Coexistence in Ecology

MONOGRAPHS IN POPULATION BIOLOGY

SIMON A. LEVIN, ROBERT PRINGLE, AND CORINA TARNITA, SERIES EDITORS

A complete series list follows the index.

Coexistence in Ecology

A MECHANISTIC PERSPECTIVE

MARK A. McPEEK

PRINCETON UNIVERSITY PRESS

Princeton and Oxford

Published by Princeton University Press
41 William Street, Princeton, New Jersey 08540
6 Oxford Street, Woodstock, Oxfordshire OX20 1TR

press.princeton.edu

Library of Congress Cataloging-in-Publication Data

Names: McPeek, Mark A., author.
Title: Coexistence in ecology : a mechanistic perspective / Mark A. McPeek.
Description: Princeton : Princeton University Press, [2022] | Series: Monographs in
    population biology ; 66 | Includes bibliographical references and index.
Identifiers: LCCN 2021036887 (print) | LCCN 2021036888 (ebook) | ISBN 9780691204871
    (paperback) | ISBN 9780691204864 (hardback) | ISBN 9780691229225 (ebook)
Subjects: LCSH: Ecology. | Coexistence of species. | BISAC: SCIENCE / Life Sciences /
    Ecology | SCIENCE / Life Sciences / Biology
Classification: LCC QH541 .M3865 2022 (print) | LCC QH541 (ebook) | DDC 577--dc23
LC record available at https://lccn.loc.gov/2021036887
LC ebook record available at https://lccn.loc.gov/2021036888

British Library Cataloging-in-Publication Data is available

Editorial: Alison Kalett & Whitney Rauenhorst
Production Editorial: Ali Parrington
Text and Jacket/Cover Design: Carmina Alvarez
Production: Danielle Amatucci
Publicity: Matthew Taylor & Charlotte Coyne
Copyeditor: Jennifer McClain

This book has been composed in Times New Roman

10 9 8 7 6 5 4 3 2 1

This book is dedicated to the memory of

Jackie Brown
(Jonathan M. Brown)
Scientist. Educator. Friend.

Contents

Acknowledgments

My career so far has spanned a very interesting time in community ecology. I started graduate school in 1982, which was in the middle of the evaluation of the centrality of resource competition as the organizing force in community ecology. Journal pages were filled with papers critiquing the validity of various types of data, reviews summarizing the results of scores of field experiments, and arguments for the validation mechanisms. As a master's student, I was a very junior author on a review of field experiments of predation that was a response and complement to similar analyses of competition (Sih et al. 1985). Following and participating in these big arguments as a graduate student defined my outlook on science. Forty years on, I hope this book is a useful contribution to rekindling these arguments, which I feel we sorely need.

Many people influenced my views on this subject. My PhD mentors, Earl Werner and Donald Hall, always preached mechanism. Robert Holt and Mathew Leibold, in hundreds of conversations, have pushed and inspired me to think about the issues of species coexistence in greater depth and broader scope. In fact, the genesis of this book is from a spark of an idea that Bob gave me while talking in his office in the Natural History Museum at the University of Kansas in 1997. Mathew pushed me harder and challenged my ideas more than anyone I've ever interacted with. My other mentors include Gary Mittelbach, Kay Gross, Susan Kalisz, Steve Tonsor, Steve Kohler, Andrew Sih, Philip Crowley, Joseph Travis, and Henry Wilbur. My graduate student colleagues, including Jonathan Brown, Robert Creed, William Resetarits, David Skelly, Thomas Miller, Oswald Schmitz, Gary Wellborn, and Josh Van Buskirk, have also been lifelong friends who have shaped and challenged me to think beyond the bounds I was imposing on myself.

I thank the following colleagues for sharing data and insights about their studies of diverse communities: Ronald Karlson, Howard Cornell, Judith Bronstein, Christopher Meyer, Jordon Casey, Mark Hay, Tyler Kartzinel, Robert Creed, Mathew Leibold, Stephen Hubbell, Robert Pringle, Meghan Duffy, and Richard Ostfeld. György Barabás helped me understand the storage effect and the role of variation to coexistence at a depth that was impossible for me before our conversations.

I thank Fran James, Donald Strong, Daniel Simberloff, Joseph Travis, Earl Werner, Peter Grant, and Thomas Schoener for recounting to me their recollections of a boat ride they took in March 1981 on the Wakulla River in the panhandle of Florida. I heard about this boat ride from Earl Werner when I was in graduate school, and the story always stuck with me (see chapter 2).

I would also like to thank Alison Kalett and her team of Ali Parrington, Jennifer McClain, Dimitri Karetnikov, Whitney Rauenhorst, and Abigail Johnson for shepherding this book through the publishing process.

Finally, I thank the many friends and colleagues who read all or parts of this book and provided valuable commentary, including Adam Siepielski, Sarah McPeek, Robert Creed, Melissa DeSiervo, Judie Bronstein, Meghan Duffy, x-Robert Holt, James Peniston, David Kikuchi, Michael Barfield, Amy Kendig, Margaret Simon, Nicholas Kortessis, and an anonymous reviewer.

And to my family—Gail, Curtis, Sarah—they all knew what I was doing. I love you all!

Coexistence in Ecology

CHAPTER 1

Introduction

A harrier glides over a field in search of voles, mice, and insects. Many of those rodents and insects themselves search for seeds while on vigil for the harrier. Other rodents, leafhoppers, grasshoppers, and caterpillars munch on the leaves and stems of the plants that produced those seeds. The plants search for mineral resources, water, and sunlight as they stand in place and suffer the consequences of losing tissue to these tormenters. However, the plants also offer up nectar and pollen for bees, butterflies, beetles, and birds as an enticement to induce pollination. Mycorrhizal fungi and the roots of the plants exchange water, nutrients, and carbohydrates. Spiders build webs in the plants' branches to ensnare various insects. Viruses and bacteria infect all these species to perpetuate their own existence. All of these interacting species form a biological community.

Current estimates suggest that 1.5–8.7 million species may currently inhabit the Earth's biosphere (May 1992, Gaston and Blackburn 2000, Mora et al. 2011). Any location where individuals of any one of those millions of species is found could be described similarly. That includes old-growth or second-growth forests, small creeks or large rivers, vernal ponds or large lakes, the abyssal ocean or intertidal coastline, or any city street or suburban neighborhood.

At every location, many species must be able to thrive with all the other species that are present there. Most have little direct effect on one another: a Cooper's hawk perched on the branch of a cherry tree probably has no direct effect on the demographic performance of that tree. The Cooper's hawk does have strong direct effects on the smaller birds and rodents on which it feeds. This reduction in the abundances of bird and rodent species reduces the food available for the other accipiters and buteos, snakes, foxes, weasels, and all the other carnivores that also feed on those species. These direct effects of the Cooper's hawk on bird and rodent abundances also indirectly benefit the cherry trees by reducing the abundances of the herbivorous rodents that feed on the cherry trees' seedling offspring, and indirectly harm the trees by reducing the abundances of birds that disperse their seeds. The bees directly benefit those cherry trees by pollinating their flowers, and in so doing indirectly benefit the caterpillars that feed on the leaves of those cherry trees. Those pollinators also indirectly harm the other trees and shrubs that compete with those cherry trees for sunlight and mineral resources and the understory plants that must deal with the allelochemicals that leach from the fallen cherry leaves and branches.

Each species at some location must deal either directly or indirectly with all the other species that are there. However, not all of those species are thriving at that site. Some species are present only because individuals continually immigrate from other areas. Some species are present now but will not be present sometime in the future. The individuals of some are vagrants. Only the species that can support populations in the face of dealing with all those other species can be considered to be thriving. These thriving species and how each deals with all the others to be able to sustain its local population are the foci of this book.

WHAT IS "COEXISTENCE" IN GENERAL TERMS?

For a community ecologist, saying that a species is "coexisting" in a community implies something much more than the colloquial meaning of the word. In general usage, coexistence simply means living together, which would suggest that all the species found in the same place would be coexisting. However, to a community ecologist, coexistence implies a very precise statement.

The word *coexistence* also typically connotes a comparative statement about a small group of species embedded in a larger community. For example, we typically might speak about two resource competitors as coexisting with one another. However, this narrow focus on only these two species in the community is too limiting a perspective. All the other species in the community must also be considered to make a definitive statement about each one's ability to coexist. Most directly, whether each of these two resource competitors will coexist depends on the dynamics of the resources over which they are competing and the dynamics of any predators, pathogens, or mutualists they may have. The dynamics of these other species will in turn depend on how they interact with other species in the community; for example, mutualists that the resources may have and additional prey that the predators may have. What determines the success or failure of each of these species are the direct impacts that the resource, predator, mutualist, and pathogen abundances have on its demographic performance. The other resource competitor only indirectly affects its competitor's success via its own impacts on these other species. Certainly, the presence or absence of a resource competitor may strongly influence the success of a particular species of interest, but this influence is mediated through the network of species interactions, and this resource competitor is just one node in that causal network. Moreover, these two resource competitors will not be able to coexist everywhere, because the conditions affecting all these other species will differ among locations.

Thus, the issue of coexistence is not a comparison of a small collection of species taken in isolation from all the other species in the community. Coexistence is a property of each species in the community. That property is its demographic performance in all the interactions with other species in the context of the abiotic environment in which it must engage. For me, the correct comparison is not limited to some subset of species, but rather pertains to whether this species can coexist with all the other species in the community.

In community ecology, the thriving species are coexisting. The simplest and most general definition of coexistence in the community ecology meaning of the word is the following:

A species is coexisting in a community if it can maintain a population in the local ecological conditions it experiences.

This sentence is simple enough to seem like a platitude, but the devil is in the details, and those details are the subject of this book.

This definition of coexistence has three important issues. The first issue is that coexistence is a statement about a single species embedded in a community and ecosystem. That community defines the web of species interactions that take place at that location, and the ecosystem context defines the regime of abiotic conditions in which those species interactions take place. However, this issue must be evaluated separately for each species in the community. Therefore, coexistence must be considered simultaneously in both granular and holistic contexts from the perspective of each species individually and all species at once.

The second issue is the meaning of the clause "maintain a population." The colloquial meaning implies merely that the population exists. However, individuals of a species may be present at a site for many different reasons. If individuals are simply vagrants that are quickly passing through, one must question whether those individuals are persistent members of the community. As geese migrate south for the winter, they stop for a few hours to days at ponds, lakes, and rivers along the way. These stopovers may have important influences on the dynamics of the local community (e.g., defecation causing significant nutrient inputs: Manny et al. 1975, Olson et al. 2005). However, these geese do not "maintain a population" at these ponds. Questions about which species are coexisting are not the same as questions about which species influence the structure of a community: as we will see, species need not be coexisting to influence community structure.

A distinction is also drawn between species that maintain a population because of the balance of local per capita birth and death rates and those that would be locally extirpated were it not for continual immigration from some other area. Again, species maintained by immigration may have important effects on local community structure: individuals of such a species still consume resources, are themselves consumed by predators, interact with mutualistic partners, and support diseases. However, their populations are present not because they are relatively successful in the local ecological conditions, but rather because their immigration rate is higher than their local rate of population decline (Shmida and Ellner 1984, Holt 1985, Pulliam 1988).

Coexistence is limited to those species that can maintain a population because of the positive balance between their per capita birth and death rates determined by local ecological conditions. This focus traces through the ideas about a species' niche (Grinnell 1917, Elton 1927, Leibold 1995), in particular G. E. Hutchinson's set-theoretic definition of the niche. Hutchinson's initial statement (1958, p. 416) is this:

> Consider two independent environmental variables x_1 and x_2 which can be measured along ordinary rectangular coordinates. Let the limiting values permitting a species S_1 to survive and reproduce be respectively x'_1, x''_1 for x_1 and x'_2, x''_2 for x_2. An area is thus defined, each point of which corresponds to a possible environmental state permitting the species to exist indefinitely.

Thus, Hutchinson's statement of the niche is a statement about where a species' per capita birth and death rates permit it to "exist indefinitely." He defines niche axes as features of the environment and considers only competition with other species as being important—which has led to much confusion about whether other species interactions, such as predation, disease, and mutualisms, are part of a species' niche (e.g., see review by Leibold 1995). My point here is not to debate the definition of a niche (that will come later), but rather to highlight the distinction that the phrase "maintain a population" is limited to the balance of per capita birth and death processes caused by local ecological conditions.

The third issue is how to define and identify "the local ecological conditions it experiences." Given that the critical issue of coexistence is whether the per capita birth and death rates of the species foster maintaining the local population indefinitely, these local ecological conditions must be evaluated with respect to those demographic rates. Specifically, what local ecological conditions influence the values of these demographic rates and how do these rates change with the changing abundances of all the species in the community, including the species of interest? These demographic rates are functions of both the local abiotic environmental regime, as Hutchinson's (1958) niche definition identified, and interactions with all the species that exist there. These include all the abiotic resources, prey, predators, competitors, mutualists, and pathogens in the community; in other words, the entire ecological milieu that influences the demography of each species. To give primacy to one particular type of species is to ignore much of this causal structure.

The central focus of this book is to explore how these demographic rates for each species in the community change because of the abiotic environmental regime and species interactions with other community members. The set of important ecological conditions may be different for every species in the community. In fact, the differences in these sets among species are typically the defining mechanistic features that permit the coexistence of each (Hutchinson 1958, MacArthur and Levins 1967, Levin 1970, Chesson 2000b). This makes our current one-size-fits-all approach to understanding coexistence inappropriate.

INVASIBILITY

How do we assess whether a species can maintain a population in a local area indefinitely? As discussed above, the mere presence of the species does not justify such a conclusion. As Hutchinson's niche definition implies, this assessment must be based on the per capita birth and death rates that shape the population dynamics of the species at that location. For clarity, I outline this issue here purely from a population ecology perspective, ignoring how interactions with other species define these demographic rates. The rest of the book incorporates those species interactions into this framework.

Because no species has increased to infinite abundance, we can infer that the abundances of all species are regulated to some degree, or at least are bounded by their ecological surroundings. The essence of population regulation is the change in the constituent demographic rates that comprise per capita population growth rate with a change in the abundances of one's own species and those of other species. A species may have only indirect impacts on its own demographic rates through how the abundances of its prey (resource limitation) or predators (predator limitation) or mutualists (mutualist augmentation) or pathogens (pathogen limitation) change as its own abundance changes, but for now I ignore the mechanisms causing these demographic relationships to exist. When viewed from this purely population ecology perspective (i.e., ignoring interactions among species), the signature of population regulation is the change in these per capita demographic rates with a species' own abundance. This implies that either per capita birth rate decreases, per capita death rate increases, or both happen simultaneously as its own abundance increases.

These ideas can be formalized mathematically. A generic model describing the change in total population growth rate caused by these demographic processes for a generic species N is

$$\frac{dN}{dt} = N[b(N,...)-d(N,...)], \qquad (1.1)$$

where $b(N, \ldots)$ and $d(N, \ldots)$ are the per capita birth and death rates, respectively, which are each functions of its own abundance (N), the abundances of all the other species in the community (the effects due to other species are represented by the ellipses because they are being ignored for now), and the abiotic environmental conditions in which these species interactions take place (typically defined by the structure and the parameter values of the function). Throughout this book, I use this continuous-time formulation of population change, but all the results presented here can also be derived in discrete-time formulations (Hassell 1978, Murdoch et al. 2003, Turchin 2003).

What are the logical implications of a population being able to "exist indefinitely," in Hutchinson's terms, given such population regulation? From a population dynamics perspective, one of three dynamical situations is implied. The first is that the community is being regulated to a stable point equilibrium in abundance, and so the population of species N is being regulated to a stable point equilibrium (Case 2000). An equilibrium abundance for N is the value where the population growth rate does not change (i.e., $dN/dt = 0$). Based on equation (1.1), that occurs when $N = 0$ or at the abundance, giving $b(N, \ldots) - d(N, \ldots) = 0$; in other words, that abundance at which per capita birth and death rates are equal (ignoring per capita immigration and emigration rates). Throughout the book, I identify such equilibrium abundances with a superscript star: N^*.

For this latter equilibrium with $N^* > 0$ to be stable, population abundance must move toward the equilibrium when it starts near the equilibrium. Thus, if the species' abundance is below this equilibrial value, the per capita birth rate of the species must be greater than its per capita death rate, and so its abundance will increase. In contrast, if its abundance is above this equilibrial value, its per capita birth rate is less than its per capita death rate, and so its abundance will decrease (fig. 1.1).

The second possible dynamical situation is that the community displays limit cycles or chaotic dynamics. These are dynamical features in which the community orbits a stationary equilibrium (i.e., an orbital attractor): the equilibrium point itself is unstable, but it is surrounded by a stable limit cycle or chaotic shell (Hirsch et al. 2012, Strogatz 2015). The cycling of hare and lynx abundances across Canada is one of the most famous examples in ecology (Elton 1942, Stenseth et al. 1997), but many other communities show analogous types of dynamics (Elton 1942, Turchin 2003, Korpimäki et al. 2004, Myers 2018). In this case, the environmental conditions remain constant, but the internal dynamics of the community cause the cycling (Krebs et al. 2013, 2018). In other words, the functions $b(N, \ldots)$ and $d(N, \ldots)$ do not change, and so the location of N^* does not change either. However, the dynamics caused by interacting with the ignored species create cycles in the abundance of N around N^*. Despite the continually changing demographic relationships, the same population regulation features hold if the

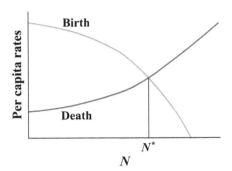

FIGURE 1.1. A stylized representation of the per capita birth and death rates for a population of some species. The birth and death rate curves are density dependent, meaning that their values change as the abundance of the population changes. The equilibrium abundance occurs at the abundance where birth and death rates are equal, and this equilibrium abundance is signified by N^*. The equilibrium pictured here is stable because per capita birth rate is greater than death rate at abundances lower than N^*, and death rate is greater than birth rate at abundances above N^*.

species is to be present: when the species reaches low abundance in the cycle, its per capita birth rate becomes greater than its per capita death rate so that it then increases in abundance; and when the species reaches high abundance in the cycle, it begins to decline because its per capita death rate becomes greater than its per capita birth rate.

The final possible dynamical situation is that temporal variability in the environmental conditions drive variability in these demographic rates for community members, which introduces variability to the dynamics of their interactions. Stochastic year-to-year weather variability can be a potent source of such demographic variability in many species (e.g., Leirs et al. 1997, Kausrud et al. 2008, Schmidt et al. 2018). One can think of this as continual change in the shapes of the per capita demographic rate relationships $b(N, \ldots)$ and $d(N, \ldots)$ caused by ecological variability through time (Tuljapurkar and Orzack 1980, Tuljapurkar 1989, Lande 1993, 2007, Lande et al. 2003). This generates temporal variation in the position of the equilibrium N^*, and this shifting position of the equilibrium drives population change. For a species to persist in this community, its long-term average per capita birth rate must be greater than its long-term average per capita death rate when at low abundance:

$$\overline{b(N \approx 0,\ldots)} > \overline{d(N \approx 0,\ldots)},$$

where the overbars signify averages (Turelli 1977, 1978b, Levins 1979, Chesson and Ellner 1989, Chesson 2000b, Schreiber et al. 2011).

All of these imply that a species will be present if its abundance on average increases when its abundance is below the equilibrium value. Taken to its extreme, this implies that the species' abundance will increase when it is extremely rare: that is, when its abundance is surely below $N^* > 0$. In fact, this is the criterion identifying coexistence for a species, which is commonly termed *invasibility* (MacArthur 1972, Holt 1977, Turelli 1978b, 1981, Chesson 2000b, Siepielski and McPeek 2010). This concept has also been referred to as *permanence* (Hutson and Law 1985, Hutson and Schmitt 1992, Law and Morton 1993, 1996, Morton and Law 1997), but I use invasibility throughout. Invasibility implies the success of the species in a number of contexts. For example, if some perturbation caused the species to be knocked to very low abundance (e.g., a disease outbreak left only a few surviving individuals of this species) or it cycles to low abundance, the species would be able to recover and increase in abundance. Alternatively, if it were initially absent, the species could successfully invade the community—hence *invasibility*. Also, note that invasibility is not evaluated with respect to any other particular species but with respect to the entire community. Thus, if invasibility is the criterion to evaluate coexistence, this argues that the concept of coexistence is a property of each species, given the community and ecosystem in which it finds itself.

A technical definition of invasibility is as follows:

A species satisfies its invasibility criterion if its per capita birth rate exceeds its per capita death rate when it is extremely rare and all other species in the community are at their demographic steady states in its absence.

Expressed mathematically, this definition becomes

$$b(N \approx 0,...) > d(N \approx 0,...) \tag{1.2}$$

(see references in the previous paragraph). Each species that can satisfy this invasibility criterion is coexisting in the community: this is the community ecologist's meaning of coexistence. Obviously, a population with few individuals will be subject to the whims of demographic stochasticity that may cause its extinction even though its expected per capita population growth rate is positive (Adler and Drake 2008). Thus, not every invasion of a potentially coexisting species or return from very low abundance will be successful, and cycling species are at greater risk of extinction when near the nadir of their abundance cycle. However, satisfying this invasibility criterion is the hallmark of every coexisting species, at least theoretically.

Biological communities do not contain only coexisting species. As mentioned above, some species that are present at a location are not sustaining themselves because of local ecological conditions, and these species can be as important to local community structure as the coexisters. By identifying which species are coexisting in a community, the species that are present but not coexisting are also being identified, namely, *co-occurring species* (Leibold and McPeek 2006). In addition to the vagrants, three other types of co-occurring but not coexisting species are present in communities.

The first type of co-occurring but not coexisting species are what Daniel Janzen called *walking dead species*. They have also been called the extinction debt (Tilman et al. 1994) and the living dead (Hanski 1998). These are species that are headed to extinction but not there yet. Species headed to extinction do not suddenly just disappear. Extinction also has a temporal dynamic based on the differences in local and average regional birth and death rates (Lewontin and Cohen 1969, Turelli 1977, Raup 1992, McPeek 2007), and these extinctions may take a very long time. For example, two million years were required to drive 12 bryozoan species in the southwestern Caribbean basin extinct; their extinction was caused by the reduction in productivity precipitated by the closing of the Isthmus of Panama and the introduction of new and presumably more competitive species via speciation (O'Dea et al. 2007, O'Dea and Jackson 2009).

The second type of co-occurring but not coexisting species are *sink species*. Sink species have their local per capita death rate exceed their local per capita birth rate, but their local population is maintained by continual immigration from other local communities in the region where they are sustaining (i.e., source populations) (Shmida and Ellner 1984, Holt 1985, Pulliam 1988). Because sink species can only emerge as a result of movement between communities from source populations to sink populations, their existence and effects on community structure are considered in chapter 9 on spatial variability.

The final type of co-occurring but not coexisting species are *neutral species* (Hubbell 1979, 2001, Hubbell and Foster 1986). When present, neutral species are a guild of ecologically identical species. For example, in eastern North America, the 8–12 *Enallagma* species found in each lake with fish display all the hallmarks of neutral species (Siepielski et al. 2010). Because they are ecologically identical, interactions with other species in the community regulate the total abundance of all guild members instead of each guild member separately (Hubbell 2001, Siepielski et al. 2010, McPeek and Siepielski 2019). As a result, the abundance of each species may change through time following a random walk within this constraint of total abundance. Neutral species are technically not coexisting species, because if one guild member is already present in the community, any additional guild member would have its per capita birth rate equal to its per capita death rate when it invades. In effect, what is coexisting in the community is the entire guild of neutral species simultaneously, even though no single member of the guild is coexisting. Neutral species are considered in chapter 10.

THE BOUNDS OF A COMMUNITY

The boundaries of a community are difficult if not impossible to define. I have worked on the communities of organisms in ponds and lakes for most of my career. Community ecologists who work in terrestrial ecosystems generally think that we aquatic types have it made: "the community" in a lake clearly ends at the water's edge, and "the community" is completely self-contained in the water of that lake.

However, even in something as discrete as a lake, multiple communities may exist, and the boundaries between them are not clear. The assemblage of crustaceans, insects, annelids, mollusks, salamanders, and fish that live clinging to and growing around the macrophytes in the littoral zone of a lake is very different from the assemblage of crustaceans, insects, and fish that live suspended in the pelagic open water zone, and the assemblage of annelids, crustaceans, insects, and mollusks that live in the mud of the benthic lake bottom zone is different still. The populations of some species span across the boundaries of these zones, but many taxa are restricted to only one. Because interactions among individuals of various species define the dynamics of a community, each of these zones might be considered separate communities of interacting organisms. Even the assemblage of organisms in each of these "zones" may not represent a single community. For example, if the macrophyte beds of a lake are disjointed, each bed may represent a separate dynamical unit with little movement of individuals between them. Individuals of many species are restricted to only one of these communities within a lake for much or all of their life cycle.

However, each of these lake zones is not an integrated and wholly separate community. Individuals of some species move between these various communities within the lake. Zooplankton migrate between the epilimnetic and hypolimnetic zones of the pelagic over the course of a day; some fish species forage in these different areas on a daily basis; some fish undergo ontogenetic habitat shifts so that they spend one life stage in one zone and another life stage primarily in another. Moreover, the water's edge is no discrete boundary. Individuals of many species, such as ducks, geese, herons, gulls, and moose, routinely forage in multiple lakes and so link the dynamics of multiple lakes. Likewise, some fish species use the stream connections to move between lakes. Does the fact that the populations of some species span these zones unite them into a single community, or are these separate communities that are linked by dispersal? Where is the line between two populations with limited dispersal versus a single continuous population? Is trying to draw a community boundary even useful if the populations of different species operate on such disparate spatial scales?

The boundaries of communities in terrestrial ecosystems are similarly obscure. The assemblage of species found in a light gap in the forest is different from that found in a patch of the continuous forest surrounding it. Some species are common to both and some are unique to each. Some species forage across multiple light gaps, while others may spend their entire lifetimes in only one patch. In the surrounding forest, the individuals of the various species that are interacting primarily with one another may be generally found over the expanse of a few hectares to square kilometers, but the forest may continuously extend over many thousands of square kilometers. Where is the boundary of this community to be drawn? Exactly the same issues emerge, but on a much smaller spatial scale, when you consider the microbiomes found in various parts of the gastrointestinal tract, the respiratory tract, and skin on your own body.

Defining the boundaries of a community or communities that develop along abiotic gradients also poses conceptual, empirical, and philosophical challenges. The plant assemblages change in a regular patterning along the altitudinal gradient on the mountains of the southern Appalachians. In the deep ravines cut by streams at the bottom of the mountain, you will find cove forests dominated by basswood, poplar, magnolias, and birches, while the dry mountaintops are dominated by oaks, hickories, and beeches (Braun 1935, 1940, Whittaker 1956). Some species can be found over much or all of this gradient, while others are restricted to particular ranges. Even salamanders that may move only a few meters in their lifetimes show distributional patterning along these altitudinal gradients (Hairston 1951, 1980, Jaeger 1971). The same issues arise over centimeters in the patterning of species distributions and community structure on the rocks of the marine intertidal (Connell 1961, Paine 1966, 1969, Menge 1976, Sousa 1979). Should the boundary of a community be placed at the distributional limits of each segregating species along the gradient or at the limits of only key species, or do the limits of the most widely distributed species set the community's boundaries?

If a biological community is not a defined and integrated unit that has no discernible boundaries, is the coexistence of species in a community something that can or should be studied? In fact, these are exactly the issues that led Ricklefs (2008, pp. 741–742) to argue that "community" is an "epiphenomenon that has relatively little explanatory power," and "coexistence can be understood only in terms of the distributions of species within entire regions" (see also Gleason 1926). If all the species present at a location are simply a sampling of all the species in the surrounding region, with each species having little to no demographic impact on any other, changes in species composition that occur as you hike up a mountain or swim from the littoral to pelagic zones of a lake would simply reflect the physiological tolerances of those various species to the local abiotic conditions.

While I completely agree with Ricklefs's arguments about the importance of regional processes and macroevolutionary dynamics in shaping local species assemblages (Ricklefs 1987, 1989, 2008, 2010) and his arguments have inspired my own work on how macroevolutionary processes shape current species diversity and the phylogenetic patterns of component taxa (McPeek and Brown 2000, 2007, McPeek 2007, 2008b), I completely disagree with the deduction that little if anything can be learned by studying coexistence in the framework of community ecology. The literature is replete with examples where adding or removing one species causes others to go locally extinct and thus permits others to locally thrive that were not previously there. Remove *Balanus balanoides* barnacles from rocks in the intertidal of Scotland, and *Chthamalus stellatus* barnacles will colonize and thrive (Connell 1961). Remove *Pisaster ochraceus* seastars from rocks in the intertidal of Washington, and as many as 25 species of barnacles, sponges, anemones, snails, chitins, urchins, and algae are replaced by a single species, *Mytilus californianus* mussels (Paine 1974). Introduce largemouth bass, *Micropterus salmoides*, to a lake in Wisconsin, and the species composition of insects, cladocerans, copepods, rotifers, and algae throughout the pelagic food web shifts dramatically (Carpenter and Kitchell 1993). This large body of results proves that interactions among species on a local scale shape not only community membership and the local abundances of those species but also larger patterns of species composition and diversity.

For me, community ecology is the study of the network of species interactions across local and regional scales. A community is an assemblage of species connected in a network of species interactions (Paine 1980, Martinez 1991, 1992, Dunne et al. 2002, Bascompte and Melián 2005). If one species directly impacts the demographic rates of another species, those two species are directly linked in this interaction network. Indirect species interactions include the effects of one species on another that passes through one or more intermediate species in the network because of changes in abundances (i.e., abundance-mediated indirect effect: Abrams 1995), or the effect that one species has on the strength of the interaction between two other species (i.e., trait-mediated indirect effect: Abrams 1995, Werner and Peacor 2003). What is included in any particular study depends on the taxa and the scope of the interaction network that are the primary foci of the researcher's interest, the linkages of those taxa within the local interaction network, and the linkages of networks among locations.

Invasibility is then a test of whether the species of interest is persistent in the local network because of the network's structure and its own local demographic success. If the species passes the test, it is a persistent member of the community. If it fails, it may be a redundant community member (i.e., neutral species), it will disappear from the network (i.e., walking dead species), or it exists primarily because of linkages to networks in other locations (i.e., sink species).

Community ecologists can ask a multitude of different questions of various scopes about these interaction networks, and different questions will require the researcher to consider the interaction network at different spatial and temporal scales. Most community ecologists are initially interested in the role of particular taxa or assemblages or trophic levels in the community, and so all questions begin with these species as the focus. Initially, only the components of the interaction network that directly impact their demographic rates and abundances will be important. Thus, for a community ecologist interested in understanding some aspects of the community ecology of *Daphnia* species found in the pelagic zone of a lake, the components of this interaction network that directly affect *Daphnia* species' demographic rates and abundances are most relevant; that is, the algal species on which they feed and the predators that feed on them. Depending on the questions being addressed, it may also be important to understand more distant parts of the interaction network that influence the abundances of these species. And so, typically, one will start with some specific component of the web and work out through various connections, because components of the interaction web that do not directly or indirectly influence the components of interest are irrelevant to study.

The extent of the network of interactions that one wants to or is forced to study is primarily determined by the questions being addressed by the researcher. This is because species composition in one part of a local food web intimately depends on the species that occupy other parts of the food web. In fact, a food web represents much (but not all) of the causal network of species interactions that constitutes a biological community. For some taxa and some types of questions, only adjacent portions of the local interaction network are necessary. For others, linkages between the networks in multiple locations are critical (Holyoak et al. 2005, Leibold and Chase 2017). I have organized this book to systematically work through the exploration of these interaction networks and how various features of the networks foster or constrain various features of communities (i.e., opportunities for new types of species to establish, consequences of species perturbations or deletions, patterns of species distributions, and abundances among communities scattered along environmental gradients).

THE WHY QUESTION

Understanding this network of species interactions is also key to understanding *why* each species is successful or unsuccessful in a community, namely, the mechanism for why it can coexist. Testing whether a species can invade a community is in principle a fairly simple exercise (but exceedingly difficult in practice for most communities), and such tests can be accomplished without any reference to why the species has a positive per capita population growth rate when it is rare. However, each coexisting species in a community is present because of its abilities to balance the conflicting demographic demands of the various species interactions it faces at its position in the food web. Understanding comes from not simply asking whether this or that species is coexisting; rather, understanding comes from asking why this species is able to coexist in this community.

One of the main problems I see today with community ecology is that we have largely gotten away from asking the why questions at a mechanistic level. This is not a new problem: Tilman (1987) actually made exactly this same point specifically about the study of resource competition; Schoener (1986) made a similar argument without focusing on a broader range of interactions, and he sketched the philosophical designs of a mechanistic research program. This book is an expanded argument for their perspectives. However, since the 1950s, coexistence has been explored primarily through the prism of a general theory of competition that is devoid of any mechanism (Gause 1934, Gause and Witt 1935, Slobodkin 1961, May 1973, Vandermeer 1975, Chesson 2000b). In fact, Chesson (2000b, p. 345) states explicitly that "these models are phenomenological: They are not defined by a mechanism of competition." The central question in that approach is, What is the effect one species has on another relative to the effect a species has on itself (Gause and Witt 1935, Slobodkin 1961, Chesson 2000b)? However, these effects are characterized devoid of any causes for those effects. As we explore in chapter 2, the justification for this analysis comes from the ability to transform simple models of two types of species interactions into a common form that subsumes the causal chain of interactions across multiple species into what appears to be a single direct interaction between two species (MacArthur 1969, 1970, Schoener 1974b, Abrams 1975, Chesson 1990, Chesson and Kuang 2008). These two types of interactions are (1) two or more consumers competing for resources and (2) two or more prey that are fed upon by the same predators. In fact, we use the language of "competition" to describe both of these interactions, namely, resource competition and apparent competition, respectively.

Although important conceptual insights have been gained and I am sure will continue to emerge from this theory, I think that relying solely on general competition theory to understand species coexistence is problematic for a number of reasons, which I explore in greater depth in chapters 2 and 11. My main reason for concern has both philosophical and practical bases. If you compare and contrast resource competition with apparent competition, you realize that the abilities and phenotypic properties that make a species successful at utilizing some resource are completely different from the abilities and phenotypic properties that make a species successful at reducing mortality from some predator—in fact, they are often in antagonistic opposition! However, when couched in the language of competition, most minds immediately gravitate to the interpretation of resource competition, which in turn focuses the mind back to this specific mechanism.

Answer these questions honestly to yourself. If I say that two species differ in competitive ability, what do you think of? Do their relative abilities to avoid a common set of predators enter your conception?

The words, language, and models we use to conceptualize nature define how we see nature and, consequently, how we measure and study nature.

As an example, consider Hutchinson's (1961) famous *paradox of the plankton*. Ponds and lakes of moderate size support dozens to hundreds of phytoplankton species (Smith et al. 2005). For example, 84 phytoplankton species were recorded in Trout Lake, Wisconsin, USA, from samples taken in 2005–2006 (data downloaded 22 August 2019 from https://portal.edirepository.org/nis/mapbrowse?scope =knb-lter-ntl&identifier=238). Hutchinson's paradox explains how so many phytoplankton species can all coexist on just a handful of limiting resources (he listed light, CO_2, phosphorus, nitrogen, and sulfur compounds, plus 14 elements), given that "natural waters, at least in the summer, present an environment of striking nutrient deficiency, so that competition is likely to be extremely severe" (Hutchinson 1961, p. 137). He offered a number of possible hypotheses for how so many species could be found together and possibly coexist. The two hypotheses that have resonated over the last 60 years or so are that either temporal variability never permits the system to reach an equilibrium and so competitive exclusion is never completed (here the species are not coexisting but simply co-occurring because they are on long transients caused by this variation: Chesson and Huntly 1997), or species really do coexist because the spatial and temporal ecological variation causes continual reversals and shuffling of the competitive dominance of species. Theoretical explorations attempting to explain the paradox of the plankton primarily approach the problem by considering such spatiotemporal variation in nutrient limitation and competitive dominance reversals (depending on the model; Armstrong and McGehee 1980, Chesson 1994, Huisman and Weissing 1999, 2001b, Huisman et al. 2001, Abrams and Holt 2002, Klausmeier 2010, Li and Chesson 2016).

However, competition for limiting nutrients is not the only ecological process that shapes the demographic rates of phytoplankton species. In fact, Hutchinson (1961) discusses two other types of species interactions, although the fact that he did so has been largely forgotten. (Note to graduate students: You actually need to carefully read classic papers yourself. Do not take other people's word for what papers say.) He devotes a paragraph to symbiosis and commensalism and offers some conjectures about how these may foster the coexistence of some species: some may benefit from acquiring vitamins that are released by others into the water column. Hutchinson (1961, p. 141) then spends exactly two sentences to suggest that predation may also foster the coexistence of some species:

It can be shown theoretically, as Dr. MacArthur and I have developed in conversation, that if one of two competing species is limited by a predator, while the other is either not so limited or is fed on by a different predator, co-existence of the two prey species may in some cases be possible. This should permit some diversification of both prey and predator in a homogeneous habit.

To my knowledge, MacArthur and Hutchinson never published these insights. We explore these conjectures explicitly in chapter 4 (see also Grover 1994).

Predation by herbivorous and omnivorous zooplankton species are actually a very likely additional set of species interactions that may foster the coexistence of many different phytoplankton species (e.g., Leibold 1989, Grover 1994, Leibold et al. 2017). For example, Trout Lake has 68 species of herbivorous or omnivorous copepods, cladocerans, and rotifers that feed on those 84 phytoplankton species (data from https://portal.edirepository.org/nis/mapbrowse?scope=knb-lter-ntl& identifier=37). As we explore in the coming chapters, enemies (i.e., herbivores, predators, omnivores, pathogens) also serve as limiting factors that foster coexistence just as limiting resources do. In fact, these 68 phytoplankton enemies plus the 19 abiotic nutrients that Hutchinson listed give a total of 87 potentially limiting factors, which are more than enough to account for 84 phytoplankton species (see Levin 1970). These numbers do not include the many herbivorous and omnivorous protozoan species that also feed on phytoplankton and the many viruses that also attack phytoplankton, which have not been enumerated in surveys of Trout Lake to my knowledge.

Thus, with great deference and all due respect to Hutchinson (whom I revere), I see no "paradox" at all, given the number and diversity of nutrients and the number and diversity of enemy species that potentially influence the per capita birth and death rates of each species in the phytoplankton assemblage of a lake. That is not to discount the contributions of spatial and temporal variation in these factors at all; we explore their contributions to fostering coexistence in chapters 8 and 9. Rather, the point of this book is to begin to synthesize a robust framework presented by dozens of theoretical ecologists over the past six decades, which simultaneously incorporates this diversity of factors that can contribute to the coexistence of the various species found in a community. The paradox only exists if we limit ourselves to considering competition for limiting resources exclusively. From what I see, community ecology's focus on a general theory of "competition" to explain and understand coexistence of every species biases us to think only in terms of limiting resources and so blinkers us from exploring the full gamut of demographic processes and limiting factors that can easily account for a diversity of species to be successful members of a community.

This focus on a general competition theory also orphans other types of species interactions outside the mainstream framework for how we understand coexistence. In its generic definition, competition is a $(-,-)$ interaction between two species, meaning that each species has a negative impact on the per capita growth rate and abundance of the other. Both resource competition and apparent competition are $(-,-)$ indirect interactions that are mediated through other species. Because predation is a general $(+,-)$ direct interaction (the two species directly interact, and the interaction has a positive effect on the predator but a negative effect on the prey), a general theory of "competition" to explain coexistence is completely agnostic in explaining the coexistence of a predator and its prey: predator and prey do not coexist because each limits its own abundance more than it limits the other species' abundance. If you want a real paradox, this is it: the direct interactions in the chains of causation that generate the indirect effects of both resource and apparent competition are all predator-prey interactions, and so this competition theory cannot explain the interactions that underlie what generates the competitive effects. The effects of pathogens, parasites, and parasitoids are similarly ignored, given that they influence their hosts in similar ways to predators influencing their prey.

Likewise, mutualisms have been largely orphaned outside the general competition theory. The general definition of a mutualism is a $(+,+)$ interaction between two species, either direct or indirect. General competition models can be converted into models of mutualisms by simply making competition coefficients positive instead of negative (e.g., Gause and Witt 1935, Vandermeer and Boucher 1978, Goh 1979, Dean 1983, Boucher 1985). However, they are subject to the "orgy of mutual benefaction" where both species increase without bounds (May 1973), and they are still only caricatures that lack any true mechanism of interaction among mutualists, just like the competition forms are.

Moreover, we need a framework for studying coexistence that explores how mutualisms alter the results of other types of interactions in which species engage. For example, imagine a field filled with plant species. From the general competition theory perspective, the natural question to ask is, How do the interspecific competitive effects between species relative to their intraspecific effects on themselves promote their coexistence? All of these plant species also have suites of enemies that depress their abundances (e.g., Reader 1992, Carson and Root 1999, 2000). Many of them also have pollinators that increase their seed yield and mycorrhizal and rhizobial symbionts that provide them with extra nutrients to grow larger (reviewed in Bronstein 2015a). These symbionts inflate the individual sizes and population abundances of their plant partners, and so differential benefits from their various mutualistic partners may decidedly shift the balance of competition among the plants. Such mutualistic effects are outside the bounds of the general competition theory framework. However, models that better capture the mechanisms of interactions among multiple species in a food web simultaneously can explore such scenarios. We do so in chapter 6.

Answering the why question is the essence of the inquiry into the mechanism of causation (Pearl and Mackenzie 2018). Testing the invasibility of a species is a necessary step to definitively evaluate whether that species is coexisting in the community. In practice, invasibility tests are almost never done because they are simply logistically impossible to perform correctly in real communities (Siepielski and McPeek 2010). However, models of community structure that allow us to probe the reasons why a species would be successful at invading can identify definitive predictions that should then be empirically tested to validate the operation of those mechanisms in promoting or retarding that species' coexistence. This is one reason why working with models that capture the essence of mechanisms is so crucial. Moreover, even if an invasibility test can be performed in a real community, the test only answers whether inequality (1.2) is true. Why it is or is not true must still be left to these same studies addressing the existence and operation of various causal mechanisms.

ASSEMBLING A MULTITROPHIC-LEVEL COMMUNITY

Analyzing a model of a causal network of interactions among multiple species at multiple trophic levels can be a very daunting task. Because of this, I have organized this book to follow a systematic approach. In this multitrophic-level food web, some of the species had to invade before others, which sets up a natural order to the assembly of a community. The algae, bacteria, and protozoa that form the foundation of the green and brown food webs of the water column in a lake must invade and become established before any of the species that base their existence on consuming them can invade. A rotifer species such as *Keratella cochlearis* cannot invade and coexist in the lake unless a resource is available to support its population. In turn, species like the predatory copepod *Mesocyclops edax* cannot invade and coexist until species such as *Keratella* establish. Thus, a logical ordering of species invasions exists. Moreover, as we will see for many of these invasions, the same kinds of criteria for invasibility emerge over and over for very different types of species at very different food web positions.

Consequently, I have structured the book around the idea of starting with an empty ecosystem that is devoid of all biological species and adding one species at a time until the multitrophic-level community is complete. In effect, this approach is organized around the idea of building community modules (Holt 1997). A *community module* is a network configuration of interacting species that denote "multispecies extensions of pair-wise interactions" (Holt 1997, p. 333). A consumer feeding on a resource is the simple pairwise interaction that forms the backbone of the community module approach. Once the criteria for each of them to coexist in the ecosystem of interest are established, we can then ask what is required for a second consumer to invade to establish a "resource competition" module, or for a second resource to invade to establish an "apparent competition" module, or for a predator of the consumer to invade and thus establish a three-trophic-level "food chain" module.

For each invading species, the criteria necessary for it to be able to invade when rare will be the first concern. Once established, each species will potentially affect the other species already present, either by altering their abundances or actually being driven extinct because of its invasion. These changes in abundance of the species already present will in turn alter the criteria that subsequent invaders must satisfy, and so how the invasion of one species affects the abilities of other species to invade is also a central issue of inquiry here.

This approach also focuses attention explicitly on the theoretical predictions that form the basis for empirical testing. What are the species interactions that directly impact the demographic rates of the invading species? How are the abundances of those species determined by the chain of indirect interactions with other species in the existing food web? After each species invasion, I will briefly discuss the critical empirical tests that are needed to both evaluate invasibility and identify signature features of the mechanisms that foster or retard that invasion.

The mathematical backbone of this analysis is a variation on the classic Rosenzweig-MacArthur model of consumer-resource interactions (Rosenzweig and MacArthur 1963, Rosenzweig 1969, 1971). This is a versatile model that can capture many of the mechanistic features of pairwise species interactions for the full gamut of mechanisms needed here. The mathematical exposition of the model is postponed until chapter 3, but I describe the general features of the model here. The basic model assumes that the resource species grows according to logistic population growth in the absence of the consumer, and the consumer has a linear or saturating functional response for feeding on the resource (Rosenzweig and MacArthur 1963, Rosenzweig 1969, 1971). These features cause this basic model to display either a stable point equilibrium or a limit cycle, depending on parameter values. Also, if a predator is added to produce a three-trophic-level food web, the community may display chaotic dynamics in addition to a point equilibrium or a limit cycle (Hastings and Powell 1991, McCann and Yodzis 1994). Thus, the model can display the full range of dynamical features we need for this analysis.

Additional mechanistic features can be easily added to the model to explore how they modify outcomes between consumer and resource (Murdoch and Oaten 1975). For present purposes, I limit this to exploring two different ways to incorporate direct self-limitation for consumers and predators into the model. Self-limitation can arise when species display such features as being cannibalistic or territorial, or interfering with the feeding of conspecifics (Beddington 1975, DeAngelis et al. 1975, Gatto 1991, Fryxell et al. 1999, Amarasekare 2002, McPeek 2012, 2014).

Finally, the Rosenzweig-MacArthur consumer-resource model is also the basis for more mechanistic models of interactions among mutualistic partners (Holland and DeAngelis 2009, 2010). The mechanism generating the benefits to partners in many mutualisms is based on trade, where each species allows its partner to consume some material that the other has in return for some benefit (Wyatt et al. 2014, Bronstein 2015b). Plants offer up nectar and pollen for pollinators to consume in return for pollination services. Other mutualistic interactions, such as between plants and mycorrhizal fungi, plants and rhizobial bacteria, or corals and zooxanthellae, involve each partner consuming some material from the other. Thus, formulation of mutualistic interactions using this consumer-resource model fundamentally characterizes their mechanism of interaction (Holland et al. 2005, Holland and DeAngelis 2010).

Of course, all models are mechanistic abstractions to some degree (Levins 1966). Models can either capture the mechanism directly (e.g., a saturating functional response) or mimic the patterns of the consequences of some mechanistic component (e.g., logistic population growth, various forms of direct consumer self-limitation) (Fryxell et al. 1999). Also, often the exact mathematical function used is not important to the outcome but rather only the functional shape: for example, various functions can be used to model a saturating functional response and all produce the same outcomes (Seo and Wolkowicz 2018). Thus, I believe that the model results presented here adequately capture major qualitative and quantitative features of coexistence mechanisms.

I am no mathematician, but over the years I have trained myself to understand the models presented here. This training came mainly from teaching an undergraduate course in community ecology where these models are the central focus. Students are typically apprehensive about their math skills. On the first day, I ask them two questions: "How many of you passed seventh-grade algebra?" and "How many of you understand what a derivative is from the first semester of calculus?" I tell them that if they can answer yes to both of these questions, they are prepared for the course. The same is true for this book.

In fact, most of what is required is to be able to draw and interpret a special kind of graph. The workhorse of our analysis will be graphs of the isoclines that result from the dynamical equations. An *isocline* is a function that maps out when a species' population growth rate is zero based on the abundances of all the species in the community. Thus, a graph of this function tells you the combinations of all species abundances where the abundance of the species in question is not changing. Moreover, if the combination of all abundances is not on the isocline, its position relative to the isocline tells you whether the species' abundance will increase or decrease. Thus, the overall dynamics of the community can be deduced from the graph of the isoclines of the interacting species (Hirsch et al. 2012, Strogatz 2015).

Biological mechanism is also embedded in the geometry of the isoclines. Each mechanistic feature of a species interaction (e.g., logistic growth, linear or saturating functional response, feeding interference, direct self-limitation) imparts a specific shape to the isoclines of the species engaged in those interactions. Also, the dynamical responses of species to one another are manifest in the geometrical relationships among the isoclines of those various species. For example, whether a consumer can invade and sustain a population on some resource species can be determined directly from the relative positions of their isoclines. A major focus for the reader should be to identify common features associated with invasibility as more species at different trophic levels are added. Increasing the dimensionality of the system by adding more mechanistic features to the model of each species and simply adding more species quickly make the mathematics quite complex and analytically intractable, particularly for someone of my modest mathematical skills. However, teaching experience tells me that understanding can emerge from considering the geometry of the isoclines in these much more complicated multispecies problems.

If this sounds like a lot of ground to cover, it is. Therefore, I cannot cover a number of important topics related to species coexistence in communities in this book. The importance of age-, size-, and stage-structure within species to promoting their own coexistence and fostering or retarding the coexistence of other species is not addressed here (e.g., Mylius et al. 2001, Rudolf and Lafferty 2011, de Roos and Persson 2013, Wollrab et al. 2013). I also do not address how adaptive behavioral shifts of species in response to one another or the resulting trait-mediated indirect effects can affect coexistence (e.g., Abrams 1992, 2010, Křivan 1998, 2000, Werner and Peacor 2003, Křivan and Schmitz 2004, Valdovinos et al. 2013, Bachelot and Lee 2018). Finally, species cannot evolve in response to one another (e.g., Slatkin 1980, Lande 1982, Taper and Case 1985, Abrams et al. 1993, Abrams and Chen 2002, McPeek 2017b, 2019a). The models I use here to explore coexistence are all easily extended to explore these issues (see the references cited here), so a thorough exploration of coexistence here sets a strong foundation for exploring these added complexities.

THEORY AND EMPIRICAL TESTING

Models are fundamental to all aspects of the scientific endeavor. However, models are constructed for many different purposes, and we must be clear what our purpose is for using any particular type of model (Levins 1966).

A *model* is a formal statement about the relationships that are deemed important among some set of variables. Some scientific models are purely statistical associations that describe covariance structures among a set of relevant variables. For example, quantitative genetic models of phenotypic associations describe the statistical associations among phenotypes based on genetic principles, but these models do not describe the causal mechanisms that map genic, genomic and developmental interactions into the phenotype. Likewise, models using artificial intelligence with big data are currently just mining statistical relationships embedded in the data trove (Pearl and Mackenzie 2018). If the covariance structures remain stable, these types of models can be excellent at making very precise predictions within the bounds of the data that are used to construct them. Yet, no matter how good its predictive powers are, this type of model is opaque to the causal mechanisms that generate the covariance structure (Woodward 2010). This is the basis of the aphorism "correlation is not causation."

Other models are derived specifically to capture relevant features of some causal network. These models are derived by first defining the causal network and stating a set of assumptions about the interactions in the network. The causal network and assumptions thereabout are meant to represent some feature of nature under study. A mathematical description of the network and the assumptions are then derived and analyzed to generate descriptors of the dynamics generated by those assumptions. Those model dynamics serve as the basis to generate predictions that can be tested in the real world to examine how well the causal model represents the feature of nature under study. If the predictions are supported by empirical testing, one can then have some level of confidence that the causal model captures important features of the real system under study. Discrepancies will require refinement of the model and then subsequent empirical testing— which is the nature of science. The models discussed in this book fall into this category.

Although the feedback loop defined by causal model development and empirical testing is the basis of the scientific method, theoreticians and empirical scientists often have a hard time communicating with one another. The gulf may not be as wide for community ecologists as it is in other scientific disciplines, but the gulf does exist. As an empirical community ecologist myself, I have tried to write this book with the empirical community ecologist in mind. However, this is a book of mathematical models, so there is no way around *doing the math*.

First, I have tried to present the models at a level that, if you exercise your seventh-grade algebra, basic calculus, and graphing skills, you will be able to rederive everything I present here. And I strongly encourage empiricists to do just that! Read this book with a pad of paper and a pencil, and do the algebra to get from model to result and isocline picture. All the insight is embedded here. You will learn that repeated issues arise in these models, and these analyses become easier with the practice of doing the analysis alongside the book as a guide.

I have also emphasized only the mathematical features that are critical for developing biological insights. Therefore, only a few exact formulas for things like equilibria are given. Instead, I simply refer to the equilibrium where it exists, namely, where the isoclines cross at a single point. To extract biological insights, that is all that's required.

Because I am encouraging understanding through graphs of isoclines, visualization of these dynamical graphs is crucial. To aid the reader, I have also constructed computer simulators for many of the community modules to be explored. Simulation modules considered in this book can be found at https://mechanismsof coexistence.com/. These simulators allow the user to examine the geometry of the isoclines for multiple parameter combinations and show the dynamics of species abundances given those geometries. Links to the corresponding simulators are found at appropriate places throughout the book.

Finally, to guide the reader in developing empirical tests of these models, each major section also contains brief expositions of what I see as some of the critical predictions and the empirical data, both observational and experimental, that would test those predictions. These are not exhaustive lists of the predictions or of the studies that could or should be done, but rather are obvious and major features that must be tested to evaluate whether the model adequately captures a community dynamic and the mechanism permitting species coexistence. Empirical research is an exercise in creativity, and so no exposition I could provide should be taken as a definitive statement about how to approach any particular model prediction. I offer these only as guides and starting points for in-depth analyses. I hope they are suggestive and inclusive of the best tests. In addition, I hope this will exercise the skills of empirical ecologists to extract the critical predictions from models. At least my students have told me this is one benefit from the approach I have taken here.

Historical Antecedents

On a sunny March afternoon in 1981, many of the world's leading community ecologists boarded boats to see the wildlife along the Wakulla River in the panhandle of Florida. They were at the Wakulla Springs Lodge (now the Edward Ball Wakulla Springs State Park) attending a conference organized by the "Florida State Mafia"—Don Strong, Dan Simberloff, Larry Abele, and their postdocs and graduate students—to discuss the evidence needed to justify the claims of what processes foster species coexistence. Each boat was piloted by a local guide who would describe the birds, snakes, turtles, fish, and other wildlife that the passengers would see on their slow float down the river and on the return journey to the spring at the river's head. The boatmen spoke with thick north Florida accents, which made their descriptions particularly resonant and memorable. In addition to wildlife, they would identify the stump on which Johnny Weissmuller stood to give his famous yell during the filming of *Tarzan's Secret Treasure* (1941) and the pool in which Ricou Browning was filmed underwater as the Gill-Man for *Creature from the Black Lagoon* (1954).

All the boatmen were well versed in the natural history of the river, and each trip was a live nature show in which they recounted facts in their slow baritone drawls about the animals that the passengers could see. "If you look to the right, you will see the limpkin, one of the rarest birds in North America; and to the left is the anhinga, the snake bird." For these community ecologists, this was heaven to be floating down a river looking at species many had never seen before. The boat trip was also a respite from the serious and often contentious discussions about the evidence needed to support the claim that species are coexisting. At one point, the boatmen drew their passengers' attention to a large white bird standing on the bank. "Now there is a white ibis. You may be wondering how it is that the white ibis can coexist with the limpkin. Well, I'll tell you how it is. It is because the ratio of their bills is 1.3 to 1."

Many of the passengers were stunned; others were impressed. These local Florida boatmen knew about Hutchinsonian ratios and were explaining bird species coexistence using the currently prevailing scientific explanation—a primary subject of their conference. (In fact, John Wiens (1984, p. 441) was so impressed that he cited a personal communication from "Sam the Boatman" about Hutchinsonian ratios in his contribution to the resulting book.) After a few seconds, all realized that the boatmen's words were an elaborate joke for them, and each boat roared with laughter. Fran James had coached the boatmen to interject Hutchinsonian ratios into their descriptions of the wading birds.

Today, few would argue that Hutchinsonian ratios are a valid metric of species coexistence, but at the time they were a primary explanation for how species were able to coexist with one another. However, the theoretical basis for understanding how communities of species are structured, and which sparked the interest in Hutchinsonian ratios, has changed very little since 1981, despite new directions in theoretical constructs about the mechanisms that foster or retard species coexistence.

Moreover, I believe that the fundamental basis of the Florida State Mafia's critique is as applicable today as it has ever been: if one is going to interpret some set of metrics about nature according to a set of theoretically derived predictions that rest on the assumption of a particular mechanism operating, then a critical test must include the evaluation of whether the assumed mechanism is actually operating in that community. The arguments at the time spiraled off into debates about null models because of specific critiques about the data being used (e.g., Strong et al. 1979, Grant and Abbott 1980, Simberloff and Boecklen 1981). However, if one claims that some species are able to coexist because of how each acquires and utilizes a shared set of resources, the first issue that should be adjudicated scientifically is that some set of resources limit their abundances and that these species actually compete for those resources today (Connell 1980, Schoener 1986, Tilman 1987). Alternatively, if one claims that a shared predator fosters the coexistence of its prey, then the first issue that should be adjudicated scientifically is that the predator actually feeds on those species to a degree that influences their demography. Once the fundamental operation of a specific mechanism has been established, then more derived predictions based on the operation of that mechanism can be evaluated. The central point is that the hypothesized mechanism and its resultant consequences should be the central focus of investigation, and species at different positions in a food web coexist in that community because of different ecological processes—one theoretical construct does not and cannot fit all species.

We see the world through the lens of the theory we use to organize our conception of how nature works. In the 1970s, Hutchinsonian ratios were all the rage because theory supposedly told empirical scientists that this was the metric to measure for understanding coexistence. My goal in writing this book is to advocate to empirical ecologists for a different and more mechanistic theoretical paradigm for querying nature about species coexistence than the one that is used by many today. This different theoretical paradigm is not new at all, and in these pages I am largely reciting the insights of others: I think of this project as a synthesis. This alternative paradigm has been on a sustained developmental trajectory since 1963 and is now very rich in depth and breadth of application. However, it has not supplanted the theoretical edifice that gave us Hutchinsonian ratios. In fact, many have used pieces of this more mechanistic body of theory to justify the current paradigm, but I think that is misguided. I hope to make a compelling case in these pages for this alternative. In this chapter, I present a history of the ideas and broad development of theory in these two different but related approaches to understanding species coexistence.

VOLTERRA, LOTKA, AND GAUSE

Every student is almost immediately introduced to the theoretical foundation of community ecology with the pioneering work of Vito Volterra and Alfred J. Lotka, but they are not taught exactly what Volterra and Lotka did. In his seminal paper on species interactions, the first model that Volterra (1926) considered was one in which two species, N_1 and N_2, compete for a single food species while living in the same environment. In his model, the parameters ε_1 and ε_2 define the rates of change in population abundance for each species, respectively, if their common food supply were always in excess for them: today, we call this parameter the *intrinsic rate of increase*. The species diminish the quantity of food at rates $h_1 N_1$ and $h_2 N_2$, respectively, and each species responds to this resource depletion differently, as signified by coefficients γ_1 and γ_2, because they have different needs for the food. These assumptions imply the following differential equations for the rates of change in their respective abundances (Volterra 1926):

$$\frac{dN_1}{dt} = N_1\left(\varepsilon_1 - \gamma_1\left(h_1 N_1 + h_2 N_2\right)\right)$$
$$\frac{dN_2}{dt} = N_2\left(\varepsilon_2 - \gamma_2\left(h_1 N_1 + h_2 N_2\right)\right) \tag{2.1}$$

This is the original model of two consumer species competing for a single resource.

In this model, only one consumer species is able to maintain a population in the long run. If the inequality

$$\frac{\varepsilon_1}{\gamma_1} > \frac{\varepsilon_2}{\gamma_2} \tag{2.2}$$

is true, N_1 will eventually drive N_2 extinct, and if inequality (2.2) is reversed, N_2 will eventually drive N_1 extinct (Volterra 1926, Lotka 1932b).

Throughout this book, I use isocline analyses (Slobodkin 1961, MacArthur 1972, Case 2000, Hirsch et al. 2012, Strogatz 2015) to explore the long-term dynamics of such models of species interactions. The *isocline* for a species is the function that defines the combinations of all species abundances at which the species in question has no change in its own abundance: that is, when $dN_i/dt = 0$. Therefore, to find the isoclines for the two species in equations (2.1), simply set each equation equal to zero, simplify, and solve for one of the species abundances so that the functions can be graphed in a state space describing the abundances of the species (i.e., the axes of the graph are N_1 and N_2). This results in

$$\frac{dN_1}{dt} = 0 : N_2 = \frac{\varepsilon_1}{\gamma_1 h_2} - \frac{h_1}{h_2} N_1$$
$$\frac{dN_2}{dt} = 0 : N_2 = \frac{\varepsilon_2}{\gamma_2 h_2} - \frac{h_1}{h_2} N_1 \quad . \tag{2.3}$$

The graph of these two functions in the two-dimensional abundance space of $N_1 - N_2$ can be used to deduce the qualitative dynamics of this simple community at all abundance combinations for the two species. As is readily apparent from equations (2.3), the isoclines for the two species are parallel lines with negative slopes in the $N_1 - N_2$ abundance space. They are parallel because their resource harvest rates h_1 and h_2 are constant across all abundance combinations. If the point represented by their current abundance $[N_1, N_2]$ is in the space below both isoclines, the abundances of both will increase because the total growth rates of both species in equations (2.1) are positive. If the point representing their current abundances is above both isoclines, the abundances of both species will decline. However, if the point representing their abundances is in the space between the two isoclines, the species with its abundance above its isocline will decrease and the species with its abundance below its isocline will increase. As a result, the species whose isocline intersects both axes at the higher abundances will drive the other extinct. Inequality (2.2) simply is a statement of which isocline is above which, since this inequality states the relationships between the points of intersection of the two isoclines on each axis (fig. 2.1).

In this model, the points where each isocline intersects its own axis—$[N_1^*, N_2^*]$ = $[\varepsilon_1/(\gamma_1 h_1), 0]$ for the N_1 isocline and $[N_1^*, N_2^*] = [0, \varepsilon_2/(\gamma_2 h_2)]$ for the N_2 isocline— are both equilibria, as well as the trivial equilibrium of both species absent at $[N_1^*, N_2^*] = [0, 0]$ (again, the superscript star signifies that the abundance is an equilibrium) (see box 2.1). If inequality (2.2) is true, $[\varepsilon_1/(\gamma_1 h_1), 0]$ is the stable equilibrium (meaning that if the two species are started at abundances near this

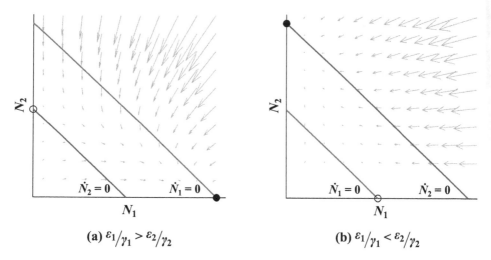

$$(a)\ {^{\varepsilon_1}}\!/\!_{\gamma_1} > {^{\varepsilon_2}}\!/\!_{\gamma_2} \qquad\qquad (b)\ {^{\varepsilon_1}}\!/\!_{\gamma_1} < {^{\varepsilon_2}}\!/\!_{\gamma_2}$$

FIGURE 2.1. The state space diagrams of the original model of two species competing for one food resource proposed by Volterra (1926). The axes in each panel are the abundances of each consumer species. The isoclines for the two species are labeled $\dot{N}_1 = 0$ and $\dot{N}_2 = 0$, respectively. The open circle is the location of the unstable equilibrium, and the filled black circle is a stable equilibrium. The grid of arrows shows how the abundances of each species change away from the isoclines, which identify the general trajectories of abundance change. In (a), N_1 is the better competitor that drives N_2 extinct; in (b), N_2 is the better competitor that drives N_1 extinct. The parameter values that are the same in both panels are $\delta_1 = \delta_2 = 1.0$, $h_1 = 0.2$, and $h_2 = 0.4$. Panel (a) also has $\varepsilon_1 = 1.0$ and $\varepsilon_2 = 0.5$, whereas (b) has $\varepsilon_1 = 0.5$ and $\varepsilon_2 = 1.0$.

equilibrium, the abundances will track to this equilibrium) and $[0,\ \varepsilon_2/(\gamma_2 h_2)]$ is unstable (meaning that if the two species are started at abundances near this equilibrium, their abundances will move away from this equilibrium) (fig. 2.1(a)); but if inequality (2.2) is reversed, the stability of these two equilibria is reversed (fig. 2.1(b)) (Lotka 1932b).

Both Volterra and Lotka also noted that, in the absence of the other species, this model simply results in logistic growth for each species of the form described by Verhulst (1838) and Pearl and Reed (1920), namely,

$$\frac{dN_j}{dt} = N_j\left(\varepsilon_j - \gamma_j h_j N_j\right). \qquad (2.4)$$

We can also analyze this model to determine the invasibility of each species. To illustrate, assume that inequality (2.2) is true, so the isoclines are as pictured in figure 2.1(a). Evaluating the invasibility of each species requires that we start the community at the long-term demographic steady state when the species is absent, and then introduce a few individuals of the species in question and evaluate

Box 2.1. In this box, I illustrate the analysis needed to come to inequality (2.2). Remember that a line intersects an axis at the point where the values of that point on all other axes are 0.0. Thus, to determine where each isocline intersects each axis, substitute 0.0 for the abundances of all species except the one for that axis in the isocline function and solve for that species. For the N_1 isocline in equations (2.3), substitute $N_1 = 0$ and then solving for N_2 gives $[N_1^*, N_2^*] = [0, \varepsilon_1/(\gamma_1 h_1)]$; substitute $N_2 = 0$ and then solving for N_1 gives $[N_1^*, N_2^*] = [\varepsilon_1/(\gamma_1 h_2), 0]$. Likewise, for the N_2 isocline in (2.3), substitute $N_1 = 0$ and then solving for N_2 gives $[N_1^*, N_2^*] = [0, \varepsilon_2/(\gamma_2 h_2)]$; substitute $N_2 = 0$ and then solving for N_1 gives $[N_1^*, N_2^*] = [\varepsilon_2/(\gamma_2 h_1), 0]$.

Since the isoclines of the two species are parallel, if the N_1 isocline intersects both axes above the N_2 isocline, this means on the N_1 axis the isocline intersection points imply

$$\frac{\varepsilon_1}{\gamma_1 h_1} > \frac{\varepsilon_2}{\gamma_2 h_1},$$

and likewise on the N_2 axis

$$\frac{\varepsilon_1}{\gamma_1 h_2} > \frac{\varepsilon_2}{\gamma_2 h_2}.$$

Note that both of these inequalities simplify algebraically to inequality (2.2).

whether their population will increase or decrease. To evaluate whether N_1 can satisfy its invasibility criterion, start the community with only N_2 present at its equilibrium of $[0, N_2^*] = [0, \varepsilon_2/(\gamma_2 h_2)]$ in the absence of N_1 (i.e., the stable equilibrium if N_1 is not present). Introducing a few N_1 individuals moves the community into the space between the two isoclines, and since N_1 has a positive per capita growth rate (i.e., $dN_1/N_1 dt > 0$) in this area of abundance space, its population will increase. Thus, N_1 can invade the community.

Likewise, to evaluate the invasibility criterion for N_2, we start the community at $[N_1^*, 0] = [\varepsilon_1/(\gamma_1 h_1), 0]$ and then introduce a few N_2 individuals. This moves the system into the space above both isoclines, where N_2 has a negative per capita population growth rate (i.e., $dN_2/N_2 dt < 0$), and the community returns to the equilibrium where only N_1 is present. Therefore, N_2 cannot invade. Volterra (1926) also extended this analysis to any number of species competing for a single resource and showed that one consumer would outcompete all others, namely, the one with the highest ε/γ_j ratio. This invasibility analysis reinforces the conclusions drawn from evaluating the stability of the various boundary equilibria where each species is absent (Law and Morton 1993, 1996, Morton and Law 1997).

The Volterra competition simulation module can be accessed at https://mechanismsofcoexistence.com/VolterraCompetition.html.

These two species could coexist with one another if a stable equilibrium existed at which both species have positive abundances. This would occur if their two isoclines could intersect at a single point with both having positive abundances. However, they can never intersect at a single point, because the isoclines of this model are parallel. Thus, the formulation of resource competition between two consumers by Volterra and Lotka (i.e., the Volterra-Lotka competition model) does not permit the two competing species to coexist. This is the result that led Hutchinson (1958, p. 417) to state, "Volterra . . . demonstrated by elementary analytic methods that under constant conditions two species utilizing, and limited by, a common resource cannot coexist in a limited system." Moreover, this result forms the basis of our intuition about niches, namely, that two species cannot occupy the same niche.

Many readers may be surprised to find that equation (2.4) is the basic model of logistic population growth, that equations (2.1) form the model of competition between two species analyzed by Volterra and Lotka, and that it is impossible for the two competitors in this model to coexist. These are not the versions of these equations that are taught to most students of ecology or used in most analyses today (e.g., Case 2000). The version of logistic population growth taught in most ecology classes today was introduced by Lotka (see Kingsland 1982 for a history of the model's early development), with the rationale being that logistic population growth is a modification of exponential population growth. Lotka (1925) reasoned that, when rare, a population initially grows at an approximately exponential rate of $dN_1/dt = b_1N_1$. However, the increasing population meets what Gause (1934) called "environmental resistance" that decreases this growth rate as a linear function of its own abundance:

$$\frac{dN_1}{dt} = b_1 N_1 \left(\frac{K_1 - N_1}{K_1} \right). \tag{2.5}$$

This is the model for logistic population growth that most students of ecology learn. In this model, b_1 is the *intrinsic rate of increase* for this species, or more directly its per capita population growth rate when rare (i.e., $N_1 \approx 0$). K_1 is interpreted as the *carrying capacity* of the environment, which is the stable equilibrial abundance of this species (i.e., $N_1^* = K_1$) if $b_1 > 0$ and no other species are present. In the coming chapters, I do not use equation (2.5) for logistic population growth because this function has a number of logical flaws that I discuss in chapter 3;

rather, I use a model of logistic population growth akin to equation (2.4). I present equation (2.5) here to orient the reader to the historical development of these ideas. However, the reader should note that equations (2.4) and (2.5) are mathematically comparable since $\varepsilon_1 = b_1$ and $\gamma_1 h_1 = b_1/K_1$.

Gause (1932, 1934) used Lotka's (1925) form of the logistic equation to define a model of competition between two species that is akin to Volterra's (1926) original model:

$$
\frac{dN_1}{dt} = b_1 N_1 \left(\frac{K_1 - N_1 - \alpha N_2}{K_1} \right)
$$
$$
\frac{dN_2}{dt} = b_2 N_2 \left(\frac{K_2 - N_2 - \beta N_1}{K_2} \right)
$$
(2.6)

In this form of the competition model, α and β are parameters that scale the effect of heterospecific abundance relative to the effect of conspecific abundance in causing population growth rate to decline as the two species' abundances increase. If $\alpha > 1$, then each N_2 individual has a greater effect on decreasing N_1's per capita population growth rate than does an N_1 individual: in other words, the strength of interspecific competition from N_2 is greater than the strength of intraspecific competition of N_1 on itself. Conversely, if $\alpha < 1$, the strength of intraspecific competition in depressing N_1's per capita population growth rate is greater than the strength of interspecific competition from N_2. Comparable statements apply to β and the relative strengths of intraspecific and interspecific competition effects on the per capita population growth rate of N_2.

The Gause competition simulation module can be accessed at https://mechanismsofcoexistence.com/GauseCompetition.html.

This is the version of the model with which most ecology students today will be familiar. Gause concluded the presentation by stating that this model coincides with that of Volterra (1926), "but it does not include any parameters dealing with the food consumption, and simply expresses the competition between species in terms of the growing populations themselves" (Gause 1934, p. 51). In other words, this model is devoid of any causal mechanism by which the species actually interact. He then used the results inferred from Volterra's and Lotka's analyses to interpret the results of his famous experiments with *Paramecium*.

Gause and Witt (1935) showed in a subsequent paper (although using a slightly different set of equations from equations (2.6)) that four different outcomes can result, depending on the specific values of α and β relative to the carrying capacities of the two species (see also Slobodkin 1961). To see these possible outcomes, the isoclines for this model must be considered. From equations (2.6), the isoclines are

$$\frac{dN_1}{dt} = 0 : N_2 = \frac{K_1}{\alpha} - \frac{1}{\alpha} N_1$$
$$\frac{dN_2}{dt} = 0 : N_2 = K_2 - \beta N_1 \qquad (2.7)$$

Clearly, these are functions describing lines in the $N_1 - N_2$ space. Therefore, their relationships relative to one another can be described by where each intersects each axis. Applying the same logic and methods used in box 2.1, the N_1 isocline intersects the N_1 axis at K_1 (i.e., substitute $N_2 = 0$ into the N_1 isocline function and solve for N_1) and intersects the N_2 axis at K_1/α (i.e., substitute $N_1 = 0$ into the function and solve for N_2). Likewise, the N_2 isocline intersects the N_1 axis at K_2/β and the N_2 axis at K_2.

If the resulting isoclines do not intersect one another at positive abundances for the two species, again the species with the isocline farther from the origin will drive the other extinct. N_1's isocline will intersect both axes above N_2's isocline, and so N_1 will drive N_2 extinct if

$$\frac{K_1}{K_2} > \alpha \text{ and } \frac{K_1}{K_2} > \frac{1}{\beta} \qquad (2.8)$$

(fig. 2.2(a)) (Slobodkin 1961). These inequalities simply state the relationships between the values of the intersection points of the two isoclines with the two axes: that is, $K_1 > K_2/\beta$ on the N_1 axis and $K_1/\alpha > K_2$ on the N_2 axis. When interpreted in their biological meaning, these inequalities show that N_1 may overwhelm N_2 in two ways. First, there is strength in sheer numbers, and so if $K_1 > K_2$, the strengths of interspecific competition for each species may be irrelevant. The relative strengths of interspecific competition become important when the carrying capacities are similar, and so their ratio is near 1.0. If $K_1 \approx K_2$, this implies that the interspecific effect of N_2 on N_1 is less than the intraspecific effect of N_1 on itself (i.e., $1 > \alpha$), and the interspecific effect of N_1 on N_2 is greater than the intraspecific effect of N_2 on itself (i.e., $\beta > 1$, which would make $1 > 1/\beta$).

Likewise, N_2's isocline intersects both axes above N_1's isocline, and so N_2 will drive N_1 extinct if

$$\frac{K_1}{K_2} < \alpha \text{ and } \frac{K_1}{K_2} < \frac{1}{\beta}, \qquad (2.9)$$

with analogous interpretations of the relative strengths of intraspecific and interspecific competition between the two species (fig. 2.2(b)) (Slobodkin 1961).

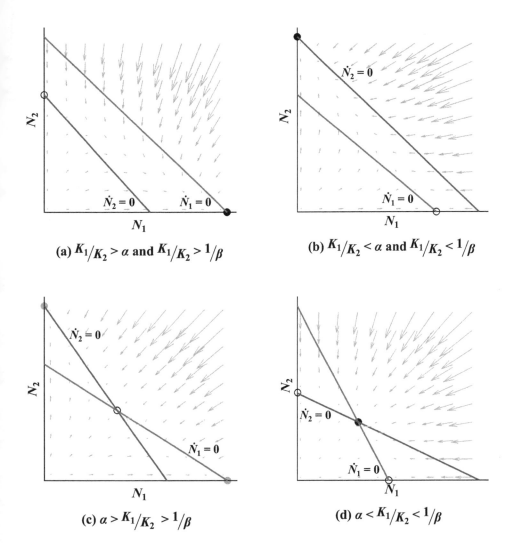

FIGURE 2.2. The state space diagram for the four possible outcomes of two species competing in the Gause (1934) model of competition. The isoclines of the two species and the stable and unstable equilibria are identified as in figure 2.1. The filled gray circles are alternative stable equilibria to which the species abundances will track depending on initial conditions, and so form a priority effect. In all panels, $b_1 = b_2 = 1.0$. In (a), $\alpha_{12} = 1.0$, $\alpha_{21} = 1.15$, $K_1 = 30$, and $K_2 = 20$. In (b), $\alpha_{12} = 1.0$, $\alpha_{21} = 1.15$, $K_1 = 20$, and $K_2 = 30$. In (c), $\alpha_{12} = 1.5$, $\alpha_{21} = 1.5$, $K_1 = 30$, and $K_2 = 30$. In (d), $\alpha_{12} = 0.5$, $\alpha_{21} = 0.5$, $K_1 = 15$, and $K_2 = 15$.

However, the two isoclines can intersect at positive abundances in this model because they are not parallel lines, and two different configurations of the isoclines can result. In one configuration, the two-species equilibrium is unstable and the consumer that has an initial abundance advantage will drive the other extinct (fig. 2.2(c) (Slobodkin 1961). This occurs when

$$\alpha > \frac{K_1}{K_2} > \frac{1}{\beta} \tag{2.10}$$

(i.e., $K_1 > K_2/\beta$ on the N_1 axis and $K_2 > K_1/\alpha$ on the N_2 axis). If $K_1 \approx K_2$, this implies that the interspecific effect of each species on the other is greater than the intraspecific effect of each species on itself (i.e., $\alpha > 1$ and $\beta > 1$). In other words, a *priority effect* results when interspecific competition is stronger than intraspecific competition for both species (and their carrying capacities are similar). In terms of invasibility, neither species can invade the community if the other is present at its long-term steady state (i.e., its carrying capacity).

The fourth and final possible configuration between the isoclines results in a stable two-species equilibrium at which the two competitors coexist (fig. 2.2(d)) (Slobodkin 1961). This occurs when

$$\alpha < \frac{K_1}{K_2} < \frac{1}{\beta} \tag{2.11}$$

(i.e., $K_2/\beta > K_1$ on the N_1 axis and $K_1/\alpha > K_2$ on the N_2 axis). This inequality can be satisfied even if interspecific competition is stronger than intraspecific competition for both species (i.e., $\alpha > 1$ and $\beta > 1$), but this requires the correct ratio of carrying capacities. However, this inequality is usually interpreted when the carrying capacities of the two species are comparable. If $K_1 \approx K_2$, this implies that the intraspecific effect of each species on itself is greater than the interspecific effect that either has on limiting the other species (i.e., $\alpha < 1$ and $\beta < 1$) (Slobodkin 1961). This is the result used to justify the typical conclusion that a species must have a greater effect in limiting its own abundance than it has in limiting the abundance of its competitor (i.e., intraspecific competition must be stronger than interspecific competition) for species to coexist (e.g., May 1973, Chesson 1994, 2000b).

While the competition coefficients in equations (2.6) can be interpreted as embodying interference competition between the species (e.g., aggression, territoriality, allelopathy) (e.g., Hutchinson and MacArthur 1959a), most discussions of this model have interpreted the competition coefficients as representing the indirect effects of resource limitation due to differences in resource utilization (as specified in Volterra's original formulation) but modeled as direct competitive effects between the consumers, which bypasses explicit resource dynamics.

At this same time, Hutchinson was developing his conceptual idea of the niche as an N-dimensional hypervolume of favorable environmental conditions and exploring the empirical patterns seen among species that are presumably coexisting. Hutchinson's own ideas about competition among species and coexistence were primarily based on empirical evidence—Gause's (1932, 1934) experiments showing competitive exclusion of *Paramecium* species in simple laboratory containers, and work by many, including himself, showing that co-occurring competitors are different (e.g., Lack 1947, Hutchinson 1951, 1959, Hutchinson and MacArthur 1959b). In his famous "Homage to Santa Rosalia," Hutchinson (1959, p. 152) marshaled empirical evidence from a number of animal groups to show that "where the species co-occur, the ratio of the larger to the small form varies from 1.1 to 1.4, the mean ratio being 1.28 or roughly 1.3"; thus the idea of Hutchinsonian ratios identifying coexisting competitors was born. This idea comported with his set-theoretic construction of the niche to infer that competitors must be different to coexist, and these differences placed them into different niches (Hutchinson 1958, p. 417):

> These findings have been extended and generalised to the conclusion that two species, when they co-occur, must in some sense be occupying different niches. The present writer believes that properly stated as an empirical generalisation, which is true except in cases where there are good reasons not to expect it to be true, the principle is of fundamental importance and may be properly called the Volterra-Gause Principle.

MacArthur and colleagues made this link between the Volterra-Lotka-Gause competition coefficients and resource limitation caused by consumers depleting those resources more explicit in the next decade. In this work, they proposed a number of different expansions of the Gause model of competition in which they attempted to explicitly explore what patterns of resource utilization would be required to permit coexistence of resource competitors. In the first, MacArthur and Levins (1964) showed how two resource competitors could coexist on two resources that are spatially segregated into different patches. They analogized this result to Slobodkin's (1961) result showing how the two competitors can coexist in Gause's model (i.e., inequality (2.11)).

MacArthur and Levins (1967) also proposed a model of competition among three consumers using Gause's formulation in which they imagined a broad and continuous spectrum of resources being available to the competitors. Their analysis began the two-decade search for the theoretical statement defining the *limiting similarity* among resource competitors to justify Hutchinson's empirical assertion of the 1.3:1 ratio. In this model, each species is able to utilize a different range of the available resource spectrum (fig. 2.3). The carrying capacity for each consumer

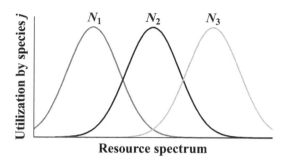

FIGURE 2.3. Resource utilization curves for three consumers feeding on a spectrum of resources. This figure is a representation of the model analyzed by MacArthur and Levins (1967).

is defined by the total area under its utilization curve, and the competition coefficients between each pair of species (e.g., α and β in equations (2.6) for each species pair) are defined by the overlap in their utilization functions. Assuming Gaussian utilization functions that were identical in shape among the species but different in location, they determined that the distance between the peaks of the utilization curves for adjacent species would have to be at least 1.1 times the standard deviation of the curves for the species to coexist along the resource spectrum. This analysis of the *species packing* problem represented the first theoretical justification for Hutchinson's empirical results of ratios of body parts and body size implying resource competition, although their analysis only made the assumption that resource utilization scales with some phenotypic feature of the consumers.

The MacArthur and Levins (1967) formulation defined a theoretically appealing approach in the attempt to link competitors' coexistence to Hutchinson's (1958) conception of the niche, by formulating the degree of *niche overlap* that was permitted while still having the resource competitors coexist. Hutchinson (1958) described the niche axes as being abiotic environmental features, such as temperature, as well as features of other species, such as the size of food. MacArthur and Levins (1967) provided a theoretical framework that made the intuitively appealing Hutchinsonian niche quantifiable and directly linked to the dynamical theory of competition.

The basic approach of MacArthur and Levins (1967) was extended and refined over the ensuing decade to determine how Volterra-Lotka-Gause competition coefficients could be quantified from other ecological situations (e.g., Schoener 1974b, Abrams 1975, 1976, 1980a, 1983a) and to determine whether a *limiting similarity* among competitors really existed that would justify Hutchinson's empirical determination of the 1.3:1 rule (e.g., May and MacArthur 1972, Turelli 1978a, 1981). However, this search was fraught with difficulty. MacArthur and Levins (1967) noted in their original analysis that, if the resource spectrum was multidimensional, no limit to competitor similarity was needed to foster coexistence: only very minute differences were needed. May and MacArthur (1972) then noted that, in a temporally constant environment, only very minute differences also permitted coexistence along a one-dimensional resource axis. Their analysis suggested the need for fluctuations in the resources to require substantial differences among competitors to permit their coexistence. Finally, Turelli (1978a, 1981) showed that even temporal variability did not limit coexisting competitors to be greater than some minimum or limiting similarity. The theoretical justification for Hutchinsonian ratios largely vanished, although some continue to explore issues about limiting similarity (Szabó and Meszéna 2006, Vergnon et al. 2013, D'Andrea and Ostling 2016, Schamp and Jensen 2019).

Real data was similarly unkind to the idea of a limiting similarity defining Hutchinsonian ratios. A number of other analyses showed how such ratios could arise merely by log-normal distributions of body size and not because of competition at all (e.g., Hespenheide 1973, Horn and May 1977, Maiorana 1978, Connor and Simberloff 1979, Strong et al. 1979, Grant and Abbott 1980, Roth 1981, Simberloff and Boecklen 1981, Wiens and Rotenberry 1981, Wiens 1982, Boecklen and NeSmith 1985, Eadie et al. 1987). Moreover, papers appeared in the literature showing that inanimate objects like frying pans, musical instruments such as recorders and stringed instruments (e.g., violin, viola, cello, bass), and tricycles and bicycles also occur in Hutchinsonian ratios (Horn and May 1977). By the mid- to late 1980s, the validity of data showing Hutchinsonian ratios to conclude that competing species were coexisting was largely rejected—the Florida State Mafia's critique of inferring process from pattern had seemingly won.

MACARTHUR RECASTS CONSUMER-RESOURCE MODELS

Along with their derivation of a model of competition between two resource competitors, Lotka and Volterra each explored a model of a predator species feeding on a prey species, which is now identified as the Lotka-Volterra predator-prey model (Lotka 1920, 1932a, Volterra 1926). Unlike the competition model, the Lotka-Volterra predator-prey model is taught to students today basically in its original form. This model was curious because the species display a neutral limit cycle in which the predator lags behind the prey by one-quarter cycle (Lotka 1920, Volterra 1926). This dynamical behavior was frequently invoked when populations of predator and prey were seen to fluctuate over time in nature. However, little was done directly with the predator-prey model for many years thereafter, with the focus for understanding species coexistence being primarily on the Volterra-Lotka-Gause model of competition.

MacArthur (1969, 1970, 1972) made another pivotal insight about coexistence of competing species when he showed how an explicit model of consumer-resource interactions like the original Lotka-Volterra predator-prey model could be recast into a model of competition of the Volterra-Lotka-Gause form. He began by imagining a set of consumer species that feed upon a set of resource species, and the consumers interact with one another only indirectly through their shared feeding on the resources. The abundance dynamics of the consumers and the resources are given by the equations

$$\frac{dN_j}{dt} = C_j N_j \left(\sum_{i=1}^{p} w_i a_{ij} R_i - T_j \right)$$

$$\frac{dR_i}{dt} = R_i \left[r_i \left(\frac{K_i - R_i}{K_i} \right) - \sum_{j=1}^{q} a_{ij} N_j \right],$$

(2.12)

where w_i is the biomass of one resource i individual, a_{ij} is the rate at which an individual consumer encounters and consumes individuals of resource i, T_j is the threshold food intake level that an individual of consumer j requires, C_j is the proportionality constant that converts the net biomass of prey eaten minus metabolic losses into new individuals of consumer j, and r_i and K_i are the intrinsic rate of increase and carrying capacity, respectively, for resource i (MacArthur 1969, 1970, 1972).

If one assumes the resource dynamics occur much faster than the consumers' dynamics, so that the resource abundances are always at equilibrium given the current consumer abundances, the resource equations can be subsumed into the consumer equations. First, consider the case in which one consumer feeds on one resource in equations (2.12). The equilibrium abundance of the resource is found by setting the resource's equation equal to zero and solving for R_1, which gives

$$R_1^* = K_1 - \frac{K_1 a_{11}}{r_1} N_1$$

(2.13)

(remember that resources are always at equilibrium). Substituting this into the consumer equation gives

$$\frac{dN_1}{dt} = C_1 N_1 \left(w_1 a_{11} \left(K_1 - \frac{K_1 a_{11}}{r_1} N_1 \right) - T_1 \right)$$

$$= N_1 \left(C_1 \left(w_1 a_{11} K_1 - T_1 \right) - \frac{C_1 K_1 w_1 a_{11}^2}{r_1} N_1 \right).$$

(2.14)

The first term is the intrinsic rate of increase for the consumer, namely, its per capita population growth rate when it invades a community containing R_1 at its equilibrium abundance in the absence of N_1 (i.e., at $R_1^* = K_1$). The second term is the rate at which N_1's per capita population growth rate declines with its increasing abundance. This is completely equivalent to logistic growth as derived by Verhulst (1838) and Pearl and Reed (1920), as in equation (2.4), with $\varepsilon_1 = C_1$ $(w_1 a_{11} K_1 - T_1)$ and $\gamma_1 h_1 = C_1 K_1 a_{11}/r_1$. Thus, one interpretation of logistic growth is as a *resource limitation* in which the resource dynamics have been subsumed into the consumer species equation.

Now consider the case in which two consumer species, N_1 and N_2, feed on a single resource species R_1 in equations (2.12). The equilibrium abundance of the resource is now

$$R_1^* = K_i - \frac{K_1 a_{11}}{r_1} N_1 - \frac{K_1 a_{12}}{r_1} N_2. \tag{2.15}$$

This equation gives the equilibrium resource abundance for all possible combinations of the two consumers' abundances. Also, note that this formulation again expresses the current resource abundance as the maximum (i.e., the resource abundance in the absence of all consumers) minus the amount removed by each consumer. Equation (2.15) is then substituted into the two consumer growth rate equations (2.12) to give the consumers' dynamics, with the resource's dynamics embedded in their equations:

$$\frac{dN_1}{dt} = C_1 N_1 \left(w_1 a_{11} \left(K_1 - \frac{K_1 a_{11}}{r_1} N_1 - \frac{K_1 a_{12}}{r_1} N_2 \right) - T_1 \right)$$

$$\frac{dN_2}{dt} = C_2 N_2 \left(w_1 a_{12} \left(K_1 - \frac{K_1 a_{11}}{r_1} N_1 - \frac{K_1 a_{12}}{r_1} N_2 \right) - T_2 \right)$$

These two equations can then be rearranged into a form that corresponds to the original Volterra-Lotka competition equations:

$$\begin{aligned}
\frac{dN_1}{dt} &= N_1 \left(C_1 \left(w_1 a_{11} K_1 - T_1 \right) - \frac{C_1 K_1 w_1 a_{11}^2}{r_1} N_1 - \frac{C_1 K_1 w_1 a_{11} a_{12}}{r_2} N_2 \right) \\
\frac{dN_2}{dt} &= N_2 \left(C_2 \left(w_1 a_{12} K_1 - T_2 \right) - \frac{C_2 K_1 w_1 a_{11} a_{12}}{r_1} N_1 - \frac{C_2 K_1 w_1 a_{12}^2}{r_1} N_2 \right)
\end{aligned} \tag{2.16}$$

The corresponding components between equations (2.1) and (2.16) are

$$\varepsilon_j = C_j \left(w_1 a_{1j} K_j - T_j \right), \gamma_j = \frac{C_j K_1 w_1 a_{1j}}{r_1}, \text{ and } h_j = a_{1j}.$$

Solving for the isoclines shows that the same result obtains in this description of the consumers' interaction as well:

$$\frac{dN_1}{dt} = 0 : N_2 = \frac{r_1 K_1 w_1 a_{11} - r_1 T_1}{w_1 a_{11} a_{12} K_1} - \frac{a_{11}}{a_{12}} N_1$$

$$\frac{dN_2}{dt} = 0 : N_2 = \frac{r_1 K_1 w_1 a_{12} - r_1 T_2}{w_1 a_{12}^2 K_1} - \frac{a_{11}}{a_{12}} N_1$$

(2.17)

As in the original Volterra-Lotka competition model (cf. equations (2.3) and (2.17)), when the consumers feed on a single resource species, their isoclines are parallel lines, and so the consumer with the isocline that intersects the abundance axes at the higher values will drive the other extinct (as in fig. 2.1).

If the two consumers feed on two or more resource species, their abundance dynamics equations with the resources embedded in them become

$$\frac{dN_1}{dt} = N_1 \left(\left(C_1 \left(\sum_{i=1}^{p} w_i a_{i1} K_i - T_1 \right) \right) - \left(C_1 \sum_{i=1}^{p} \frac{K_i w_i a_{i1}^2}{r_i} \right) N_1 - \left(C_1 \sum_{i=1}^{p} \frac{w_i a_{i1} K_i a_{i2}}{r_i} \right) N_2 \right)$$

$$\frac{dN_2}{dt} = N_2 \left(\left(C_2 \left(\sum_{i=1}^{p} w_i a_{i1} K_i - T_2 \right) \right) - \left(C_2 \sum_{i=1}^{p} \frac{w_2 a_{i2} K_i a_{i1}}{r_i} \right) N_1 - \left(C_2 \sum_{i=1}^{p} \frac{K_i w_2 a_{i2}^2}{r_i} \right) N_2 \right)$$

(2.18)

where p is the number of resource species. Now, the resulting isoclines are not parallel, since the species will differ in their abilities to capture the various resources and may cross in any of the four configurations, analogous to figure 2.2.

This model typically is then recast by embedding the functions of each term into single parameters. For this model, this recasting is simple, because these terms are all simple functions of only model parameters. Thus, equations (2.16) and (2.18) can both be recast as

$$\frac{dN_1}{dt} = N_1 \left(r_1 - \alpha_{11} N_1 - \alpha_{12} N_2 \right)$$

$$\frac{dN_2}{dt} = N_2 \left(r_2 - \alpha_{21} N_1 - \alpha_{22} N_2 \right)$$

(2.19)

where the parameters in equations (2.19) represent the corresponding terms in equations (2.16) or (2.18). The first term $r_j (= C_j (\sum_{i=1}^{p} w_i a_{ij} K_i - T_j))$ is the *intrinsic rate of increase* for each competitor: the population growth rate when the consumer invades a community that contains the resources at their equilibrium abundances in the absence of all consumers (K_i's). The *competition coefficients* $\alpha_{jk} (= C_j \sum_{i=1}^{p} K_i w_i a_{ij} a_{ik} / r_i)$ quantify the indirect effects of species k on species j mediated through their shared direct effects on the resource abundances. Given this formulation, these two competitors will coexist if

$$\frac{\alpha_{11}}{\alpha_{21}} > \frac{r_1}{r_2} > \frac{\alpha_{12}}{\alpha_{22}}. \tag{2.20}$$

This result has been derived a number of times in the context of equations (2.19) directly (e.g., inequality (6.25) of May 1973, inequality (4) of Vandermeer 1975) or in the context of MacArthur's recasting of the consumer-resource model of (2.18) (e.g., inequality (2) of Abrams 1975, inequality (13) of Chesson 1990). Schoener (1974b) also constructed competition coefficients from other forms of the base model.

The MacArthur recasting competition simulation module can be accessed at https://mechanismsofcoexistence.com/MacArthurCompetition.html.

The above analyses present MacArthur's classic recasting of a model of resource competition among multiple consumers feeding on multiple resources into the framework of the Volterra-Lotka-Gause competition model that subsumes the resource dynamics into parameters of the model. Others have also explored the consequences of this recasting (Abrams 1975, 1980a, Chesson 1990, Letten et al. 2017, McPeek 2019b). This recasting undergirds the typical interpretation of the Volterra-Lotka-Gause competition models in the context of only resource competition, as in Gause's (1934) original statement. The intraspecific and interspecific competitive effects are typically interpreted as quantifying the degree of resource limitation that each consumer indirectly generates on themselves and on the other consumers (e.g., Chesson 2000b, Letten et al. 2017, Saavedra et al. 2017).

This interpretation, though, is completely unwarranted. It is surely true that the demography of every single species is influenced by more than one single type of species interaction. In fact, other types of species interactions can provide the same features as resource competition in this recasting. For example, Holt (1977) showed how the indirect interactions among a set of prey species that are mediated through a shared predator can result in the same experimental results as when the abundances of resource competitors are manipulated. He dubbed this indirect interaction *apparent competition* because of this correspondence in experimental results between the different interactions. This interaction is discussed in greater depth later in this chapter and more formally in chapter 3. Here, I use it only as an illustration in the framework of MacArthur's recasting. MacArthur's recasting can be applied if the consumers in equations (2.12) also experience some degree of direct self-limitation (Chesson and Kuang 2008). For example, one can expand the threshold food intake to be negatively density dependent, $T_j(N_j) = T_j + G_j N_j$, so that each consumer has a higher minimum food threshold for subsistence with greater consumer abundance (perhaps the consumers develop greater levels of stress with conspecific crowding). With this addition to the model, the consumers' abundances can be incorporated into the resource equations by making the assumption that the consumer abundances are always at equilibrium with respect to the current resource abundances (Chesson and Kuang 2008).

To illustrate, consider a single consumer species feeding on two resource species from equations (2.12) with the addition of $T_j (N_j) = T_j + G_j N_j$. Recasting the resource equations to include the consumer's dynamics gives

$$
\begin{aligned}
\frac{dR_1}{dt} &= R_1 \left(\left(r_1 - \frac{a_{11} T_1}{G_1} \right) - \left(\frac{r_1}{K_1} + \frac{w_1 a_{11}^2}{G_1} \right) R_1 - \left(\frac{w_2 a_{11} a_{21}}{G_1} \right) R_2 \right) \\
\frac{dR_2}{dt} &= R_2 \left(\left(r_2 - \frac{a_{21} T_1}{G_1} \right) - \left(\frac{w_1 a_{11} a_{21}}{G_1} \right) R_1 - \left(\frac{r_2}{K_2} + \frac{w_2 a_{21}^2}{G_1} \right) R_2 \right)
\end{aligned}
\tag{2.21}
$$

These equations, too, can be interpreted in the Volterra-Lotka-Gause competition context,

$$
\begin{aligned}
\frac{dR_1}{dt} &= R_1 \left(r_1 - \alpha_{11} R_1 - \alpha_{12} R_2 \right) \\
\frac{dR_2}{dt} &= R_2 \left(r_2 - \alpha_{21} R_1 - \alpha_{22} R_2 \right)
\end{aligned}
\tag{2.22}
$$

where parameters in equations (2.22) correspond to the respective terms in parentheses in equations (2.21).

Both of these indirect interactions have been recast into Volterra-Lotka-Gause competition, and in this context coexistence of these species requires that both satisfy inequality (2.20). A little algebra shows that satisfying this inequality in each case reflects the satisfaction of the criteria that would be derived if the consumer-resource models were analyzed directly (presented in chapter 3). However, the reasons that the species satisfy inequality (2.20) in the two cases are fundamentally different. Recasting these interactions into Volterra-Lotka-Gause competition coefficients of "intraspecific versus interspecific competitive effects" relative to their intrinsic rates of increase can identify if two or more species are coexisting, but the reason why they are coexisting—the mechanism—is left to the imagination, and that imagination typically gravitates to the resource limitation interpretation regardless of the true mechanism permitting their coexistence.

The r-α competition simulation module can be accessed at https://mechanismsofcoexistence.com/rAlphaCompetition.html.

In effect, MacArthur's recasting converts a model of causation into a model of correlation. Equations (2.12) are a causal model of multiple consumers feeding on multiple resources, whereas equations (2.19) and (2.22) are correlational models devoid of mechanism because the mechanism is embedded inside their parameters. Just as a correlation coefficient results from the underlying network of causation determining the values of two variables, the Volterra-Lotka-Gause competition coefficients result from the underlying causal network of species interactions. Consequently, just as correlations are not explanations for causation, Volterra-Lotka-Gause competition coefficients shed no light on the causal mechanisms of coexistence. These correlational models will predict the correct correlated change in species abundances if the assumptions of the two underlying causal models are met. However, these models will fail to predict the correct change in species abundances if the assumptions of the two underlying models are not met (Reynolds and Brassil 2013, O'Dwyer 2018). Moreover, simply demonstrating that the model predicts the right outcome provides no strong test of any causal mechanism whatsoever.

Understanding the mechanism of how species coexist has important ramifications for how we understand community structure. With resource competition, the critical issue is whether each consumer has adequate resources for it to have a positive per capita population growth rate when rare. With apparent competition, the critical issue is whether the mortality rate each resource experiences from the consumer is low enough for it to have a positive per capita population growth rate when rare. Moreover, how other species are affected by these species are fundamentally different in these two cases: the addition of each consumer depresses resource levels for subsequent invading species, whereas the addition of each resource inflates consumer levels for subsequent invading species. In the words of Elton (1927), the "occupation" of each of these species and what makes each species successful in the community are fundamentally different issues. As I said in chapter 1, we view and interpret the world through the lens of the theoretical constructs we use.

Moreover, this recasting can only be accomplished under very restrictive assumptions—in other words, when the assumptions of the two underlying causal models used as justification fail. The primary assumption is that the subsumed species are instantaneously at equilibrium with the abundances of the species in question, which can be problematic for many realistic biological situations (Reynolds and Brassil 2013, O'Dwyer 2018). Also, the consumer dynamics cannot be subsumed into the resource dynamics unless the consumers are assumed to experience some level of direct intraspecific density dependence in addition to resource limitation (e.g., Chesson and Kuang 2008). The approach also relies on the fact that all per capita demographic processes are modeled as linear functions (Abrams 1980a, 1980b). The self-limitation in resources is modeled as logistic, which is a linear per capita effect, and the functional responses of consumers feeding on the resources and their numerical responses to that feeding are also linear (MacArthur 1969, 1970, 1972, Chesson 1990). Other linear models of resource self-limitation may be used, but they also suffer those same difficulties (O'Dwyer

2018). If any of these demographic processes are nonlinear, the competition coefficients are not simply functions of the parameters of the underlying models, as in equations (2.18) and (2.21), but rather they change with the abundances of the species that are subsumed into the dynamical equations (Abrams 1980a, 1980b). Moreover, the pairwise competition coefficient between two species changes in the presence of other resources and competitors (e.g., Wilbur 1972, Neill 1974).

Many different topics have been introduced to the discussion over the years to accommodate these anomalies of underlying nonlinearities, namely, *higher-order interactions* and "*diffuse*" *competition* (May 1973, Case and Bender 1981, Abrams 1983b, 2006, 2009a, Billick and Case 1994, Chesson 2000b, Stump 2017). These anomalies are accommodated by adding terms to the Volterra-Lotka-Gause competition models. These additions are all meant to account for the empirical consequences that the pairwise competition coefficients measured between species are not static numbers, but rather change with the abundances of the species and in the presence of other types of species (e.g., Wilbur 1972, Neill 1974, Smith-Gill and Gill 1978). This makes interpretation of these results highly problematic for understanding community structure. Yet, we continue to use derivatives of the Volterra-Lotka-Gause competition model as the basis for our understanding and our guide for structuring our empirical inquiries about species coexistence in nature. These additions also provide little to no insight for understanding the mechanisms that foster or retard coexistence among the species of interest. Such modifications to a correlational model try to make it fit the data better without any insight into causal mechanisms.

These fundamentally problematic issues with the Volterra-Lotka-Gause competition framework as the way of understanding species coexistence have apparently been forgotten. Many have now characterized this approach, the resulting equations (2.19) and (2.22), and the criteria for coexistence they define, namely, inequality (2.20), as modern coexistence theory (Mayfield and Levine 2010, HilleRisLambers et al. 2012, Letten et al. 2017, Saavedra et al. 2017, Barabás et al. 2018). Moreover, this approach has formed the theoretical basis for many recent field experimental studies of coexistence, particularly for assemblages of annual plants (e.g., Adler et al. 2006, 2010, 2013, 2018, Harpole and Suding 2007, Sears and Chesson 2007, Angert et al. 2009, Levine and HilleRisLambers 2009, Godoy et al. 2014, Godoy and Levine 2014, Kraft et al. 2015, Staples et al. 2016). The general predictions for coexisting species from this are as follows: (1) each species has a positive per capita population growth rate when it is rare in the community (i.e., invasibility); (2) each species' per capita population growth rate declines as its frequency increases (i.e., $\alpha_{ii} > 0$); (3) each species limits its own abundance more than it limits the abundances of other species in the community (i.e., $\alpha_{ii} > \alpha_{ji}$); and (4) the decline in per capita population growth rate with relative frequency and the differences in intraspecific versus interspecific effects are caused by trade-offs among coexisting species (Chesson 1990, 1991, 2000b, Adler et al. 2007, Levine and HilleRisLambers 2009, Mayfield and Levine 2010, HilleRis-Lambers et al. 2012, Levine et al. 2017, Saavedra et al. 2017, Barabás et al. 2018).

Personally, I have always found it inexplicable why this recasting of interactions among species is needed—why we turn causal models into correlational models—when analyses of the models that cogently capture the bases of the causal mechanisms defining the direct and indirect interactions among all the relevant species are just as simple to accomplish. Moreover, understanding whether a species is able to coexist in a particular community because of its resource acquisition abilities or because of its predator avoidance abilities has important ramifications for what trade-offs are crucial for its ecological success, and these trade-offs cannot be deduced by simply collapsing all species interactions into their relative intraspecific versus interspecific "competitive" effect sizes. Exploring species coexistence in the Volterra-Lotka-Gause competition framework places an unnecessary veil over the mechanisms that foster or retard coexistence. So let us work with the models that more directly capture the mechanisms of species interactions.

THE CONSUMER-RESOURCE APPROACH

Almost every species on the face of the Earth simultaneously is a consumer of individuals or at least tissues or products of other species and is itself at threat of being consumed by individuals of other species. As MacArthur's recasting made clear, resource competition is not a direct interaction between two competing species, but rather is an indirect interaction between two consumers feeding on one or more resources. However, the interactions among consumer and resource species can be modeled directly. In fact, another intellectual track through theoretical ecology does model the causal interactions among species more mechanistically, instead of trying to abstract these interactions into the Volterra-Lotka-Gause model. I use the phrase "more mechanistically" deliberately to represent that, at some level, these models are still abstractions of the specifics of particular interactions. However, this line of modeling does try to capture the main features of the mechanisms of interaction between species embedded in some food web structure, and does incorporate many of the nonlinearities inherent in species interactions.

One of the first attempts at modeling such an interaction between two species was undertaken by Sir Ronald Ross (1911) on the dynamics of malaria in a human population. Ross was awarded the 1902 Nobel Prize for Physiology or Medicine for his work proving that mosquitoes were the vector for the malaria parasite.

With this model, he attempted to predict when the disease would persist or die out in a population of humans and, if persisting, the proportions of uninfected, affected but asymptomatic, and sick individuals. This paper inspired Lotka (1912) to publish a more general model of disease dynamics that took into account the diseased portion of the population, death from the disease, and recovery. Although both of these models accounted for only the dynamics of humans, they grounded the idea that individuals of one species injuring or killing individuals of another for their own benefit can play a huge role in the dynamics of both.

Lotka (1920) then published a note on the curious dynamics, which resulted in a very simple model of a single predator species feeding on a single prey species. The model predicted that the two species would cycle through "indefinitely continued oscillations," with the predator lagging one-quarter cycle behind the prey (Lotka 1920, 1932a). Volterra published work on the same model and verified Lotka's analysis (Volterra 1926). This was the birth of the study of predator-prey or, more generally, consumer-resource (including plant-herbivore, host-pathogen, victim-enemy) dynamics.

From this point to the early 1960s, the study of predator-prey dynamics was undertaken mainly by scientists interested in more practical matters, such as crop pest control (e.g., Nicholson and Bailey 1935) or natural resources (particularly fisheries) management (e.g., Ricker 1954, Beverton and Holt 1957). These models were developed as difference equations and were also capable of displaying the cycling dynamics that Lotka and Volterra had demonstrated in their differential equation models.

One important contribution made in this context that moved into the broader basic ecological modeling endeavor was the realization that predators cannot continue to consume their prey at constant rates across a huge range of prey abundances; rather, their feeding rate would saturate at a high but constant level at high prey abundances—the *saturating functional response*. Holling (1959a) explored these relationships in his work on small mammals feeding on sawflies, which were themselves pests feeding on pines in southwestern Ontario forests. Holling showed in experiments that a predator's feeding rate would not increase linearly with increasing prey availability, as the Lotka-Volterra predator-prey model assumed, but rather would asymptote to a maximum rate at high prey abundances (Holling 1959a, 1959b, 1966). He also derived a function to describe this asymptotic relationship—the saturating functional response equation (Holling 1959b). At the same time, Ivlev (1955) was performing similar studies on fish feeding, and he used a different function to describe the saturating functional response. The saturating functional response is the source of a fundamental nonlinearity that makes the Volterra-Lotka-Gause competition framework problematic (Abrams 1980a, 1980b).

The study of models of predator-prey interactions was reinvigorated in mainstream theoretical ecology with the publication of Rosenzweig and MacArthur's (1963) graphical analysis of the interaction between predator and prey, in which the prey population has logistic dynamics in the absence of the predator and the predator has a saturating functional response for feeding on the prey (Rosenzweig and MacArthur 1963). Their graphical analyses of the isoclines resulting from such a system showed that the consumer and resource could coexist either at a stable point equilibrium or in a stable limit cycle, depending on the relative geometries of the isoclines (Rosenzweig and MacArthur 1963, Rosenzweig 1969, 1971). Their analyses renewed interest in the dynamics that could result from studying the mechanisms of interactions between predator and prey directly (for general reviews, see Royama 1971, Murdoch and Oaten 1975). Also, such cycling caused by internal dynamics is not possible in the Volterra-Lotka-Gause competition model.

To my mind, two papers defined the way forward for the more mechanistic approach to understanding coexistence outside the confines of Volterra-Lotka-Gause competition. The first of these was Robert Holt's contrasting the interactions among prey species that are fed upon by the same predator—what he called *apparent competition*—with the interactions among multiple predators feeding on the same prey that defines resource competition (Holt 1977). Interestingly, he began with MacArthur's consumer-resource model (i.e., equations (2.12)), but took these equations at face value and asked the reverse question of what is required for multiple prey species to coexist with one predator that feeds on all of them to some degree. In this community, the prey interact with one another only indirectly through their shared effects on inflating the predator's abundance. (This scenario is analyzed explicitly in chapter 3.)

Holt (1977) showed that one prey could drive another extinct if it inflated the predator's abundance to a sufficiently high level. The moniker "apparent competition" comes from the fact that increasing one prey's abundance should decrease the other prey's abundance, just as is expected for resource competition. This paper illustrated that the importance of the indirect effects among species mediated through shared predators are just as consequential as those mediated through shared resources. It also showed the ambiguity of interpretation for the results of density manipulation experiments alone, which were taken as the gold standard for demonstrating resource competition among species at the time.

The other defining paper was David Tilman's (1980) explicit analysis of consumers feeding on different types of resources (see also Tilman 1982). In addition to showing how an explicit model of multiple consumers feeding on multiple resources could be recast into the simpler Volterra-Lotka-Gause competition model that only accounted for the consumers, MacArthur also presented an analysis in which interactions among the consumers could be understood simply by following resource abundances and ignoring consumer abundances (see fig. 2.14 in MacArthur 1972). Tilman (1980) greatly expanded MacArthur's approach to include many different types of resources, both abiotic nutrients and biotic species, defined by their effects on the population dynamics of the consumers through the shapes of their isoclines. His analysis showed that coexistence of two consumers on two of any of the defined resource types always required the same qualitative properties for the consumers; namely, that each consumer must have a greater proportional impact on the resource that is more limiting to its own abundance. Moreover, his models allowed the dynamics of both resource competition among the consumers and apparent competition among the resources to shape the ultimate outcome of the species interactions.

These papers were the vanguard of a parallel body of theory that has explored an almost bewildering variety of configurations of species interactions in food webs (e.g., Rosenzweig and MacArthur 1963, Levin 1970, Rosenzweig 1971, MacArthur 1972, DeAngelis et al. 1975, Murdoch and Oaten 1975, Abrams 1976, Levine 1976, Holt 1977, Hassell 1978, Hsu et al. 1978a, Case and Casten 1979, Armstrong and McGehee 1980, Pimm 1982, Tilman 1982, Vance 1985, Schoener 1986, Hastings and Powell 1991, DeAngelis 1992, Grover 1994, 1997, Holt et al. 1994, Leibold 1996, Holt and Polis 1997, Diehl and Feißel 2000, Cantrell and Cosner 2001, Holland et al. 2002, Křivan 2003, Křivan and Eisner 2003, Murdoch et al. 2003, Neubert et al. 2004, Briggs and Borer 2005, Rudolf and Antonovics 2005, Amarasekare 2008a, Holland and DeAngelis 2010, Kuang and Chesson 2010, McCann 2011, Jones et al. 2012, McPeek 2014, 2019b, Revilla 2015). This body of theory explores the consequences of interactions in various configurations of consumers and their resources, including algae, plants, and bacteria foraging for inorganic nutrients; herbivores foraging for algae and plants, predators foraging for herbivores and other predators,;and omnivores foraging for various types of species. These various configurations of interacting species have been termed *community modules* (Holt 1997). The community modules approach is more mechanistic in that it models the dynamics of various configurations of interacting species explicitly, while incorporating features into the models that attempt to capture a flavor of the processes involved in those interactions (i.e., various forms of functional responses for consumers feeding on resources; various dietary requirements; additional forms of self-limitation besides what arises from resource limitation, such as stress, territoriality, and feeding interference). This permits the full consequences of both direct and indirect interactions among the species to shape community dynamics.

No systematic synthesis of this literature of more mechanistic species interaction models has been made. The goal of this book is to begin such a synthesis. My approach involves building a multitrophic-level food web by adding one species at a time within the framework of these models, starting with no species present at all. I follow an organized pattern of adding species to all possible positions in the trophic structure of a food web, both to add multiple species to the same position in one module (e.g., adding multiple resource competitors) and to add a new type of species to move from one module type to another (e.g., adding a top predator to add a trophic level to a food chain).

With each new addition, I characterize what is required for this species to invade and coexist in the community of species that are already present. For each invasion, I also compare how different mechanistic properties of the interactions among species (e.g., linear or saturating functional responses, intraspecific feeding interference within predator populations, additional forms of density-dependent self-limitation besides resource limitation) influence the potential success of the invader and how each property alters the criteria for subsequent invaders. Invaders include not only resources and consumers but also mutualists and pathogens. I use a specific model form to illustrate the main features associated with each invasion, but I also highlight the commonalities that emerge from different mathematical representations of particular properties or that cause major changes to the predictions.

In addition, what I hope to show to the reader is that the fundamental mechanisms that permit species at different positions in a food web to coexist in a community are different. Thus, a single criterion, such as inequality (2.20), cannot be applied to understanding coexistence for every species in a community. However, this diversity of criteria is not cause for consternation, because the set of criteria is not infinitely large: the systematic analysis of coexistence criteria across the different positions in a food web yields a relatively small and easily intelligible set of criteria, and this analysis identifies clear guides for empirical analyses of these different types of species. The differences among species across a food web are what drive the differences in their natural histories. My goal in the following is to identify a way to a *mechanistic natural history*.

Building a Two-Trophic-Level Food Web

Swim onto the coral reef around the island of Mo'orea in the Indian Ocean, French Polynesia, and you will be surrounded by hundreds of fish species that feed on other organisms that inhabit the reef (Siu et al. 2017). These fishes have fantastically diverse diets. For example, two damselfish species, *Chromis viridis* and *Dascyllus flavicaudus*, feed on calanoid, cyclopoid, and harpacticoid copepods, amphipods, isopods, ostracods, polychaete annelids, shrimp, crabs, tanaids, sponges, stomatopods, gastropods, tunicates, cnidarians, bryozoans, chaetognaths, echinoderms, hemichordates, nemerteans, platyhelminth flatworms, peanut worms, xenacoelomorphs, and a diversity of other fishes (Leray et al. 2019). Other fishes across the many taxonomic groups represented on this reef have similarly diverse diets (Casey et al. 2019). Their diets are different in the quantitative proportions of the various prey they take, but all have substantially diverse and greatly overlapping diets (Casey et al. 2019, Leray et al. 2019).

Stride onto the savanna in Laikipia, Kenya, and you will be similarly surrounded by consumers that feed on a diverse array of species. For example, the mammalian herbivore megafauna on this savanna contains over two dozen species, including twelve species of gazelles and antelopes, two zebras, a giraffe, a warthog, two rhinos, a buffalo, a camel, a porcupine, two rabbits, two hyrax, a baboon, and an elephant, plus five domesticated species. These herbivores collectively feed on more than 200 plant species from 54 families, with each herbivore species feeding on 19–57 plant species at any one time (Kartzinel et al. 2019). For example, the two zebra species, *Equus grevyi* and *E. quagga*, feed on 45 plant species in 9 families but focus mainly on 30 grasses and 6 legumes (Kartzinel et al. 2015). As with the coral reef fishes, these savanna mammalian herbivores have quantitatively different but substantially overlapping and diverse diets (Kartzinel et al. 2015, 2019).

Resource competition has always held a central place in the conception of community ecology for how communities are structured. However, our model conceptions of resource competition typically have considered only one or two limiting resources explicitly conceived (Stewart and Levin 1973, Hsu et al. 1977, 1978a, 1978b, Tilman 1980, 1982, Letten et al. 2017), or a spectrum of resources along a single axis of phenotypic variation that ignores species identity (MacArthur and Levins 1967, May and MacArthur 1972, Roughgarden 1974, Turelli 1978b, 1981) or completely ignores the resource base (Gause 1934, Gause and Witt 1935, Slobodkin 1961, Vandermeer 1970, 1975, Chesson 2000b). These models predict that consumers should be differentiated onto different components of the resource base—either explicitly (e.g., Tilman 1982) or implicitly (e.g., Gause and Witt 1935, Chesson 2000b). However, as the fishes on coral reefs and the large ungulates on African savannas show, many if not most locally co-occurring consumers broadly overlap in diet over the diverse resource bases of dozens to hundreds of species available to them. Certainly, differences in their diets are discernible, but the mere fact that their diets are different does not identify whether they are different in ways that promote their coexistence. We need to expand our theoretical explorations to consider how and why consumers may coexist when more than two resources or resources that differ in more than a single trait are available to them.

Moreover, not only do the consumers interact with one another through their shared effects on resource abundances, the resources interact with one another through their shared effects on consumer abundances (Holt 1977). Differences in productivities among the resources can shape which consumers can coexist, but through those effects on the consumers, these same productivity differences can cause certain resources to be unable to coexist. The final constellation of consumers and resources that coexist in a community is defined by the network of these direct and indirect effects at both of these trophic levels, and we can understand the consequences of these mechanisms only if the models we use explicitly include them.

THE FIRST RESOURCE INVADES

The foundations of any food web are the resource species at its base. These species may have autotrophic metabolisms (e.g., algae, plants, many bacteria and archaea) that permit them to harvest energy from the environment in the form of sunlight (i.e., photoautotrophs) or chemical energy (i.e., chemoautotrophs) along with other needed materials (e.g., inorganic nutrients, water), or they may have heterotrophic metabolisms (e.g., mainly bacteria and archaea) that scavenge energy from inorganic and organic materials. Throughout, I identify this class of species as *resources* and symbolize them as R_i, where the subscript i identifies the particular species being referenced ($i = 1,2,3, \ldots, p$). The R_i also serve as the state variables for their abundances in models.

Imagine an ecosystem at some location that is completely lacking of any living species. This ecosystem has some set of abiotic environmental properties, including the availability of water; the salinity and pH of that water; the thermal regime

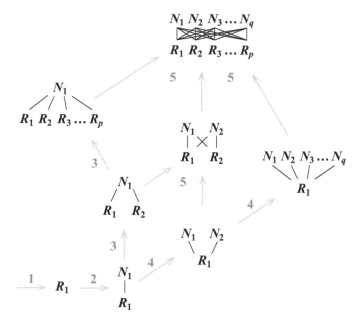

FIGURE 3.1. A road map of the species invasions that are considered in this chapter. For each community module, R_i represents a resource species and N_j represents a consumer species. A line between two species identifies a trophic link where the species above consumes the species below. Arrows show transitions between modules that result from the invasion by different types of species. Numbers are associated with each arrow so that different types of transitions can be identified in the text.

over the course of a day, season, and year; the concentrations of inorganic nutrients; the parent materials and constitution of the soil; and so forth. Some large subset of these abiotic factors will shape the demographic rates of the first resource species that colonizes this ecosystem (invasion 1 in fig. 3.1). These demographic rates are the per capita birth and death rates of the species defined by the environmental conditions in which it finds itself.

When R_1 colonizes, its initial abundance will be small, with only a few individuals constituting its initial abundance (i.e., $R_1 \approx 0$). If this species is to thrive and maintain a population in this ecosystem, its per capita population growth rate at this small abundance must be positive: in short, its per capita birth rate must be greater than its per capita death rate so that its abundance increases when it is rare. This is the recurring issue in this larger analysis. As the species' abundance increases, it may alter the environmental conditions of the ecosystem in ways that either decreases its per capita birth rate or increases its per capita death rate or both, until eventually its abundance increases to a level at which its per capita birth rate equals its per capita death rate. At this point, the species' abundance will stop increasing.

This biological description of population growth for this species can be described mathematically. (For much of this book, I utilize the continuous-time framework of differential equations to model population dynamics. However, this entire book could also be presented using discrete-time difference equation models to accomplish the same analyses (Hassell 1978, Case 2000, Turchin 2003). I use the continuous-time framework because this is the more common framework used in much of the ecological literature.) The classic model describing this biological scenario is logistic population growth (Verhulst 1838, Pearl and Reed 1920, Gause 1934, Slobodkin 1953, 1961). The per capita population growth rate of R_1 is then given by

$$\frac{dR_1}{R_1 dt} = c_1 - d_1 R_1, \tag{3.1}$$

where c_1 is the intrinsic rate of increase and d_1 is the rate at which per capita population growth rate decreases with increasing abundance. Both of these model parameters are determined by how the phenotypic properties of R_1 interact with the abiotic environmental conditions that it experiences in the ecosystem. R_1 is both the identifier for the species and the variable describing its abundance (table 3.1 lists

TABLE 3.1. Summary descriptions of the state variables and parameters used in the models presented in this chapter

State variable	Description
i	Subscript indexing resource species
p	Total number of resource species available to colonize community
R_i	Population abundance of resource i
j	Subscript indexing consumer species
q	Total number of consumer species available to colonize community
N_j	Population abundance of consumer j

Parameter	Description
c_i	Intrinsic rate of increase for resource i
d_i	Strength of intraspecific density dependence for resource i
a_{ij}	Attack coefficient for consumer j feeding on resource i
b_{ij}	Conversion efficiency for consumer j feeding on resource i
h_{ij}	Handling time for consumer j feeding on resource i
f_j	Intrinsic death rate for consumer j
g_j	Strength of direct intraspecific density dependence on death rate for consumer j
z_j	Strength of feeding interference for consumer j

all the model state variables and parameters used in this book). Another species with different phenotypic properties would have different values for these parameters in this ecosystem, and this species would have different parameters if it were invading another ecosystem with different abiotic conditions. This dual determination of model parameters is an essential idea to keep in mind throughout this analysis.

Logistic population growth is not a mechanistic characterization of population regulation, but rather a nonmechanistic descriptor of population dynamics. However, as we saw in chapter 2, various mechanisms regulating the population, either resource limitation or predator limitation, may be embedded in the terms of the logistic growth equation (Schaffer and Leigh 1976, Schaffer 1981). Moreover, logistic population growth can be interpreted as a simpler form of a more complex model in which both per capita birth and death rates of the species depend on the abundance of the species (Hairston et al. 1970, Pianka 1972). To illustrate, imagine that the per capita birth and death rates of R_1 are both linear functions of R_1's abundance: birth rate $= \beta_0 + \beta R_1$ and death rate $= \delta_0 + \delta R_1$, where β_0 and δ_0 are the intercepts of the respective rates (i.e., the respective rates when $R_1 = 0$) and β and δ are the slopes of these relationships. The resulting population dynamics equation is then

$$\frac{dR_1}{R_1 dt} = (\text{per capita birth rate}) - (\text{per capita death rate})$$
$$= (\beta_0 + \beta R_1) - (\delta_0 + \delta R_1)$$
$$= (\beta_0 - \delta_0) - (\delta - \beta) R_1$$

Comparing this equation to equation (3.1) makes the demographic interpretation of the parameters of logistic growth more apparent. c_1, the intrinsic rate of increase in equation (3.1), is the representation of the per capita birth rate minus the per capita death rate when R_1 is rare (i.e., $c_1 = \beta_0 - \delta_0$). Likewise, the rate at which the per capita population growth rate changes with abundance is a function of the rates of change in the underlying demographic rates (i.e., $d_1 = \delta - \beta$). Thus, logistic population growth is a representation that can be interpreted both as encapsulating some underlying mechanism of population regulation (akin to MacArthur's embedding the population dynamics of one species into that of another) and as a general descriptor of underlying demographic rates.

The one-resource simulation module can be accessed at https://mechanisms ofcoexistence.com/twoLevels_1D_1R.html.

As the biological scenario above implies, whether R_1 can invade this ecosystem and maintain a population depends on whether it has a positive per capita population growth rate when rare. Mathematically, this is a question of whether equation (3.1) is positive when $R_1 = 0$:

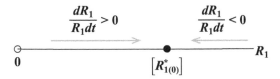

FIGURE 3.2. The isocline configuration for the invasion of an empty ecosystem by a single resource species. In this case, the isocline is a single point at the equilibrium abundance of the resource. If the invading resource has positive per capita population growth rate when at low abundance (i.e., near zero, which will be its abundance when it invades), its abundance will increase along this axis until it reaches the demographic equilibrium at $R^*_{1(0)}$, where its average per capita birth and death rates balance (i.e., a point on its isocline). Conversely, if population abundance is above $R^*_{1(0)}$, per capita growth rate will be negative, which will decrease its abundance. This stable equilibrium abundance is identified with a filled black circle. Arrows above the axis show the change in population abundance along different ranges of the axis of resource abundance. $R_1 = 0$ in this case is, therefore, an unstable equilibrium, which is signified by an open circle.

$$\left. \frac{dR_1}{R_1 dt} \right|_{R_1 = 0} > 0.$$

From equation (3.1), R_1 will increase when rare if $c_1 > 0$ (substitute $R_1 = 0$ into equation (3.1) and evaluate when the resulting equation is positive). $c_1 > 0$ is, therefore, the *invasibility criterion* for R_1 to be the first species to invade this ecosystem. Thus, if R_1's per capita birth rate exceeds its per capita death rate when it is rare (i.e., $c_1 = \beta_0 - \delta_0 > 0$), it will be able to invade and maintain a population in this ecosystem, and it will not be able to do so if this is not true (i.e., if $c_1 < 0$).

If it can increase when rare, its per capita growth rate will decline as its abundance increases until this rate reaches zero. At this point, the population has reached its *equilibrium abundance*. The equilibrium abundance is found by setting equation (3.1) equal to zero and solving for R_1. Thus, the equilibrium for R_1 is

$$R^*_{1(0)} = \frac{c_1}{d_1}. \tag{3.2}$$

The superscript asterisk identifies this specific abundance as an equilibrium, and the parenthetical zero in the subscript identifies that no other species are present at this equilibrium.

This is also a stable equilibrium, because population abundance moves to this value if the population abundance is started above or below this value (fig. 3.2). Mathematically, a stable equilibrium is identified as having the partial derivative of the total growth rate with respect to population abundance at this equilibrium that is negative (Case 2000):

$$\left. \frac{\partial}{\partial R_1} \left(\frac{dR_1}{dt} \right) \right|_{R_1 = R_{1(0)}^*} < 0.$$

Evaluating equation (3.1) thus gives

$$\left. \frac{\partial}{\partial R_1} \left(\frac{dR_1}{dt} \right) \right|_{R_1 = R_{1(0)}^*} = \left. \frac{\partial}{\partial R_1} \left(c_1 R_1 - d_1 R_1^2 \right) \right|_{R_1 = R_{1(0)}^*} = \left. \left(c_1 - 2 d_1 R_1 \right) \right|_{R_1 = R_{1(0)}^*} = c_1 - 2 d_1 \frac{c_1}{d_1} = -c_1.$$

When $c_1 > 0$, so that $-c_1 < 0$, $R_{1(0)}^*$ is a stable equilibrium. Thus, satisfying the invasibility criterion for R_1 also ensures that the equilibrium is stable.

The final issue that is important to evaluate is the *feasibility* of the equilibrium, meaning the criterion that must be satisfied for the equilibrium abundance to be nonnegative: populations cannot have negative abundances. It should be easy for the reader to see that $R_{1(0)}^* > 0$ if $c_1 > 0$ (from the formulation of equation (3.1), d_1 is assumed to be positive). Thus, the invasibility criterion for when R_1 can increase when rare is identical to the criterion for its equilibrium abundance to be feasible and to be stable! This equality of invasibility, feasibility, and stability criteria is a property of many of the models we will evaluate in the coming pages.

These results suggest a number of important inferences about the invasion of this first resource species into the ecosystem. Biologically, if the species can satisfy the invasibility criterion for this ecosystem when it is the first to invade, this species can maintain a population on the abundances of nutrients and abiotic conditions found in this ecosystem in its absence. Additionally, the strength of density dependence (d_1 in equation (3.1)) that regulates its population after it invades plays no role whatsoever in determining whether it can invade. The strength of this density dependence only determines its equilibrium abundance after it invades. Thus, the strength of density dependence does not influence whether this species will exist in the community. The strength of density dependence in this species might influence whether other species can invade and coexist by determining the resource's abundance, but that is another species' problem.

Logistic population growth described by equation (3.1) is equivalent to the form originally derived by Verhulst (1838), by Pearl and Reed (1920), and by Volterra (1926) in his original exposition of competition within and between two species (i.e., equation (2.4), where $c_1 = \varepsilon_1$ and $d_1 = \gamma_1 h_1$). This formulation is also consistent with many mechanistic underpinnings that result in a function of this form (Schoener 1973).

Moreover, I prefer equation (3.1) to the more typical formulation of logistic population growth introduced by Lotka (1925) and given in equation (2.5) for four reasons. First, in equation (2.5), if $b_1 < 0$, per capita population growth rate increases with the resource's abundance, and K_1 becomes an unstable equilibrium

(Gabriel et al. 2005). If the population would somehow become larger than K_1 with $b_1 < 0$, the population would increase toward infinity. Therefore, the classic formulation of logistic population growth does not adequately describe situations in which the species cannot maintain a population (i.e., sink or walking dead species). Also, in the classic formulation, the strength of density dependence is b_1/K_1, the intrinsic rate of increase divided by the equilibrium population abundance, which yields little direct biological intuition about the strength of density dependence. In addition, the equilibrium abundance of the species is not specified as a parameter of the model in equation (3.1), but rather emerges as a balance of the parameters describing per capita growth rate when rare and the strength of density dependence. Finally, the formulation of equation (3.1) also better matches a series of historically important community models that were discussed in chapter 2 (e.g., MacArthur and Levins 1967, MacArthur 1969, May 1973).

Many other functional forms could have been used to describe the population dynamics of R_1 in this ecosystem by itself, including other formulations for logistic population growth or other nonlinear forms describing the growth rate of a single population (Gilpin and Ayala 1973, Schoener 1973). Other mathematical forms are frequently used to model species such as R_1 when they are bacteria, for example, the Monod or Michaelis-Menten equations (Michaelis and Menten 1913, Monod 1949). Alternatively, instead of a species, R_1 could also be considered a depletable abiotic resource, such as various phosphorus and nitrogen compounds, silica, and light, and other mathematical forms are used (e.g., Stewart and Levin 1973). All of these mathematical forms could be used to model a basal resource, because they all share the same qualitative features: all result in a stable equilibrium abundance when in the absence of all consumers, and all can be depleted to lower abundances by consumers. Moreover, any of these could be used as the basis of resource dynamics in the models to be considered in the coming chapters, and all the same qualitative insights would be derived. I refer to R_1 as a species and use equation (3.1) throughout to describe its dynamics, although one can interpret R_1 as an abiotic, depletable resource or a biological species, both of which are therefore potentially limiting resources.

Throughout this book, I focus on logistic population growth for the basal resources, and I do not consider things like *Allee effects*, where species have per capita population growth rates that increase with their own abundance (Allee and Bowen 1932, Lande et al. 2003). For the purposes of invasibility, the only issue is whether the species has a positive per capita growth rate when it is rare, not the form of density dependence (i.e., how population growth rate changes with abundance). Certainly, some species may have negative intrinsic rates of increase coupled with a strong Allee effect that would permit it to maintain a population were it to somehow reach abundance where its per capita population growth rate is positive (e.g., Eskola and Parvinen 2007, Kramer et al. 2009, McLellan et al. 2010). However, for our purposes, these more complicated scenarios add little additional insight about invasibility.

Even though the community in this section only contains one species, the model of logistic population growth presents a number of important testable predictions about the mechanisms that permit this species to coexist in this ecosystem. Obviously, the test that is first and foremost on the list is to evaluate the invasibility criterion of the species. The mere presence of a species at a location does not mean that it is coexisting there. A species may be present simply because it has not gone extinct yet (a walking dead species); or a species may be maintained locally by continual immigration from somewhere else (a sink species). The discriminating feature between coexisting species and walking dead/sink species is that a coexisting species has $c_1 > 0$ because of the balance between its local per capita birth and death rates, and walking dead and sink species have $c_1 < 0$.

Typically, experimental manipulations of R_1's abundance will be required to estimate c_1. For example, one experiment would be to manipulate R_1 from a very low abundance to well above its typical abundance (*abundance* here meaning the number of individuals in a standard-sized enclosure or area) and measure per capita birth and death rates in each abundance treatment level. The results of this experiment could then be used to test a number of important issues concerning the mechanisms permitting R_1 to support a population in this ecosystem. First, functions fit to these per capita birth and death rates across abundance treatments describe how these demographic rates change with R_1's abundance (e.g., as in fig. 1.1). The species' invasibility criterion is satisfied if its per capita birth rate is higher than its per capita death rate in the very low abundance treatments and specifically when the functions are projected to $R_1 = 0$. This is the critical test for whether the species is coexisting or is a walking dead or sink species.

The results of this experiment can also be used to evaluate a rarely tested hypothesis that nevertheless is the basis of much of our theoretical analyses, namely, whether the species' "natural" abundance is at or near a stable equilibrium (see fig. 1.1). An equilibrium is the abundance at which per capita birth and death rates balance. Therefore, the curves fitted to the per capita birth and death rates across the experimental abundance gradient should cross near the typical abundance of the species at that site. Also, the equilibrium is stable if birth rate is higher than death rate below this crossing point, and death rate is higher than birth rate above this crossing point. For species with simple life histories (e.g., algae, bacteria, annual plants), simple manipulations of abundance are sufficient for this type of study. However, for species with complex life cycles, more complicated designs that integrate demographic rates across life cycles may be needed to fully evaluate these issues (e.g., McPeek and Peckarsky 1998, Caswell 2001).

Quantifying the functional relationships between the per capita demographic rates and abundance also identifies how density-dependent mechanisms act to regulate R_1's abundance. Per capita birth rate may decline or per capita death rate may increase, or both may change, with increasing abundance. Knowing the demographic rates that are density dependent also identifies the key demographic variables to quantify in subsequent tests to identify the mechanisms of population regulation and coexistence.

Identifying the mechanistic causes of these birth and death rate relationships and, consequently, the sources of density dependence are also critical for understanding *why* a species is coexisting in a community. Static features of the environment (e.g., average temperature, relative humidity, salinity) will influence the elevations of the demographic curves and thus the difference between birth and death rates at any given abundance. With only a single species present, density dependence will have to come from either depletion of some abiotic resource (e.g., water, light, mineral and organic nutrients) or from some mechanism of direct self-limitation (e.g., allelopathy, waste product accumulation, territoriality, aggression, cannibalism). Observational data and experimental manipulations of the possible sources of these demographic drivers to see which cause changes in the demographic rates of the species will identify these key environmental features that permit coexistence and limit the species abundance.

The efficacy of these factors can be further tested by making comparisons across sites. For example, imagine that some mineral nutrient, such as nitrate, is identified as a key environmental factor that limits the per capita birth rate of species at one site: in the logistic model, this would be equivalent to nitrate availability being a surrogate for c_1. If nitrate supply is a general factor limiting the distribution and abundance of this species, the species should be absent from sites where nitrate availability is extremely low because it cannot meet its invasibility criterion (i.e., $c_1 < 0$). Where the species is present, its abundance should increase with increasing nitrate supply across sites. This observational study is equivalent to evaluating $R_{1(0)}^*$, the equilibrium abundance of the species, across a gradient of c_1.

Although not crucial here, understanding the traits of the species that make it successful or unsuccessful in this environment are invaluable for developing a mechanistic understanding of species distributions and abundances (Schoener 1986). This is because the per capita birth and death rates of a species in a given ecosystem are determined by how their traits interact with their ecological environment. Obviously, multiple environmental factors also determine the demographic rates of the species. Some influence only the elevations of the rates (i.e., the intercepts of the demographic functions that determine c_1), whereas others generate density dependence (i.e., how the demographic functions change as abundance changes to determine d_1 or how they make the relationships nonlinear).

In this case, with only one species present, all of these environmental factors will be abiotic environmental factors. In the coming pages when more species are introduced, these environmental factors will also include other species and their products. The outline of studies here does not change when other species are also influencing the demographic rates of the species of interest. The list of limiting factors simply lengthens to include them. However, knowing which factors— abiotic and biotic—influence the demography of a species and what traits influence their success in those interactions is critical for determining the circuitous paths of indirect effects that determine whether a species will coexist and that shape the overall community and food web structure.

TWO TROPHIC LEVELS

Once the first species establishes a population in the ecosystem by itself, this provides opportunities for other species who cannot live by themselves, namely, heterotrophs that must consume other species (e.g., algivores, herbivores, bacterivores, carnivores, and all manner of predators), to colonize and exploit the established resource species. This adds a trophic level to the food web in the ecosystem. It also is the most stark illustration that the establishment of one species in a community opens ecological opportunities for other species to now colonize that were not available before. This section considers the opportunities and constraints placed on having multiple species invade to fill functional roles at these two trophic levels.

The First Consumer Invades

Now that a resource species has established a population in the ecosystem, a consumer of that resource can now attempt to invade and potentially establish a simple food chain of two species (invasion 2 in fig. 3.1). Throughout, I refer to the species that feed on resource species as *consumers* and identify a pool of them as N_j, with $j = 1,2,3, \ldots ,q$, again with N_j both identifying the species and being the variable describing its abundance. The basic model of a consumer feeding on a resource that I use is a modified version of the classic Rosenzweig-MacArthur model (Rosenzweig and MacArthur 1963, Rosenzweig 1969, 1971):

$$\frac{dN_1}{N_1 dt} = \frac{b_{11} a_{11} R_1}{1 + a_{11} h_{11} R_1} - f_1 - g_1 N_1$$

$$\frac{dR_1}{R_1 dt} = c_1 - d_1 R_1 - \frac{a_{11} N_1}{1 + a_{11} h_{11} R_1}$$

$$(3.3)$$

This model has simply added predation by N_1 on R_1 to equation (3.1). R_1 displays logistic population growth in the absence of the consumer, with c_1 and d_1 having the same meanings.

The consumer has a saturating functional response for feeding on the resource—the last term in the equation for R_1—with an attack coefficient of a_{11} and handling time of h_{11} (Holling 1959a, 1959b). The attack coefficient quantifies the rate at which consumer individuals capture resource individuals when the resource is at very low abundance. I use Holling's disc equation for the saturating functional response because it can represent both linear and saturating forms: linear has $h_{11} = 0$ and saturating has $h_{11} > 0$. The handling time describes how long it takes an individual consumer to capture and process one resource individual and sets the maximum rate at which consumers can kill and consume resources. Other mathematical forms could be used for the functional response: for example, the Michaelis-Menten form, the Monod function, the Ivlev function, or various trigonometric functions (Monod 1949, Ivlev 1955, Jassby and Platt 1976, Jeschke et al. 2002). Luckily, all the various functions used for saturating functional responses have very similar consequences for the shapes of resource and consumer isoclines, and thus all give very similar dynamics (i.e., stable point equilibria, stable limit cycles, or chaotic dynamics) depending on the same characteristics of the shapes of the resource and consumer isoclines around points where they intersect (Hesaaraki and Moghadas 2001, Fussmann and Blasius 2005, Seo and Wolkowicz 2018). The consumer's numerical response (i.e., per capita birth rate due to feeding on the resource) is simply the functional response times a conversion efficiency parameter b_{11} that specifies the number of consumer offspring produced per resource eaten.

The consumer also has a per capita intrinsic death rate f_1. In addition, the consumer may also limit its own abundances in a negatively density-dependent fashion through the consequences of direct interactions among conspecifics and independent of the effects of resource limitation (Tanner 1966). This term ($-g_1N_1$) can be interpreted as representing a decrease in the per capita birth rate or increase in the per capita death rate of the consumer with increasing consumer abundance (Gilpin 1975, Gatto 1991, Caswell and Neubert 1998, Fryxell et al. 1999, Amarasekare 2002, 2008a, Neubert et al. 2004, McPeek 2012, 2014). Mechanisms causing such direct density-dependent effects include territoriality and despotic habitat filling (Pulliam and Danielson 1991, McPeek, Rodenhouse, et al. 2001, Both and Visser 2003, López-Sepulcre and Kokko 2005), cannibalism (Fox 1975, Polis 1981, Van Buskirk and Smith 1991, Rudolf 2007), opportunity costs and physiological stress due to agonistic interactions among conspecifics (Marra et al. 1995, Lochmiller 1996, McPeek, Grace, et al. 2001, Glennemeier and Denver 2002, McPeek 2004), and mate finding and competition for mates (Zhang and

Hanski 1998, Bauer et al. 2005, M'Gonigle et al. 2012). All can generate direct intraspecific density dependence in species at any trophic level by either reducing their per capita birth rates or increasing their per capita death rates as their own abundances increase, and these occur independent of their foraging abilities. For simplicity, I assume that these relationships are linear with rate g_1, but any functional shape with a negative first derivative of demographic rate with respect to abundance will give qualitatively similar results.

Consumers might also reduce their own feeding rates because of agonistic interactions while foraging, and these effects of feeding interference can be incorporated directly into the functional responses of the species (Beddington 1975, DeAngelis et al. 1975). While these various ways of incorporating direct intraspecific density dependence into consumer dynamics have quantitative differences on outcomes, their qualitative effects on community membership and structure are very similar (DeAngelis et al. 1975, McPeek 2012, 2014, 2019b). I will consider this mechanism of feeding interference shortly.

The dynamics of model (3.3) can be intuitively understood by considering the relationships between the *isoclines* for the two species. The isocline of a species maps all combinations of abundances for every species in the community at which that particular species' population growth rate is zero. Again, the function describing the shape and position of the isocline is found by setting the species' population growth rate equation equal to zero and solving for one of the species abundances so that they can be graphed in a state space of all species in the model. For equations (3.3), the corresponding isoclines are

$$\frac{dN_1}{dt} = 0 : R_1 = \frac{f_1 + g_1 N_1}{a_{11}\left(b_{11} - h_{11}\left(f_1 + g_1 N_1\right)\right)}$$

$$\frac{dR_1}{dt} = 0 : N_1 = \frac{c_1}{a_{11}} + \frac{c_1 a_{11} h_{11} - d_1}{a_{11}} R_1 - d_1 h_{11} R_1^2$$

Mapping the isoclines of every species in the community into the state space described by abundance axes of all species allows one to graphically describe the dynamics of the community and identify important abundance combinations (i.e., equilibria) (Hirsch et al. 2012, Strogatz 2015). Figure 3.3 shows the isoclines for the community in equations (3.3) for various parameter combinations.

To build intuition, first consider the simplest case in which the consumer has a linear functional response ($h_{11} = 0$) and experiences no direct self-limitation ($g_1 = 0$) (fig. 3.3(a)). Note that this two-dimensional representation of the $N_1 - R_1$ abundance space simply has an added dimension, namely, N_1, to the depiction in figure 3.2. Adding another species to the community adds an additional dimension to the abundance space and reveals more about the dynamics of the system, but it does not change anything about the reduced space that was considered before. Therefore, what we learn about simple systems is the foundation for understanding more complex systems. For these parameters, the R_1 isocline is a line that intersects the R_1 axis at $R_{1(0)}^* = c_1/d_1$ (i.e., when $N_1 = 0$) (cf. figs 3.2 and 3.3(a)) and intersects the N_1 axis at c_1/a_{11} (i.e., when $R_1 = 0$). If R_1 is below its isocline (i.e., the point $[R_1, N_1]$ is in the space between the origin and the R_1 isocline), R_1 increases; biologically, R_1 increases in this area of abundance space because the mortality inflicted by the consumer population is not substantial enough to prevent population growth. If it is above its isocline (i.e., if the R_1 isocline is between the origin and the point $[R_1, N_1]$), R_1 decreases because the consumer is abundant enough to inflict substantial mortality to cause R_1 to decrease in abundance.

The N_1 isocline is a straight vertical line that intersects the R_1 axis at $f_1/(a_{11}b_{11})$ and is parallel to the N_1 axis because N_1 has no direct effect on its own demographic rates in this case (fig. 3.3(a)). If N_1 is to the right of its isocline (i.e., if the N_1 isocline is between the origin and the point $[R_1, N_1]$), N_1 increases; biologically, this means that R_1's abundance is large enough for the consumer population to grow. If N_1 is to the left of its isocline (i.e., if the N_1 isocline is not between the origin and the point $[R_1, N_1]$), N_1 decreases because the R_1 abundance is not sufficient to support the N_1 population.

An equilibrium with both species present in the community exists if the isoclines for the two species intersect at a point where both have positive abundances (the point $[R_{1(N_1)}^*, N_{1(R_1)}^*]$ in fig 3.3(a)), and this equilibrium is stable, meaning that the species abundances will move toward this equilibrium from any point in this space where both have positive abundances (Case 2000, Murdoch et al. 2003, McCann 2011). As a result, the equilibrium with R_1 by itself $[R_{1(0)}^*, 0]$ is now unstable, meaning that the system will move away from this equilibrium (Haygood 2002). A little algebra is needed to solve for the exact equations defining $R_{1(N_1)}^*$ and $N_{1(R_1)}^*$. These two quantities are found by substituting one isocline into another and solving for each abundance. For present purposes, these exact equations are not useful to our discussion, so I do not present them here. Readers interested in the exact equations for these equilibrium values should consult the appropriate papers cited above and below for specific community modules.

The criterion for N_1 being able to invade the community with R_1 at its demographic steady state in N_1's absence is also apparent in this figure. When N_1 invades, the abundances will be at a point just above $[R_{1(0)}^*, 0]$. N_1's population will increase if this point is to the right of the N_1 isocline. Thus, the invasibility criterion for N_1 in this community is for the N_1 isocline to intersect the R_1 axis below $[R_{1(0)}^*, 0]$. Mathematically, this implies

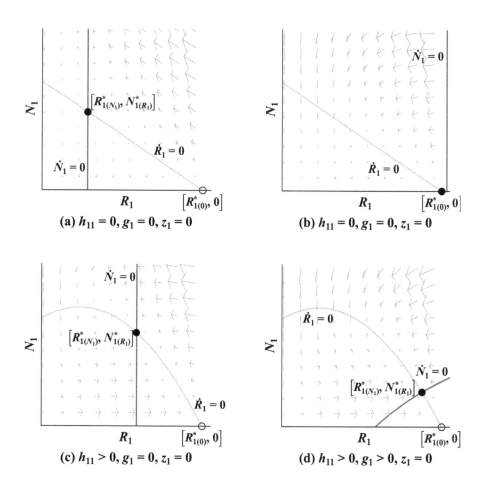

(a) $h_{11} = 0, g_1 = 0, z_1 = 0$

(b) $h_{11} = 0, g_1 = 0, z_1 = 0$

(c) $h_{11} > 0, g_1 = 0, z_1 = 0$

(d) $h_{11} > 0, g_1 > 0, z_1 = 0$

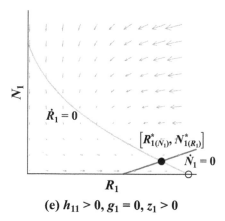

(e) $h_{11} > 0, g_1 = 0, z_1 > 0$

$$\frac{f_1}{a_{11}b_{11}} < \frac{c_1}{d_1}, \tag{3.4}$$

which can also be derived analytically by substituting $R_{1(0)}^*$ into equation (3.3) and querying when $dN_1/N_1 dt > 0$:

$$\left.\frac{dN_1}{N_1 dt}\right|_{R_1 = R_{1(0)}^*} = b_{11}a_{11}R_{1(0)}^* - f_1 > 0$$

$$b_{11}a_{11}R_{1(0)}^* > f_1 \qquad .$$

$$\left(\frac{c_1}{d_1} = \right)R_{1(0)}^* > \frac{f_1}{b_{11}a_{11}}$$

Biologically, this means that the resource population is abundant enough so that the consumer's per capita birth rate is higher than its per capita death rate when it invades (i.e., $b_{11}a_{11}R_{1(0)}^* > f_1$). Also, note that this is the criterion that the two isoclines intersect at positive abundances for both species.

Alternatively, N_1 cannot invade and coexist with R_1 if inequality (3.4) is not satisfied, meaning that the N_1 isocline intersects the R_1 axis above $[R_{1(0)}^*, 0]$ (fig. 3.3(b)). Biologically, this means that the consumer's per capita death rate is higher than the per capita birth rate it can generate by feeding on the resource when it invades. In other words, sufficient resources do not exist in this ecosystem to

---

FIGURE 3.3. Various isocline configurations for a single consumer (N_1) invading an ecosystem with one resource (R_1). In each panel, the R_1 isocline is labeled $\dot{R}_1 = 0$ and the N_1 isocline is labeled $\dot{N}_1 = 0$. Unstable equilibria having at least one species present are identified by open circles (the trivial equilibrium at the origin is not identified), and stable equilibria are identified by filled black circles. The grid of arrows shows how the abundances of each species change away from the isoclines, which identify the general trajectories of abundance change. Panels (a) and (b) show the isocline system when N_1 has a linear functional response ($h_{11} = 0$) and experiences no direct self-limitation ($g_1 = 0$ and $z_1 = 0$). The two-species equilibrium is feasible in (a) because the isoclines intersect at positive abundances for both species, and does not exist in (b) because the isoclines do not intersect at positive abundances for both species. Panel (c) shows the same isocline system when the consumer has a saturating functional response ($h_{11} > 0$). The two-species equilibrium is stable if the isoclines intersect on the portion of the resource isocline with a negative slope (as pictured), but this equilibrium is unstable and has an attractor for a stable limit cycle around it if the isoclines intersect on the portion of the resource isocline with a positive slope (not pictured). Panel (d) shows the isocline system when the consumer has both a saturating functional response ($h_{11} > 0$) and experiences some level of direct self-limitation (i.e., $g_1 > 0$), and in (e) the consumers display feeding interference ($z_1 > 0$). Unless otherwise stated, the parameter values in all panels are as follows: $c_1 = 1.0$, $d_1 = 0.02$, $a_{11} = 0.25$, $b_{11} = 0.1$, $h_{11} = 0.0$, $f_1 = 0.35$, $g_1 = 0.0$, and $z_1 = 0.0$. Panel (b) has $f_1 = 1.35$, (c) has $h_{11} = 0.15$, (d) has $h_{11} = 0.15$ and $g_1 = 0.05$, and (e) has $h_{11} = 0.15$ and $z_1 = 0.2$.

support a consumer population of even one individual. In this case, $[R^*_{1(0)}, 0]$ remains the stable equilibrium.

Mechanistically, it is critical to realize that coexistence of the consumer in this community is not simply a question of whether any prey are available for it to eat. Understanding coexistence requires understanding the balance between the per capita birth and death rates of the consumer. Whether the resource and consumer isoclines intersect at positive abundances for both species will depend on how the abiotic features of the ecosystem and the phenotypic properties of the species determine the parameters in inequality (3.4).

Greater productivity of the ecosystem potentially means a higher value of c_1, which would shift the R_1 isocline up on both axes. Many features of the ecosystem will influence the attack coefficient a_{11} of the consumer (e.g., water turbidity, structural complexity), which will affect the position of the consumer's isocline on the resource axis. Still other ecosystem properties will influence the consumer's intrinsic death rate f_1 and the strength of density dependence experienced by the resource d_1, which will also change the positions of the isoclines in abundance space. For example, in an ecosystem in which the consumer has a very low intrinsic death rate (small f_1), a low resource abundance can sustain its population. However, in an ecosystem in which the consumer has a high intrinsic death rate (large f_1) because of higher levels of physical stressors (e.g., Wilson and Keddy 1986, Emery et al. 2001, Callaway et al. 2002, Thuiller et al. 2004, He and Bertness 2014), a substantially larger resource abundance will be necessary to sustain its population.

Likewise, the traits of the interacting species define how these environmental features are translated into per capita demographic rates for the species because of their interactions. Obviously, the traits of the two interacting species shape the attack coefficient (a_{11}) and conversion efficiency (b_{11}) values. For example, lions and cheetahs that can run faster will have higher attack coefficients on zebra and impala species, and conversely faster zebras and impalas will mean lower attack coefficients from the lions and cheetahs (Wilson et al. 2018). Bill size of a Darwin's finch and the diameter of seeds it eats will both influence the attack coefficient of the finch feeding on the seeds (Grant and Grant 1982, 2006, Schluter and Grant 1984). Competition among phytoplankton taxa will be influenced by their ability to actively take up phosphorus-containing compounds (McKew et al. 2015, Lin et al. 2016, Zhang et al. 2019). The mechanistic reasons *why* a consumer and resource species can or cannot coexist are found in how their traits operate in the environmental conditions in which their interaction takes place and determine their respective birth and death rates.

A saturating functional response ($h_{11} > 0$) does quantitatively alter the consumer's invasibility criterion, but the qualitative features of the criterion are unchanged (fig. 3.3(c)). The points where the R_1 isocline intersects the two axes are unchanged with $h_{11} > 0$, but the R_1 isocline becomes convex up, and can have a maximum point higher than the point at which it intersects the N_1 axis with a sufficiently large handling time: what Rosenzweig (1969) called the "hump." The point where the N_1 isocline intersects the R_1 axis also shifts with $h_{11} > 0$, which is now $f_1/(a_{11}(b_{11} - f_1 h_{11}))$: a larger handling time for the consumer causes its isocline to intersect the R_1 axis at a higher value. Thus, with $h_{11} > 0$, the invasibility criterion for N_1 becomes

$$\frac{f_1}{a_{11}\left(b_{11} - f_1 h_{11}\right)} < \frac{c_1}{d_1}. \tag{3.5}$$

This makes intuitive biological sense because feeding satiation means that the consumer has a lower per capita capture rate at higher resource abundances, which in turn makes its invasibility criterion more stringent. Because the N_1 isocline is still parallel to its own axis, the equilibrium abundance of the resource is equivalent to the intersection point of the N_1 isocline with the R_1 axis, and is still less than R_1's abundance in N_1's absence (fig. 3.3(a)–(c)):

$$\left(R^*_{1(N_1)} =\right)\frac{f_1}{a_{11}(b_{11} - f_1 h_{11})} < \frac{c_1}{d_1}\left(= R^*_{1(0)}\right).$$

A nonzero handling time for the consumer also shifts the location of their joint equilibrium (fig. 3.3(c)). Because a larger value of h_{11} shifts the N_1 isocline to a higher value on the R_1 axis, the resource's equilibrium abundance will always be higher with a higher consumer handling time. This increase in resource equilibrium abundance may be important for subsequent invaders, which means that a saturating functional response of one consumer can influence which other species might subsequently invade. However, the consumer's equilibrium abundance may increase or decrease depending on the relative changes in the position of the N_1 isocline and bowing of the R_1 isocline (Rosenzweig and MacArthur 1963, Rosenzweig 1969, 1971, Fussmann and Blasius 2005, Seo and Wolkowicz 2018).

The stability of the equilibrium at $[R^*_{1(N_1)}, N^*_{1(R_1)}]$ also can change because of $h_{11} > 0$. If the new equilibrium point is on the descending part of the resource isocline, this equilibrium point is globally stable, but if this equilibrium point is on the ascending part of the resource isocline, the two species will enter a stable limit cycle around this attractor (Rosenzweig and MacArthur 1963, Rosenzweig 1969, 1971, Murdoch et al. 2003, Fussmann and Blasius 2005, McCann 2011, Seo and Wolkowicz 2018). In both cases, the consumer and resource are coexisting, but the temporal dynamics of their abundances are different. How limit cycles can influence coexistence is considered in greater detail in chapter 8. For now, it is sufficient to understand that the consumer and resource in both of these dynamics are coexisting, since they satisfy their invasibility criteria and persist in the community (Law and Morton 1993, 1996).

If the consumer limits its own abundance through some mechanism of direct intraspecific density dependence (i.e., $g_1 > 0$), its ability to invade the community when rare is unchanged. With $g_1 > 0$, the point where the N_1 isocline intersects the R_1 axis is unchanged because the effect it has on its own demography when it is rare is essentially nil, but the N_1 isocline now bends away from the N_1 axis (fig. 3.3(d)) (Gilpin 1975, Gatto 1991, Caswell and Neubert 1998, Neubert et al. 2004, McPeek 2012, 2017b). Thus, inequality (3.5) still defines its invasibility criterion with $g_1 > 0$. Because the consumer now limits its own abundance to a lower value, the resource's equilibrium abundance increases because the two isoclines now intersect at a resource abundance that is not equivalent to where the N_1 isocline intersects the R_1 axis. This also tends to increase the stability of the equilibrium: with $g_1 > 0$, equilibria resulting from the isoclines intersecting to the left of the peak in the R_1 isocline may now be stable because of the shape of the N_1 isocline (Gatto 1991, McPeek 2012).

The consumer species may also limit its own abundance through feeding interference. Feeding interference occurs when consumer individuals reduce their own feeding rate because of interactions among the consumer individuals while they are feeding. Feeding interference is typically introduced into these models by using a Beddington-DeAngelis functional response to describe the attack rate of consumers on resources (Beddington 1975, DeAngelis et al. 1975, Cantrell and Cosner 2001). Incorporating feeding interference in the form of a Beddington-DeAngelis functional response (and omitting the direct density dependence term used in equations (3.3)) into the interaction between N_1 and R_1 gives

$$
\begin{aligned}
\frac{dN_1}{N_1 dt} &= \frac{b_{11} a_{11} R_1}{1 + a_{11} h_{11} R_1 + z_1 N_1} - f_1 \\
\frac{dR_1}{R_1 dt} &= c_1 - d_1 R_1 - \frac{a_{11} N_1}{1 + a_{11} h_{11} R_1 + z_1 N_1}
\end{aligned}
\tag{3.6}
$$

where z_1 scales the rate at which the feeding interference depresses the feeding rate of N_1 on R_1 with N_1's abundance. The isocline functions for these equations are

$$
\begin{aligned}
\frac{dN_1}{dt} = 0 : R_1 &= \frac{f_1 (1 + z_1 N_1)}{a_{11} (b_{11} - f_1 h_{11})} \\
\frac{dR_1}{dt} = 0 : N_1 &= \frac{c_1 + (c_1 a_{11} h_{11} - d_1) R_1 - d_1 h_{11} R_1^2}{a_{11} - z_1 (c_1 - d_1 R_1)}
\end{aligned}
$$

As with the linear intraspecific density dependence term, feeding interference in the form of a Beddington-DeAngelis functional response does not alter the point at which the N_1 isocline intersects the R_1 axis, but the N_1 isocline bends away from the N_1 axis (fig. 3.3(e)). Consequently, the invasibility criterion for N_1 remains inequality (3.5), and feeding interference increases the equilibrium abundance of R_1: a higher value of z_1 causes the N_1 isocline to bend more steeply away from its own axis and thus increases $R^*_{1(N_1)}$. However, the shape of the R_1 isocline also changes because of consumer feeding interference. The R_1 isocline still intersects the R_1 axis at c_1/d_1, but higher values of z_1 cause the R_1 isocline to become concave up, even with $h_{11} > 0$, and to intersect the N_1 axis at higher values. Moreover, if $c_1/a_{11} > 1/z_1$, the R_1 isocline does not intersect the N_1 axis at all, but rather an asymptote occurs vertically at $R_1 = (c_1z_1 - a_{11})/(d_1z_1)$ (i.e., where the denominator of the R_1 isocline equation is zero), thus making low resource abundances act like a prey refuge (DeAngelis et al. 1975). All of these shape changes to the isoclines make the two-species equilibrium more stable if it exists (DeAngelis et al. 1975). (Because feeding interference (i.e., $z_1 > 0$) and other forms of direct self-limitation (i.e., $g_1 > 0$) have qualitatively identical effects on coexistence, I primarily explore the effects of self-limitation in more species-rich communities using the $g_1 > 0$ formulation because of its greater mathematical simplicity.)

Whether the consumer is directly self-limiting (i.e., $g_1 > 0$ or $z_1 > 0$) or not determines how the equilibrium abundances of the two species are expected to change along environmental gradients. For example, imagine various locations that differ in the productivity for the resource, which can be characterized as locations that differ in c_1. If c_1 is too low, only the resource will be present ($c_1 < 0.59$), and it will be more abundant in locations where c_1 is higher ($c_1 > 0.59$ in fig. 3.4). For a consumer that is not self-limiting at sites where it can invade, its abundance will be higher in locations where c_1 is higher, but the resource's abundance is constant (fig. 3.4(a)). Remember that the resource's abundance is set by the position of the vertical consumer isocline, which does not change with c_1 (see fig. 3.3(a)–(c)). The consumer's abundance increases because the intersection points of the R_1 isocline increase on both axes with higher values of c_1. However, because the consumer's isocline bends away from its own axis with both forms of consumer self-limitation, both the resource and consumer abundances are higher at locations with higher c_1 if the consumer is self-limiting (fig. 3.4(b)–(c)). (See box 3.1 for a general interpretation of direct self-limitation terms in these models.)

The one-consumer/one-resource simulation module can be accessed at https://mechanismsofcoexistence.com/twoLevels_2D_1R_1N.html.

Thus, for this consumer to invade and coexist, it must have a positive per capita population growth rate when rare, but no other restrictions limit its membership in this community. Whether this consumer can satisfy its invasibility criterion

Box 3.1. The inclusion of direct self-limitation for the consumer in these models illustrates the heart of the mechanisms that regulate species in a community, how limiting factors emerge, and how those processes are represented in models.

First, consider the per capita growth rate equation for the consumer in equations (3.3) with $g_1 = 0$. In this case, the per capita growth rate of the consumer does not directly depend at all on its own abundance: specifically, N_1 does not appear on the right side of the equation with $g_1 = 0$. The consumer's per capita growth rate directly depends only on the resource abundance, and so it is only *resource limited*. This is the only limiting factor for this consumer. The resource's abundance does depend on the consumer's abundance (N_1 does appear in the resource's per capita growth rate equation). As the consumer abundance increases, the resource abundance decreases because of the greater mortality inflicted. As in all of these models, the consumer equilibrates at an abundance where its per capita birth rate is equal to its per capita death rate:

$$\frac{b_{11}a_{11}R^*_{1(N_1)}}{1+a_{11}h_{11}R^*_{1(N_1)}} = f_1.$$

The two species equilibrate because N_1 depresses R_1 to the point where this is true: $R^*_{1(N_1)}$.

One common way to introduce intraspecific density dependence to the consumer is with the term $-g_1 N_1$ in equations (3.3) (Gilpin 1975, Gatto 1991, Caswell and Neubert 1998, Fryxell et al. 1999, Amarasekare 2002, 2008a, Neubert et al. 2004, McPeek 2012, 2014). Such a term may be a close representation of some mechanism. For example, if the consumer is cannibalistic, this term would represent a linear functional response for consumer individuals feeding on conspecifics (Rudolf 2007):

$$\frac{b_{11}a_{11}R^*_{1(N_1)}}{1+a_{11}h_{11}R^*_{1(N_1)}} = f_1 + g_1 N^*_{1(R_1)}.$$

Alternatively, it may encapsulate the effects of some process,for example, stress responses to the presence of conspecifics (Marra et al. 1995, Lochmiller 1996, McPeek, Grace, et al. 2001, Glennemeier and Denver 2002) that causes the consumer's per capita mortality rate to increase with its own abundance. Schoener (1973) also developed models for interference competition among conspecifics by various mechanisms and showed that such a term commonly emerges. Alternatively, this term may represent a diminution of the birth rate of the consumer because of opportunity costs or stress responses caused by conspecific abundance or competition for mates (Zhang and Hanski 1998, Bauer et al. 2005, M'Gonigle et al. 2012):

$$\frac{b_{11}a_{11}R^*_{1(N_1)}}{1+a_{11}h_{11}R^*_{1(N_1)}} - g_1 N^*_{1(R_1)} = f_1.$$

Feeding interference among the consumers that is introduced via a Beddington-DeAngelis functional response (equations (3.6)) reduces the consumer's per capita birth rate by reducing its feeding rate on the resource:

$$\frac{b_{11}a_{11}R^*_{1(N_1)}}{1+a_{11}h_{11}R^*_{1(N_1)}+z_1 N^*_{1(R_1)}} = f_1.$$

In all three of these formulations with consumer direct self-limitation, the abundances of two species appear in their per capita growth rate equations, which identify two limiting factors in determining their per capita growth rate: resource limitation and direct self-limitation. The consumer equilibrium abundance is now defined by the balance between these two processes.

The common feature that various forms of direct consumer self-limitation introduces is to cause the consumer's isocline to bend away from its own axis (fig. 3.3(d)–(e)). Territoriality, which regulates the population by imposing a maximum on the number of breeding individuals in the population (Lande 1987, Both and Visser 2003, López-Sepulcre and Kokko 2005, Berestycki and Zilio 2019), also causes the consumer isocline to asymptote at a consumer abundance set by this maximum (e.g., breeders plus floaters). This is the common effect that all forms of direct self-limitation impose on the geometry of the consumer isocline.

Clearly, many of these same mechanisms may influence the demography of other consumer species in the community as well. For example, some birds defend their breeding territories against both conspecifics and heterospecifics (Robinson and Terborgh 1995). The release of allelopathic chemicals may also have detrimental demographic effects on both conspecifics and heterospecifics (Rice 1984, Thacker et al. 1998, Inderjit and Duke 2003, Loh and Pawlik 2014). For now, only the effects on conspecifics are considered. I take up how the differential effects of such mechanisms on conspecifics and heterospecifics may influence the coexistence of closely related species in chapter 10.

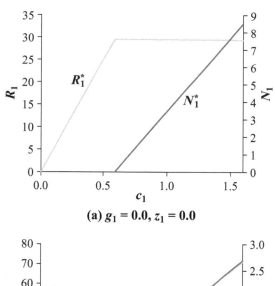

(a) $g_1 = 0.0$, $z_1 = 0.0$

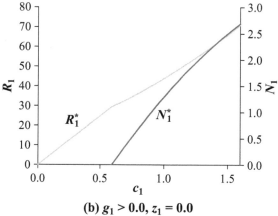

(b) $g_1 > 0.0$, $z_1 = 0.0$

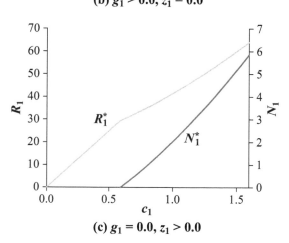

(c) $g_1 = 0.0$, $z_1 > 0.0$

depends critically on the equilibrium abundance of the resource before it invades, which determines whether the consumer's isocline intersects the resource axis above or below the equilibrium abundance of the resource in its absence. This will be the recurring theme of this analysis. *The successful invasion of the next species will depend critically on the abundances of the species already present in the community.* A saturating functional response reduces the scope of invasion for a consumer because it requires a higher resource abundance to have positive population growth when rare (inequality (3.5) instead of inequality (3.4)), but it increases the scope for subsequent invaders because the successful consumer does not depress the resource abundance as far as a consumer would with similar parameters but a linear functional response. Also, the effect of direct self-limitation via feeding interference or some other mechanism generating direct intraspecific density dependence in the consumer has no effect on its own ability to invade and coexist in a community; but it will strongly influence the invasibility criteria that subsequent species must meet, because such self-limitation reduces the ability of the consumer species to depress resource abundances, as well as depressing its own abundance.

The first issue that must be empirically adjudicated for N_1 in evaluating its coexistence is that it actually consumes R_1. If N_1 is a consumer of R_1, one should be able to readily verify that this a component of its diet. If the contents of N_1 individuals' guts are examined, are R_1 individuals found in those contents? Can N_1 individuals be observed consuming R_1 individuals? If R_1 is some abiotic molecule, do isotopically labeled variants of the molecule become incorporated into the consumer; for example, is isotopically labeled phosphate taken up by an alga species? Also, as the abundance of N_1 increases, the abundance of R_1 should decline as it is drawn down by consumption. For example, Passarge et al. (2006) performed a series of experiments on phosphorus and light limitation in algae in which they grew single species of algae in batch culture. They proved that each of these depletable resources were limiting to the growth of all the algae species by showing that as each alga species increased in abundance, the abundance of phosphorus and light declined. As a basis for this mechanism, if one species is supposedly

---

FIGURE 3.4. The equilibrium abundances of a single consumer feeding on a single resource at different points along an environmental productivity gradient that determines the intrinsic growth rate of the resource c_1 for different parameter combinations. Panel (a) illustrates equilibrium abundances when the consumer has no direct self-limitation. In (b), the consumer experiences direct self-limitation through a density-dependent death rate ($g_1 > 0$), and in (c), the consumers have feeding interference ($z_1 > 0$). The parameter values are as follows in all panels: $d_1 = 0.02$, $a_{11} = 0.25$, $b_{11} = 0.1$, $h_{11} = 0.15$, and $f_1 = 0.35$. Panel (b) has $g_1 = 0.05$, and (c) has $z_1 = 0.2$.

present in a community because it feeds on and is limited by another, one should be prepared to marshal evidence that the consumer's abundance is actually limited by the resource in question.

Once N_1 is confirmed to consume R_1, the next issue is to determine that this interaction influences the demographics of both species. Observational studies can suggest such relationships by demonstrating higher prey abundances in areas where their predators are absent. For example, spiders are 10 times more abundant on islands lacking their anole predators (Schoener and Toft 1983, Polis and Hurd 1995). More definitively, an experiment that manipulates the presence/absence of N_1 should be able to demonstrate that R_1 has a higher per capita death rate and a lower abundance when N_1 is present as compared to when N_1 is absent. For example, when I began working on the coexistence of *Enallagma* and *Lestes* damselfly species with different predators, the first experiment I did was to show that each species has a higher per capita death rate and thus a lower abundance in the presence of both fish and dragonflies (McPeek 1990b, Stoks and McPeek 2003b). Likewise, an experiment that demonstrates the demographic rates of N_1 are more favorable (e.g., higher per capita feeding and thus birth rate or lower per capita death rate) when resource levels are experimentally increased confirms that the consumer's abundance is limited by the resource. It is important to realize that demonstrating these facts about the interaction between R_1 and N_1 says nothing about whether N_1 is coexisting; both sink and walking dead species as well as coexisting species would show the same relationships to their critical resources.

These preliminary studies do, however, form the basis for interpreting the results of invasibility studies for N_1. Experiments evaluating the invasibility for N_1 will follow the same design as described for R_1 in the previous section. A gradient of N_1 abundance treatments should be established from very low to greater than its normal abundance, with its per capita birth and death rates as the response variables to be quantified (e.g., Godoy and Levine 2014). Ideally, this density gradient would be established in experimental units where N_1 has first been removed and the remaining species and abiotic resources have been allowed to approach their new demographic steady states. Again, N_1 is coexisting if its per capita growth rate is positive at the lowest abundances.

Another important response variable in this experiment is the resource's abundance across the N_1 abundance gradient: if population regulation of N_1 occurs primarily by resource limitation, R_1's abundance should decrease as N_1's abundance increases. This prediction comes directly from examining the R_1 isocline. The R_1 isocline identifies R_1's equilibrium abundance if N_1's abundance is fixed at various values: from this relationship, as N_1's treatment abundance increases, the abundance of R_1 is predicted to decrease, except at the highest N_1 abundances if the R_1 isocline has a hump (see fig. 3.3). Moreover, the abundance to which R_1 is depressed at equilibrium by N_1 (i.e., $R^*_{1(N_1)}$) can also be assessed in this experiment by quantifying R_1 abundances around the N_1 treatment abundances that are near natural for N_1. Comparisons of R_1 abundances across the N_1 gradient to its naturally occurring abundances will indicate how realistically the experimental conditions mimic their real populations.

Another important experiment will be to assess the minimum resource level at which N_1 can support a population, and so empirically determining where the N_1 isocline intersects the R_1 axis (i.e., $f_1/(a_{11}(b_{11} - f_1 h_{11}))$). Tilman (1982) described this point as the consumer's "R-star," the level to which the consumer would depress the resource. However, if the consumer experiences some form of direct self-limitation, R_1's equilibrium abundance with the consumer $R^*_{1(N_1)}$ will not be the same as the minimum R_1 abundance at which N_1 has a positive population growth rate (i.e., $R^*_{1(N_1)} \neq f_1/(a_{11}(b_{11} - f_1 h_{11}))$) if $g_1 > 0$ or $z_1 > 0$; see fig. 3.3(d)–(e)). This experiment has treatments across a wide range of R_1 abundances and a low and constant number of N_1 individuals at each treatment level. Again, the response variable is N_1's per capita population growth rate across this gradient. For example, Tilman et al. (1981) performed a series of experiments measuring the per capita growth rates of the diatoms *Asterionella formosa* and *Synedra ulna* across silica gradients, an essential nutrient for their growth. In addition to demonstrating the relationship between resource availability and per capita growth rate, they also performed these experiments across a range of temperatures to show how the diatoms' demographic rates change with this environmental condition.

This experiment is in effect mapping the qualitative shape of the N_1 isocline. The point along the R_1 gradient at which N_1's per capita population growth rate is zero is the R_1 abundance at which the N_1 isocline intersects the R_1 axis. Moreover, the difference between this R_1 abundance and the R_1 abundance from the N_1 gradient experimental treatment, where N_1 is at its natural abundance (i.e., an estimate of $R^*_{1(N_1)}$ when N_1's birth and death rates balance), is a measure of the degree to which mechanisms of direct self-limitation prevent N_1 from depressing resource abundances to its fullest possible extent.

Observational studies and experimental manipulations that quantify the species abundances along environmental gradients also provide insights about the importance of direct self-limitation in addition to resource limitation. For example, along the range of the environmental gradient that influences c_1, the resource's intrinsic rate of increase, and where the consumer and resource can coexist, $R^*_{1(N_1)}$ does not increase with an increase in c_1 if the consumer is not directly self-limiting (i.e., $g_1 = 0$ and $z_1 = 0$); all the increased resource productivity should be immediately translated into the consumer's abundance (fig. 3.4(a)). However, if the consumer is self-limiting to some degree, both $R^*_{1(N_1)}$ and $N^*_{1(R_1)}$ are predicted to increase with increasing c_1 (fig. 3.4(b)–(c)). Imagine sampling the abundances of an algal species and a zooplankton species that feeds on that alga in a series of lakes that vary in the supply of total phosphorus to the water column (e.g., McQueen et al. 1986). If the zooplankton species (N_1 in this case) is resource limited (by the algal abundance, which in this case is R_1) and also directly limits its own abundance via some mechanism, both the algal and zooplankton abundances should positively correlate with total phosphorus concentration across the lakes. However, if the zooplankton species is only limited by algal abundance, the zooplankton's abundance should increase with total phosphorus concentration across the lakes, but the algal abundance should be flat. Thus, observational studies relating a resource to consumer abundances along productivity gradients for the resource actually provide a test of whether the consumer is self-limiting by some mechanism. Obviously, this is not a definitive test for consumer self-limitation, but it should serve as an indicator that such mechanisms, in addition to resource limitation, should be examined. Subsequent studies should then explore the mechanisms of direct self-limitation.

At this point, studies of the functional and numerical responses and the traits involved in defining the parameters are not directly relevant to issues of coexistence. However, such studies will become important when other species are added and for understanding the causes and consequences of the mechanisms of coexistence. What consumer traits determine its ability to pursue and subdue resource individuals, and what traits of the resource are important in deterring consumer attacks? Does the consumer have a linear or saturating functional response in natural settings? Studies that address such questions enhance the richness of understanding for the results derived above. They also provide a foundation for making predictions on the relative performance abilities of new species to be added at each trophic level.

An Even More Mechanistic Basal Trophic Level

Throughout this book, I characterize the basal species in the food web as biotic species and assume that those species do not directly interact with one another. However, even if those species are photoautotrophs, they must still harvest various commodities from their environment (e.g., light, water, inorganic nutrients), which means that they may compete for these commodities and so interact indirectly through these shared resources as well, even if they do not directly interact.

This means that an even more mechanistic description of the food web would include equations describing the dynamics of these abiotic resources and the indirect competition of the basal biological species for those shared resources (DeAngelis 1992, Grover 1997, Schmitz 2010, Calcagno et al. 2011, Hunter 2016).

Many different functional forms can be used to describe the dynamics of abiotic nutrients (e.g., phosphorus, nitrogen, and silica-containing compounds) and other commodities (e.g., light, water), such as the Michaelis-Menten, Monod, or chemostat equations (Michaelis and Menten 1913, Monod 1949, Stewart and Levin 1973, Hsu et al. 1981, Tilman 1982). All of these various mathematical forms have the same general properties as the logistic equation: each describes a steady state amount that is present in the environment if no consumption occurs, and each describes the renewal rate if the current amount is below that steady state.

If the basal species are characterized as abiotic resources and one of these alternative mathematical forms is used to describe their dynamics, qualitatively identical conclusions emerge. For example, Hsu et al. (1977) developed a model in which they imagined multiple species of microorganisms competing for a single nutrient in a continuous-culture chemostat. Their model assumed a simple model of nutrient renewal and that the microorganisms take up the abiotic nutrient R from the environment according to Monod/Michaelis-Menten dynamics:

$$\frac{dN}{dt} = \frac{mRN}{(a+R)} - DN$$

$$\frac{dR}{dt} = \left(R^{(0)} - R\right)D - \frac{mRN}{y(a+R)}$$

where $R^{(0)}$ is the input concentration of the abiotic resource, D is the dilution rate of the chemostat, m is the maximum specific growth rate of N, y is the cell growth yield for N, and a is the Michaelis-Menten half-saturation constant. Also, see Stewart and Levin (1973) for an alternative model of microorganisms grown by consuming mineral nutrients in a chemostat thatcomes to the same qualitative outcomes. Tilman (1977) developed the same model for phytoplankton taking up mineral nutrients from the water. The model he analyzed was slightly different, assuming that multiple nutrients are available to the consumer and that the consumer's growth rate is determined by the most limiting nutrient at the time. For these models, in the absence of consumers, the nutrient equilibrates at a constant level (i.e., in this case $R^*_{(0)} = R^{(0)}$), and the consumer and resource either come to a stable equilibrium or enter a limit cycle, depending on parameter values, if the consumer has a positive per capita growth rate when it is rare and $R^*_{(0)} = R^{(0)}$ (Hsu et al. 1977, 1978a, 1978b, Tilman 1977, 1980, 1982, Hsu 1980). They went on to show that all the qualitative conclusions developed in the rest of this chapter about resource and apparent competition among two resources and two consumers emerge from analyzing this model as well (see also Grover 1997). Tilman (1980, 1982) also derived similar results from models assuming a variety of different types of nutrients. This general mathematical framework is also used to model nutrient cycling in food webs (DeAngelis 1992).

These are not the only approaches that have been taken to show how mineral nutrient uptake (i.e., consumption) determines the population dynamics of phytoplankton and plants. One model that is frequently used for phytoplankton is the Droop model, which imagines that the phytoplankton growth rate is determined by the degree to which phytoplankton cells have met their internal quota of a limiting nutrient or nutrients (Droop 1974). Imagine that a phytoplankton species has an asymptotically maximum growth rate of μ_{max}, and the species has a minimum quota of some limiting nutrient Q_{min} that it needs to have a positive per capita population growth rate. In this model, the consumer's growth rate is modeled as depending on the nutrient content inside its cell and not the nutrient content in the surrounding environment. The per capita growth rate of the phytoplankton population is then

$$\mu = \mu_{max}\left(1 - \frac{Q_{min}}{Q}\right),$$

where Q is the current quantity of the limiting nutrient in the phytoplankton population. Note its analogous mathematical form and properties to Lotka's logistic growth equation (see equation 2.5). Ågren (1988) developed a similar model for terrestrial plant growth based on internal nutrient content. Sterner and Elser (2002) used these models as the basis of their even more mechanistic stoichiometric theory to understand how the requirements for these mineral nutrients in the biochemistries of various taxa shape patterns of diversity within communities.

Thus, any of these frameworks could be used as the descriptors for the basal trophic level in this book. If the basal trophic level is taken to be limiting abiotic nutrients (e.g., phosphorus and nitrogen compounds) and other limiting factors (e.g. light, water), herbivores (e.g., grasshoppers feeding on plants in a prairie or zooplankton feeding on phytoplankton in a lake) would represent the third "trophic level" in these models—depletable resources, such as light, water, and inorganic nutrients, at the basal level; primary producers at the second level; and herbivores at the top level. As we will see in the coming chapters, a consistent pattern for understanding coexistence requirements emerges for different positions in the food web across various mathematical formulations. However, many resources, whether they are considered abiotic depletable resources or biological species, serve as the limiting factors for those species that consume them. For Hutchinson's (1961) description of the plankton, he listed light, "inorganic media containing a source of CO_2, inorganic nitrogen, sulphur, and phosphorus compounds and a considerable number of other elements (Na, K, Mg, Ca, Si, Fe, Mn, B, Cl, Cu, Zn, Mo, Co and V)" as those factors limiting phytoplankton diversity.

In what follows, I typically refer to the resources as *species*, but the reader should keep in mind that qualitatively identical results obtain and all the same criteria emerge if the "resources" are considered to be abiotic depletable resources and the "consumers" are phytoplankton or plants that consume them, and alternative models describing their dynamics (such as abiotic resource renewal (e.g., Hsu et al. 1977) or logistic growth) result in qualitatively similar conclusions.

Apparent Competitors

The next invading species to consider are a series of additional resource species. Imagine now that a number of the resource species from the pool defined above can potentially invade the community of one resource and one consumer. The resulting per capita population growth equations for this set of species are

$$\frac{dN_1}{N_1 dt} = \frac{\sum_{i=1}^{p} a_{i1} b_{i1} R_i}{1 + \sum_{i=1}^{p} a_{i1} h_{i1} R_i} - f_1 - g_1 N_1$$

$$\frac{dR_i}{R_i dt} = c_i - d_i R_i - \frac{a_{i1} N_1}{1 + \sum_{i=1}^{p} a_{i1} h_{i1} R_i}$$

$$(3.7)$$

If any of these additional resources can invade and coexist with R_1 and N_1, they will form an apparent competition module (invasion 3 in fig. 3.1) (Holt 1977).

First, consider the invasion of a single additional resource species R_2. Figure 3.5 shows the isocline systems for representative parameter combinations of this apparent competition module. By adding a new species, we are again increasing the dimensionality of our isocline analysis by one, but all the features of the coexistence between R_1 and N_1 still apply in this new system. The graphs on the right in panels (a)–(c) of the figure show the plane where $R_2 = 0$ through the three-dimensional systems shown on the left in (a)–(c): *the $N_1 - R_1$ face.* Compare these graphs to those in figure 3.3.

If the consumer has a linear functional response on all resources ($h_{i1} = 0$) and experiences no direct self-limitation ($g_1 = 0$), the isoclines for all the species are planes in the three-dimensional abundance space (fig. 3.5(a)). Each resource's isocline still intersects its own axis at $R_{i(0)}^* = c_i/d_i$ and the N_1 axis at c_i/a_{i1}, but each is parallel to the other resource's axis, because neither resource directly affects the demographic rates of the other. The N_1 isocline also still intersects each resource axis at $f_1/(a_{i1} b_{i1})$ but is parallel to its own axis. R_2 is invading a community that is

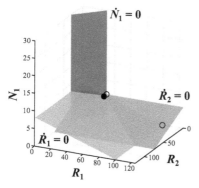
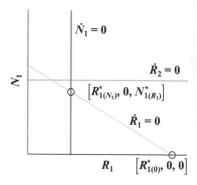

(a) $h_{i1} = 0$, $g_1 = 0$

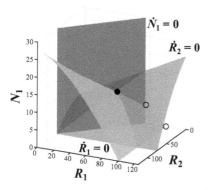
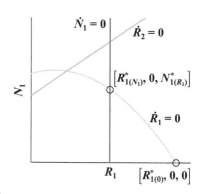

(b) $h_{i1} > 0$, $g_1 = 0$

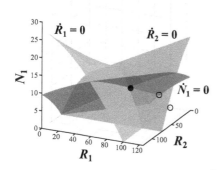
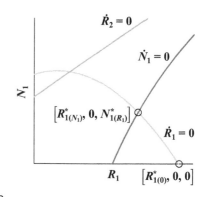

(c) $h_{i1} > 0$, $g_1 > 0$

at $[R^*_{1(N_1)}, 0, N^*_{1(R_1)}]$ in the $N_1 - R_1$ face (fig. 3.5(a), right). R_2 will be able to invade if its isocline passes above this point in the $N_1 - R_1$ face: remember that a resource's abundance will increase if the current location in abundance space is below its isocline. The two resources will coexist with N_1 if R_1 can also pass the analogous invasibility test of its isocline passing above the equilibrium $[0, R^*_{2(N_1)}, N^*_{1(R_2)}]$ in the $N_1 - R_2$ face (i.e., *mutual invasibility*) (Holt 1977). If so, the community will come to the new equilibrium at the point where all three isoclines intersect $[R^*_{1(R_2,N_1)}, R^*_{2(R_1,N_1)}, N^*_{1(R_1,R_2)}]$ (i.e., the filled black circle in fig. 3.5(a), left). At this new equilibrium, N_1's abundance will have increased because it now feeds on two resource species instead of one, and R_1's abundance will have decreased because N_1 is more abundant. Because increasing the abundance of one resource causes a decrease in the abundance of the other resource, Holt (1977) dubbed this interaction "apparent" competition, which is an indirect effect of one resource on another mediated by their shared effects of directly inflating the consumer abundance.

In contrast, if R_2 can invade, but R_1's isocline passes below $[0, R^*_{2(N_1)}, N^*_{1(R_2)}]$ in the $N_1 - R_2$ face, R_1 will be driven extinct because R_2 inflates the consumer's abundance above the level at which R_1 can maintain a population (Holt 1977).

Holt (1977) showed that many resource species can potentially invade and coexist with a single consumer species for this model. I identify the species according to the ranking of where their isoclines intersect the N_1 axis, such that

---

FIGURE 3.5. Various isocline configurations for an apparent competition module of one consumer (N_1) that feeds on two resources (R_1 and R_2). Panel (a) shows the isocline system when the consumer has linear functional responses and no direct self-limitation, (b) for the consumer having saturating functional responses, and (c) for the consumer having saturating functional responses and some level of direct self-limitation. The R_1 isocline is labeled $\dot{R}_1 = 0$, the R_2 isocline is labeled $\dot{R}_2 = 0$, and the N_1 isocline is labeled $\dot{N}_1 = 0$. In each panel, the left graph shows the full three-dimensional representation of the isoclines, and the right graph shows the $N_1 - R_1$ face (i.e., the plane where $R_2 = 0$). In the graphs on the right, the stable equilibrium point before R_2 invades is labeled $[R^*_{1(N_1)}, 0, N^*_{1(R_1)}]$. This equilibrium becomes unstable when R_2 invades if R_2's isocline passes above this point (as shown) in this face, which means that the consumer abundance is low enough at this two-species equilibrium for R_2 to have a positive population growth rate when it invades. If R_2's isocline passes below this point (not shown), R_2 cannot invade because the consumer abundance is high enough to cause a negative population growth rate for R_2 at invasion. In the full three-dimensional figures, the unstable equilibria in the $N_1 - R_1$ face are identified with open circles, and the stable three-species equilibrium is identified with a filled black circle. Unless otherwise stated, the parameter values in all panels are as follows: $c_1 = 4.0$, $c_2 = 3.0$, $d_1 = d_2 = 0.04$, $a_{11} = a_{21} = 0.5$, $b_{11} = b_{21} = 0.1$, $h_{11} = h_{21} = 0.0$, $f_1 = 1.5$, and $g_1 = 0.0$. Panel (b) has $h_{11} = h_{21} = 0.04$, and panel (c) has $h_{11} = h_{21} = 0.04$ and $g_1 = 0.03$.

$$\frac{c_1}{a_{11}} > \frac{c_2}{a_{21}} > \frac{c_3}{a_{31}} > \ldots > \frac{c_p}{a_{p1}}. \tag{3.8}$$

No matter their order of invasion, the same set of resources will coexist with N_1 after all have attempted to invade; but identifying this coexisting set is best understood by considering sequential invasions starting with R_1 (Holt 1977). First, allow R_1 and N_1 to come to their equilibrium and then allow R_2 to invade. If R_2 can invade, the three species will coexist, because R_1's isocline must pass above $[0, R^*_{2(N_1)}, N^*_{1(R_2)}]$, given the species ordering specified by inequality (3.8) (fig. 3.5(a), left). Because each resource's isocline is parallel to all the other resource axes with $h_{i1} = 0$, this means simply that R_2 can invade if

$$\frac{c_2}{a_{21}} > N^*_{1(R_1)}, \tag{3.9}$$

which is its invasibility criterion (fig. 3.5(a), right). The new consumer abundance will be inflated by the presence of both resource species (i.e., $N^*_{1(R_1,R_2)} > N^*_{1(R_1)}$). By the same logic, R_3 can then invade and coexist if $c_3 / a_{31} > N^*_{1(R_1,R_2)}$, R_4 can invade and coexist if $c_4 / a_{41} > N^*_{1(R_1,R_2,R_3)}$, and so on until the resource in the rank order of inequality (3.8) is reached that cannot satisfy its invasibility criterion (Holt 1977). Also, with the addition of each new resource species, the abundances of the resources already present will decrease (e.g., $R^*_{1(N_1)} > R^*_{1(R_2,N_1)} > R^*_{1(R_2,R_3,N_1)}$).

If the consumer has a saturating functional response ($h_{11} > 0$), the same qualitative invasibility criteria define which resources can coexist with the consumer, but specifying an exact criterion analytically is difficult. However, it is easy to understand graphically using isoclines. A saturating functional response makes each resource's abundance dependent on the abundances of the others in the community, because consumer satiation by one resource reduces the realized attack rate of the consumer on the other resources. As a result, each resource isocline still intersects the N_1 axis at c_i/a_{i1}, but it now has a positive slope in each $N_1 - R_i$ face for the other resources (fig. 3.5(b)). As a result, consumer satiation makes it easier for each resource to have its isocline pass above the consumer abundance for invasibility. Thus, the saturating functional response of the consumer makes the invasibility criteria for each resource less stringent, and so may permit more resources to coexist. However, the order specified in inequality (3.8) no longer defines which ones necessarily coexist. The community may come to a stable point equilibrium or limit cycle, depending on the relative shapes of the isoclines at this equilibrium (Křivan and Eisner 2006). If a stable equilibrium results, the consumer will increase and resources already present will decrease with each successful invasion. If, however, limit cycles result, the addition of subsequent resources may decrease or increase the average abundances of the resources already present (Abrams et al. 1998).

If the consumer also experiences some degree of direct self-limitation ($g_1 > 0$), all the same issues apply, but the invading resource will have an even easier time because the consumer's abundance is lower still (fig. 3.5(c)). Again, this is because the consumer's isocline bends away from its own axis along both resource axes. The consumer's abundance will again increase with each new invading resource, but not as much as with no direct consumer self-limitation. Consequently, each resource's abundance is depressed to a lesser degree with the addition of each resource. The same effects occur if the consumer's direct self-limitation is manifest as feeding interference in the functional responses (McPeek 2017b, 2019b). Thus, direct self-limitation in the consumer increases the scope of coexistence for resources that are apparent competitors by decreasing the consumer's abundance.

The apparent competition simulation module can be accessed at https://mechanismsofcoexistence.com/twoLevels_3D_2R_1N.html.

In this apparent competition module, the invading resource species must simply satisfy an invasibility criterion to coexist. Also, once established, each new resource species will influence the invasibility criterion that subsequent invaders must satisfy, because of how their phenotypic and demographic properties influence the consumer's abundance. For example, a resource that experiences a greater level of intraspecific density dependence itself (i.e., higher d_i) will cause the consumer to equilibrate at a lower abundance, which will permit poorer apparent competitors to subsequently invade; but the level of intraspecific density dependence has no consequence for whether the resource can itself invade and coexist with other resources that are already in the community. Whether a species can invade is irrelevant to how that species might affect either the community or the success of subsequent invaders once it is established.

Considering the geometry of the isoclines for these species also suggests other important insights about the assembly of a community that might not be readily apparent. For example, one can easily position the consumer isocline so that it will coexist with two or more resources but would be unable to maintain a population on any one of the resources alone. Remember that for the consumer to coexist with only one resource, its isocline must intersect the resource axis below $R^*_{i(0)}$ (see inequality (3.5) and fig. 3.3). Consider the situation of a consumer feeding on two resources that cannot support a population on either alone (fig. 3.6). In the $R_2 - R_1$ face, the resource isoclines are each parallel to the other resource axis, and the consumer isocline is a straight line between its intersection points on the two resource axes (figs. 3.5(a) and 3.6). The consumer will be able to invade and coexist with these two resources, even though it cannot coexist with either alone, because its isocline passes below the point of intersection between the resource isoclines in the $R_2 - R_1$ face: this is the invasibility criterion for this consumer (fig. 3.6(b)). Consequently, the consumer can only invade this community once both

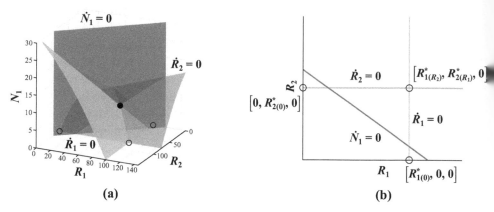

(a)   (b)

FIGURE 3.6. An isocline configuration for an apparent competition module in which the consumer (N_1) coexists in the system if both resources (R_1 and R_2) are present, but cannot coexist if only one of the two resources is present. All isocline and equilibrium points are as shown in figures 3.3 and 3.5. The parameter values are as follows: $c_1 = 4.0$, $d_1 = 0.04$, $c_2 = 3.0$, $d_2 = 0.04$, $a_{11} = 0.4$, $a_{21} = 0.5$, $b_{11} = b_{21} = 0.1$, $h_{11} = h_{21} = 0.03$, $f_1 = 1.95$, and $g_1 = 0.0$.

resources are established, which implies a definitive order for community assembly. Both the multidimensional nature of many invasibility criteria and the ordering of community assembly will be recurring issues as food webs become more complex.

Contrast the equilibrium abundances of two resources and one consumer species at locations along an environmental gradient that differ in the value for c_1 but are identical for all other parameters (fig. 3.7). For example, these sites differ in the availabilities of some limiting nutrient for R_1, but this nutrient does not influence the abundances of the other species. If c_1 is too low, R_1 cannot invade and coexist because of the combined effects of low nutrient availability and mortality imposed by the consumer ($c_1 < 0.16$ in the figure). At intermediate values of c_1, all three species coexist, but R_2's abundance is lower and N_1's abundance is higher at sites that are more productive for R_1 ($0.16 < c_1 < 1.8$). Finally, at sites where c_1 is high, R_1's abundance inflates N_1's abundance to a level at which R_2 cannot invade and coexist ($c_1 > 1.8$).

All the issues addressed in the empirical testing discussions in the previous sections apply here. Are all these resources in N_1's diet, and do they each make some material contribution to N_1's population dynamics? Likewise, can all the resource species support populations in this community in the absence of N_1 (i.e., $R_{i(0)}^* > 0$ for each)?

The invasibility experiment for each resource species should follow the same basic strategy as outlined in the previous sections. The basic design would

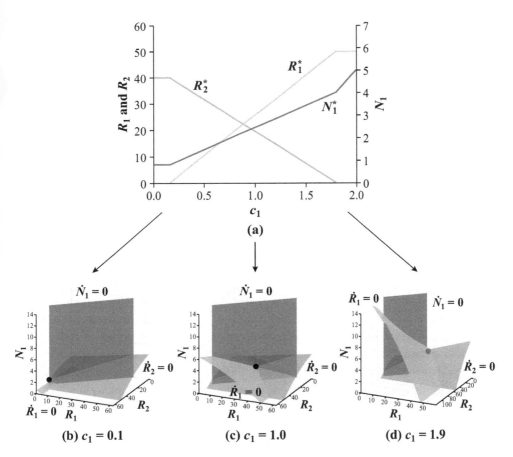

FIGURE 3.7. The equilibrium abundances and isocline configurations of a single consumer (N_1) feeding on two resources (R_1 and R_2) at different points along an environmental productivity gradient that determines the intrinsic growth rate of R_1 (i.e., a c_1 gradient). Panel (a) shows the species equilibrium abundances along this gradient. Panels (b)–(d) show the isocline configurations at three different points along this gradient. The isocline diagrams are as described in figure 3.6. The parameter values are as follows: $c_2 = 1.0$, $d_1 = d_2 = 0.02$, $a_{11} = 0.4$, $a_{21} = 0.5$, $b_{11} = b_{21} = 0.1$, $h_{11} = h_{21} = 0.05$, $f_1 = 1.0$, and $g_1 = 0.0$.

manipulate the abundance of each species over a large range, from very low to above natural, with the other species near their natural abundances after deleting the species in question. Again, the per capita birth and death rates of the manipulated species are the key response variables to evaluate its growth rate when rare and how that growth rate changes with its own abundance. Invasibility is satisfied if the resource has a positive per capita growth rate at its lowest abundance treatment. The higher abundance treatments again would show the form of density dependence. If one cannot determine the consumer's abundance in the absence of the resource in question, the resource abundance gradients may be crossed in a factorial experimental design with a few lower consumer abundances (e.g., 0.5×, 0.75×, and 1× the natural consumer abundance), so as to cover the possible consumer abundances into which the resource would be invading. The invasibility of the consumer should also be evaluated in the presence of all resource species together and on each resource separately to evaluate the diversity of resources needed to support the consumer population (cf. figs. 3.5 and 3.6).

While simple in principle, invasibility is difficult if not impossible to test in practice. First, the species must be removed from the community, and then the community must be allowed to come to its new steady state. Only then should the species be reintroduced to test for its per capita growth rate when rare. Clearly, this is not possible in most communities: imagine trying to remove a single fish species from a pond or lake while leaving all other species in place. In fact, empirical tests of invasibility have only been performed in relatively simple communities, mainly in laboratory microcosm experiments (Siepielski and McPeek 2010). Thus, for most communities, more indirect evidence must be mustered to claim coexistence.

Even if an invasibility test can be performed, the invasibility experiments only address whether each species is coexisting, as opposed to being a sink or walking dead species. However, the invasibility experiments do not test *why* each species is able to maintain a population in the community—the *mechanism of coexistence*. To test the mechanism of coexistence, one must elucidate the demographic circuitry of the food web and probe the ecological performance strengths and weaknesses of the various species that pass their invasibility criteria. For this community module, the test of the mechanism promoting or retarding their coexistence is the apparent competition experiment that Holt (1977) described and for which the module is named. The abundance of each resource species should be manipulated over a gradient, and the performances of the consumer and the other resources evaluated in response (as in fig. 3.7). Obviously, the simplest gradient would be the presence and absence of each resource species (e.g., Schmitt 1987, Karban et al. 1994, Bonsall and Hassell 1997, Muller and Godfray 1997, Cronin 2007, Orrock et al. 2008). For example, imagine examining a community containing R_1, R_2, and N_1. In such experiments, the abundance of the manipulated species is held constant within an experimental replicate, and the other species are allowed to respond as they will. This is the experimental equivalent of changing the conditions that would alter the equilibrium abundance of the manipulated species and how the other species respond. For example, if R_1 is held experimentally constant at a particular abundance, this treatment mimics placing the community at the corresponding position on the c_1 gradient in figure 3.7, since c_1 only affects R_1's abundance. In one set of treatments, R_1 is manipulated over an abundance gradient from low to well above natural, and R_2 and

N_1 are present initially at their natural abundances. The ideal experiment would allow R_2 and N_1 to respond in abundance, with N_1 predicted to increase and R_2 predicted to decrease with increasing R_1. The reciprocal set of treatments manipulating R_2 and quantifying the responses of R_1 and N_1 fill out the experimental design (imagine the corresponding figure to fig. 3.7 for a gradient of c_2).

Understanding the mechanism is critical. For example, competitive responses among plants in density manipulation experiments are almost invariably interpreted as resulting from competition for nutrients or light (e.g., Goldberg et al. 2001, Levine and HilleRisLambers 2009, Kraft et al. 2015). Certainly, resource manipulation experiments do show that nutrient limitation and so resource competition is one possible circuit for connections between plant species (e.g., Goldberg and Miller 1990, Tilman 1993, Harpole et al. 2011). However, the herbivores that feed on these same plants also substantially depress the fitnesses of these same plants and link plant species via apparent competition (e.g., Reader 1992, Carson and Root 1999, 2000, Borer et al. 2014). Responses of herbivores to plant density manipulations may drive the abundance responses of competitor plants just as much as changes in nutrient levels.

One of the important consequences of a saturating functional response in this mechanism is to reduce the stringency of the invasibility criterion for invading apparent competitors (see fig. 3.5(b)–(c)). A critical test for this component is to measure the per capita death rate inflicted by the consumer on one resource species at low abundance across an abundance gradient of the other resource species. For example, consider the right panel of fig. 3.5(b). The R_2 isocline has a positive slope in the $N_1 - R_i$ face because predator satiation causes the per capita death rate imposed by N_1 on R_2 to decrease as R_1's abundance increases. The experiment to test this important component of the mechanism is to create a treatment gradient of R_1 from very low abundance to greater than natural, and to have a natural abundance of N_1 and a very low abundance of R_2 in every replicate across the R_1 abundance gradient. The critical response variable is the per capita death rate of R_2, which is predicted to decrease with increasing R_1 abundance. The reciprocal design quantifying the per capita death rate of R_2 from N_1 predation along an R_1 abundance gradient completes the test of this important mechanistic component caused by saturating functional responses. This experiment should be augmented by quantifying the parameters of N_1's saturating functional response for feeding on the two resources, which of course also validates that N_1 actually has a saturating functional response for feeding on these resources.

Functional studies to understand how the resources' traits determine their relative apparent competitive abilities reinforce these experimental inquiries about ecological performance. For example, what traits influence the relative susceptibilities of the resources to the consumer (i.e., their various a_{i1} values), their relative intrinsic growth rates (i.e., their various c_i values), and their relative strengths of density dependence (i.e., their various d_i values)?

Resource Competitors

Return to the simple food chain of one consumer feeding on one resource and now consider the pool of potential invading consumers (invasion 4 in fig. 3.1). This series of invasions queries the conditions required for multiple consumers to coexist while feeding on a single resource. This is the community module that Volterra (1926) originally considered (see equations (2.1) and fig. 2.1)), but here I include the explicit dynamics of all the interacting species. The equations describing their joint population dynamics are then

$$\frac{dN_j}{N_j dt} = \frac{a_{1j} b_{1j} R_1}{1 + a_{1j} h_{1j} R_1} - f_j - g_j N_j$$

$$\frac{dR_1}{R_1 dt} = c_1 - d_1 R_1 - \sum_{j=1}^{q} \frac{a_{1j} N_j}{1 + a_{1j} h_{1j} R_1}$$

(3.10)

With linear functional responses ($h_{1j} = 0$) and no direct self-limitation ($g_j = 0$), the isocline of each consumer intersects the R_1 axis at $f_j/(a_{1j} b_{1j})$ as above and is parallel to all the consumer axes, including its own (fig. 3.8(a)). This is because

---

FIGURE 3.8. Various isocline configurations for a resource competition module with two consumers (N_1 and N_2) that feed on one resource (R_1). The panels show the isocline system when (a) the consumers have linear functional responses and no direct self-limitation, (b) the consumers have saturating functional responses, (c) the consumers have saturating functional responses and N_1 has some level of direct self-limitation, and (d) both consumers have saturating functional responses and some level of direct self-limitation. Panels (a)–(c) show the full three-dimensional representation of the isoclines on the left and the $N_1 - R_1$ face (i.e., the plane where $N_2 = 0$) on the right. (The $N_1 - R_1$ face for (d) is the same as for (c) and is not shown.) In the graphs on the right, the stable equilibrium point before N_2 invades is labeled $[R^*_{1(N_1)}, 0, N^*_{1(R_1)}]$. This equilibrium becomes unstable when N_2 invades if N_2's isocline passes to the left of this point in this face (as shown in (c)), meaning that the invading consumer has adequate resources available for positive population growth rate. If N_2's isocline passes to the right of this two-species equilibrium (as in (a) and (b)) insufficient resources are available for N_2 to invade. In the full three-dimensional graphs, the unstable equilibria in the $N_1 - R_1$ face are identified with open circles, and the stable three-species equilibrium is identified with a filled black circle. Unless otherwise stated, the parameter values in all panels are as follows: $c_1 = 4.0$, $d_1 = 0.04$, $a_{11} = 0.5$, $a_{12} = 0.25$, $b_{11} = b_{21} = 0.1$, $h_{11} = h_{21} = 0.0$, $f_1 = f_2 = 1.0$, and $g_1 = g_2 = 0.0$. Panel (b) has $h_{11} = h_{21} = 0.04$, (c) has $h_{11} = h_{21} = 0.04$ and $g_1 = 0.125$, and (d) has $h_{11} = h_{21} = 0.04$, $g_1 = 0.125$, and $g_2 = 0.05$.

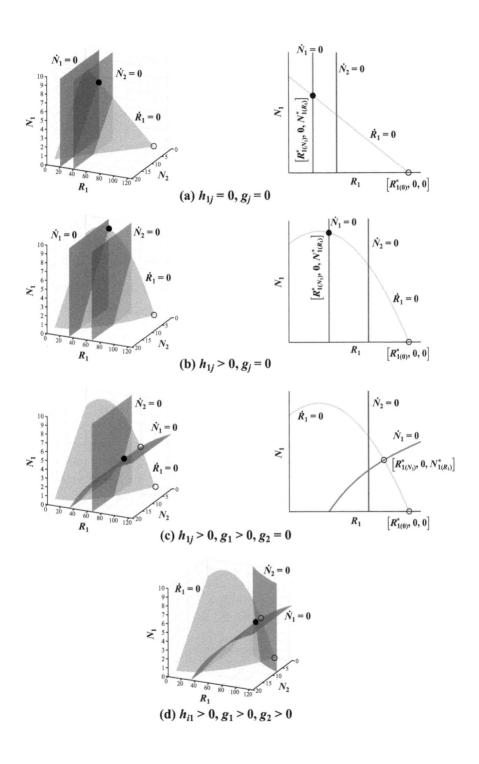

(a) $h_{1j} = 0, g_j = 0$

(b) $h_{1j} > 0, g_j = 0$

(c) $h_{1j} > 0, g_1 > 0, g_2 = 0$

(d) $h_{i1} > 0, g_1 > 0, g_2 > 0$

no consumer directly influences the per capita demographic rate of any other consumer, including itself. The interactions among the consumers, including the effects on themselves, are all *indirect effects*: intraspecific and interspecific resource competition are both indirect effects mediated through the resource responses to consumer abundances. As a result, the isoclines of the consumers are all parallel to one another, and if all intersect the R_1 axis at unique points (i.e., their isoclines are not coincident), one consumer species will depress the resource abundance to a level at which no other consumer in the pool can maintain a population (Hsu et al. 1977, 1978a, Tilman 1980, 1982, Grover 1997, McPeek 2012).

Identify the species according to the points where they intersect the R_1 axis from lowest to highest, such that

$$\frac{f_1}{a_{11}b_{11}} < \frac{f_2}{a_{12}b_{12}} < \frac{f_3}{a_{13}b_{13}} < \ldots < \frac{f_q}{a_{1q}b_{1q}}. \tag{3.11}$$

Because each consumer will increase if the resource abundance is above its isocline, each has an invasibility criterion of

$$\frac{f_j}{a_{1j}b_{1j}} < R^*_{1(N_k)} \tag{3.12}$$

for the situation when R_1 is already coexisting with N_k. If N_j meets this invasibility criterion, it will invade, and subsequently depress R_1's abundance below the level at which N_k can maintain a population. Consequently, the consumer species that intersects the R_1 axis at the lowest value, in this case N_1, will depress R_1's abundance lower than any other consumer can stand. If all these species eventually invade, the resulting community would then consist only of N_1 feeding on R_1. All of this is simply an illustration of the *R*-star rule, where the consumer with the lowest *R*-star—the lowest intersection point on the resource axis—will drive all other consumers extinct, given the assumptions of $h_{1j} = 0$ and $g_j = 0$ (Hsu et al. 1977, Tilman 1980, 1982). The reader should note that this result is another way to derive Volterra's (1926) original conclusion based on equations (2.1) where he ignored the resource dynamics.

Saturating functional responses ($h_{1j} > 0$) do not alter this qualitative outcome that only one consumer can coexist with R_1 if the community comes to a stable point equilibrium (fig. 3.8(b)). In this case, the consumers are ordered according to

$$\frac{f_1}{a_{11}(b_{11} - f_1 h_{11})} < \frac{f_2}{a_{12}(b_{12} - f_2 h_{12})} < \frac{f_3}{a_{13}(b_{13} - f_3 h_{13})} < \ldots < \frac{f_q}{a_{1q}(b_{1q} - f_q h_{1q})}, \tag{3.13}$$

but because every consumer isocline is still parallel to all consumer axes, N_1 will depress R_1's abundance to a level at which no other consumer can maintain a population, since

$$R^*_{1(N_j)} = \frac{f_j}{a_{1j}\left(b_{1j} - f_j h_{1j}\right)}. \tag{3.14}$$

Thus, the point at which each consumer's isocline intersects the resource axis is still a direct measure of its relative competitive ability for this resource, even with saturating functional responses.

In contrast, if the resulting dynamic with one or more consumers is a limit cycle, two consumers may coexist on this single resource (Koch 1974b, Hsu et al. 1978b, Armstrong and McGehee 1980, Abrams 2004, Xiao and Fussmann 2013). A necessary condition for two consumers to coexist in a limit cycle is that the average resource abundance over the trajectory of the limit cycle must be greater than the R-star value of the poorer resource competitor (Hsu et al. 1978a, Armstrong and McGehee 1980). Thus, cycling removes the constraint that the number of available resource species limits the number of coexisting consumer species. However, I defer a more in-depth discussion of the mechanisms permitting their coexistence in a limit cycle until chapter 8.

For a second consumer species from this pool to be able to invade and coexist with R_1 and N_1 at a stable point equilibrium, N_1 must limit its own abundance through some form of direct self-limitation (i.e., $g_1 > 0$ in equations (3.10) or feeding interference in a Beddington-DeAngelis functional response, with $z_1 > 0$ as in equations (3.6)) (Vance 1984, McPeek 2012, 2017b). Remember that both forms of self-limitation cause N_1's isocline to bend away from its own axis but still intersect the R_1 axis at $f_1/(a_{11}(b_{11} - f_1 h_{11}))$ (see figs. 3.3(d)–(e) and 3.5(c)). As a result, with some form of consumer self-limitation,

$$\frac{f_1}{a_{11}\left(b_{11} - f_1 h_{11}\right)} < R^*_{1(N_1)}, \tag{3.15}$$

and it is now possible for a poorer resource competitor to still satisfy its invasibility criterion (fig. 3.8(c)). For three species, this is true if the poorer competitor's isocline passes below the two-species equilibrium $[R^*_{1(N_1)}, 0, N^*_{1(R_1)}]$ in the $N_1 - R_1$ face (fig. 3.8(c), right). Since each consumer isocline is parallel to the axes of all other consumers (i.e., they only interact indirectly through the resource), this simply requires that its isocline intersect the R_1 axis between the two quantities in (3.15):

$$\frac{f_1}{a_{11}\left(b_{11} - f_1 h_{11}\right)} < \frac{f_j}{a_{1j}\left(b_{1j} - f_j h_{1j}\right)} < R^*_{1(N_1)}. \tag{3.16}$$

Coexistence of the poorer competitor simply requires that the superior competitor experience a sufficient level of direct self-limitation to increase the resource's abundance enough for inequalities (3.16) to be true.

Because direct self-limitation causes the consumer isocline to bend away only from its own axis, all consumer isoclines remain parallel to all other consumer axes besides their own. As a result, the poorer competitor can itself experience any level of direct self-limitation and still coexist with the better competitor (fig. 3.8(d)). The pair of consumers that can jointly depress R_1's abundance to the lowest level and mutually invade will coexist, although they may not be the two with the lowest intersection points on the R_1 axis (McPeek 2012, 2017b). In other words, Tilman's (1982) R-star rule no longer applies in this case. If both of these consumers (call them N_1 and N_2) experience some degree of direct self-limitation, a third consumer can invade if it can satisfy the invasibility criterion $f_j / (a_{1j}(b_{1j} - h_{1j}f_j)) < R^*_{1(N_1,N_2)}$. Invasions of consumers will continue until no remaining consumer species in the available pool of potential invaders can satisfy its invasibility criterion given the consumers already present in the community; that is, all remaining consumers that cannot invade have $f_y / (a_{1y}(b_{1y} - h_{1y}f_y)) > R^*_{1(N_1,...,N_j)}$. Also, as with apparent competition, the order of invasion by the various consumers has no effect on the set of consumers that will ultimately coexist (McPeek 2012).

The resource competition simulation module can be accessed at https://mechanismsofcoexistence.com/twoLevels_3D_1R_2N.html.

Here again, whether a consumer can invade is determined solely by whether it has a positive per capita population growth rate when it is rare. In this case, what directly determines this is whether it has adequate resources to foster population growth when its population abundance is near zero and not the abundances of other consumers. Obviously, the other consumers determine $R^*_{1(N_1,...,N_j)}$, but these are only indirect effect for the invader. If the invading consumer can satisfy its invasibility criterion at $R^*_{1(N_1,...,N_j)}$, it will invade no matter if the resource level is set by other consumers or other factors besides consumers. Moreover, the level of intraspecific density dependence it experiences at equilibrium after it invades, both directly from some form of self-limitation or indirectly via resource limitation, relative to its effect on other species is immaterial to whether it will coexist in this resource competition module.

The properties of other species already present in the community do strongly influence whether any additional consumer can invade. For a poorer resource competitor to invade, all the consumers that are already coexisting and that are better resource competitors must experience adequate levels of direct self-limitation to have the resource equilibrate at a level that permits the poorer competitor to invade. For example, in figure 3.8(c), if N_2 is a better resource competitor than all potential consumers who could subsequently invade this community, none of them will be able to invade because $f_2 / (a_{12}(b_{12} - h_{12}f_2)) = R^*_{1(N_1,N_2)}$. In contrast, in figure 3.8(d), where N_2 does have some degree of self-limitation, a third consumer could potentially invade because $f_2 / (a_{12}(b_{12} - h_{12}f_2)) < R^*_{1(N_1,N_2)}$.

Such invasions of subsequent consumers could continue if the consumer before them also had some degree of self-limitation and they can meet their invasibility criterion of $f_{j+1} / (a_{1j+1}(b_{1j+1} - h_{1j+1}f_{j+1})) < R^*_{1(N_1, N_2, \ldots, N_j)}$.

This model also makes some seemingly counterintuitive predictions for how equilibrium consumer abundances will change across communities arrayed along various environmental gradients (fig. 3.9). First, consider how the three species' abundances change across communities that would result along a resource productivity gradient of c_1. If only the better resource competitor N_1 is self-limiting, only N_2's abundance—the poorer resource competitor—increases with increasing productivity across communities on the portion of the gradient where all three species coexist ($c_1 > 1.89$ in fig. 3.9(a)). However, if both consumers are self-limiting, both consumers increase in abundance across communities with higher resource productivities (fig. 3.9(b)). Similar patterns are apparent when more than two consumers coexist on a single resource as well. Naively, a researcher might think that the better competitor should do better and the poorer competitor should do worse if resource abundance is increased. However, if only one resource limits the abundances of multiple consumers, environments that generate more productivity in the resource will increase the abundances of all consumers that are self-limiting to some degree without harming any coexisting competitor. In the next section, we explore what is required for competitive replacement along resource gradients.

It is interesting to note that multiple consumers can also coexist on a single resource if each consumer has its own specialized disease or specialist predator (Grover 1994), but I defer considering this in depth until chapters 4 and 7.

Replacement of the consumers is predicted to occur along potential gradients of intrinsic death rates for the consumers. For example, imagine two consumers competing for one resource along an altitudinal gradient where one consumer's intrinsic death rate increases with altitude but the other's does not (e.g., classic studies of salamander distributions along altitudinal and habitat gradients: Jaeger 1970, 1971, Hairston 1980, 1986). Figure 3.9(c) illustrates the equilibrium abundances of the three species along such a gradient for f_1. Imagine that f_1 increases with altitude for N_1, but this stressor does not alter N_2's intrinsic death rate. In communities with low values of f_1, N_1 is the better resource competitor; but even though it is self-limiting, it causes R_1's abundance to equilibrate at levels where N_2 cannot invade and coexist ($f_1 < 0.46$). In communities with intermediate values of f_1, all three species coexist with N_1—still the better resource competitor—but N_1 is more abundant and N_2 is less abundant in communities with higher f_1 ($0.46 < f_1 < 1.13$). In communities where f_1 is too high, N_1 becomes the poorer resource competitor; and since N_2 is not self-limiting, N_1 cannot maintain populations in those communities along the gradient ($f_1 > 1.13$). A similar pattern of consumer

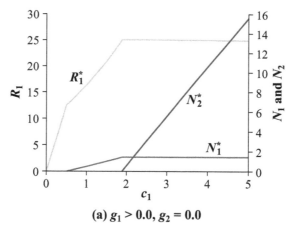

(a) $g_1 > 0.0, g_2 = 0.0$

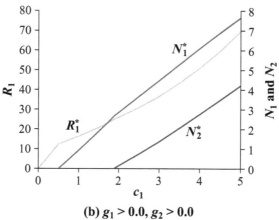

(b) $g_1 > 0.0, g_2 > 0.0$

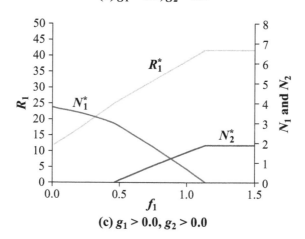

(c) $g_1 > 0.0, g_2 > 0.0$

species replacement along this gradient also occurs if both consumers are self-limiting, although their zone of coexistence along the gradient is broader.

As with a single consumer and a single resource, the first issue to empirically adjudicate is whether the consumers actually eat the resource and, if so, whether the availability of the resource limits their abundances. I cannot count how many times I have sat through seminars in which the speaker claimed that multiple species were competing for some resource, and the speaker had absolutely no evidence to show that the species in question were competing for or even limited by anything. For resource competition to occur, some resource must be limiting to the abundances of the consumers of that resource. The critical test of this is to manipulate the abundance of the resource and determine whether the demographic rates and therefore the abundances of the consumers change (e.g., Tilman et al. 1981, Passarge et al. 2006).

Invasibility experiments that query the coexistence status of each consumer also follow the same design described in previous sections. The critical issue with evaluating invasibility for this community module is the resource abundance that the consumer experiences in the experiment. If the resource's abundance dynamics respond quickly to consumer abundances, simply establishing consumer density treatments may be adequate. However, if resource abundance responds slowly, care must be given to having the resource abundance at the start of the experiment at the level it would be in the absence of the consumer in question but with the other consumers present. In this latter case, treatments would have to begin with only the other consumers present and allow these to approach equilibrium before the consumer in question is introduced at low abundance. The resource abundance should also be monitored if at all possible throughout the course of the experiment to assure that the conditions are appropriate for evaluating invasibility.

The test of this mechanism is the classic consumer density manipulation experiment, in which the abundance of each consumer is manipulated and the responses of the other consumers quantified (Connell 1983a, Schoener 1983). The expectations for this experiment are that the manipulation of one consumer's

---

FIGURE 3.9. The equilibrium abundances of two consumers (N_1 and N_2) feeding on one resource (R_1) at different points along an environmental productivity gradient that determines the intrinsic growth rate of R_1 (i.e., a c_1 gradient in (a) and (b)) and an environmental gradient that determines the intrinsic death rate of consumer 1 (i.e., an f_1 gradient in (c)). In (a), only N_1 is directly self-limiting, and in (b) and (c) both consumers are directly self-limiting. Unless otherwise stated, the parameter values in all panels are as follows: $c_1 = 2.0$, $d_1 = 0.04$, $a_{11} = 0.5$, $a_{12} = 0.25$, $b_{11} = b_{21} = 0.1$, $h_{11} = h_{21} = 0.04$, $f_1 = f_2 = 0.5$, and $g_1 = g_2 = 0.0$. Panels (a) and (c) have $g_1 = 0.125$, and (b) and (c) have $g_2 = 0.125$.

abundance will cause the abundances of other consumers to change in opposition to the manipulation: increase N_1's abundance and N_2 is expected to decrease, and decrease N_1's abundance and N_2 is expected to increase. It is important to remember that this experiment is equivalent to changing the other demographic forces that impinge on the manipulated consumer and allowing the other consumers and the limiting resource to change as they may. So, for example, such density manipulations are akin to moving N_1 to a different position on the f_1 gradient in figure 3.9(c) and determining whether the abundances of R_1 and N_2 change as predicted by this model.

Manipulating the abundance of the resource also provides important information that allows the researcher to discriminate this community module from others we consider later. For example, increasing the productivity for the basal resource (e.g., nitrogen addition in a terrestrial plant-herbivore system or phosphorus addition to an aquatic alga-zooplankton system) should increase the abundances of species at both trophic levels (fig. 3.9(a)–(b)). This may seem rather trivial at this point—make the world better for the basal resource and the resource and its consumers all do better (or at least not worse)—but in the next section and chapter, we will see examples where this is not so. The result described here will only occur if one resource is limiting to the competing consumers.

When one thinks of the resource competitive ability of a particular species, the mind is generally drawn to the ability of the consumer to capture, consume, and utilize the resource. However, the species that is most effective at harvesting the resource is not necessarily the best resource competitor. One of the critical insights that comes from exploring these more mechanistic models is that the competitive ability of each consumer is defined by a balance between demographic forces: $f_j/(a_{1j}(b_{1j} - h_{1j}f_j))$. The denominator of this ratio quantifies the resource harvesting and utilization ability of the consumer, and the numerator quantifies the other demographic forces—that do not involve resource harvest rate—against which the benefits of consuming the resource must be balanced in this case. We typically think of resource competitive ability as being defined by the resource level to which the consumer can depress the resource in isolation, which is what this value represents: Tilman's (1982) R-star. For example, Passarge et al. (2006) could predict the winner of pairwise competitive interactions between algal species based on single-species growth experiments in which they measured the degree to which each algal species depressed phosphorus or light, the limiting resources in these experiments.

However, the level to which the consumer can depress the resource is determined by the balance of these two demographic forces (Tilman 1982). In fact, a number of classic empirical studies of plant distributions show that plant species segregate due to resource competition along stressor gradients (e.g., Grace and

Wetzel 1981, Wilson and Keddy 1986, Emery et al. 2001). Consequently, the relative rank of consumers at utilizing the resource (i.e., only considering the denominator of equation (3.14)) cannot be assumed to represent their rank in resource competitive ability. Certainly, the fact that one species of Darwin's finch has a bill that is best at utilizing some seed will influence its ability to compete for those seeds, but this does not make that species the best competitor for that seed. Studies to understand why one species is a better resource competitor than another cannot simply focus on resource utilization abilities alone.

The ratio defining competitive ability is also important in another context. Multiple consumers can only coexist on a single resource if those that can drive the resource to a lower abundance also limit their own abundances in some way (assuming they do not have their own enemies, which is considered in later chapters). Therefore, if the consumers compete for a single resource, the mechanism of their coexistence must involve the better competitors also limiting their own abundances through some form of direct self-limitation (see fig. 3.8(c)–(d)).

For example, it is important to remember that MacArthur (1958) documented five *Setophaga* warbler species that tend to forage in different parts of trees in the conifer forests of northern New England, USA. He did not conclusively demonstrate that these five species compete for resources or that their partial segregation for feeding on different areas of trees causes them to segregate onto different resources. In fact, his analysis considered a broad set of natural history features of these five species that might limit their population abundances throughout their lifetimes (e.g., mortality during migration, survival on wintering grounds, diseases, nest predation, temporal variability of food, and territoriality). As MacArthur noted, all five of these species defend breeding/feeding territories. On the Birds of North America website, three of the species are described as defending their territories against only conspecifics, one defending against conspecifics and heterospecifics, and one being unclear about the specificity of defense (Cornell Laboratory of Ornithology, https://birdsna.org/Species-Account/bna/home, accessed 10 October 2019). Such territoriality serves as an additional limiting factor for each species that would permit many additional warbler species to coexist in this forest. In fact, MacArthur (1958, p. 609) acknowledges this in the paper:

> From the ecological point of view, the distinction is of very great importance, however, for, as G. E. Hutchinson pointed out in conversation, if each species has its density (even locally) limited by a territorial behavior which ignores the other species, then there need be no further differences between the species to permit them to persist together.

If such interaction types are evident, the critical test of their importance for the mechanism of coexistence is the gap between where the isocline of each intersects the resource axis and the abundance at which the resource equilibrates in their presence (i.e., $f_j/(a_{1j}(b_{1j}-h_{1j}f_j))$ versus $R^*_{1(N_j)}$). To quantify $f_j/(a_{1j}(b_{1j}-h_{1j}f_j))$, each consumer would be placed at low abundance over a gradient of resource abundances and in the absence of all other consumers, with the response variable being the consumer's per capita growth rate. The results of this experiment would identify the minimum resource abundance at which each consumer could support a population (i.e., satisfy its invasibility criterion), which is the point where each consumer's isocline intersects the resource axis (see fig. 3.8). A second experiment that places each consumer by itself with the resource and allows them to come to equilibrium would be done to quantify $R^*_{1(N_j)}$. The difference between these two quantities for each consumer measures the degree to which mechanisms of self-limitation prevent each consumer from depleting resources to its full extent and thus permits other consumers to support populations on this single resource. The results of these two experiments would also validate and reinforce the invasibility experiment results by showing that resource availability is the key environmental factor determining each consumer's success in the community and by quantifying the performance of each consumer over a gradient of resource availabilities. These results would also allow the prediction of consumer composition across sites that differ in the total availability of the limiting resource.

Multiple Consumers Eating Multiple Resources

Now that we have considered requirements for having multiple species coexist at each of two trophic levels when only one species is present at the other trophic level, we now take up the issue of having multiple species at both trophic levels (invasion 5 in fig. 3.1). For much of the history of community ecology, this has been the fundamental problem, given that coexistence of resource competitors has been the central focus (e.g., Hutchinson 1958, 1959, MacArthur and Levins 1967, MacArthur 1969, 1972, May 1973, Tilman 1982, Letten et al. 2017, McPeek 2019b). However, the typical problem has been to consider only two resources (e.g., Stewart and Levin 1973, Hsu et al. 1978a, 1978b, Tilman 1980, 1982) or resource species arrayed along a single trait dimension (e.g., MacArthur and Levins 1967, May and MacArthur 1972). Our models need to begin to explore the mechanisms of coexistence that are available when more than two resources that differ in multiple trait dimensions are available to consumers, like in the open water of Trout Lake, or on the Mo'orea coral reef, or on the Laikipia savanna. Here, I offer a modest start to this process by explicitly considering up to four resource species and projecting the inferences derived here to more than four at each.

With multiple consumers feeding on multiple resources, the dynamical system of equations becomes

$$
\frac{dN_j}{N_j dt} = \frac{\sum_{i=1}^{p} a_{ij} b_{ij} R_i}{1 + \sum_{i=1}^{p} a_{ij} h_{ij} R_i} - f_j - g_j N_j
$$

$$
\frac{dR_i}{R_i dt} = c_i - d_i R_i - \sum_{j=1}^{q} \frac{a_{ij} N_j}{1 + \sum_{k=1}^{p} a_{kj} h_{kj} R_k}
$$

(3.17)

This appears to be a daunting set, but we have already considered all the necessary analytical features of this system.

Begin by considering the simplest system with two resources and two consumers, with linear functional responses ($h_{ij} = 0$) and no direct self-limitation ($g_j = 0$). This module forms the basis for classic analyses that have guided our intuition about species coexistence of resource competitors (MacArthur 1969, Tilman 1980, 1982). All four isoclines are four-dimensional planes with these parameters, and they must intersect at a single point for a four-species equilibrium to exist. However, this equilibrium may be stable or unstable, depending on the relative resource competitive abilities of the consumers and the relative apparent competitive abilities of the resources. Figure 3.10(a)–(c) provides the three-dimensional perspectives on this four-dimensional system when the isoclines intersect to give a stable four-species equilibrium. Also note the similarities to the previous analyses with only one species at one of the two trophic levels (cf. figs. 3.5(a) and 3.8(a)).

First, consider the consumer isoclines. Because neither consumer species directly influences its own or any other consumer's abundance, both consumer isoclines are parallel to both consumer axes. Thus, only their positions on the resource axes are needed to determine whether they intersect. To intersect at positive resource abundances, each consumer must be the better competitor for one of the two available resources (fig. 3.10(d)). If N_1 is the better competitor for R_1 (i.e., the N_1 isocline intersects the R_1 axis below that of N_2) and N_2 the better competitor for R_2 (i.e., the N_2 isocline intersects the R_2 axis below that of N_1), the inequality

$$
\frac{a_{11} b_{11}}{a_{12} b_{12}} > \frac{f_1}{f_2} > \frac{a_{21} b_{21}}{a_{22} b_{22}}
$$

(3.18)

is satisfied (see box 3.2), which is simply based on the relationships among the intersection points of the consumer isoclines with the resource axes (McPeek 2019b). Tilman's (1980, 1982) classic analysis of two consumers feeding on two resources identifies this competitive trade-off required for consumer invasion and

Box 3.2. In this box, I illustrate the analysis needed to come to inequality (3.18). The consumer isoclines are found by setting each consumer equation in (3.17) equal to 0 and then simplifying. For two consumers and assuming all $h_{ij} = g_j = 0$, the resulting isoclines are

$$\frac{dN_1}{dt} = 0 : R_2 = \frac{f_1}{a_{21}b_{21}} - \frac{a_{11}b_{11}}{a_{21}b_{21}} R_1$$

$$\frac{dN_2}{dt} = 0 : R_2 = \frac{f_2}{a_{22}b_{22}} - \frac{a_{12}b_{12}}{a_{22}b_{22}} R_1.$$

Both are lines in $R_1 - R_2$ space. The N_1 isocline intersects the N_1 axis at $f_1/(a_{11}b_{11})$ (remember to substitute $N_2 = 0$ into the N_1 isocline and solve for N_1) and the N_2 axis at $f_1/(a_{21}b_{21})$. Likewise, the N_2 isocline intersects the N_1 axis at $f_2/(a_{12}b_{12})$ and the N_2 axis at $f_2/(a_{22}b_{22})$.

For these isoclines to intersect one another at positive abundances for both species, the isocline of one species must intersect the N_1 axis above the other species, but intersect the N_2 axis below the other species. If we define N_1 as the better competitor for R_1, this means

$$\frac{f_1}{a_{11}b_{11}} < \frac{f_2}{a_{12}b_{12}},$$

which can be rearranged to

$$\frac{f_1}{f_2} < \frac{a_{11}b_{11}}{a_{12}b_{12}}.$$

For the isoclines to intersect, this means that N_2 must be the better competitor for R_2, which implies

$$\frac{f_2}{a_{22}b_{22}} < \frac{f_1}{a_{21}b_{21}}$$

$$\frac{a_{21}b_{21}}{a_{22}b_{22}} < \frac{f_1}{f_2}.$$

These two inequalities resulting from the relative positions of the intersection points of the isoclines on the two axes can then be combined (since they both involve the ratio f_1/f_2) to give

$$\frac{a_{11}b_{11}}{a_{12}b_{12}} > \frac{f_1}{f_2} > \frac{a_{21}b_{21}}{a_{22}b_{22}},$$

which is inequality (3.18).

coexistence (see also Hsu et al. 1978a, Fox and Vasseur 2008, Kleinhesselink and Adler 2015, Letten et al. 2017). Coexistence requires that "each [consumer] species consumes proportionately more of the resource that more limits its own growth" (Tilman 1982, p. 75).

However, coexistence of all four species also depends critically on the apparent competitive abilities of the two resources to interact with the consumers. First, each resource must be able to invade when each consumer is present with the other resource (i.e., the invasibility criterion for each resource in an apparent competition module as depicted in fig. 3.5(a) for each consumer-resource pairing). In addition, given the resource competitive relationships between the two consumers specified by inequality (3.18), the resource isoclines must intersect at the four-species equilibrium so that R_2 is a better apparent competitor against N_1 and R_1 is a better apparent competitor against N_2 (fig. 3.10(e)). Remember that being a better apparent competitor means that the resource can support a population at a higher consumer abundance because its isocline intersects the consumer axis at a higher value. Thus, to coexist, each resource must be more impacted by the consumer that is the better resource competitor for it. This relationship between the resource isoclines at the equilibrium means

$$\frac{a_{11}}{a_{21}} > \frac{c_1 - d_1 R^*_{1(R_2, N_1, N_2)}}{c_2 - d_2 R^*_{2(R_1, N_1, N_2)}} > \frac{a_{12}}{a_{22}} \tag{3.19}$$

is satisfied, which is simply a statement about the relative intersection points of the resource isoclines with the consumer axes in a plane through the four-species equilibrium that is perpendicular to both resource axes and intersects the resource axes at their equilibrium abundances (McPeek 2019b). This inequality ensures that the intersection between the two consumer isoclines intersects the intersection between the two resource isoclines to form a feasible four-species equilibrium. (For those familiar with Tilman's (1982) graphical formulation of this problem, inequality (3.19) is the criterion that ensures the resource supply point be in the wedge formed by the consumers' consumption vectors.)

If inequality (3.19) is not satisfied because the ratio of the resource productivities is outside the range formed by the ratios of the attack coefficients (but $a_{11}/a_{21} > a_{12}/a_{22}$ is true), only one consumer will coexist with the two resources. If the resource productivities ratio is larger than both attack coefficient ratios, only N_1 will coexist with the resources because R_1's productivity inflates N_1's abundance to a level that depresses N_2's more critical resource below a level where it can persist. If the resource productivities ratio is smaller than both attack coefficient ratios, only N_2 will coexist with the two resources for analogous reasons. (These

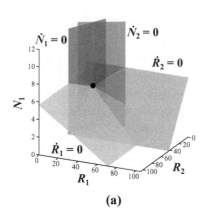

(a)

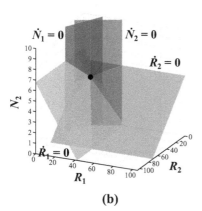

(b)

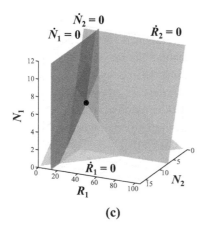

(c)

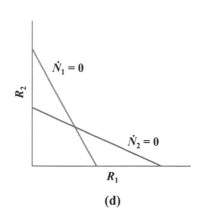

(d)

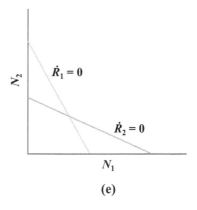

(e)

situations correspond to the supply point being outside the consumption vector wedge in Tilman's (1982) graphical formulation. See figure 3 in McPeek (2019b) for corresponding isocline representations of these.)

If one of these inequalities is reversed (i.e., each resource is more impacted by the consumer that is the poorer resource competitor for it), the four-species equilibrium is unstable (Hsu et al. 1978a, Tilman 1980, 1982, Letten et al. 2017, McPeek 2019b). In this case, each consumer can coexist alone with the two resources, and each consumer cannot invade the community if the other consumer is already present. In other words, this situation represents a *priority effect*, such that whichever consumer is the first to invade will coexist with the two resources, and in so doing will prevent the other consumer from being able to invade.

This full range of possible outcomes and this same set of qualitatively identical conclusions are reached for similar models that assume basal abiotic resources instead of biotic species as resources (Tilman 1982). So no matter whether the scenario imagined is phytoplankton species consuming abiotic resources or herbivore species consuming plants, the conditions for coexistence are qualitatively the same.

Saturating functional responses do not qualitatively alter the criteria for invasibility and coexistence, but they will alter some quantitative aspects of the invasibility criteria and obviously the equilibrium abundances that result. With all $h_{ij} > 0$ in equations (3.17), inequality (3.19) is still the criterion to have the resource isoclines intersect at positive consumer abundances and produce a four-species equilibrium (McPeek 2019b). However, inequality (3.18) is modified somewhat to incorporate nonzero handling times:

$$\frac{a_{11}\left(b_{11} - f_1 h_{11}\right)}{a_{12}\left(b_{12} - f_2 h_{12}\right)} > \frac{f_1}{f_2} > \frac{a_{21}\left(b_{21} - f_1 h_{21}\right)}{a_{22}\left(b_{22} - f_2 h_{22}\right)}, \tag{3.20}$$

which again is simply the statement that N_1 is the better competitor for R_1 and N_2 the better competitor for R_2 (McPeek 2019b). Thus, all the same qualitative relationships hold for stable coexistence of the four species, for one consumer dominating because of a difference in resource productivities, or for priority effects between the two consumers. The only other difference that saturating functional responses cause

---

FIGURE 3.10. The isocline configuration for a community module in which two consumers with linear functional responses (N_1 and N_2) feed on two resources (R_1 and R_2). Neither consumer has direct self-limitation. Each panel shows a different perspective on the same isocline system: (a) the $N_1 - R_1 - R_2$ subspace, (b) the $N_2 - R_1 - R_2$ subspace, (c) the $N_1 - N_2 - R_1$ subspace, (d) the $R_1 - R_2$ face, and (e) the $N_1 - N_2$ plane at $R^*_{1(R_2,N_1,N_2)}$ and $R^*_{2(R_1,N_1,N_2)}$. All isocline and equilibrium points are as shown in figures 3.6 and 3.8. The parameter values are as follows: $c_1 = c_2 = 4.0$, $d_1 = d_2 = 0.04$, $a_{11} = a_{22} = 0.5$, $a_{12} = a_{21} = 0.25$, $b_{11} = b_{21} = 0.1$, $h_{11} = h_{21} = 0.0$, $f_1 = f_2 = 1.0$, and $g_1 = g_2 = 0.0$.

are that they reduce the stridency of resource invasibility criteria, because their iso-clines increase along the axes of other resources (fig. 3.11(a); see also fig. 3.5(b)).

In some areas of parameter space, the resulting dynamics for the community are limit cycles or chaos if the consumers have saturating functional responses. As with consumers competing for one resource, if the dynamics of the system is a limit cycle, the number of consumers that can coexist is not limited by the number of available resources; I postpone consideration of the consequences of limit cycles on coexistence until chapter 8.

The two-consumers/two-resources simulation module can be accessed at https://mechanismsofcoexistence.com/twoLevels_3D_2R_2N.html.

Thus, with linear or saturating functional responses and no direct self-limitation, two consumers can coexist with two resources at a stable point equilibrium if each pair of species can satisfy both their individual invasibility criteria and the interspecific trade-offs for each pair of species at the two trophic levels. These trade-offs require (1) each consumer to be a better resource competitor for one of the resources, and (2) each resource must be the poorer apparent competitor against the consumer that is the better resource competitor for it (Hsu et al. 1978a, Tilman 1980, 1982, Letten et al. 2017). Moreover, at most two consumers can coexist with two resources at a stable point equilibrium if $g_j = z_j = 0$ (Tilman 1977, 1980, 1982, Hsu et al. 1978a, 1981).

The changes in equilibrium abundances along various gradients are also different when multiple consumers coexist with multiple resources. Again, consider the

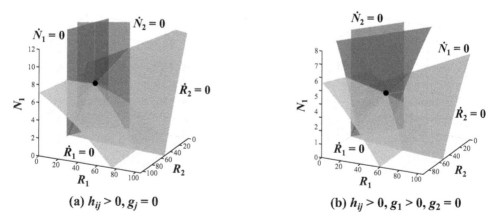

(a) $h_{ij} > 0, g_j = 0$ (b) $h_{ij} > 0, g_1 > 0, g_2 = 0$

FIGURE 3.11. Various isocline configurations for a community module in which two consumers with saturating functional responses (N_1 and N_2) feed on two resources (R_1 and R_2). All isocline and equilibrium points are as shown in figure 3.10. Parameters are also as in figure 3.10, except $h_{11} = h_{21} = 0.01$ in both (a) and (b) and $g_1 = 0.18$ in (b).

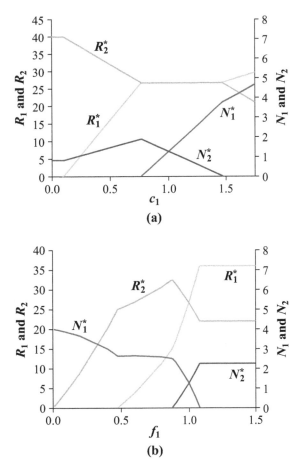

FIGURE 3.12. The equilibrium abundances of two consumers (N_1 and N_2) feeding on two resources (R_1 and R_2) at different points along an environmental productivity gradient that determines the intrinsic growth rate of R_1 (i.e., a c_1 gradient in (a)) and an environmental gradient that determines the intrinsic death rate of consumer 1 (i.e., an f_1 gradient in (b)). The parameter values in both (a) and (b) are as follows: $c_1 = c_2 = 1.0$, $d_1 = d_2 = 0.02$, $a_{11} = a_{22} = 0.5$, $a_{12} = a_{21} = 0.25$, $b_{11} = b_{21} = 0.1$, $h_{11} = h_{21} = 0.05$, $f_1 = f_2 = 1.0$, and $g_1 = g_2 = 0.0$.

communities that develop at different points along a productivity gradient encapsulated in c_1 for R_1, but all the other species have the same parameter values in all communities (fig. 3.12(a)). (In Tilman's (1980, 1982) formulation of this problem, such a gradient represents a shift in the positions of the supply point for the community without changing the consumer isoclines or the consumption vectors.) N_1 is the better resource competitor for R_1, N_2 is the better resource competitor for R_2, and the four species can coexist stably on some range of the gradient. In communities with the lowest values of c_1, only R_2 and N_2 are present: in this range, the mortality inflicted by N_2 on R_1 prevents R_1 from invading ($c_1 < 0.09$ in fig. 3.12(a)). In communities with the next higher range of c_1 values, R_1 can invade and coexist, but the combined abundances of R_1 and R_2 are not sufficient to support N_1 ($0.09 < c_1 < 0.77$). All four species coexist in communities in the next higher range of productivity, with the relative abundances of the two consumers shifting to increasing relative frequency of N_1 at higher c_1 values in this range ($0.77 < c_1 < 1.47$). However, in this range of four-species coexistence, the abundances of the two resources do not change. At the highest productivities, N_1 depresses R_2 abundance below a level at which N_2 can support a population on the combined abundances of R_1 and R_2 ($1.47 < c_1$).

In experiments in which the diatoms *Asterionella formosa* and *Synedra ulna* competed against one another under various abundance combinations of silica and phosphate, Titman (1976) demonstrated exactly this replacement along the resource gradient and showed that his studies of their abilities to grow on the two resources in single consumer experiments predicted almost exactly the range along the resource gradient in which they would coexist. Thus, the relative productivities of the two resources can result in food chains of only one resource and one consumer, apparent competition modules with one resource coexisting on both resources, or both consumers coexisting with both resources.

A similar diversity of outcomes results in communities that develop at different points along stress gradients for each consumer (e.g., the gradient of f_1 depicted in fig. 3.12(b)). If the differences in the intrinsic death rates of the two consumers make one a substantially better competitor than the other, only the better resource competitor coexists with the two resources in apparent competition modules. In the range of the gradient where all four species coexist, both the consumers and the resources shift in relative abundance with different levels of stress for N_1.

With three or more resources present in the community but no consumer self-limitation ($g_j = 0$ and $z_j = 0$), the number of coexisting consumers at a stable point equilibrium cannot exceed the number of coexisting resources on which they feed (Hsu et al. 1978a, Tilman 1980, 1982, Letten et al. 2017). However, the trade-offs in resource and apparent competitive abilities that are strictly required for coexistence with two resources present are not required for coexistence of multiple consumers feeding on three or more resources (McPeek 2019b). Three consumers can coexist if each is the best competitor for one of three available resources (fig. 3.13(a)–(b)). However, a consumer that is not the best competitor for any available resource can outcompete all others to coexist by itself on three resources (fig. 3.13(c)–(d)). Alternatively, a consumer that is not the best competitor for any of the available resources may coexist with consumers that are superior competitors for the available resources (fig. 3.13(e)–(f)).

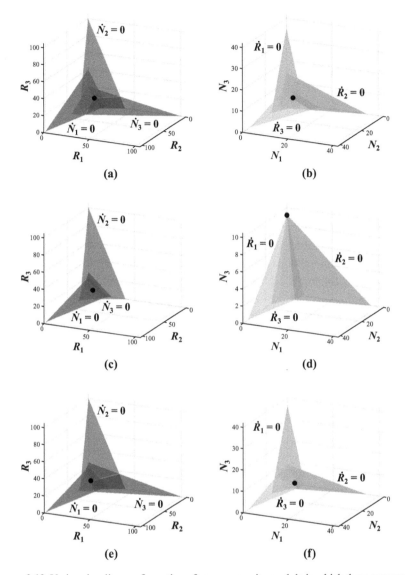

FIGURE 3.13. Various isocline configurations for a community module in which three consumers (N_1, N_2, and N_3) feed on three resources (R_1, R_2, and R_3) for three different parameter combinations. The isoclines for each parameter combination are illustrated by a pair of isocline graphs ((a) with (b), (c) with (d), (e) with (f)), where (a), (c), and (e) show the three consumer isoclines in the $R_1 - R_2 - R_3$ subspace and (b), (d), and (f) show the three resource isoclines in the $N_1 - N_2 - N_3$ subspace. In all panels, the parameters are as follows: $c_i = 4.0$, $d_i = 0.04$, $b_{ij} = 0.1$, $h_{ij} = 0.0$, $f_j = 1.0$, and $g_j = 0.0$. In (a) and (b), $a_{11} = a_{22} = a_{33} = 0.5$, $a_{31} = a_{12} = a_{23} = 0.25$, and $a_{21} = a_{32} = a_{13} = 0.1$. In (c) and (d), $a_{11} = a_{31} = a_{22} = 0.5$, $a_{12} = 0.25$, $a_{21} = a_{32} = 0.1$, and $a_{13} = a_{23} = a_{33} = 0.4$. In (e) and (f), $a_{11} = a_{31} = a_{22} = 0.5$, $a_{12} = 0.25$, $a_{21} = a_{32} = a_{13} = 0.1$, and $a_{23} = a_{33} = 0.4$.

The apparent stridency of the competitive trade-off that each must be a supe-rior competitor on one resource if two resource species are present simply results from the fact that each consumer is faced with a binary choice if only two resources are present. As the number of resource species in the community increases, the possible constellation of competitive abilities that permit consumers to coexist also increases (McPeek 2019b). Thus, when dozens of resources are available to the next higher trophic level, a huge constellation of patterns of differences among the consumers will permit them all to coexist, with each having broad, general-ized, but somewhat different diets: exactly what is seen in the diets of fish on the Mo'orea reef or the mammalian herbivores on the Laikipia savanna.

The three-consumers/three-resources simulation module can be accessed at https://mechanismsofcoexistence.com/twoLevels_4D_3R_3N.html.

Also, as when only one resource is present, consumer self-limitation ($g_j > 0$ or $z_j > 0$) also permits more consumer species to coexist than the number of available resource species at a stable point equilibrium (McPeek 2019b). Remember that consumer self-limitation makes a consumer's isocline bend away from its own axis, and so it cannot depress resource abundances as low as it could were it not self-limiting (see fig. 3.3(d) for the single-resource case and fig. 3.11(b) for the two-resource case). Consequently, any consumer whose isocline passes below the joint resource equilibrium abundances in the resource abundances face (as in fig. 3.6(b)) can invade and coexist with the currently coexisting assemblage of con-sumers and resources. Consumer self-limitation reduces the degree to which each consumer can depress resource abundances, and those unutilized resources are then available to any consumer that can support a population at those resource abundance levels.

A strong example of the importance of such consumer self-limitation is appar-ent in the diversity of forest tree assemblages (Harms et al. 2000, HilleRisLambers et al. 2002, Comita et al. 2010, LaManna et al. 2017). For example, Comita et al. (2010) found that conspecific abundance predicted seedling survival rates for 180 tree species on Barro Colorado Island, Panama, but the effect of heterospecific seed abundance was generally negligible. Thus, these 180 species, which are pre-sumably competing for a small number of resources, all show demographic pat-terns consistent with direct self-limitation. Comita and colleagues ascribed these strong patterns of intraspecific density dependence to the actions of host-specific natural enemies, such as herbivores and soil pathogens, and I explore the possible influence of pathogens in chapter 7; but a number of intraspecific mechanisms may also contribute to this apparent self-limitation.

So far in this section, the presentation has focused on having consumers invad-ing an apparent competition module. The other route to initially achieve two consumers and two resources coexisting is to have a second resource invade a resource competition module. However, remember that the only way to have multiple consumers coexist on a single resource is for the consumers to directly limit their own abundances (e.g., fig. 3.8(c)–(d)) or for the entire community to display a limit cycle or chaotic dynamics (to be discussed in chapter 8). Given the considerations in this chapter to this point, it should be obvious that the only constraint to whether a second, third, or fourth resource could invade is whether the invader has a positive population growth rate, given the abundances of the consumers already present in the community.

All the same empirical studies as described in the previous section (e.g., invasibility, consumer abundance manipulations) with only one available resource are relevant here, and so I will not repeat those study descriptions. Here, I focus on additions to these and other tests that arise because of the presence of multiple resource species.

The first issue to resolve is how many resource species are present in the community that influence the demographies of the consumers in question. This number will determine the maximum number of consumer species that can coexist, unless the consumers are also directly self-limiting. For example, with 84 algal species in Trout Lake, 84 algivorous zooplankton species could potentially coexist from only feeding on them. Likewise, if more consumer species are present than limiting resources, one immediately knows that other mechanisms in addition to resource limitation influence the consumer abundances. Therefore, knowing the number of limiting resources is crucial to understanding the mechanisms that structure the consumer trophic level. Obviously, the best information on the number of resources that influence the consumer trophic level are from diet information on the consumers: for example, observational studies of feeding, gut content analyses, and DNA barcoding of gut contents (e.g., Kartzinel et al. 2015, 2019, Casey et al. 2019, Leray et al. 2019). Manipulations of the resource abundances will then identify which are limiting to the consumers and thus which are important for further studies.

Experimental studies to estimate $f_j/(a_{1j} (b_{1j} - h_{1j}f_j))$ for each consumer on each limiting resource (described in the last section) will provide fundamental information about the strategies employed by the consumers to coexist. If only two resources are present, only two consumers should be able to coexist in the community if neither directly limits its own abundance (Hsu et al. 1978a, Tilman 1980, 1982). Moreover, with only two available resources, each consumer must be better at consuming the resource that more limits its own abundance (Hsu et al. 1978a, Tilman 1980, 1982, Letten et al. 2017). However, with more than two limiting resources, many different configurations of relative resource utilizations by the consumers will permit their coexistence (e.g., fig. 3.13) (McPeek 2019b). Given that many co-occurring consumers have broad and overlapping diets, the kinds of predictions derived from simple models of one and two species do not apply. We need to expand and broaden our conceptual mechanistic studies to many more resource species being present. Which consumers are specialists that are superior competitors for one of the resources but less limited by the others, and which consumers are generalists that require a greater diversity of resources to support their population? These studies identifying the relative resource competitive abilities of consumers also lay the foundation for functional mechanistic studies of how the traits of consumers (i.e., traits determining resource uptake, processing, assimilation, and utilization rates) make them successful and permit their segregation onto different subsets of the limiting resources.

Manipulations of the limiting resources also provide critical insights about the mechanisms of indirect effects and how these indirect effects shape the diversity of the consumer guild of species. One of the seeming conundrums of manipulating resources is that greatly increasing the abundance of one limiting resource often leads to the extirpation of some consumers. Why would greatly augmenting a limiting resource not alleviate competition and make all consumers thrive? This is the result that occurs if multiple consumers coexist on only one limiting resource (see fig. 3.9(a)–(b)). In contrast to this expectation, the typical result of supplementing a limiting resource to the bottom of a food web is that many algae and plant species go extinct (e.g., Goldberg and Miller 1990, Tilman 1993, Foster and Gross 1998, Leibold 1999, Grover and Chrzanowski 2004, Clark et al. 2007, Harpole and Tilman 2007, Clark and Tilman 2008).

For example, Goldberg and Miller (1990) manipulated the abundances of nitrogen, phosphorus, potassium, and water in an old-field assemblage of 12 summer annual plant species. Relative to the models considered here, the abiotic nutrients and water are the resources in the models and the plants are the consumers. Goldberg and Miller found a strong shift in relative biomass toward dominance by ragweed, *Ambrosia artemisiifolia*, in plots that were watered and smaller increases in biomass by *A. artemisiifolia* to the addition of nitrogen. The additions of phosphorus and potassium had no effect. This experiment accomplished two goals relative to testing the models analyzed here. First, this design determined which of these five potential resources were limiting to the plant assemblage. Second, and more interesting, it demonstrated that increasing the abundance of one resource to a consumer guild that is simultaneously limited by multiple resources will shift the abundances in favor of one of the consumers.

This shift in limiting factors was proved in another experiment performed on a diverse grassland plant assemblage (Hautier et al. 2009). A diverse grassland assemblage was grown in experimental plots for four years, and then treatments of nitrogen addition and understory light supplementation were performed in a cross-factored design and maintained for two additional years. Treatments with only nitrogen supplementation had the lowest understory light levels and lost on average two species as compared to controls. In contrast, species richness in the treatment with nitrogen and understory light supplementation was no different from the controls. The interpretation of these results was that nitrogen addition alone made light level limiting and some plant species were lost because of this shift in limiting factors for some species, and augmentation of understory light prevented these extinctions.

As most of these resource addition experiments demonstrate, increasing one limiting resource typically favors one consumer species (an alga or plant in this case), and other species decrease in relative abundance or are driven extinct outright because of newly created shortages of other limiting resources because of the increased abundance of the consumers that were most limited by the manipulated resource (see fig. 3.12(a)). These experiments manipulate the apparent competitive abilities of the resources to affect consumer abundances, and so depress the abundances of other resources to levels that may drive other consumers locally extinct. In other words, consumers are being driven extinct because of apparent competition among the resources. Thus, such manipulations of the productivity of one specific limiting resource provides fundamental information about whether only that resource limits the consumers or multiple resources are limiting simultaneously (cf. figs. 3.9(a)–(b) and 3.12(a)).

The combination of such studies will reinforce understanding of the web of interactions among consumers and resources. For example, consider the three consumers competing for three resources depicted in figure 3.13(a)–(b), where each consumer is the best competitor on a different resource. Increasing the abundance of one resource (e.g., R_1 in this example) greatly inflates the abundance of the consumer that is the best competitor for that resource (N_1 in this example) and most depresses the abundance of the consumer that is the poorest competitor for that resource (N_3) (see fig. 3.13 caption for parameter values for these species). The same is true for manipulations where specialist and generalist consumers can coexist (fig. 3.13(e)–(f)). Consumer abundances are negatively affected in reverse order to their rank competitive abilities. Thus, the results of the studies estimating $f_j/(a_{1j} (b_{1j} - h_{1j} f_j))$ for each consumer are the basis for predicting which consumers should benefit and which should be harmed by increasing a single limiting nutrient, and how consumer abundances should respond along multidimensional resource gradients (Tilman 1982).

Also, as discussed above, resource competitive ability is defined by not only the ability of the consumer to harvest and utilize resources (i.e., the denominator in $f_j/(a_{1j} (b_{1j} - h_{1j} f_j)))$ but also the per capita death rates of consumers (i.e., the numerator). This insight is critical for interpreting the segregation of resource competitors across environmental stress gradients (e.g., fig. 3.12(b)) (e.g., Jaeger 1970, 1971, Hairston 1980, 1986, Grace and Wetzel 1981, Wilson and Keddy 1986, Emery et al. 2001). If the consumer species differ in resource utilization abilities, they must also differ in their susceptibility to the stressor to segregate along the stressor gradient; otherwise, the same species would be superior along the entire gradient. Segregation requires the species that are better able to utilize the nutrients to be more susceptible to the stressor, and so these species dominate the more benign end of the gradient. The species that dominate the stressful end of the gradient must be poorer at utilizing the resources but have lower mortality in areas of high stressor levels.

For example, four perennial plants segregate into a sharp zonation pattern in the tidal salt marshes of New England, USA (Emery et al. 2001). At low sites in the tidal gradient, water logging and soil anoxia are important stressors, but high sites are only inundated with water for brief periods in the tidal cycle so that stresses are reduced at this end of the gradient (Bertness and Ellison 1987). Plant productivity, and so presumably nutrient availability, are similar all along the gradient (Bertness and Ellison 1987, Bertness 1991b). The species (*Spartina patens, Juncus gerardi*, and *Distichlis spicata*) that dominate the lowest stressor (i.e., high tidal) areas competitively exclude *Spartina alterniflora*, which dominates the high stressor end of the gradient (Bertness and Ellison 1987, Bertness 1991a, 1991b). Presumably, *Spartina alterniflora* is the poorer competitor for the resources along the gradient, but it also has the lowest detrimental consequences of soil anoxia and inundation by salt water, which is what makes it the best resource competitor overall at the high stress end of the gradient. Moreover, the boundaries between plant distributions are set by the points at which their isoclines change relative positions on the resource axes to reverse their competitive dominance along the gradient. Interestingly, fertilizing at these boundaries causes the species toward the more stressful end of the gradient to encroach up the gradient into the area dominated by the next species (Emery et al. 2001). Again, this experimental result will only occur if multiple resources limit the consumer abundances and permit them to coexist: increasing the abundance of one limiting resource will only shift the dominance among the consumers if the consumers are competing for multiple resources.

Understanding the coexistence of each species at this mechanistic level also allows for more sophisticated predictions to be made concerning broader patterns of community structure. For example, what types of species additions and deletions should occur across real (i.e., speciose) communities arrayed along some environmental gradient? Intuitively, you can imagine that as you move among communities along a gradient of some basal resource, the number of coexisting species should increase because more of the basal resource is available at higher positions on the gradient. This is typically seen at the lower end of gradients in productivity, or in some limiting nutrient like phosphorus or nitrogen compounds for plants in prairies and forests, phytoplankton in ponds and lakes, consumers that feed on those basal resource species, or higher trophic levels (e.g., Leibold 1999, Mittelbach et al. 2001, Grover and Chrzanowski 2004, Irigoien et al. 2004). However, at the higher ends of these gradients, species richness tends to decline as the basal resource abundance is increased further, just as seen in the experiments described above (Tilman 1993, Leibold 1999, Grover and Chrzanowski 2004, Harpole and Tilman 2007, Clark and Tilman 2008, Harpole et al. 2011). Thus, over the broad range of a gradient of basal resources or productivity, species richness tends to have a hump-shaped relationship (Leibold 1999, Mittelbach et al. 2001, Grover and Chrzanowski 2004, Irigoien et al. 2004).

What types of species are added and deleted at various positions along such a gradient? Although the decline in species richness at high productivity and resource availability seems counterintuitive, the critical issue to remember is that all these species depend on multiple resources below them and so each is influenced by the balance of available resources (Elser et al. 2000, 2007, North et al. 2007, Harpole et al. 2011), and that some consumers of those resources have more skewed dependencies (i.e., more specialized) and others have broader dependencies (i.e., more generalized). As an example, consider the two-trophic-level community of four consumers feeding on four resources depicted in figure 3.14. N_1 is the best resource competitor for both R_1 and R_2 but is the worst competitor for R_3 and R_4. N_2 is the best resource competitor for R_3 and N_3 is the best resource competitor for R_4. N_4 is the best resource competitor for none of the resources individually, but rather is a generalist that has intermediate resource competitive abilities on all the resources. How will these species change in distribution and abundance along various productivity gradients for the available resources?

Figure 3.14(b)–(c) shows the changes in the eight species individually and their collective abundances at each trophic level along a gradient of productivity for R_3 (i.e., increasing c_3). For communities at very low values of c_3, R_3 is absent because its productivity is not high enough to permit it to coexist given the abundances of the consumers that are present (i.e., $c_3 < 0.475$). Consequently, both N_2 (the best competitor for R_3) and N_4 (the generalist feeder) are also absent in this range. At a position on the gradient where R_3's abundance is high enough in concert with the other available resources, N_2—the specialist that feeds primarily on R_3—can invade the community and support a population (i.e., $c_3 > 1.35$). At still higher productivity values, N_4—the generalist that feeds on all the resources but is not the best on any—can invade and coexist (i.e., $c_3 > 1.95$). Thus, on this lower

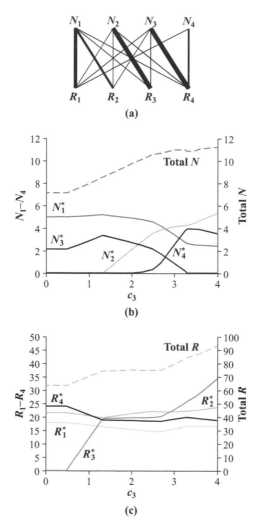

FIGURE 3.14. The equilibrium abundances of four consumers ($N_1 - N_4$) feeding on four resources ($R_1 - R_4$) at different points along an environmental productivity gradient that determines the intrinsic growth rate of R_3 (i.e., a c_3 gradient). Panel (a) shows the connections of feeding relationships among the eight species, with the width of each line connecting a consumer to a resource proportional to the magnitude of the attack coefficient between them. Panel (b) shows the equilibrial abundances of each consumer species at each point along the gradient (solid lines) and the summed total abundance of all four consumers (dashed line). Likewise, (c) shows the equilibrial abundances of each resource species (solid lines) and the summed total abundance of all four resources (dashed line). The parameter values in all panels are as follows: $c_i = 2.0$ (except c_3), $d_i = 0.04$, $b_{ij} = 0.1$, $h_{ij} = 0.05$, $f_j = 1.0$, $g_j = 0.0$, $a_{11} = 0.5$, $a_{21} = 0.4$, $a_{31} = a_{41} = 0.1$, $a_{12} = 0.12$, $a_{22} = 0.03$, $a_{32} = 0.7$, $a_{42} = 0.1$, $a_{13} = 0.03$, $a_{23} = 0.12$, $a_{33} = 0.2$, $a_{43} = 0.7$, $a_{14} = 0.2$, $a_{24} = 0.25$, $a_{34} = 0.3$, and $a_{44} = 0.3$.

segment of the gradient, consumer species richness increases with increasing productivity of R_3, with the specialists tending to invade first and then generalist consumers invading.

At still higher productivity values, consumer species start to be driven extinct because of apparent competition among the resources that drive essential components of their resource base to lower values. The first to go extinct at higher productivity values is N_3, the specialist on R_4, because of a subtle decrease in R_4's abundance (i.e., $c_3 > 3.3$). Finally, at very high values of c_3, the generalist N_4 cannot coexist because the combination of the other resources besides R_3 are decreased to levels at which N_4 cannot maintain a population (results not shown so that relationships at lower productivity values are clear). On this upper portion of the gradient, consumer species richness declines with increasing productivity of R_3. Thus, along the entire gradient, consumer species richness is hump-shaped.

This same pattern for the types of species replacements occurs along productivity gradients for the other resource species if each is manipulated individually. In general, the pattern I have found along these model gradients from examining a number of various species combinations is as follows: At the lower values of each gradient, the consumers that most depend on that resource to support their population are absent. In a species-rich community, these are more specialized species that are the better competitors on that resource but poor competitors on the other available resources, because the community would not be species-rich if one species were the dominant competitor on most available resources. Species added at higher positions on the gradient tend to be more generalized in their abilities to acquire and utilize the resources but not exceptional competitors on any single resource.

Conversely, the species that are lost at intermediate positions on the gradient are typically species that are more specialized on other resources that are depleted by the other consumers that are favored at those gradient positions. For example, in figure 3.14, N_3 is the specialist that is driven extinct first because N_2 also is an adequate feeder on R_4, and so the most important resource to maintaining N_3 is depleted by the consumer that is most favored by having high productivity of R_3. Finally, at the highest productivities, generalist consumers begin to be lost.

Thus, considering these more mechanistic models generates specific predictions for what types of species should be added and lost along single resource or productivity gradients to generate a hump-shaped relationship for species richness (e.g., Leibold 1999, Mittelbach et al. 2001, Grover and Chrzanowski 2004). On the ascending portion of the curve, more specialized consumers that are better competitors for the gradient resource should be added, while more generalized consumers should be added near the peak of the curve. The curve should take the downturn at the peak because consumers are more specialized on resources other than the gradient resource in question. Finally, the descending portion of the curve past the peak should decline primarily because of the loss of more generalized feeders.

This is one example showing that the more mechanistic approach to understanding species coexistence pays huge dividends for second-order questions of community structure. To identify species in these various categories, studies of their resource competitive relationships on the various available resources (i.e., orderings of $f_j/(a_{ij} (b_{ij} - h_{ij} f_j))$) provide essential tests of the predictions stated in the previous few paragraphs about species gains and losses along these environmental gradients. Conversely, one could also take the data for the species richness relationship along a gradient and test the competitive abilities of species that are gained and lost in various ranges along the gradient for these predicted relationships as well: for example, are the species that are gained at the low end of the gradient more specialized in their competitive ability to the gradient resource, or are the species that are lost at the very highest end of the gradient subordinate and more generalized in their competitive abilities on all resources? Moreover, this example points the way to extending these predictions to real communities containing dozens of species at each trophic level. Simply considering four to five species per trophic level can generate patterns of community structure and species replacements along environmental gradients that are seen in real and diverse communities.

SUMMARY OF MAJOR INSIGHTS

- Invasibility is proximally determined by the direct effects of species interactions on the invader. Indirect effects influence the magnitudes of these direct effects.
- A saturating functional response makes invasibility more difficult for consumer species but easier for resource species.
- The magnitude of intraspecific density dependence that a species experiences has no influence on whether it can coexist. However, greater direct self-limitation in one species does increase the scope for invasibility for other species.
- Rigid interspecific trade-offs in performance capabilities only constrain coexistence in communities with two species per trophic level.
- The number of species at the resource trophic level sets the maximum limit on the number of consumers that can coexist if the consumers are not directly self-limiting.

- Because direct self-limitation also counts as a limiting factor, the maximum number of consumer species that can coexist at a stable point equilibrium is equal to the number of resources on which they feed plus the number of consumers that are self-limiting.
- Competitive replacements of consumers along an environmental productivity gradient require that multiple resources limit the abundances of the consumers.

Adding a Third (and a Fourth and a Fifth . . .) Trophic Level

Swim out to the middle of Trout Lake in northern Wisconsin, USA, and although the water looks clear, you will be surrounded by hundreds of species. As noted in chapter 1, a total of 84 phytoplankton species alone were recorded in the water that surrounds you in monthly samples taken in 2005–2006. Hutchinson (1961) suggested that the potential coexistence of so many phytoplankton species was a paradox because they must compete for a much smaller number of limiting resources. How many of those phytoplankton species are limited primarily by the abundance of some set of nutrients and not also fed upon by one or more of the 68 herbivorous and omnivorous zooplankton species (i.e., rotifers, cladocerans, copepods) that also are swimming in those waters with you? These zooplankton also feed on the bacteria, fungi, and protist species found in those waters. This planktonic food web does not end there. These zooplankton species are themselves fed upon to varying degrees by other zooplankton, such as predatory protists, rotifers (*Ploesoma lenticulare*), copepods (*Mesocyclops edax*), and cladocerans (*Leptodora kindti, Polyphemus pediculus*), as well as phantom midges (*Chaoborus flavicans, C. punctipennis*) and 18 species of zooplanktivorous fish (Carpenter and Kitchell 1993). The planktivorous fish are fed upon by 11 species of piscivorous fish (Carpenter and Kitchell 1993), and all these fish species are fed upon by kingfishers, herons, egrets, mergansers, scaup, loons, ospreys, and bald eagles.

A similarly diverse food web can be found among the macrophytes that grow in the littoral zone of a lake such as Trout Lake, although no littoral food web has been characterized in such taxonomic detail as the pelagic. While the macrophytes provide little of the biomass that flows through the littoral food web (James et al. 2000), these plants serve as the surfaces and structure that harbor the organisms of the web. Like the pelagic food web, highly diverse and species-rich groups of bacteria, algae, and protists form the base of the littoral food web. Feeding on these are speciose groups of scrapers and filter feeders, including snails, annelids, crustaceans, many different insect orders (mayflies, caddisflies, dipterans), and anurans, with most of these having tens to hundreds of species in a single macrophyte bed.

Preying on these are diverse sets of crustaceans, bryozoans, insects (hemipterans, beetles, odonates), fish, and salamanders. Also, as in the pelagic, these are all typically fed upon by larger fish, as well as snakes, turtles, frogs, and wading birds.

Community ecology is not simply the study of consumers competing for resources. Certainly, all individuals of all species need to harvest resources from their environment to support their populations, and so resource competition clearly plays an important role in structuring communities. However, individuals of most species also face the prospect on a daily basis of being a resource for some other species, either losing tissue as when a plant's leaf is eaten by an herbivore or outright dying as when a predator kills a prey. A community is a food web, and we cannot ignore all but one major type of interaction in which species must engage to coexist.

Each of these scores of species may be a coexisting member of this food web. We know that if a major top predator is either added to or deleted from a lake food web, major changes in species richness and community structure happen at every trophic level (Elser and Carpenter 1988, Carpenter and Kitchell 1993), which implies strong direct and indirect connectivity across trophic levels that shape coexistence up and down the food web. We also know that some species occupying similar functional positions in these food webs are able to coexist with one another because they are differentially successful in the various species interactions in which they engage. Some algae are more susceptible to grazing, while other algae are better able to compete for nutrients (e.g., Tilman 1977, Leibold 1989, Grover and Chrzanowski 2004). Likewise, some zooplankton species coexist because of their differential abilities to utilize resources and avoid their own predators (e.g., Leibold 1990, 1991, Tessier and Leibold 1997, Gliwicz and Wrzosek 2008).

In this chapter, we consider what is required to add trophic levels to a food web, what is required to add multiple species to each trophic level, how the requirements are different at the different trophic levels, and how these issues propagate up and down the food web. Obviously, we will not be constructing models of food webs with hundreds of species at each trophic level, but, hopefully, this exercise will allow us to develop a sense from this chapter and the next about what processes are required for adding so many potentially coexisting species and so many trophic levels to a food web. If these models capture important features of the causal mechanisms that permit such diverse communities of species in nature, we should be able to glean their essences here.

THREE TROPHIC LEVELS

With only two trophic levels, each species is either at the bottom or at the top of the food web. Adding species that are predators of the consumers creates no new

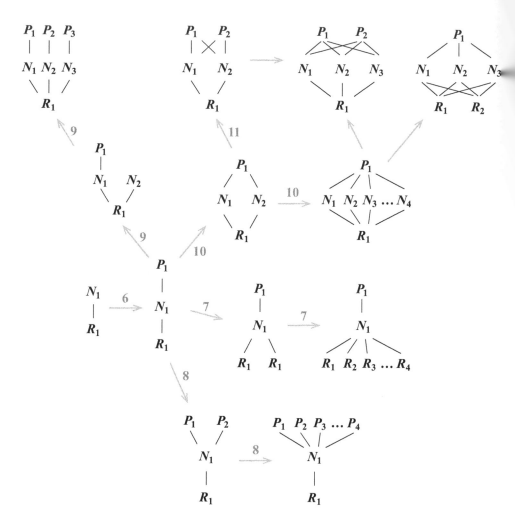

Figure 4.1. A road map of the species invasions that are considered in this chapter. For each community module, R_i represents a resource species, N_j a consumer species, and P_k a predator species. A line between two species identifies a trophic link where the species above consumes the species below. Arrows show transitions between modules that result from the invasion by different types of species. Numbers are associated with each arrow so that different types of transitions can be identified in the text.

conceptual problems for understanding species coexistence. In fact, much of the understanding that derives from considering species at two trophic levels translates directly to three or four or five. Really, the only new issue arises for the consumers in the middle, because they now face two simultaneous challenges to invasion and coexistence: obtaining adequate resources and suffering mortality from predation.

TABLE 4.1. Summary descriptions of the additional state variables and parameters used in the models presented in this chapter

State variable	Description
k	Subscript indexing predator species
s	Total number of predator species available to colonize community
P_k	Population abundance of resource i

Parameter	Description
m_{ik}	Attack coefficient for predator k feeding on consumer j
n_{jk}	Conversion efficiency for predator k feeding on consumer j
l_{jk}	Handling time for predator k feeding on consumer j
x_k	Intrinsic death rate for predator k
y_k	Strength of direct intraspecific density dependence on death rate for predator k

We begin with the community module having one consumer and one resource, and now permit a predator P_1 that feeds exclusively on the consumer to potentially invade to make a three-trophic-level food chain (invasion 6 in fig. 4.1). The dynamical equations for this situation are

$$\frac{dP_1}{P_1 dt} = \frac{n_{11}m_{11}N_1}{1+m_{11}l_{11}N_1} - x_1 - y_1 P_1$$

$$\frac{dN_1}{N_1 dt} = \frac{b_{11}a_{11}R_1}{1+a_{11}h_{11}R_1} - \frac{m_{11}P_1}{1+m_{11}l_{11}N_1} - f_1 - g_1 N_1 \quad , \tag{4.1}$$

$$\frac{dR_1}{R_1 dt} = c_1 - d_1 R_1 - \frac{a_{11}N_1}{1+a_{11}h_{11}R_1}$$

where the consumer and resource parameters are as in chapter 3. The parameters for the predator feeding on the consumer are as follows: m_{11} is the attack coefficient, n_{11} is the conversion efficiency, and l_{11} is the handling time. Like the consumer, the predator has an intrinsic death rate x_1, and the predator may limit its own abundance directly with rate y_1 (table 4.1). The isoclines for these species are

$$\frac{dP_1}{dt} = 0 : N_1 = \frac{x_1 + y_1 P_1}{m_{11}\left(n_{11}-l_{11}\left(x_1+y_1 P_1\right)\right)}$$

$$\frac{dN_1}{dt} = 0 : P_1 = \left(\frac{b_{11}a_{11}R_1}{1+a_{11}h_{11}R_1} - f_1 - g_1 N_1\right)\frac{1+m_{11}l_{11}N_1}{m_{11}}.$$

$$\frac{dR_1}{dt} = 0 : N_1 = \frac{c_1}{a_{11}} + \frac{c_1 a_{11} h_{11} - d_1}{a_{11}} R_1 - d_1 h_{11} R_1^2$$

If the predator can successfully invade, a third trophic level would be added to the food web. With linear functional responses ($h_{11} = l_{11} = 0$) and no direct self-limitation ($g_1 = y_1 = 0$) for either the consumer or predator, the P_1 isocline is a plane that is parallel to both the P_1 and R_1 axes and intersects the N_1 axis at $x_1/(m_{11}n_{11})$ (fig. 4.2(a), left). As when only the consumers are present, the R_1 isocline intersects the R_1 axis at $R_{1(0)}^* = c_1/d_1$ and the N_1 axis at c_1/a_{11}, but because the predator does not consume the resource, the R_1 isocline is parallel to the P_1 axis (cf. fig. 3.3(a)). The N_1 isocline also still intersects the R_1 axis at $f_1/a_{11}b_{11}$) and is parallel to the N_1 axis, but it tilts away from the P_1 axis (as in fig. 4.2(a), left).

For the predator to be able to invade, its isocline must pass below the $[R_{1(N_1)}^*, N_{1(R_1)}^*, 0]$ equilibrium in the $N_1 - R_1$ face (fig. 4.2(a), right). Because the P_1 isocline is parallel to the R_1 axis, this sets the invasibility criterion for P_1 as

$$\frac{x_1}{m_{11}n_{11}} < N_{1(R_1)}^*. \tag{4.2}$$

Note that this invasibility criterion is exactly the same as that faced by the consumer invading the community when only R_1 was present (i.e., inequality (3.4) and fig. 3.3(a)). This criterion is also derived in exactly the same way by asking what makes its per capita growth rate positive when $P_1 \approx 0$:

FIGURE 4.2. Various isocline configurations for a three-trophic-level food chain module in which one predator (P_1) feeds on one consumer (N_1), which in turn feeds on one resource (R_1). The isocline system is shown (a) when the consumer and predator have linear functional responses and no direct self-limitation, (b) when the consumer and predator have saturating functional responses, and (c) when the consumer and predator have saturating functional responses and the predator also has some level of direct self-limitation. The R_1 isocline is labeled $\dot{R}_1 = 0$, the N_1 isocline is labeled $\dot{N}_1 = 0$, and the P_1 isocline is labeled $\dot{P}_1 = 0$. In (a) and (b), the left graph shows the full three-dimensional representation of the isoclines, and the right graph shows the $N_1 - R_1$ face (i.e., the plane where $P_1 = 0$). (The $N_1 - R_1$ face for (c) is the same as for (b) and so is not duplicated.) In the graphs on the right in (a) and (b), the stable equilibrium point before P_1 invades is labeled $[R_{1(N_1)}^*, N_{1(R_1)}^*, 0]$. This equilibrium becomes unstable when P_1 invades if P_1's isocline passes below this point (as shown), which means that the consumer abundance is high enough at this two-species equilibrium for P_1 to have a positive population growth rate when it invades. If P_1's isocline passes above this point (not shown), P_1 cannot invade because the consumer abundance is not high enough to support a predator population. In the full three-dimensional graphs, the unstable equilibria in the $N_1 - R_1$ face are identified with open circles, and the stable three-species equilibrium is identified with a filled black circle. The parameter values in all panels are as follows: $c_1 = 4.0$, $d_1 = 0.04$, $a_{11} = 0.5$, $b_{11} = 0.1$, $f_1 = 0.5$, $g_1 = 0.0$, $m_{11} = 0.25$, $n_{11} = 0.1$, and $x_1 = 0.15$. Panel (a) has $h_{11} = l_{11} = 0.0$ and $y_1 = 0.0$, (b) has $h_{11} = l_{11} = 0.05$ and $y_1 = 0.0$, and (c) has $h_{11} = l_{11} = 0.05$ and $y_1 = 0.01$.

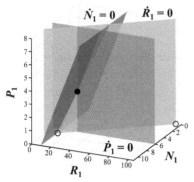
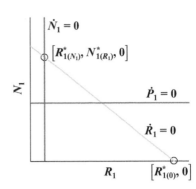

(a) $h_{11} = 0$, $l_{11} = 0$, $y_1 = 0$

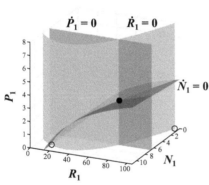
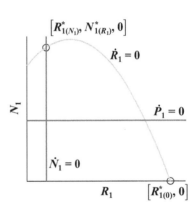

(b) h_{11} and $l_{11} > 0$, $y_1 = 0$

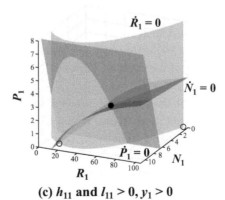

(c) h_{11} and $l_{11} > 0$, $y_1 > 0$

$$\left. \frac{dP_1}{P_1 dt} \right|_{P_1=0} = n_{11} m_{11} N^*_{1(R_1)} - x_1 > 0.$$

Biologically, this means that the consumer is at an adequate abundance for the predator population to increase when rare. If P_1 can invade, the community will come to a stable point equilibrium at $[R^*_{1(N_1,P_1)}, N^*_{1(R_1,P_1)}, P^*_{1(R_1,N_1)}]$: the equilibrium abundance of the intermediate-trophic-level consumer decreases (i.e., $N^*_{1(R_1)} > N^*_{1(R_1,P_1)}$), and the resource's abundance increases (i.e., $R^*_{1(N_1)} < R^*_{1(N_1,P_1)}$). This is the theoretical result that predicts the *trophic cascade* described by Hairston et al. (1960) and demonstrated in Wisconsin lakes (Carpenter et al. 1985, Carpenter and Kitchell 1993). This result also illustrates the basic *top-down* effect of a predator on a food chain.

Adding a saturating functional response for the predator ($l_{11} > 0$) shifts the position of the predator isocline on the N_1 axis to a higher value (i.e., $x_1/(m_{11} (n_{11} - l_{11}x_1)))$, but it remains a plane parallel to the R_1 and P_1 axes (fig. 4.2(b), left), just as previously with the consumer. Therefore, its invasibility criterion is

$$\left. \frac{dP_1}{P_1 dt} \right|_{P_1=0} = \frac{n_{11} m_{11} N^*_{1(R_1)}}{1 + m_{11} l_{11} N^*_{1(R_1)}} - x_1 > 0$$

$$\frac{x_1}{m_{11}(n_{11} - x_1 l_{11})} < N^*_{1(R_1)}$$

(4.3)

With the consumer also having a saturating functional response ($h_{11} > 0$), in addition to the convex-up shape of the R_1 isocline, the N_1 isocline is convex up along the R_1 axis as well, because of its saturating feeding on the resource. The community in this case will come to a stable point equilibrium $[R^*_{1(N_1,P_1)}, N^*_{1(R_1,P_1)}, P^*_{1(R_1,N_1)}]$ if the equilibrium is on the decreasing side of the R_1 isocline with respect to the $N_1 - R_1$ face (fig. 4.2(b), right). If the equilibrium is on the ascending side of the R_1 isocline, the system will display a stable limit cycle or chaotic dynamics around this attractor (Rosenzweig 1973, Hastings and Powell 1991, McCann and Yodzis 1994). Again, discussion of the cases of limit cycles or chaos are deferred until chapter 8.

Just as in chapter 3 with the consumer, predator self-limitation ($y_1 > 0$) causes its isocline to tilt away from its own axis, which decreases its equilibrium abundance, increases the consumer's abundance, and decreases the resource's abundance (fig. 4.2(c)). As with the consumer, predator direct self-limitation does not change its invasibility criterion. This predator self-limitation also reduces the likelihood of limit cycles and chaos, and creates a greater range of parameter space giving stable point equilibria (Wollkind 1976). Thus, as with the consumer invading a community with only one resource present, here the only requirement for the top predator to invade and coexist is that it be able to increase when rare. This is possible if the consumer is abundant enough for the invading predator to have a per capita birth rate higher than its per capita death rate and so have an overall positive per capita population growth rate (inequality (4.3)).

The food chain simulation module can be accessed at https://mechanismsof coexistence.com/threeLevels_3D_1R_1N_1P.html.

The changes in species' abundances along environmental gradients for this three-trophic-level food chain show some of the classic patterns of community structure. First, consider the differences in equilibrium abundances in communities that develop at different points along a gradient of resource productivity c_1. If neither the consumer nor the predator has any direct self-limitation (e.g., $g_1 = 0$ and $y_1 = 0$), the species at the top of the food chain increases in abundance with increasing c_1, but the abundance of the species at the trophic level below the top does not change because the P_1 isocline is parallel to the R_1 and N_1 axes (fig. 4.3(a)). Thus, on the range of the gradient where only R_1 and N_1 can support populations, N_1 increases with increasing c_1, but R_1 remains unchanged ($0.54 < c_1 < 1.59$). Likewise, where all three species can support populations, P_1 and R_1 increase with increasing c_1, but N_1 remains unchanged ($c_1 > 1.59$). Alternatively, if the consumer and predator do have some level of direct self-limitation, the abundances of the species at every trophic level increase with increasing resource productivity (fig. 4.3(b)) (Gleeson 1994).

These shifts in the equilibrium abundances in both cases are caused by how the value of c_1 determines the position of the R_1 isocline (fig. 4.3(c)–(d)). Oksanen and colleagues (1981) originally described this alternating pattern of abundance change at adjacent trophic levels for this model, and they presented data from a number of real communities that show this pattern, although their examples pool all species on a given trophic level and not a food chain of one species per trophic level.

Increasing abundances of all trophic levels with increasing productivity have also been identified in a number of other systems, but again species pooling within each trophic level dominates these data (e.g., McQueen et al. 1986, Mittelbach et al. 1988, Leibold 1989, Abrams 1993). These patterns have been called the *bottom-up* effects in food web structure.

The pattern of equilibrium abundance change is different along an environmental gradient that would cause differences in the predator's intrinsic death rate. Regardless of whether P_1 and N_1 experience any direct self-limitation, in communities where the intrinsic death rate of P_1 is higher, the equilibrium abundances of R_1 and P_1 are lower and N_1's abundance is higher. Biologically, this occurs

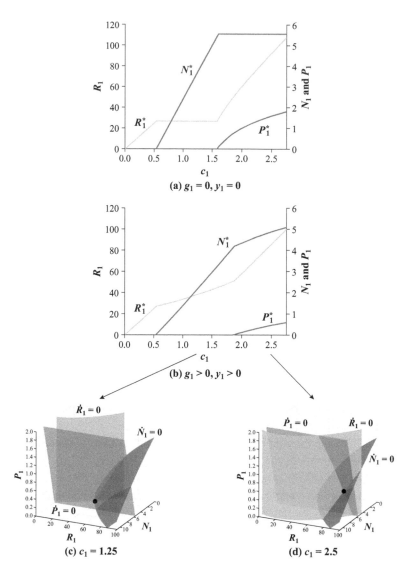

FIGURE 4.3. The equilibrium abundances and isocline configurations of one predator (P_1) feeding on one consumer (N_1), which in turn is feeding on one resource (R_1) along ((a) and (b)) an environmental productivity gradient that determines the intrinsic growth rate of R_1(i.e., a c_1 gradient) and ((e) and (f)) an environmental gradient that determines the intrinsic death rate of the predator (i.e., an x_1 gradient). Panels (a) and (e) show the species equilibrium abundances along these gradients when both consumer and predator have no self-limitation, and (b) and (f) show the species equilibrium abundances when both consumer and predator are self-limiting. Panels (c)–(d) and (g)–(h) show the isocline configurations for two points along the gradient for (b) and (f), respectively. The isocline diagrams are as described in figure 4.2. The parameter values are as follows: $c_1 = 2.75$, $d_1 = 0.02$, $a_{11} = 0.25$, $b_{11} = 0.1$, $h_{11} = 0.05$, $f_1 = 0.5$, $m_{11} = 0.4$, $n_{11} = 0.1$, $l_{11} = 0.05$, and $x_1 = 0.2$. Panels (a) and (e) also have $g_1 = y_1 = 0.0$, and (b) and (f) also have $g_1 = y_1 = 0.05$.

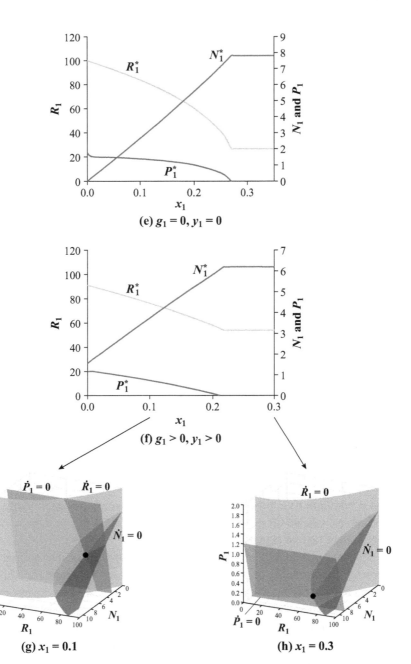

(e) $g_1 = 0, y_1 = 0$

(f) $g_1 > 0, y_1 > 0$

(g) $x_1 = 0.1$

(h) $x_1 = 0.3$

because higher mortality in P_1 causes it to equilibrate at a lower abundance, which in turn reduces mortality on N_1 to allow its abundance to be higher, which in turn increases the mortality imposed on R_1. From the geometric perspective, the equilibrium abundances change as the P_1 isocline shifts as x_1 changes (fig. 4.3(g)–(h)). Note once again that this is the mechanism that generates trophic cascades from adding or removing the top predator or top-down effects in food webs, although, again, these larger cascades pool species within a given trophic level (Hairston et al. 1960, Carpenter and Kitchell 1987, 1993, Persson 1999, Schmitz and Suttle 2001, Shurin et al. 2002).

At this point, we have a three-trophic-level food chain, in which P_1 feeds on N_1 and N_1 feeds on R_1. Remember that exactly the same issues face N_1 invading a community with only R_1 to feed upon, and P_1 invading a community with only N_1 to feed upon. By induction, it should be obvious that if we now wanted to add a fourth trophic level—a predator of P_1—that species would also face exactly the same issues. Namely, is P_1 at high enough abundance when this new top predator invades so that the top predator has adequate resources to increase in population size? Once this fourth trophic level is established, a predator of this top predator may be able to become established to create a fifth trophic level. A species being added at the top of a food chain faces the same problem, no matter how many trophic levels are below it.

Empirically testing the issues raised in this section would largely follow the types of approaches discussed in the chapter 3 section entitled "The First Consumer Invades," given that a new top predator faces the same issues. What evidence supports the fact that P_1 actually eats N_1 and that the abundance of each species has a demonstrable impact on the demographic rates of the other? What is the minimum abundance of N_1 at which P_1 can support a population, and so can invade this community? Does P_1 display any mechanism of direct self-limitation that would limit its ability to depress N_1's abundance? Is this community at equilibrium; namely, do the per capita birth and death rates of these three species balance at their normal abundances?

Trophic cascades are now also a substantial component of interpreting the systemwide consequences of depression or loss of major top predators. The experimental studies performed at the North Temperate Lakes Long-Term Ecological Research program in the 1980s and 1990s were instrumental in verifying the causal linkages of how removing or adding a single species at the top of the food web could propagate through all trophic levels (Carpenter and Kitchell 1993). These studies are now used to interpret the consequences of other conservation management strategies. For example, the reintroduction of wolves to Yellowstone National Park caused a substantial decrease in the elk population, which in turn has allowed alder and cottonwood trees to repopulate riparian areas of the park (Ripple and Beschta 2012, Ripple et al. 2015, Beschta and Ripple 2016). Similar responses were seen with the reintroduction of sea otters to Pacific coastal areas to reinvigorate kelp forests by reducing urchin densities, a major herbivore of kelp (Estes and Palmisano 1974). Opposite effects have been seen because of overfishing in coastal ecosystems as well (Estes and Palmisano 1974).

The patterns of how abundances at various trophic levels change along environmental gradients also provide critical insights into the population regulatory mechanisms operating on each species, and how those direct and indirect effects shape overall community and food web structure. Observational and experimental studies of how the abundances of species at the various trophic levels change along productivity or predator stress gradients have been critical features of studies for decades (e.g., Oksanen et al. 1981, McQueen et al. 1986, Mittelbach et al. 1988). Typically, though, these studies have pooled all species at a given trophic level and not considered individual species. Moreover, some systems show relationships that are more consistent with no direct self-limitation in the predator trophic level (e.g., fig. 4.3(a)), and others show relationships more consistent with direct self-limitation at the predator level (fig. 4.3(b)). Studies to determine whether predator self-limitation causes these differences among systems in generating these patterns would be quite informative.

ADDING SPECIES TO THE THREE TROPHIC LEVELS

Now that a three-trophic-level food chain has been established with one species at each level, we take up the problems faced by species invading at each of these trophic levels. The basic problem that each species faces is whether it has a positive per capita population growth rate when it invades—as always. However, what is required of each species to solve this problem is quite different at each trophic level.

Multiple Resources Invading a Three-Trophic-Level Food Chain

Once a three-trophic-level food chain is established, a new species may be an invading resource at the base (invasion 7 in fig. 4.1). If we permit a pool of p resources that can potentially invade with $i = 1,2,3, \ldots , p$, our model becomes

$$\frac{dP_1}{P_1 dt} = \frac{n_{11} m_{11} N_1}{1 + m_{11} l_{11} N_1} - x_1 - y_1 P_1$$

$$\frac{dN_1}{N_1 dt} = \frac{\sum_{i=1}^{p} b_{i1} a_{i1} R_i}{1 + \sum_{i=1}^{p} a_{i1} h_{i1} R_i} - \frac{m_{11} P_1}{1 + m_{11} l_{11} N_1} - f_1 - g_1 N_1. \tag{4.4}$$

$$\frac{dR_i}{R_i dt} = c_i - d_i R_i - \frac{a_{i1} N_1}{1 + \sum_{i=1}^{p} a_{i1} h_{i1} R_i}$$

As with every invasion, the success of the invader will depend on whether its per capita population growth rate is positive when it invades. In this case, that rate will be determined by its intrinsic rate of increase combined with the mortality rate imposed by the consumer N_1.

Interestingly, if the consumer and the predator both have no direct self-limitation, the addition of a second-resource species does not affect the equilibrium abundance of N_1 because its abundance is set solely by the top predator, and this is true for both linear and saturating functional responses (Rosenzweig 1971, Holt 1977, Oksanen et al. 1981). To understand this point, consider again the three-trophic-level isocline system in figure 4.2(a). Since the P_1 isocline is a plane perpendicular to the N_1 axis for both types of functional responses, the consumer abundance is fixed by the position of the predator isocline regardless of any resource abundances (i.e., the predator isocline is parallel to all resource axes because it does not directly affect any of their abundances here). Therefore, adding resource species at the bottom of the food web is transduced through the consumer to increase the predator's abundance without changing the consumer's abundance (Holt 1977). Any resource with $c_i / a_{i1} > N^*_{1(R_1, ..., P_1)}$ can invade because $N^*_{1(R_1, ..., P_1)}$ is insensitive to the identity of the resources present. This is true with both linear and saturating consumer functional responses.

The two-resource food chain simulation module can be accessed at https://mechanismsofcoexistence.com/threeLevels_3D_2R_1N_1P.html.

In contrast, if the predator does experience some form of direct self-limitation (e.g., feeding interference or costs such that $y_1 > 0$) so that its isocline now tilts away from its own axis toward the consumer axis (fig. 4.2(c)), additional resource species will cause both the consumer and the predator to increase in abundance. Consequently, the resources will in this case experience apparent competition among themselves because additional resource species will inflate the consumer's abundance, albeit more weakly as compared to when the top predator is absent. In this case, all the issues discussed in chapter 3 for apparent competition apply here as well, but the presence of the predator makes these issues less stringent.

Consequently, this module would require all the same empirical tests as outlined in chapter 3, "Apparent Competitors."

Multiple Predators Invading a Three-Trophic-Level Food Chain

New species may also invade at the top of the food chain (invasion 8 in fig. 4.1). This requires us to consider what is required for two or more predators to coexist while feeding on one consumer:

$$\frac{dP_k}{P_k dt} = \frac{n_{1k} m_{1k} N_1}{1 + m_{1k} l_{1k} N_1} - x_k - y_k P_k$$

$$\frac{dN_1}{N_1 dt} = \frac{b_{11} a_{11} R_1}{1 + a_{11} h_{11} R_1} - \sum_{k=1}^{s} \frac{m_{1k} P_k}{1 + m_{1k} l_{1k} N_1} - f_1 - g_1 N_1. \tag{4.5}$$

$$\frac{dR_1}{R_1 dt} = c_1 - d_1 R_1 - \frac{a_{11} N_1}{1 + a_{11} h_{11} R_1}$$

Assume that a pool of s predator species is available to invade with $k = 1, 2, 3, \ldots, s$.

Luckily, this problem is identical to adding multiple consumers to feed on a single resource species (i.e., invasion 4 in fig. 3.1), and so all the same conclusions apply (see chapter 3, "Resource Competitors"). Likewise, issues of whether multiple predators can coexist when competing for feeding on multiple consumers have identical consequences as when consumers compete for multiple resources (invasion 5 in fig. 3.1). And just as adding a single top predator at the top of a food chain to add a trophic level, competition among multiple predators at the top trophic level—no matter the number of trophic levels below it—occurs with the same mechanisms, constraints, and dynamics. With no direct self-limitation (all $y_k = 0$), only one predator—the one whose isocline intersects the N_1 axis at the lowest $x_1/(m_{11} n_{11})$—can coexist, but direct self-limitation in the predators can permit multiple predators to coexist while feeding only on N_1. This module would also require all the same empirical approaches and tests as outlined in chapter 3, "Resource Competitors."

The only species that can have resource competition as the sole interspecific interaction influencing their population dynamics and coexistence with other species are those at the top of a food web, and resource competition at the top of the food web is identical regardless of how many trophic levels are below them. Consequently, *this is the primary reason why focusing solely on resource competition as the structuring force for any guild of species at any level in the food web is ludicrous.*

The two-predator food chain simulation module can be accessed at https://mechanismsofcoexistence.com/threeLevels_3D_1R_1N_2P.html.

Multiple Consumers Invading a Three-Trophic-Level Food Chain

The final kinds of species that could invade a three-trophic-level food chain are more consumers. First, consider the addition of a single consumer that feeds on the basal resource but is completely invulnerable to the predator (invasion 9 in fig. 4.1). For example, Leibold (1989) used a model of this type to explore the conditions that permit two algae species that compete for a shared resource to coexist, with one also being fed upon by a zooplankton grazer and the other being invulnerable to the grazer.

Adding this second consumer that is not eaten by the predator to the food chain in equations (4.1) produces the following set of equations:

$$\frac{dP_1}{P_1 dt} = \frac{n_{11} m_{11} N_1}{1 + m_{11} l_{11} N_1} - x_1 - y_1 P_1$$

$$\frac{dN_1}{N_1 dt} = \frac{b_{11} a_{11} R_1}{1 + a_{11} h_{11} R_1} - \frac{m_{11} P_1}{1 + m_{11} l_{11} N_1} - f_1 - g_1 N_1$$

$$\frac{dN_2}{N_2 dt} = \frac{b_{12} a_{12} R_1}{1 + a_{12} h_{12} R_1} - f_2 - g_2 N_2 \tag{4.6}$$

$$\frac{dR_1}{R_1 dt} = c_1 - d_1 R_1 - \frac{a_{11} N_1}{1 + a_{11} h_{11} R_1} - \frac{a_{12} N_2}{1 + a_{12} h_{12} R_1}$$

In the scenario considered here, the second invading consumer is competing for the resource with the consumer that is already present (Inouye 1980, Leibold 1989).

The problem to be solved is where N_2's isocline can be placed relative to N_1's isocline in figure 4.2 and have all four species coexist. Because N_2 does not directly interact with N_1 or P_1, its isocline intersects the R_1 axis at $f_2/(a_{12} (b_{12} - f_2 h_{12}))$ and is parallel to both the N_1 and P_1 axes (fig. 4.4(a)). Thus, the N_2 isocline is a vertical plane in the $P_1 - N_1 - R_1$ subspace. If N_2 is a better competitor than N_1 for R_1 (i.e., $f_2/(a_{12} (b_{12} - f_2 h_{12})) < f_1/(a_{11} (b_{11} - f_1 h_{11}))$ so that the N_2 isocline intersects the R_1 axis at a lower value than the N_1 isocline), N_2 will invade and drive the resource below the level where N_1 can support a population. In this case, both N_1 and P_1 would go extinct, and the food web would collapse to a two-trophic-level food chain of N_2 feeding on R_1.

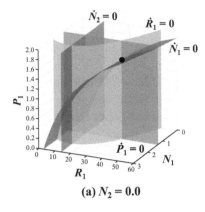

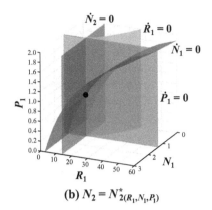

(a) $N_2 = 0.0$

(b) $N_2 = N^*_{2(R_1, N_1, P_1)}$

FIGURE 4.4. The isocline configurations for a community module containing two consumers (N_1 and N_2) feeding on one resource (R_1), and a predator (P_1) that feeds only on consumer 1 (N_1). The isocline diagrams are as described in figure 4.2. This is a three-dimensional representation in the $P_1 - N_1 - R_1$ subspace of a four-dimensional system. Because the N_2 dimension is omitted, the R_1 isocline changes position as the N_2 abundance changes. Panel (a) shows the isocline configuration when $N_2 = 0$, and (b) shows the isocline configuration when the community reaches its four-species equilibrium with $N_2 = N^*_{2(R_1, N_1, P_1)}$. The parameter values for this figure are as follows: $c_1 = 2.0$, $d_1 = 0.04$, $a_{11} = 1.0$, $b_{11} = b_{12} = 0.1$, $h_{11} = h_{12} = 0.04$, $f_1 = f_2 = 0.25$, $g_1 = g_2 = 0.0$, $m_{11} = 0.8$, $m_{21} = 0.0$, $n_{11} = 0.1$, $l_{11} = 0.04$, $x_1 = 0.1$, and $y_1 = 0.0$.

Coexistence of the four species requires that the N_2 isocline intersect the R_1 axis above the N_1 isocline but at a value lower than $R^*_{1(N_1, P_1)}$, that is,

$$\frac{f_1}{a_{11}(b_{11} - f_1 h_{11})} < \frac{f_2}{a_{12}(b_{12} - f_2 h_{12})} < R^*_{1(N_1, P_1)}, \qquad (4.7)$$

so that it has an adequate resource abundance to increase (fig. 4.4(a)). If this inequality is satisfied so that it can invade and increase, the R_1 isocline shifts position in the $P_1 - N_1 - R_1$ subspace as N_2 increases until the new four-species equilibrium is reached. (The R_1 isocline only appears to move because we are only considering a three-dimensional subspace of the full four dimensions, and the isocline's intersection points on the axes depend on the dimension not in view.)

This analysis shows that the second consumer must be a poorer resource competitor than the consumer that is also suffering mortality from the predator, but the two consumers must be similar enough in competitive ability to permit N_2 to survive on the available resource abundance. If the predator were not present and N_1 experienced no direct self-limitation ($g_1 = 0$), N_2 could not invade because N_1

would depress the resource abundance below a level at which N_2 could support a population (see chapter 3, "Resource Competitors"). If N_2's isocline intersects the R_1 axis above $R^*_{1(N_1,P_1)}$ (i.e., the right inequality in (4.7) is not true), an adequate resource abundance is not available for N_2 to invade.

Thus, a specialized predator can permit two consumers to coexist on one resource, if the predator feeds on the consumer that is the better competitor. Grover (1994) greatly extended this analysis to show that any number of consumers can potentially coexist while competing for a single resource if each consumer has its own specialized predator. Each predator depresses its consumer's abundance so that the consumer reduces its impact on the resource abundance. This in turn leaves the resource abundance at an adequate level to support other consumers.

It is very important to realize that predation does not "alleviate" resource competition among the consumers. Some consumers will be excluded if resource abundance is not sufficient to support their population. Moreover, altering the environment to increase the resource's abundance (e.g., increasing c_1) will alter the abundances of the consumers. Predation and resource competition are not alternatives for regulating the consumers: they operate simultaneously to shape species richness and structure (Menge and Sutherland 1976). Instead of damping or removing resources as limiting factors of their prey (i.e., removing R_1 as a limiting factor for N_1 and N_2 in this case), adding a predator to the community adds a limiting factor that permits additional species on the trophic level on which they feed. This is also the simplest model example showing why adding predators can enhance species richness at lower trophic levels (e.g., Paine 1966, Harper 1969, Paine 1969, 1974, Lubchenco 1978, Olff and Ritchie 1998, Hillebrand et al. 2007, Borer et al. 2014, Leibold et al. 2017).

Specialist predators thus act in analogous fashion to direct self-limitation to permit multiple consumers to coexist on a single resource (cf. chapter 3, "Resource Competitors"). This is because both processes have similar effects on the geometry of the consumer isoclines that permit the consumer isoclines to intersect. Remember that two or more consumers cannot coexist on one resource if their isoclines are parallel (e.g., see fig. 3.8(a)–(b)). Direct self-limitation in a consumer causes its isocline to bend away from its own axis (as in fig. 3.8(c)–(d)), whereas a predator feeding on a consumer causes the consumer's isocline to bend away from the predator axis (as in fig. 4.4). These effects in the better resource competitor will cause its isocline to intersect the isoclines of poorer competitors in multi-dimensional space, and so possibly permit consumers that are poorer competitors to invade and coexist. If the consumers have both specialized predators and experience direct self-limitation ($g_j > 0$, or $z_j > 0$ if the consumers have Beddington-DeAngelis functional responses), their abundances will be further limited, and so even more consumers may be permitted to coexist on a single resource.

As we will see in chapter 7, specialized pathogens can also have the same effect. For example, a number of studies of forest tree assemblages have ascribed the pattern of strong intraspecific density dependence but little heterospecific density dependence in seedling survival across dozens to hundreds of co-occurring tree species to the actions of host-specific herbivores and pathogens (e.g., Harms et al. 2000, HilleRisLambers et al. 2002, Comita et al. 2010). Although the mathematics is extremely complex for these situations, the geometry of the isoclines reveals the simple causes of coexistence are shared among these various types of species interactions.

DIAMOND/KEYSTONE PREDATION

Consumers may also invade that are fed upon by the top predator (invasion 10 in fig. 4.1). These invaders face a new challenge, because they must now satisfy invasibility and coexistence criteria that depend on the abundances of both the resource and the predator that is already present in the community, and these two abundances are linked because of the transduction of biomass through the consumer already present. This is often called a *diamond module* because of the shape of the diagram of interactions among the species (fig. 4.1), or a *keystone module* after Robert Paine's (1966) pioneering work on such a community module in intertidal food webs (Holt et al. 1994, Leibold 1996). The following equations describe such a community module with multiple intermediate-trophic-level consumers, one basal resource species, and one top predator:

$$\frac{dP_1}{P_1 dt} = \frac{\sum_{j=1}^{q} n_{j1} m_{j1} N_j}{1 + \sum_{j=1}^{q} m_{j1} l_{j1} N_j} - x_1 - y_1 P_1$$

$$\frac{dN_j}{N_j dt} = \frac{b_{1j} a_{1j} R_1}{1 + a_{1j} h_{1j} R_1} - \frac{m_{j1} P_1}{1 + \sum_{j=1}^{q} m_{j1} l_{j1} N_j} - f_j - g_j N_j. \qquad (4.8)$$

$$\frac{dR_1}{R_1 dt} = c_1 - d_1 R_1 - \frac{a_{11} N_1}{1 + a_{11} h_{11} R_1}$$

For a second consumer N_2 to invade a three-trophic-level food chain, this linkage causes an interspecific trade-off that must be satisfied in order to coexist with the consumer already present. The second consumer in the last section (i.e., equations (4.6)) faced the same trade-off, but the trade-off for the second consumer in equations (4.8) is more stark. In the last section, coexistence of all four species required that the consumer that is the better resource competitor must also be the one that has a specialist predator. Because the second consumer was invulnerable to the predator, the focus only needed to be on the consumers' relative resource competitive abilities. Here, both consumers simultaneously must compete for the resource and suffer mortality imposed by the predator.

Begin by considering the situation with all linear functional responses ($h_{1j} = l_{j1} = 0$) and no consumer or predator self-limitation ($g_j = y_1 = 0$). In this situation, all the isoclines are planes. Like a resource invading at the base of the food web, the invading consumer must have a positive population growth rate based on the birth and death rates given the resource and predator abundances at $[R^*_{1(N_1,P_1)}, N^*_{1(R_1,P_1)}, P^*_{1(R_1,N_1)}]$ (as in fig. 4.2(a)). This problem has now become four-dimensional; the most informative perspective on the isoclines is to ignore the P_1 axis and focus on the subspace defined by the consumer and resource axes (fig. 4.5(a)). The R_1 isocline intersects all three of these axes, since each of these species affects R_1's abundance. The P_1 isocline intersects each consumer axis at $x_1/m_{j1}n_{j1}$ and is parallel to the R_1 axis. The positions of neither of these isoclines depend on P_1's abundance, and the positions of these two isoclines do not change in this subspace view regardless of P_1's abundance. For the four-species equilibrium $[R^*_{1(N_1,N_2,P_1)}, N^*_{1(R_1,N_2,P_1)}, N^*_{2(R_1,N_1,P_1)}, P^*_{1(R_1,N_1,N_2)}]$ to exist, the R_1 and P_1 isoclines must intersect one another in this view, which means that the P_1 isocline must intersect one or both of the consumer axes below where the resource isocline intersects the axes.

In contrast, the positions of the consumer isoclines in this view of the $N_1 - N_2 - R_1$ subspace depend on P_1's abundance (see fig. 2 in McPeek 2014). Each consumer isocline is parallel to both consumer axes and intersects the R_1 axis at $R_1 = ((f_j/a_{1j}b_{1j})) + m_{ji}P_1$. Thus, each consumer isocline's minimum intersection point on the R_1 axis is its R-star value (see chapter 3, "Resource Competitors"), and this intersection point increases at a rate equal to P_1's attack coefficient on it with increasing P_1 abundance (McPeek 2014). For the four-species equilibrium to exist, the two consumer isoclines must be coincident on the R_1 axis at some abundance of $P_1 > 0$ that also causes them to intersect the line of intersection between the R_1 and P_1 isoclines (fig. 4.5(a)). For this to be true, P_1 must inflict greater per capita mortality (i.e., have a larger attack coefficient) on the consumer that is the better competitor for the resource (i.e., the consumer whose isocline intersects the R_1 axis at the lower value in the absence of P_1). In other words, the isocline of the better resource competitor moves faster up the R_1 axis as P_1's abundance increases. If we designate N_1 as the better resource competitor, these relationships are true if

$$\frac{m_{11}}{m_{21}} > \frac{a_{11}b_{11}}{a_{12}b_{12}} > \frac{f_1}{f_2} \tag{4.9}$$

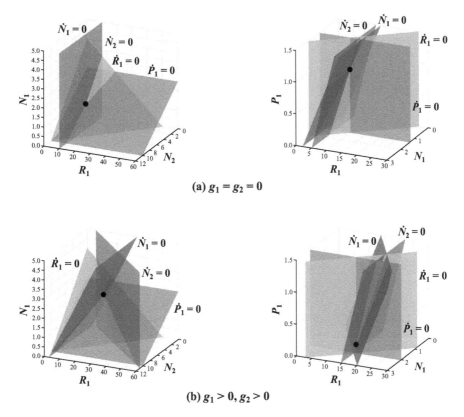

FIGURE 4.5. The isocline configurations for a diamond/keystone community module containing a predator (P_1) that feeds on two consumers (N_1 and N_2), which in turn both feed on one resource (R_1). The isocline diagrams are as described in figures 3.8 and 4.2. In (a), neither consumer is self-limiting, and in (b), both consumers are self-limiting. In each panel, the left graph shows this four-dimensional system in the $N_1 - N_2 - R_1$ subspace, and the right graph shows it in the $P_1 - N_1 - R_1$ subspace. The parameter values are as follows: $c_1 = 2.0$, $d_1 = 0.04$, $a_{11} = 0.5$, $a_{12} = 0.2$, $b_{11} = b_{12} = 0.1$, $h_{11} = h_{12} = 0.0$, $f_1 = f_2 = 0.1$, $m_{11} = 0.4$, $m_{21} = 0.1$, $n_{11} = n_{21} = 0.1$, $l_{11} = l_{21} = 0.0$, $x_1 = 0.1$, and $y_1 = 0.0$. Panel (a) has $g_1 = g_2 = 0.0$, and (b) has $g_1 = 0.3$ and $g_2 = 0.1$.

(Vance 1978, Holt et al. 1994, Leibold 1996, McPeek 1996, 2014). The point at which the two consumer isoclines are coincident on the R_1 axis thus defines the equilibrium abundances of R_1 and P_1 in the four-species equilibrium. This constraint also identifies that, at most, two consumers will be able to coexist with the predator and resource (i.e., the isoclines of all coexisting consumers must be coincident at the same point on the R_1 axis).

Inequality (4.9) states an important trade-off that the consumers must satisfy in order to coexist: the consumer that is the better resource competitor must also experience disproportionately greater mortality from the predator. Thus, the predation pressure on the better competitor is what permits the poorer competitor to coexist in the system, as Paine (1966, 1974) first articulated and showed experimentally. Note that the coexisting consumers in the last section were simply the logical extreme of this trade-off where the poorer competitor experienced no predation at all. However, the key insight here is that the predation pressures on the consumers do not alleviate resource competition between them, as many have interpreted predation to do. Rather, each consumer balances the conflicting demands of resource and apparent competition in different ways: one must be the better resource competitor, and the other is the better apparent competitor (i.e., less susceptible to the predator).

The other criterion that must be true for the four-species equilibrium to be feasible and stable is

$$\frac{m_{11}n_{11}}{a_{11}} > \frac{x_1}{c_1 - d_1 R^*_{1(N_1,N_2,P_1)}} > \frac{m_{21}n_{21}}{a_{12}}. \tag{4.10}$$

If this inequality is true and assuming the relationship defined by inequality (4.9) is also true, the consumer isoclines will be coincident at a point where they also intersect the line of intersection between the R_1 and P_1 isoclines (Holt et al. 1994, Leibold 1996, McPeek 2014). Satisfying both inequalities, therefore, means that each consumer can invade the three-species food chain containing the other consumer. This inequality identifies a second trade-off between the predator and the resource that must be satisfied to ensure that the predator and resource isoclines intersect at positive consumer abundances: the predator must be a better resource competitor for the consumer whose abundance is more inflated by the resource, which is a statement of the resource competitive abilities of the predator and the apparent competitive abilities of the resource (as defined in chapter 3), even though neither is competing with another species (Holt et al. 1994, Leibold 1996, McPeek 1996, 2014, 2017b).

If inequality (4.10) is not true because the inner ratio is larger than both outer ratios (i.e., P_1 has a large intrinsic death rate relative to the resource productivity), only the consumer that is the better competitor (N_1 in this case) will coexist with R_1 and P_1. Biologically, this means that the predation pressure on the consumer that is the better competitor is not great enough to increase R_1's abundance (i.e., $R^*_{1(N_1,P_1)}$) enough to permit the consumer that is the poorer competitor to invade and coexist. In contrast, if this inner ratio is smaller than both outer ratios, only the consumer that is the poorer resource competitor (i.e., N_2 in this case) will coexist with R_1 and P_1. Biologically, this outcome means that the predator abundance when coexisting with only the consumer that is the poorer competitor (i.e., $P^*_{1(R_1,N_2)}$) is high enough to prevent the consumer that is the better competitor from invading the three-species food chain and coexisting.

If one of these inequalities is reversed so that the consumer that is less susceptible to the predator is also the consumer for which the predator is a better resource competitor, the four-species equilibrium is unstable (Holt et al. 1994, Leibold 1996, McPeek 1996, 2014). In this case, both consumers can coexist in the three-species food chain by themselves, but the other consumer cannot invade and displace or coexist with the consumer that is already present. This is another example of ecological conditions giving a *priority effect*.

Saturating functional responses for the consumers and predator do not qualitatively change these conditions for coexistence (Grover and Holt 1998). With saturating functional responses, inequality (4.9) for the consumers becomes

$$\frac{f_1}{a_{11}(b_{11}-f_1 h_{11})} < \frac{f_2}{a_{12}(b_{12}-f_2 h_{12})} \text{ and } \frac{a_{11}b_{11}\big/\left(1+a_{11}h_{11}R^*_{1(N_1,N_2,P_1)}\right)}{a_{12}b_{12}\big/\left(1+a_{12}h_{12}R^*_{1(N_1,N_2,P_1)}\right)} < \frac{m_{11}}{m_{21}}, \quad (4.11)$$

which still states that the consumer that is the better resource competitor (left inequality) must experience disproportionate mortality from the predator relative to the benefit it derives from feeding on the resource (right inequality). In addition, assuming that inequalities (4.11) are satisfied, for the four isoclines to intersect at a single point equilibrium that is either stable or attracts a stable limit cycle, the inequality

$$\frac{m_{11}\left(n_{11}-l_{11}x_1\right)\left(1+a_{11}h_{11}R^*_{1(N_1,N_2,P_1)}\right)}{a_{11}} > \frac{x_1}{c_1-d_1 R^*_{1(N_1,N_2,P_1)}}$$

$$> \frac{m_{21}\left(n_{21}-l_{21}x_1\right)\left(1+a_{12}h_{12}R^*_{1(N_1,N_2,P_1)}\right)}{a_{12}}, \quad (4.12)$$

which is the comparable criterion to inequality (4.10) with saturating functional responses, must also be satisfied. Still, with saturating functional responses and no consumer self-limitation (i.e., $g_j = 0$), at most two consumers can coexist with a single resource and a single predator, because the only limiting factors on the consumer abundances are the resource and predator (Levin 1970). Again, more than two consumers can coexist if the equilibrium is an attractor of a stable limit cycle, which is considered in chapter 8.

The diamond/keystone predation simulation module can be accessed at https://mechanismsofcoexistence.com/threeLevels_3D_1R_2N_1P.html.

If both consumers directly limit their own abundances ($g_j > 0$), their coexistence no longer depends on satisfying the trade-off that one is better at avoiding the predator and the other is better at utilizing the resource (McPeek 2014, 2017b). Now each consumer's isocline tilts away from its own axis, and so they intersect over a broad range of resource and predator abundances instead of just a single combination of the two (fig. 4.5(b)). Moreover, direct consumer self-limitation causes the resource abundance to increase and the predator abundance to decrease, which now permits more than two consumers to invade and coexist (McPeek 2014, 2017b). Each successive invading consumer must be able to have a positive population growth rate at the resource and predator abundances when it invades, but it also faces no specific trade-off, as in the case with all $g_j = 0$. The resource abundance decreases and the predator abundance increases with each successive invading consumer, until no other potential invader can satisfy its own invasibility criteria (McPeek 2014). Here again, direct self-limitation permits more coexisting species than limiting resources and predators: any consumer that has a positive per capita population growth rate at the current resource and predator abundances can invade and coexist with the consumer species already present.

Consumers also replace one another in the food web along environmental gradients because of the shifting importance of resource availability and predator mortality (see also Leibold 1996). Consider the equilibrium abundances in communities potentially containing a resource, two consumers, and a predator along a productivity gradient of c_1 (fig. 4.6). Consumer species replacements follow the same pattern along such gradients whether the consumers and predator are directly self-limiting or not. At low productivities, the consumer that is the better resource competitor (N_1 in this case) can coexist with only the resource at very low productivity ($c_1 < 1.27$) or with the resource and predator at slightly higher productivity ($1.27 < c_1 < 1.55$) (fig. 4.6(a)–(b)). In the range of productivities where all four species can coexist ($1.55 < c_1 < 2.6$ in fig. 4.6(a) and $1.55 < c_1 < 4.5$ in fig. 4.6(b)),

---

FIGURE 4.6. The equilibrium abundances and isocline configurations of one predator (P_1) feeding on two consumers (N_1 and N_2), which in turn both feed on one resource (R_1) along an environmental productivity gradient that determines the intrinsic growth rate of R_1 (i.e., a c_1 gradient). In (a), the consumers and the predator are not self-limiting, and in (b), the consumers and the predator are self-limiting. Panels (c) and (d) show the isocline configurations for two points along the gradient for (b). The isocline diagrams are as described in figure 4.5. The parameter values are as follows: $c_1 = 2.0$, $d_1 = 0.04$, $a_{11} = 1.0$, $a_{12} = 0.2$, $b_{11} = b_{12} = 0.1$, $h_{11} = h_{12} = 0.04$, $f_1 = f_2 = 0.25$, $m_{11} = 0.8$, $m_{21} = 0.1$, $n_{11} = n_{21} = 0.1$, $l_{11} = l_{21} = 0.04$, and $x_1 = 0.1$. Panel (a) has $g_1 = g_2 = y_1 = 0.0$, and (b)–(d) have $g_1 = g_2 = y_1 = 0.05$.

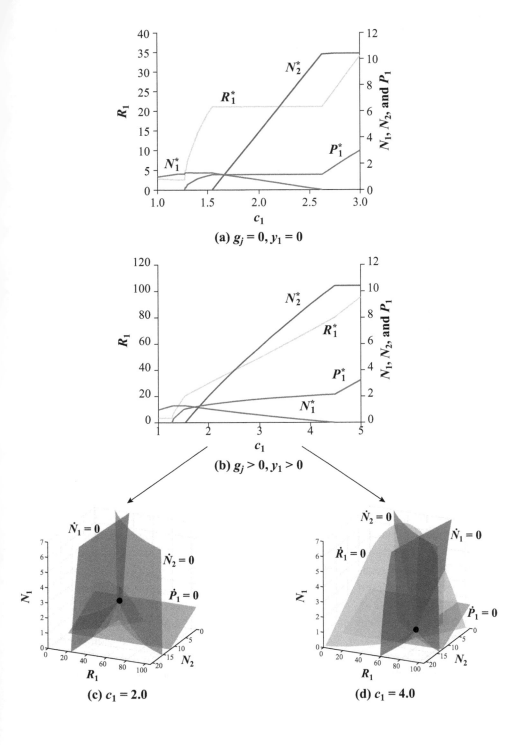

(a) $g_j = 0, y_1 = 0$

(b) $g_j > 0, y_1 > 0$

(c) $c_1 = 2.0$

(d) $c_1 = 4.0$

the consumer that is the poorer resource competitor (N_2 in this case) is at higher relative abundance in communities with higher c_1. The shift in consumer relative abundance occurs because the trade-off they face favors resource competitive ability at lower productivities but favors apparent competitive ability (i.e., predator avoidance ability) at higher productivities. In communities at the highest productivity end of the gradient, the better resource competitor cannot support a population, and only the poorer resource competitor coexists with the resource and predator (fig. 4.6(a)–(b)). If the predator is not directly self-limiting ($y_1 = 0$), the R_1 and P_1 abundances are constant across all communities where the four species coexist (fig. 4.6(a)). However, if the predator is directly self-limiting ($y_1 > 0$), R_1 and P_1 abundances increase with increasing c_1 regardless of the constellation of consumers with which they coexist (fig. 4.6(b)).

Analogous patterns of consumer invasion and replacement occur for communities arrayed along a gradient of the predator's intrinsic death rate (results not shown). If the predator's intrinsic death rate is high and so its equilibrium abundance is low, the consumer that is the better resource competitor is favored. If the predator's intrinsic death rate is low and so its abundance is higher, the consumer that is the better apparent competitor is favored.

If multiple consumers can potentially invade and the consumers are not self-limiting so that only two can coexist simultaneously, species replacements along these environmental gradients follow a sequential pattern. Because the same results obtain whether the top predator does or does not feed on the basal resource, I defer discussing this replacement until chapter 5.

All the results developed in this section for consumers at the intermediate level of a three-trophic-level food web also apply directly to intermediate trophic levels in food webs with any number of trophic levels. What is important is that species face the dual problems of finding enough food while being fed upon by species higher in the food web.

The important new features that must be explored empirically for these types of community modules are how each consumer species balances the conflicting demands of predator avoidance and resource acquisition. Obviously, I am quite partial to these types of community modules because much of my empirical research has explored how different species settle this trade-off to coexist in the freshwater habitats of North America (Kohler and McPeek 1989, McPeek 1990a, 1990b, 1998, McPeek and Peckarsky 1998, Stoks and McPeek 2003a, 2003b, 2006, Stoks et al. 2003). Studies quantifying the causes of differences in predator susceptibility and resource acquisition abilities should be coupled with resource and predator density manipulations to explore the mechanisms that permit each species to balance these conflicting demands and thus satisfy their invasibility criteria.

Another set of critical empirical tests would manipulate the presence and abundance of the predator and the abundance of the basal resource. For example, if the predator is removed from the system, do the consumers that appear to be the better resource competitors (based on the results of other studies) now have a stronger resource competitive effect on the consumers that are weaker competitors? For example, overfishing on coral reefs has resulted in shifts from a dominance of space by corals to macroalgae (Hughes 1994, McManus et al. 2000). This implies that the macroalgae were more susceptible to fish grazing than corals and superior competitors for space than corals, and so with reduced fish abundances the superior competitors then displaced the species that were less susceptible to predation. Experimental tests confirm that these resource and apparent competitive relationships among coral and macroalgae are as these models would predict: that the macroalgae suffer disproportionate mortality from grazers relative to corals, and algal grazers increase algal diversity but maintain algal total abundance that does not overtop corals (Carpenter 1986, Hay 1986, Burkepile and Hay 2010, Rasher et al. 2012, McCook et al. 2014, Adam et al. 2015). Likewise, experimentally inflating the resource abundance should increase the abundances of the predator and consumers that are weaker competitors and decrease the abundances of the consumers that are the better competitors (fig. 4.6).

One important feature that these field studies highlight is that the effects of higher trophic levels on species richness at lower trophic levels can differ. For example, in a review of a number of field experiments, Proulx and Maxumder (1998) found that grazing tended to decrease species richness when nutrient supplies to the plants that were being grazed were relatively low, but the presence of grazers tended to increase species richness when nutrient supply to plants was relatively high. As the results of the models considered here show, consumers (plants in this case) that are better resource competitors have the advantage at low positions on a nutrient supply/productivity gradient, but consumers that are better apparent competitors against their enemies have the advantage at more productive positions on such gradients (fig. 4.6). Thus, grazing may tend to exclude species at low productivity sites because those species typically balance in favor of nutrient acquisition. In contrast, species at high productivity sites are already better able to deal with their enemies, and increasing the intensity of grazing only decreases their abundances, making more nutrients available for additional species.

The role of direct self-limitation in the consumer, and the mechanism underlying it, is also important to explore empirically in this community module. One prediction that these analyses suggest is that if the consumers are only weakly self-limiting, species richness at the consumer trophic level should be low and the consumers should have ecological performances and trait combinations that strongly reflect the trade-off between resource acquisition and predator avoidance. In contrast, if the guild of consumers all have mechanisms operating that also limit their own abundances (territoriality, cannibalism, allelopathy, stress responses to conspecific densities, etc.), the species richness of coexisting guild members can be much higher and their ecological performances and trait combinations are not constrained by the trade-off. Comparative analyses of different consumer guilds could be undertaken to test these predictions.

More and More New Species

We typically use these more mechanistic models to work with only a handful of species. However, they do point the way to how we might get to the dozens to hundreds of species living together on a coral reef, in the open water or littoral vegetation of a lake, or on a savanna. This is true for both the number and types of species that can be added to a community to construct a diverse food web.

The reader by now should have noted that if consumers and predators do not contribute to directly limiting their own abundances, each trophic level has a maximum richness of coexisting species if the dynamics of the community take it to a stable point equilibrium. That limit is the number of species at other trophic levels that generate some degree of limitation on their numbers: Levin's (1970) *limiting factors*. For predators at the top trophic level, the number of limiting factors is the number of consumers on which they can feed; and for consumers at the intermediate trophic level, the number of limiting factors is the number of resources on which they feed plus the number of predators that feed on them. This is also one issue that contributes to why species richness tends to be higher at lower trophic levels. For example, with no predators above them, the maximum number of possible coexisting consumers is equal to the number of available resource species; with one resource and one predator, a maximum of two consumers can coexist at a stable equilibrium. However, with a trophic level above and below, species that directly shape the per capita growth rates from both above and below constitute the limiting factors at a given trophic level.

Obviously, this constraint on species richness does not apply to the basal trophic level in these models, because logistic growth (i.e., direct self-limitation) is assumed for them. This is a consideration that would argue for the basal level in these food web models to represent the depletable inorganic nutrients that nourish the plants, and the plants would then be the consumers of those basal nutrient resources. However, as the development here and in the coming chapters shows, this is not a problem because the same issues apply to all intermediate trophic levels in the food web. Therefore, we would simply need to shift our focus to the plant species as simultaneously experiencing resource competition for these limiting nutrients and suffering mortality and fitness decrements via herbivory as inter-mediate-trophic-level "consumers" in the interaction network of the food web.

Given these ideas, now consider the full model with multiple species at each of three trophic levels:

$$\frac{dP_k}{P_k dt} = \frac{\sum_{j=1}^{q} n_{jk} m_{jk} N_j}{1 + \sum_{j=1}^{q} m_{jk} l_{jk} N_j} - x_k - y_k P_k$$

$$\frac{dN_j}{N_j dt} = \frac{\sum_{i=1}^{p} b_{ij} a_{ij} R_i}{1 + \sum_{i=1}^{p} a_{ij} h_{ij} R_i} - \sum_{k=1}^{s} \frac{m_{jk} P_k}{1 + \sum_{j=1}^{q} m_{jk} l_{jk} N_j} - f_j - g_j N_j. \qquad (4.13)$$

$$\frac{dR_i}{R_i dt} = c_i - d_i R_i - \sum_{j=1}^{q} \frac{a_{ij} N_j}{1 + a_{ij} h_{ij} R_i}$$

Again, this set of equations may look daunting, but we already have all the tools to analyze what we need. The community depicted in figure 4.5 has one basal resource, two consumers, and one predator coexisting. For this module and with no self-limitation in the consumers or the predator ($g_j = y_1 = 0$), no more consumers could invade and coexist since that trophic level has two limiting factors (i.e., R_1 and P_1, which are the only two species' abundances that appear in the per capita growth equations of N_1 and N_2), but another predator could potentially invade and coexist since the predator trophic level has two limiting factors (i.e., N_1 and N_2) but only one species present.

In fact, invasion by a second predator P_2 (invasion 11 in fig. 4.1) faces the same set of problems as a second consumer invading an apparent competition module with two resources and one consumer already present (see chapter 3, "Multiple Consumers Eating Multiple Resources"). The problem is simply translated up a trophic level. With no predator self-limitation ($y_k = 0$), the isoclines of both predators are independent of their own axes and that of R_1, since neither consumes the resource. Therefore, for the two predators to coexist, their isoclines must intersect at positive consumer abundances in the $N_1 - N_2$ face (fig. 4.7(c)). This occurs when the predators' ecological performance capabilities satisfy

$$\frac{m_{11}(n_{11} - x_1 l_{11})}{m_{12}(n_{12} - x_2 l_{12})} > \frac{x_1}{x_2} > \frac{m_{21}(n_{21} - x_1 l_{21})}{m_{22}(n_{22} - x_2 l_{22})}, \qquad (4.14)$$

which is exactly analogous to inequality (3.20) that two consumers must satisfy to each coexist while feeding on two resources. Again, this inequality implies the

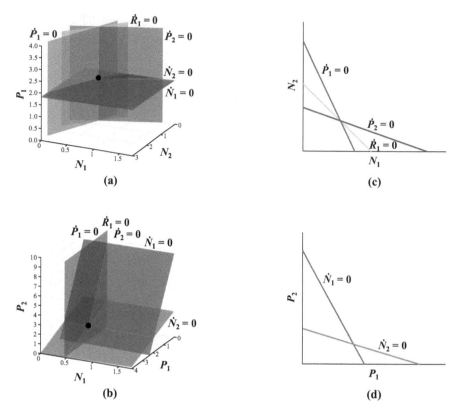

FIGURE 4.7. The isocline configurations for two predators (P_1 and P_2) that feed on two consumers (N_1 and N_2), which in turn feed on one resource (R_1). Panels (a) and (b) present two different three-dimensional perspectives on this five-dimensional isocline system when at the five-species equilibrium. In (b), the resource and both predator isoclines are completely coincident. Panel (c) presents the predator and resource isoclines in the $N_1 - N_2$ face, and (d) presents consumer isoclines in the $P_1 - P_2$ plane through the five-species equilibrium. All isoclines and equilibria are as described and identified in figures 4.4 and 4.6, with the P_2 isocline labeled as $\dot{P}_2 = 0$. The parameter values are as follows: $c_1 = 2.0$, $d_1 = 0.02$, $a_{11} = 1.0$, $a_{12} = 0.2$, $b_{1j} = n_{jk} = 0.1$, $h_{1j} = l_{jk} = 0.04$, $f_j = 0.25$, $g_j = 0.0$, $m_{11} = 0.8$, $m_{12} = m_{21} = 0.2$, $m_{22} = 0.3$, $x_1 = 0.05$, $x_2 = 0.03$, $y_k = 0.0$.

trade-off that each predator must be the better resource competitor for feeding on one of the two available consumers (fig. 4.7(a) and (c)). (Sorry, but this is where the designations of species to trophic levels finally becomes somewhat confusing.)

Likewise, stable coexistence of all five species also requires that the two consumer isoclines intersect at positive predator abundances and that each consumer is the better apparent competitor against the predator that is the better resource competitor for the other consumer. This is analogous to the result defined for resources in inequality (3.19), which shows that the issues of coexistence at a given position generalizes across trophic levels with similar species interaction structures. This means that, for the species depicted in figure 4.7, N_2 must be the better apparent competitor against P_1 and N_1 must be the better apparent competitor against P_2 (remember that the better apparent competitor is the prey that intersects its predator axis at the highest point) (fig. 4.7(c)–(d)). If the competitive relationships between the predators satisfy inequality (4.14), then the apparent competitive relationship among the consumers requires

$$\frac{m_{11}}{m_{21}} > \frac{\dfrac{b_{11}a_{11}R_1^*}{1+a_{11}h_{11}R_1^*} - f_1}{\dfrac{b_{12}a_{12}R_1^*}{1+a_{12}h_{12}R_1^*} - f_2} > \frac{m_{12}}{m_{22}}. \tag{4.15}$$

Again, this inequality is analogous to the comparable inequality (3.19) for two resources to each coexist with two consumers.

Adding species to the top trophic level of a food web is the same regardless of how many trophic levels are below. The only issues for species at the top of the food web are the diversity and types of species that form the resources/prey base supporting the top trophic level. And only species at the top have resource competition as the primary indirect interaction influencing their coexistence.

From these results, the reader should see a general principle emerging; namely, that additional species can invade and coexist at a given trophic level once new species invade and coexist at adjacent trophic levels both below and above. *Diversity begets diversity—the remarkable diversity we see in real biological communities is a consequence of itself in many ways.* Thus, once the community module of two predators, two consumers, and one resource is established, the opportunity exists for a third consumer with the appropriate set of ecological skills that permit it to have a positive per capita growth rate when it invades and to therefore coexist (results not shown). Once the third consumer is present in the community, a third predator with the right ecological properties can invade and coexist with the three consumers. In fact, all the various constellations of ecological performance capabilities illustrated in figure 3.13 can be constituted among three predators feeding on three consumers in this model (results not shown). And once a third predator is established, a fourth consumer can potentially invade and coexist.

Likewise, a third consumer can invade and coexist, after a second resource species invades a diamond/keystone module (i.e., one resource, two consumers, one predator) and becomes established (fig. 4.1). In this case, three consumers can coexist because three limiting factors—two resources and one predator—limit their abundances (fig. 4.8). However, those three consumers can have various ecological abilities to utilize the resources and to evade the predator that permit

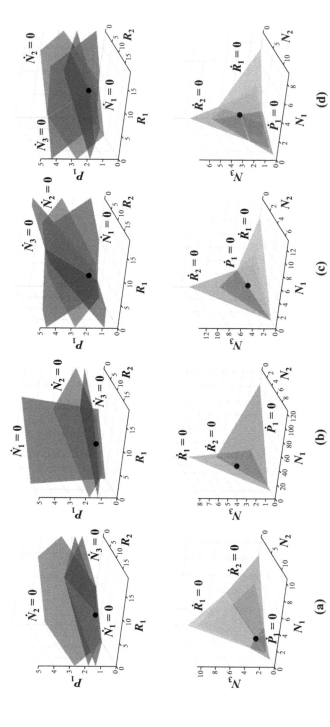

FIGURE 4.8. The isocline configurations for a predator (P_1) that feeds on three consumers (N_1, N_2, N_3), which in turn feed on two resources (R_1 and R_2). The top graph in each panel shows the N_1, N_2, and N_3 isoclines in the $P_1 - R_1 - R_2$ subspace, and the bottom graph shows the P_1, R_1, and R_2 isoclines in the $N_1 - N_2 - N_3$ subspace. The isocline diagrams are as described in figure 4.6. The parameter values are as follows: $c_i = 3.0$, $d_i = 0.2$, $b_{ij} = 0.1$, $h_{ij} = 0.0$, $f_j = 0.2$, $g_j = 0.0$, $x_1 = 0.1$, $y_1 = 0.01$, $n_{j1} = 0.1$, and $l_{j1} = 0.00$. Panel (a) has $a_{11} = a_{23} = 0.3$, $a_{12} = a_{13} = a_{21} = a_{22} = 0.25$, $m_{11} = m_{31} = 0.2$, and $m_{21} = 0.1$. Panel (b) has $a_{11} = a_{21} = 0.2$, $a_{12} = a_{22} = 0.25$, $a_{13} = a_{23} = 0.3$, $m_{11} = 0.01$, $m_{21} = 0.15$, and $m_{31} = 0.29$. Panel (c) has $a_{11} = a_{23} = 0.2$, $a_{12} = a_{22} = 0.4$, $a_{13} = a_{21} = 0.6$, $m_{j1} = 0.2$, $m_{21} = 0.15$, and $m_{31} = 0.29$. Panel (d) has $a_{11} = 0.2$, $a_{21} = 0.8$, $a_{12} = a_{22} = 0.25$, $a_{13} = a_{23} = 0.4$, $m_{11} = 0.3$, $m_{21} = 0.1$, and $m_{31} = 0.3$.

coexistence. The only real constraint to coexistence is that no single consumer can be the best competitor on both resources and the best apparent competitor against the predator. Thus, again we see that the operation of a stringent trade-off between limiting factors—either between competitive ability for two resources or between resource acquisition and predator avoidance abilities—is relaxed when more than two limiting factors constrain coexistence at a given trophic level. Moreover, the diversity of ecological performance capabilities that will permit coexistence also increases with the number of limiting factors at each trophic level. Some of these involve trade-offs and some do not.

However, as was true for simpler community modules, this general rule of the maximum number of coexisting species at one trophic level being equal to the number of species at adjacent trophic levels that limit their abundances in a density-dependent fashion only applies when consumers and predators do not directly limit themselves as well (Levin 1970). Again, direct self-limitation also counts as a limiting factor in the community; so if these species also generate some form of direct self-limitation, nothing about community dynamics limits species richness, since adding a new species to a trophic level adds another limiting factor to that trophic level (Levin 1970). Also, as we will see in chapter 8, certain forms of temporal variability in population regulation (e.g., endogenously generated or exogenously forced cycling) also can permit more species than limiting factors, if those species also have specific performance capabilities.

In general, species invading at lower trophic levels face the added problem of mortality from above, and so they must garner enough of their own resources while mitigating mortality from their predators to sustain a population. In other words, they must balance the conflicting demands of obtaining enough of the resources on which they feed while reducing the mortality they experience from their enemies, and many different constellations of ecological performance characteristics may permit balances that foster coexistence. This is true for plants and algae that compete for limiting nutrients and are fed upon by herbivores and for intermediate-trophic-level consumers that feed on species below them and are fed upon by species above them. Only the species at the top of the food web that have no enemies of their own (and this means no pathogens as well, as we will see in chapter 7) are limited exclusively by resources from below. Again, the number of species that can stably coexist at a point equilibrium is set by the number of species that directly limit their abundances at adjacent trophic levels, both above and below, and the diversity of ecological performance characteristics that will permit coexistence greatly increases as the number of those limiting factors increases above two. Thus, the number and diversity of limiting factors that impact coexistence at a given trophic level is a critical, defining feature that is needed to understand community structure. However, even when temporal variability permits additional species (as we will see in chapter 8), the number and diversity of limiting factors still influences both the number and properties of species that can coexist. Identifying the full set of limiting factors, both above and below, for the set of species of interest is a fundamental piece of information for understanding why they can coexist.

One important pattern identified in large-scale studies of species diversity that is consistent with this view is that local species richness tends to increase with regional species richness. The species in local communities can be thought of as sampled from a regional species pool. Cornell and Lawton (1992) hypothesized two possible relationships between local and regional species richness, based on the importance of local species interactions limiting community membership. They reasoned that if local species interactions are important in determining community membership, a plot of local species richness against regional species richness should show an asymptote at higher regional richness—a pattern they identified as *saturated*. The other pattern, in which local species richness increases as a direct proportion to the size of the regional species pool, they identified as *proportionate sampling*. They assumed that niche space would be limiting to the number of species that could coexist locally, but they identified a number of mechanisms (e.g., neutral species, habitat fragmentation and disturbance, spatiotemporal heterogeneity, fugitive/tramp species) that might make local richness increase directly with regional richness instead of displaying an asymptote.

In fact, most studies have found that communities are not saturated with species (e.g., as in Terborgh and Faaborg 1980), but rather local species richness increases linearly with increasing regional species pool size (e.g., Cornell 1985, Cornell and Lawton 1992, Hawkins and Compton 1992, Cornell and Karlson 1996, Caley and Schluter 1997, Karlson and Cornell 1998, 2002, Srivastava 1999, Shurin et al. 2000, Shurin and Allen 2001, Karlson et al. 2004, Witman et al. 2004, Cornell et al. 2007, Cornell and Harrison 2013). Although the factors that Cornell and Lawton (1992) identify surely do contribute to these proportional sampling relationships, I would suggest that the fact that diversity at one trophic level fosters diversity at adjacent trophic levels may also play a major role in creating this pattern. Most of these studies included heterogeneous assemblages of species at multiple trophic levels, and so the models considered here can account for increases in diversity across trophic levels. Adding a species at one trophic level increases "niche space" available for species at adjacent trophic levels. "Niche space" is not some finite quantity that is defined outside the community, but rather is a direct function of the species that are present. However, no study to my knowledge has stratified their sampling to allow one to evaluate whether these patterns hold across multiple trophic levels.

Various patterns in the fossil record also support this interpretation of the accumulation of species diversity at one trophic level fostering diversity at adjacent trophic levels. Across the Phanerozoic, genus-level diversity, which is assumed to be correlated with species-level diversity, in the marine invertebrate fossil record is strongly positively correlated with proxies for the intensity of predation (e.g., trace fossil marks, such as drill holes left on prey skeletons) (Huntley and Kowalewski 2007). More directly, North American assemblages of large carnivores and herbivores over the past 25 million years show the same pattern of increasing local species richness with increasing regional richness, and continent-wide carnivore and herbivore species richnesses are positively correlated across this time span (Van Valkenburgh and Janis 1993).

The development of the models in this chapter ultimately illustrate why Hutchinson's (1961) paradox of the plankton is no paradox at all, once one realizes that phytoplankton species have limiting factors both above and below them. In fact, only after enumerating and mechanistically studying the diversity of limiting factors can one assess how many species can potentially coexist and what properties those species must have to be successful and thrive in the community.

THE IMPORTANCE OF ENUMERATING LIMITING FACTORS

The fundamental property of each coexisting species is that it satisfies its invasibility criterion; namely, were the species to be extirpated from the community, it could reinvade because its per capita birth rate would be greater than its per capita death rate as it reinvades the community in which all the other species are at their demographic steady state. In other words, the density-dependent demographic forces acting on each coexisting species would permit it to increase in this community were it to be perturbed to low abundance.

To really understand why each coexisting species can increase were it made rare requires knowledge about the abundances of the limiting factors that directly determine its per capita birth and death rates, and the indirect interactions with other species that are mediated through those limiting factors. The limiting factors are those abiotic resources and species that directly determine the per capita birth and death rates of a species (Levin 1970). Abiotic limiting factors are things such as light, phosphate, nitrate, and silica oxide, which have abundances and are depleted by the actions of individuals from the local environment. In a fully mechanistic model, the dynamics of the abundances for these abiotic limiting factors would also be modeled just as species are modeled (DeAngelis 1992, Schmitz 2010, Hunter 2016). (This is why I have typically characterized the models here as "more mechanistic" because they capture the same features; but the dynamics of abiotic limiting factors are assumed to be subsumed into the logistic terms of basal resource species, and the interactions among resource species are ignored for simplicity. See chapter 3 for a fuller discussion of this issue.) Likewise, species whose abundances have direct impacts (i.e., directly appear in the equations) on the per capita birth and death rates of a particular coexisting species are also limiting factors for that species.

It is the combined effect of these direct interactions with its limiting factors that determine whether the species can satisfy its invasibility criterion.

A species will coexist in a community when its abiotic and biotic resource abundances are sufficiently high and its enemy abundances (i.e., herbivores, consumers, predators, pathogens) are sufficiently low when it is rare so that its per capita birth rate exceeds its per capita death rate.

Even if all species in a given functional guild or trophic level have identical sets of limiting factors, each coexisting species will experience different demographic impacts from those limiting factors. Thus, whether a particular resource is sufficiently high enough or an enemy is sufficiently low enough to permit its coexistence will differ for each species. This is why I argue that coexistence is best understood as a species-specific property for each member of a community, and is not a pairwise analytic reality that converts indirect interactions with other guild or trophic-level members into direct interactions.

The idea of coexistence as a pairwise comparison between indirectly interacting species derives from wishing to ignore these limiting factors and so to focus only on the indirect interactions among guild members. Surely, understanding these indirect interactions is a critical feature of the full story for why a species can coexist, but not to the exclusion of understanding the direct impacts on each species that they mediate. These indirect effects are important because of the direct impacts these other species have on each of the shared limiting factors for the species in question. Two consumers will both be able to coexist on a single limiting resource if neither can depress the resource abundance below a level at which the other will

---

FIGURE 4.9. The equilibrium abundances of four predators ($P_1 - P_4$) feeding on four consumers ($N_1 - N_4$), which in turn feed on four resources ($R_1 - R_4$) at different points along an environmental productivity gradient that determines the intrinsic growth rate of R_3 (i.e., a c_3 gradient). (a) The connections of feeding relationships among the 12 species, with the width of each line connecting one species to another being proportional to the magnitude of the attack coefficient between them. (b)–(d) The equilibrium abundances of each of the 12 species (solid lines) in communities along the gradient and the total abundances at each trophic level (dashed lines). All parameters for the resources and consumers are as in figure 3.14. The parameters for the predators are as follows: $n_{jk} = 0.1$, $l_{jk} = 0.02$, $x_k = 0.1$, $y_k = 0.0$, $m_{11} = 0.5$, $m_{21} = 0.09$, $m_{31} = 0.09$, $m_{41} = 0.1$, $m_{12} = 0.11$, $m_{22} = 0.5$, $m_{32} = 0.1$, $m_{42} = 0.11$, $m_{13} = 0.08$, $m_{23} = 0.12$, $m_{33} = 0.28$, $m_{43} = 0.31$, $m_{14} = 0.27$, $m_{24} = 0.25$, $m_{34} = 0.17$, and $m_{44} = 0.12$.

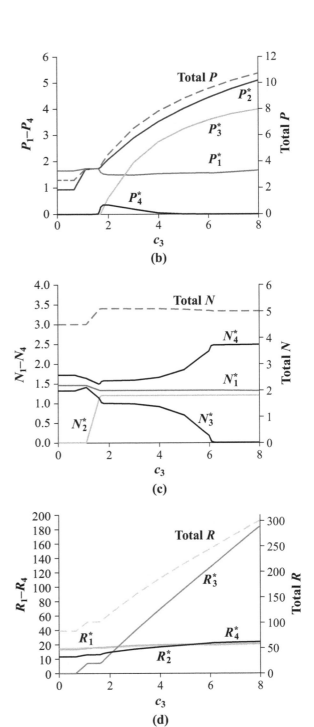

have a positive per capita growth rate when rare; or two resource species can coexist while being fed upon by a shared consumer if neither resource inflates the consumer abundance above which the other would have a positive per capita growth
rate when rare. The importance of studying mechanism is understanding which
combination of limiting factors is sufficient to permit a particular species to coexist. Whether two guild members may or may not coexist together in a community
depends on all the other species in the community that also influence the levels of
those limiting factors; so simply focusing on the guild members to the exclusion of
the limiting factors provides no causal understanding of the community.

As an example for a more complex food web, add a third trophic level of four
predators ($P_1 - P_4$) to the two-trophic-level food web depicted in figure 3.14. The
feeding relationships among the consumers and resources are the same as in figure
3.14. At the added trophic level, P_1 is the best resource competitor for N_1 and P_2 is
the best resource competitor for N_2. P_3 is the best resource competitor for both N_3
and N_4 and, finally, P_4 is a generalist that is not the best resource competitor for any
of the consumers (fig. 4.9). All predators are different only in their attack coefficients on the various consumers. And as in figure 3.14, the equilibrium abundances
of the 12 species along a gradient of productivity for only R_3 (i.e., a gradient of c_3)
is shown. Changes in species composition and abundances at the lower trophic
levels in the three-trophic-level community are similar to those in the two-trophic-
level system, with a few notable and instructive exceptions. The most notable
exception is that N_4, the generalist and subordinate competitor at that trophic level,
is present all along the gradient. This is because the equilibrium resource abundances are higher in the three-trophic-level community (cf. resource abundances in
figs. 3.14 and 4.9), and so N_4 can meet its invasion criteria in communities all along
the gradient. These higher resource abundances all along the gradient also allow N_2
to invade at a lower c_3 value, and N_3 can exist in communities with a higher c_3 value.

The effects on the predator trophic level are somewhat different in this example. This is because the generalist is in most conflict for prey with the predator that
is most affected by the effects of c_3. Thus, P_4, the generalist and subordinate
resource competitor for feeding on all the consumers, exists in communities on a
narrow range of the c_3 gradient (fig. 4.9). Because the predators' diets are largely
segregated onto different consumers, their abundances change along the gradient,
but none of them go extinct.

This community is also instructive for the possible results generated in another
type of common experimental manipulation—species deletion experiments (e.g.,
Díaz et al. 2003). Because they are largely dependent on specific consumers, P_1 or
P_2 go extinct if N_1 or N_2, respectively, are deleted. In contrast, deleting N_3 causes
the generalist P_4 to go extinct because P_3 has an alternative prey. However, deleting the generalist N_4 at the consumer level causes both P_2 and P_3 to go extinct.

Similar consequences occur for deleting one of the predator species. In this community, all 12 species coexist if all the basal resources have similar intrinsic birth rates (i.e., all c_i equal). Some deletions have only quantitative effects: deleting P_2 or P_4 causes changes in abundances of the other species, but no species go extinct. However, deleting other top predators also has cascading effects at that trophic level or lower, what Pimm (1982) called *species deletion stability or instability*. For example, deleting P_1 causes P_2 to go extinct. This secondary extinction occurs because of the propagation of the complex web of indirect effects that propagate through the food web: removing P_1 permits N_1's abundance to increase, which in turn causes P_4's abundance to increase, and it is resource competition with the generalist P_4 that drives P_2 extinct when P_1 is deleted. Deleting P_3 from this community causes the greatest change, with four species, N_2, N_3, P_2, and P_4, all going extinct as a consequence. Understanding the web of species interactions is the only way to understand and potentially predict the cascading consequences of species loss from communities, and why consumers and predators increase species richness at lower trophic levels (Hay 1986, Collins et al. 1998, Proulx and Maxumder 1998, Bakker et al. 2006, Hillebrand et al. 2007, Borer et al. 2014, Leibold et al. 2017).

The diversity of limiting factors is also important in setting constraints on the abilities of species in a guild to coexist. For example, the axiom derived from Tilman's (1980, 1982) analysis of two consumers competing for two resources—that each consumer must be more limited by a different resource—only applies when two resources limit the consumer trophic level (McPeek 2019b). If three resources are available, three coexisting resources may each be the best competitor for a different resource (see fig 3.13(a)–(b)), or one of the consumers may be a subordinate competitor for all the resources (see fig. 3.13(e)–(f)). For the three consumers to coexist, the constraint must hold that no single consumer can be the best competitor for all three resources (i.e., it is then impossible to have the three consumer isoclines intersect), but a number of possible configurations of their abilities may permit them to coexist. Analogous considerations govern multiple intermediate-trophic-level consumers coexisting (see figs. 4.4–4.8). As the number of limiting factors increases, the constellation of possible consumer abilities that permit coexistence also increases. Thus, with dozens of predators feeding on multiple consumers that in turn feed on dozens of resources (as on the Mo'orea coral reef or the Laikipia savanna) species are expected to have broad and varied diets that are different from one another, if they are to coexist, but no necessary pattern for how they should be different is required.

However, parsing guilds of species into collections that are more or less limited by various subsets of their limiting factors has great conceptual and practical implications. For example, first consider the simplest case of two consumers coexisting with two resources. If one of the resources is supplemented in some way (e.g., increasing the productivity of that resource), the consumer that is the better competitor for that resource will benefit most from this supplement. As a result, the increase in this consumer will then depress the abundance of the other resource through its greater feeding pressure and so may depress the other resource to a level at which the other consumer cannot coexist. Extending this to a much more diverse community, augmenting one resource may indirectly depress the abundances of consumers that are not most limited by that resource—even to the point of extinction (see figs. 3.14 and 4.9). Exactly the same can be said for augmenting the abundance of a consumer or predator, perhaps through immigration from some other communities or in more benign environments where their intrinsic death rates are lower; the resources that are the best apparent competitors against that consumer will be less affected by consumer augmentation, and so they will be less susceptible to extinction. Thus, understanding the general patterns of limitation among the set of limiting factors allows for general predictions about responses to changes in the levels of limiting factors.

This has important practical implications as well. One of the serious human environmental impacts for the past century has been the input of limiting phosphorus and nitrogen-containing compounds into freshwater and estuarine systems (e.g., Scavia and Chapra 1977, Lowe and Keenan 1997, Jeppesen et al. 2000, Brett and Benjamin 2007, Conley et al. 2009). One of the major consequences of the influx of excessive nutrients in runoff is the loss of species diversity (e.g., Ludsin et al. 2001, Worm et al. 2002, Silliman and Bertness 2003, Maskell et al. 2010). The causes of this biodiversity loss are predicted to be exactly the mechanisms highlighted here (e.g., figs. 3.14 and 4.9), namely, shifts in the abundances of limiting factors that are caused by propagation of indirect effects through the food web. Consumers that are most limited by the supplemented nutrients will increase most strongly to the added influx and, as a consequence, drive down other limiting nutrients on which they previously had less impact. This in turn will cause other consumers that are more dependent on these alternative nutrients to be depressed, some to the point of extinction. Thus, understanding broad patterns of resource limitation among a diverse array of consumers will help managers understand and potentially mitigate the consequences of shifts in abundances among their limiting factors. Overexploitation of species higher in the food web (e.g., overfishing) can similarly result in extinctions at other food web positions (e.g., Estes et al. 1989, 1998, Jackson et al. 2001). For example, the loss of biodiversity in kelp forests due to the extirpation of sea otters results from the removal of a strong limiting factor—namely, sea otter predation—on urchins (Estes and Palmisano 1974).

One final insight derived from these considerations that has important practical implications is the fact that a single predator cannot drive a single prey species extinct in isolation. In all the models of consumer-resource interactions, if only one resource species is present, the predator cannot drive the prey to extinction. The interaction may become unstable in the sense of entering a limit cycle, which may increase the probability of the resource going extinct probabilistically when at low abundance in the cycle, but equilibria with the resource at zero abundance are not stable. The applied implication of this result is that alternative prey must be present for the consumer to drive a resource species extinct. The alternative resources inflate and support the consumer population at a level where it may drive some of the resource species extinct via apparent competition. Moreover, the collateral effects on other resource species are necessary. This insight should guide conservation efforts that are trying to mitigate the consequences of introduced predators (e.g., brown tree snakes in Guam, pythons in Florida) or the biocontrol efforts where an enemy is introduced to try to control some pest species (DeBach 1964, Mills and Getz 1996, Klapwijk et al. 2016). The apparent competitive interactions among species mediated through the predator cannot be eliminated but may be mitigated in beneficial directions with forethought.

SUMMARY OF MAJOR INSIGHTS

- Invasibility for species at the top trophic level, regardless of how many trophic levels are below, is identical to that discussed for consumers in chapter 3.
- Invasibility at intermediate trophic levels requires that a species simultaneously have adequate resources and not suffer undue mortality from enemies (i.e., herbivores, predators, parasitoids).
- Rigid interspecific trade-offs in performance capabilities only constrain which intermediate-level consumers can coexist when only one predator and one resource are present.
- Specialist predators have the same effect as direct self-limitation on community structure.
- Diversity begets diversity. Adding a species at one trophic level creates a new opportunity for a species to be added at an adjacent trophic level.
- The maximum number of species that can coexist at a given trophic level is determined by the number of species (or abiotic nutrients) on which they feed at the trophic level below and the number of predators that feed on them at the trophic level above, if no species on that trophic level is also self-limiting.

Omnivory in a Food Web

The diet diversity of many species requires them to be able to process and digest quite different types of biological materials, namely, plant and animal tissues. The colloquial term for this is *omnivory*: an omnivore consumes both plants and animals as part of its diet. Many species in all ecosystems are omnivores. Many of the rotifers and copepods in the Trout Lake plankton are omnivores. Many of the fishes on the Mo'orea reef, such as triggerfish and butterfly fish, feed on both algae and the diverse animal life available to them (Casey et al. 2019). Even for hummingbirds, which are stereotypically thought of as exclusively nectar feeders, the majority of their diet is actually insects (Lucas 1893, Young 1971). And when you sit down to eat a dinner of steak, potatoes, and vegetables, you are such an omnivore.

Community ecologists, however, have a broader definition of omnivory: feeding on more than one trophic level without regard to the taxonomic composition of the diet (e.g., Pimm and Lawton 1978, Diehl and Feißel 2000, Rudolf 2007, Gellner and McCann 2012). Clearly, simultaneously feeding on heterotrophic animals and autotrophic plants is by definition feeding on more than one trophic level. However, animals higher in the food web also feed across trophic levels. Spiders on the Konza Prairie ensnare both herbivorous and predatory insects in their webs. The planktivorous fish in Trout Lake feed on both *Chaoborus* midges and the cladocerans, copepods, and rotifers that are the prey of those midges (Carpenter and Kitchell 1993). Even in small ponds, large dragonflies like *Tramea lacerate* feed on smaller dragonflies like *Erythemis simplicicollis*, and all these dragonflies feed on smaller damselflies like *Enallagma aspersum* and *Ischnura verticalis* (Wissinger and McGrady 1993). Such omnivory has also been termed *intraguild predation*, because the predator and some of its prey are also competing for shared resources (Polis et al. 1989, Polis and Holt 1992, Holt and Polis 1997).

Even some organisms that are photosynthetic can be consumers, which is termed *mixotrophy*. In the plankton of lakes and oceans, dinoflagellates both photosynthesize and engulf smaller organisms (Stoecker et al. 2017, Edwards 2019). In terrestrial ecosystems, carnivorous plants, such as pitchers, flytraps, and sundews, are obvious examples.

Omnivory is a common feature of food webs in all ecosystems (Polis and Strong 1996). As the concept of intraguild predation highlights, omnivory puts additional constraints on the coexistence of predator and prey at adjacent trophic levels, particularly for the prey. The prey must be able to obtain enough resources while simultaneously suffering mortality inflicted by its resource competitor (Holt and Polis 1997). However, as we will see, omnivory also creates the opportunity for more species to coexist at higher trophic levels, which to my knowledge has not been a recognized feature of omnivory to now.

ADDING AN OMNIVORE TO A FOOD CHAIN

In its most basic form, omnivory or intraguild predation is an interaction network involving three species—a resource, a consumer, and a predator. The consumer feeds on the resource, and the predator feeds on the consumer, just as in a three-trophic-level food chain. However, the predator also feeds on the resource (Polis et al. 1989). This network of interactions makes an intraguild predation module, which is a bundle of direct and indirect effects among these three species: the predator is both a predator and resource competitor of the consumer; the consumer is both prey and a resource competitor of the predator; and the resource is both prey and an apparent competitor of the consumer. All of these direct and indirect effects among the species play a role in determining whether each can coexist with the other two.

The basic model of omnivory/intraguild predation, with saturating functional responses and all species potentially having some degree of self-limitation, is

$$\frac{dP_1}{P_1 dt} = \frac{w_{11}v_{11}R_1 + n_{11}m_{11}N_1}{1 + v_{11}u_{11}R_1 + m_{11}l_{11}N_1} - x_1 - y_1 P_1$$

$$\frac{dN_1}{N_1 dt} = \frac{b_{11}a_{11}R_1}{1 + a_{11}h_{11}R_1} - \frac{m_{11}P_1}{1 + v_{11}u_{11}R_1 + m_{11}l_{11}N_1} - f_1 - g_1 N_1, \qquad (5.1)$$

$$\frac{dR_1}{R_1 dt} = c_1 - d_1 R_1 - \frac{a_{11}N_1}{1 + a_{11}h_{11}R_1} - \frac{v_{11}P_1}{1 + v_{11}u_{11}R_1 + m_{11}l_{11}N_1}$$

TABLE 5.1. Summary descriptions of the additional parameters used in the models presented in this chapter

Parameter	Description
v_{ik}	Attack coefficient for predator k feeding on resource i
w_{ik}	Conversion efficiency for predator k feeding on resource i
u_{ik}	Handling time for predator k feeding on resource i

where v_{11}, w_{11}, and u_{11} are, respectively, the attack coefficient, the conversion efficiency, and the handling time for the predator feeding on the resource (table 5.1), and all the other parameters are as in chapters 3 and 4. The corresponding isocline functions are

$$\frac{dP_1}{dt} = 0 : N_1 = \frac{(x_1 + y_1 P_1)(1 + u_{11} v_{11} R_1) - w_{11} v_{11} R_1}{m_{11}(n_{11} - l_{11}(x_1 + y_1 P_1))}$$

$$\frac{dN_1}{dt} = 0 : P_1 = \left(\frac{b_{11} a_{11} R_1}{1 + a_{11} h_{11} R_1} - f_1 - g_1 N_1 \right) \frac{1 + v_{11} u_{11} R_1 + m_{11} l_{11} N_1}{m_{11}}.$$

$$\frac{dR_1}{dt} = 0 : P_1 = \left(c_1 - d_1 R_1 - \frac{a_{11} N_1}{1 + a_{11} h_{11} R_1} \right) \frac{1 + v_{11} u_{11} R_1 + m_{11} l_{11} N_1}{v_{11}}$$

With saturating functional responses (i.e., $h_{11} > 0$, $u_{11} > 0$, $l_{11} > 0$), the population dynamics of each species depends directly on the abundances of all other species, which makes analytical calculations torturous. This is why we focus primarily on the geometry of the isoclines to understand the conditions for invasibility. Consequently, much of the presentation that follows focuses on situations where all the functional responses are linear. However, as we know from chapters 3 and 4, saturating functional responses do not qualitatively alter outcomes if stable point equilibria result, and so considering the linear functional response situations can get us far.

Therefore, consider the coexistence of these three species when all functional responses are linear (i.e., $h_{11} = u_{11} = l_{11} = 0$) and neither the consumer nor the predator directly limits its own abundances (i.e., $g_1 = y_1 = 0$). Start with the community of R_1 and N_1 already coexisting with one another and P_1 invading (invasion 12 in fig. 5.1). The isocline system for these species is similar to that of the three-trophic-level food chain (see fig. 4.2(a)), with two important exceptions. Because the predator eats the resource, the R_1 isocline now intersects the P_1 axis at c_1/v_{11}, and continues to intersect the N_1 axis at c_1/a_{11} and the R_1 axis at $R_{1(0)}^* c_1/d_{11}$ (fig. 5.2(a)). Also, the P_1 isocline now intersects the R_1 axis at $x_1/(v_{11} w_{11})$ as well as the N_1 axis at $x_1/(m_{11} n_{11})$, but is still parallel to the P_1 axis. The N_1 isocline is unchanged. The three isoclines also remain planes.

For the predator to invade and coexist with the consumer at a stable point equilibrium, the P_1 isocline must intersect the line of intersection between the R_1 and N_1 isoclines with a particular configuration. The line of intersection between the R_1 and N_1 isoclines runs from the point $[\hat{R}_{1(R_1,N_1)}, 0, \hat{P}_{1(R_1,N_1)}]$ in the $P_1 - R_1$ face to the point $[R_{1(N_1)}^*, N_{1(R_1)}^*, 0]$ in the $N_1 - R_1$ face (fig. 5.2(b)–(c)). (I use the notation $\hat{R}_{1(R_1,N_1)}$ to identify the value of a point of intersection between two isoclines that is not an equilibrium.) Thus, if the P_1 isocline is between $[R_{1(N_1)}^*, N_{1(R_1)}^*, 0]$ and the

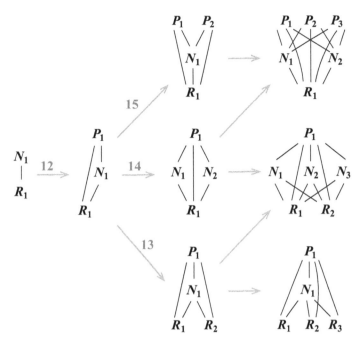

FIGURE 5.1. A road map of the species invasions that are considered in this chapter. For each community module, R_i represents a resource species, N_j a consumer species, and P_k a predator species. A line between two species identifies a trophic link where the species above consumes the species below. Note that the predators here are intraguild predators feeding on both consumers and resources. Arrows show transitions between modules that result from the invasion by different types of species. Numbers are associated with each arrow so that different types of transitions can be identified in the text.

origin in the $N_1 - R_1$ face, the predator can invade and coexist at a stable equilibrium. If the P_1 isocline passes above $[R_{1(N_1)}^*, N_{1(R_1)}^*, 0]$, the combined abundances of both R_1 and N_1 are not enough to support P_1's population.

One way this will be true is if the P_1 isocline intersects the N_1 isocline below the R_1 isocline in the $N_1 - R_1$ face (fig. 5.2(c)):

$$N_{1(R_1)}^* > \widehat{N}_{1(N_1, P_1)}\Big|_{P_1 = 0}$$

$$\frac{c_1 a_{11} b_{11} - d_1 f_1}{a_{11}^2 b_{11}} > \frac{x_1 a_{11} b_{11} - v_{11} w_{11} f_1}{a_{11} b_{11} m_{11} n_{11}} \tag{5.2}$$

If this is true, the combined abundances of R_1 and N_1 will give P_1 a positive per capita population growth rate when rare (i.e., when P_1's abundance begins very

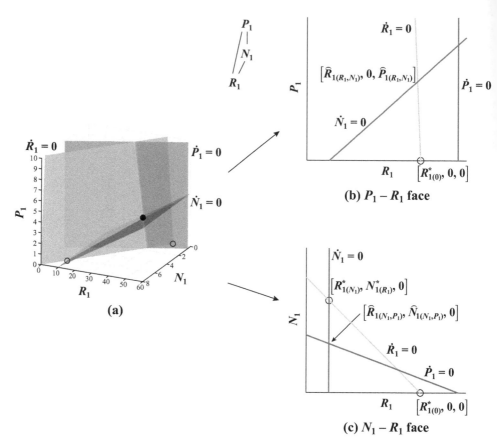

FIGURE 5.2. An isocline configuration for an intraguild predation/omnivory community module containing a consumer (N_1) that feeds on a resource (R_1) and a predator (P_1) that feeds on both the consumer and the resource, in which the three species coexist at a stable equilibrium and P_1 can support a population by feeding only on N_1 but cannot support a population by feeding only on R_1. The full three-dimensional representation of the isoclines is given in (a). Panel (b) shows the $P_1 - R_1$ face and (c) shows the $N_1 - R_1$ face. The isoclines are identified as in figure 4.2. Unstable equilibria are identified with open circles, and the stable equilibrium with a filled black circle. The parameter values are as follows: $c_1 = 2.0$, $d_1 = 0.04$, $a_{11} = 0.25$, $b_{11} = 0.1$, $h_{11} = 0.0$, $f_1 = 0.25$, $g_1 = 0.0$, $m_{11} = 0.25$, $n_{11} = 0.1$, $l_{11} = 0.0$, $v_{11} = 0.015$, $w_{11} = 0.1$, $u_{11} = 0.0$, $x_1 = 0.1$, and $y_1 = 0.01$.

near $[R^*_{1(N_1)}, N^*_{1(R_1)}, 0]$). Note that this is exactly the same criterion for its invasibility when P_1 was not an "intraguild" predator. Now that P_1 eats R_1, the P_1 isocline is not parallel to the R_1 axis. In the situation depicted in figure 5.2, P_1 could coexist in the three-trophic-level food chain without consuming R_1 since its isocline also intersects the N_1 axis below $N^*_{1(R_1)}$, but now it has a very low attack coefficient on

R_1 such that its isocline intersects the R_1 axis well above $R^*_{1(N_1)}$. However, P_1 could not invade the community if N_1 were absent, because P_1 cannot support its population on R_1 alone since the P_1 isocline intersects the R_1 axis above $R^*_{1(0)}$ (fig. 5.2(a)). An intrepid community ecologist studying these species in the wild would find that P_1 was sustained by feeding on N_1 and that R_1 is simply an incidental component of P_1's diet.

The strengths of the various direct and indirect effects between the species are also critical to understanding coexistence in this situation. P_1 imposes substantial mortality directly on N_1 and so depresses its abundance. In contrast, N_1 is a much better resource competitor than P_1 for R_1, which is identified by N_1's isocline intersecting the R_1 axis well below where P_1's isocline intersects (fig. 5.2). Consequently, the combined effects of P_1 on N_1 from direct mortality and competition via resource depression are not enough to drive N_1 extinct. Finally, R_1 is a much better apparent competitor against P_1 because its isocline intersects the P_1 axis at a value that is substantially greater than the value where it intersects the N_1 axis. These are all necessary attributes for these three species to coexist.

The three species can also coexist at a stable equilibrium if the predator can support a population on the resource alone. Figure 5.3(a)–(c) shows the isocline perspectives for such a case. Here, P_1 could still support its population from feeding on N_1 alone, because its isocline still intersects the N_1 axis below $N^*_{1(R_1)}$ (fig 5.3(c)). P_1 can also support its population solely on R_1 as well, because its isocline intersects the R_1 axis below $R^*_{1(0)}$ (i.e., $x_1 / (v_{11}w_{11}) < R^*_{1(0)}$) (fig. 5.3(b)). In this case, the three species coexist because the P_1 isocline passes to the right of the intersection point between the R_1 and N_1 isoclines in the $P_1 - R_1$ face, which ensures that the three isoclines intersect at a feasible and stable three-species equilibrium (fig. 5.3(a)). Because the P_1 isocline is a vertical line in the $P_1 - R_1$ face, this requires that the P_1 isocline intersect the R_1 axis such that

$$\widehat{R}_{1(R_1,N_1)} < R^*_{1(P_1)}. \tag{5.3}$$

In this case, the inequality is

$$\frac{c_1 m_{11} + f_1 v_{11}}{a_{11} b_{11} v_{11} + d_1 m_1} < \frac{x_1}{v_{11} w_{11}}.$$

Inequality (5.3) is in fact the invasibility criterion for N_1 into the community where R_1 and P_1 are already present. For N_1 to have a positive per capita population growth rate when rare, its isocline must intersect the R_1 isocline at a lower resource abundance. Thus, the three species will coexist at a stable equilibrium only if inequality (5.3) is satisfied. Obviously, the N_1 isocline must also intersect the R_1 axis below $R^*_{1(0)}$ (as originally developed in chapter 3). The community

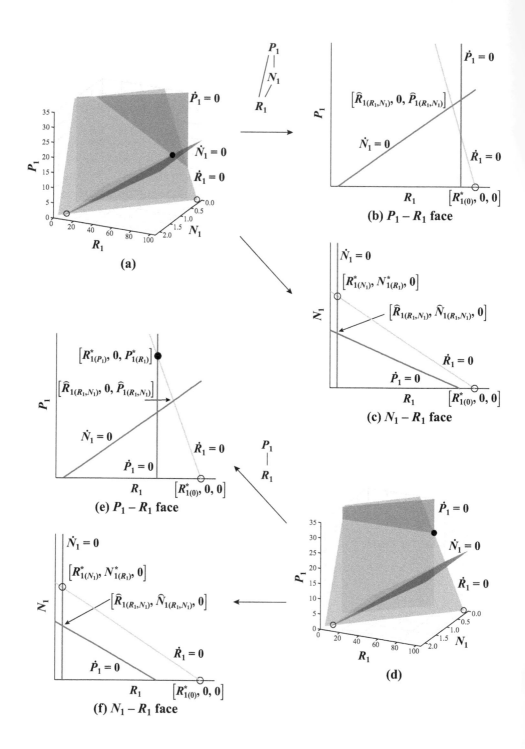

ecologist studying this system would conclude that both components of P_1's omnivorous diet were important to maintaining its population.

The only difference between the species depicted in figure 5.2 and those in figure 5.3(a)–(c) is that the predator in figure 5.2 cannot support its population solely on R_1 and the predator in figure 5.3(a)–(c) can. In both, the consumer is the better competitor for the resource, and the predator in each does not impose enough mortality on the consumer to drive it extinct at the resource abundance level that results (Holt and Polis 1997).

If inequality (5.3) is not satisfied and the R_1 isocline passes above the point of intersection of the R_1 and N_1 isoclines, P_1 drives the resource abundance to a level at which N_1 cannot maintain a population, even though N_1 may still be the better resource competitor (fig. 5.3(d)–(f)). Now the community will come to a stable equilibrium with only P_1 and R_1 present at $[R^*_{1(P_1)}, 0, P^*_{1(R_1)}]$. If the predator can directly depress resource abundance below $\hat{R}_{1(R_1,N_1)}$, the combined mortality from the predator and the low resource abundance result in the consumer having a negative per capita population growth rate when rare, even though the consumer may be able to support a population on that resource level alone. Conversely, *if the predator cannot support a population on the resource alone, it cannot drive the consumer extinct.* Only the combined direct and indirect effects of the predator on the consumer can prevent the consumer from invading and coexisting in this community (Holt and Polis 1997).

The three species can also coexist if P_1 can support a population on R_1 but cannot support a population on N_1 (fig. 5.4(a)–(c)). The range of resource abundances at which N_1 could survive with P_1 (i.e., the range from $\hat{R}_{1(R_1,N_1)}$ to $R^*_{1(0)}$) has increased because the slope of the N_1 isocline in the $P_1 - R_1$ face increases when P_1 derives less benefit from feeding on N_1 (i.e., smaller $m_{11}n_{11}$). Thus, the balance of demographic forces on N_1 has shifted to greater importance of the per capita birth rate from feeding on the resource and less importance of per capita mortality from the predator. In this case, the ecologist studying this community would consider the consumer to be the less important component of the predator's diet.

FIGURE 5.3. Two isocline configurations for an intraguild predation/omnivory community module containing a consumer (N_1) that feeds on a resource (R_1) and a predator (P_1) that feeds on both the consumer and the resource in which (a)–(c) the predator can support a population on the resource alone and the three species coexist at a stable equilibrium, and (d)–(f) the predator drives the consumer extinct because it outcompetes the consumer for the resource so that only the predator and resource can coexist. All isoclines and equilibria are identified as in figure 5.2. The parameter values common to both examples are as follows: $c_1 = 1.0$, $d_1 = 0.01$, $a_{11} = 0.5$, $b_{11} = 0.1$, $h_{11} = 0.0$, $f_1 = 0.25$, $g_1 = 0.0$, $m_{11} = 0.25$, $n_{11} = 0.3$, $l_{11} = 0.0$, $w_{11} = 0.1$, $u_{11} = 0.0$, $x_1 = 0.09$, and $y_1 = 0.01$. Panels (a)–(c) have $v_{11} = 0.01$ and (d)–(f) have $v_{11} = 0.013$.

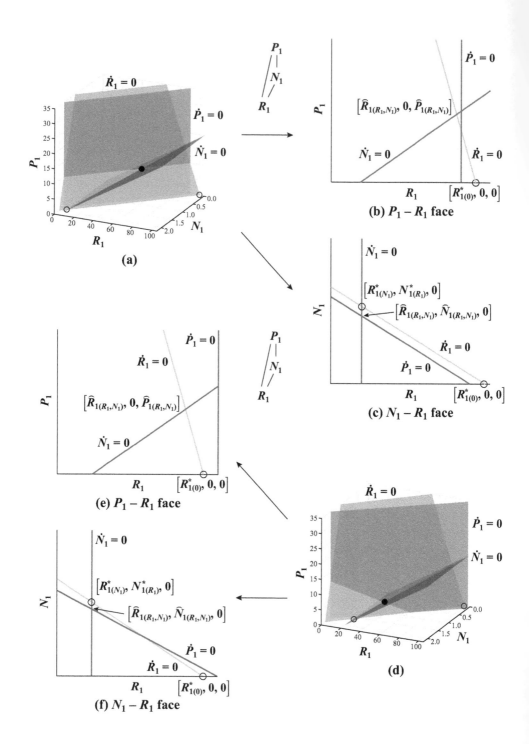

(a)

$$\left[\widehat{R}_{1(R_1,N_1)}, 0, \widehat{P}_{1(R_1,N_1)}\right]$$

$$\left[R^*_{1(0)}, 0, 0\right]$$

(b) $P_1 - R_1$ face

$$\left[R^*_{1(N_1)}, N^*_{1(R_1)}, 0\right]$$

$$\left[\widehat{R}_{1(R_1,N_1)}, \widehat{N}_{1(R_1,N_1)}, 0\right]$$

$$\left[R^*_{1(0)}, 0, 0\right]$$

(c) $N_1 - R_1$ face

$$\left[\widehat{R}_{1(R_1,N_1)}, 0, \widehat{P}_{1(R_1,N_1)}\right]$$

$$\left[R^*_{1(0)}, 0, 0\right]$$

(e) $P_1 - R_1$ face

(d)

$$\left[R^*_{1(N_1)}, N^*_{1(R_1)}, 0\right]$$

$$\left[\widehat{R}_{1(R_1,N_1)}, \widehat{N}_{1(R_1,N_1)}, 0\right]$$

$$\left[R^*_{1(0)}, 0, 0\right]$$

(f) $N_1 - R_1$ face

Finally, it is possible that the three species will coexist even though P_1 cannot support a population on either R_1 or N_1 alone (fig. 5.4(d)–(f)). Here, the combined abundances of both R_1 or N_1 are needed to sustain the predator. Obviously, in this case, both are essential components of P_1's diet.

The intraguild predation module reveals one final interesting combination of species properties. If neither inequality (5.2) nor (5.3) is satisfied but P_1 can coexist with R_1, then these species will display *priority effects* so that the first species to invade the community with R_1 already present will coexist with R_1 and prevent the other from invading (fig. 5.5) (Holt and Polis 1997). Each two-species food chain is a locally stable equilibrium, with the three-species equilibrium being unstable. Thus, if P_1 is the first to invade, the mortality imposed by P_1 relative to the birth rate resulting from feeding on R_1 at their joint equilibrium is inadequate for N_1 to have a positive per capita population growth rate when it invades (fig. 5.5(b)). Conversely, if N_1 is the first to invade, the combined abundances of R_1 or N_1 at their joint equilibrium is inadequate for P_1 to have a positive per capita population growth rate when it invades (fig. 5.5(c)).

Saturating functional responses ($h_{ij} > 0$, $l_{jk} > 0$, $u_{ik} > 0$) for the consumer and resource have identical effects as were seen for the three-trophic-level food chain (see fig. 4.2(c)). Because a saturating functional response makes the point of intersection between the consumer's or predator's isocline and its prey's axis increase, saturating functional responses make satisfying a species' invasibility criterion more difficult by requiring a higher prey abundance to support a positive population growth rate when the species is rare.

Moreover, the isoclines of species at intermediate trophic levels with saturating functional responses bend away from the axes of their predators. Thus, with $h_{11} > 0$, the N_1 isocline in an intraguild module curves away from the P_1 axis. This makes the ability of N_1 to satisfy inequality (5.3) more difficult when P_1 can support a population on R_1 alone. Figure 5.6 shows the isocline configurations for the same parameter combination as in figure 5.3(a)–(c) but with saturating functional

---

FIGURE 5.4. Two isocline configurations for an intraguild predation/omnivory community module containing a consumer (N_1) that feeds on a resource (R_1) and a predator (P_1) that feeds on both the consumer and the resource in which (a)–(c) the three species coexist at a stable equilibrium and the predator can support a population on the resource alone but not on the consumer alone, and (d)–(f) the three species coexist at a stable equilibrium but the predator cannot support a population alone on either the consumer or the resource. All isoclines and equilibria are identified as in figure 5.2. The parameter values common to both examples are as follows: $c_1 = 1.0$, $d_1 = 0.01$, $a_{11} = 0.5$, $b_{11} = 0.1$, $h_{11} = 0.0$, $f_1 = 1.0$, $g_1 = 0.0$, $m_{11} = 0.25$, $l_{11} = 0.0$, $v_{11} = 0.01$, $w_{11} = 0.1$, $u_{11} = 0.0$, and $y_1 = 0.01$. Panels (a)–(c) have $n_{11} = 0.2$ and $x_1 = 0.09$, and (d)–(f) have $n_{11} = 0.25$ and $x_1 = 0.11$.

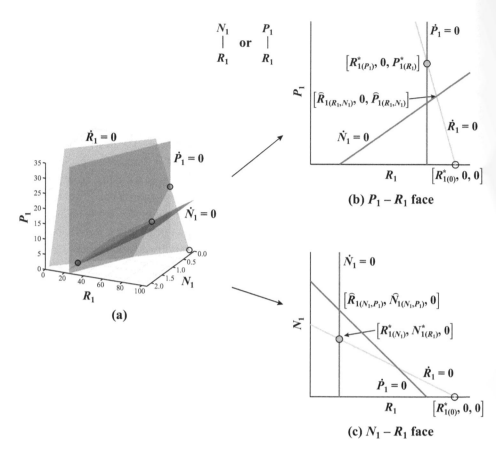

FIGURE 5.5. An isocline configuration for an intraguild predation/omnivory community module containing a consumer (N_1) that feeds on a resource (R_1) and a predator (P_1) that feeds on both the consumer and the resource in which the three-species equilibrium is unstable and each of the two-species equilibria are locally stable, giving a priority effect. All isoclines and equilibria are identified as in figure 5.2. The parameter values common to both examples are as follows: $c_1 = 1.0$, $d_1 = 0.01$, $a_{11} = 0.5$, $b_{11} = 0.1$, $h_{11} = 0.0$, $f_1 = 1.0$, $g_1 = 0.0$, $m_{11} = 0.25$, $n_{11} = 0.1$, $l_{11} = 0.0$, $v_{11} = 0.01$, $w_{11} = 0.1$ $u_{11} = 0.0$, $x_1 = 0.08$, and $y_1 = 0.01$.

responses. Unlike figure 5.3(a), N_1 now cannot invade and coexist with P_1, even though R_1's equilibrium abundance is higher and P_1's equilibrium abundance is lower than when both had linear functional responses.

Direct self-limitation in N_1 and P_1 causes their isoclines to bend away from their own axes. As we saw in chapters 3 and 4, such self-limitation does not affect the per capita population growth rate of a species when it is invading, but it can strongly influence whether others will be able to meet their invasibility criteria because of

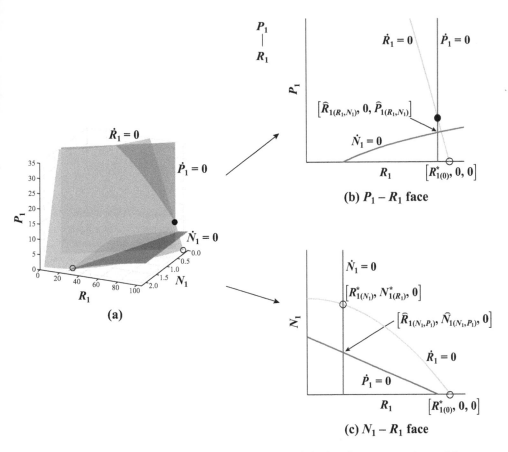

FIGURE 5.6. An isocline configuration for an intraguild predation/omnivory community module containing a consumer (N_1) that feeds on a resource (R_1) and a predator (P_1) that feeds on both the consumer and the resource, and both consumer and resource have saturating functional responses. In this case, only the predator and resource can coexist. All isoclines and equilibria are identified as in figure 5.2. All parameter values are the same as in figure 5.3(a)–(c), except $h_{11} = l_{11} = u_{11} = 0.02$.

its effect on species abundances. For example, if P_1 is self-limiting (i.e., $y_1 > 0$) and can support its population on N_1 alone, the self-limitation increases $R^*_{1(P_1)}$ in inequality (5.3), and so makes invasion and coexistence of N_1 easier (fig. 5.7).

The basic intraguild predation (IGP) simulation module can be accessed at https://mechanismsofcoexistence.com/omnivory_3D_1R_1N_1P_IGP.html.

All the results above describe areas of parameter space in which stable point equilibria result. With saturating functional responses, the dynamics of an

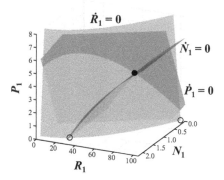

FIGURE 5.7. An isocline configuration for an intraguild predation/omnivory community module containing a consumer (N_1) that feeds on a resource (R_1) and a predator (P_1) that feeds on both the consumer and the resource. In this case, both consumer and resource have saturating functional responses, and they also are directly self-limiting. All three species coexist at a stable equilibrium. All isoclines and equilibria are identified as in figure 5.2. All parameter values are the same as in figure 5.6, except $g_1 = 0.03$ and $y_1 = 0.01$.

intraguild predation module may be a limit cycle or chaos, just as outlined in chapter 4. However, the diversity and strength of indirect effects among the species in this simple community module cause small areas of parameter space that can generate limit cycles and chaos even with all linear functional responses (Tanabe and Namba 2005, Hsu et al. 2015). One such area of parameter space is where P_1 has a moderately large attack coefficient and a moderate conversion efficiency for N_1, but P_1 has a disproportionately large attack coefficient and a disproportionately small conversion efficiency for R_1 (Tanabe and Namba 2005, Hsu et al. 2015). Thus, while it is possible to have chaotic dynamics with linear functional responses here, the parameters giving them do not seem biologically very reasonable.

All of the issues addressed in the empirical testing sections in chapters 3 and 4 apply here as well. What evidence identifies that the omnivore actually consumes the consumer and resource species, and does this consumption have a material demographic impact on them? Does this omnivore have a positive per capita population growth rate when at low abundance in treatments where the other community members have been allowed to approach their steady state abundances in its absence? Are its per capita birth and death rates roughly equal when the omnivore is at or near its natural abundance? Does the omnivore have an essentially linear or saturating functional response at normal consumption rates? What other mechanisms also influence the demography of the omnivore (e.g., feeding interference, stress responses to conspecific crowding, territoriality)?

In addition to these issues, understanding the mechanisms promoting the coexistence and determining the local demographic steady state abundance of the omnivore requires additional information. Specifically, as this section highlights, one of the critical issues is whether the omnivore can support a population on each species found in its diet separately. Feeding experiments should be done in which the omnivore is placed at low abundance and allowed to feed on only a single prey type that is held constant at the natural abundance of the prey in the absence of all of its enemies. This experiment should then be repeated for each component species in its diet. This set of experiments tests where the omnivore's isocline intersects each prey species' abundance axis relative to the abundances of the various prey in the omnivore's absence. These experimental results establish which components of the omnivore's diet are essential to its persistence in a given community and which are less critical or only incidental.

These experiments will also provide valuable mechanistic insights into understanding patterns of species composition and abundance among communities along environmental gradients in nature. For example, in locations that favor high abundance for R_1 (e.g., high c_1), the high productivity of R_1 may inflate the omnivore's abundance to a level that will drive N_1 extinct (Diehl and Feißel 2000, Mylius et al. 2001, Amarasekare 2008a). This occurs because the R_1 isocline increases in the $P_1 - R_1$ face as c_1 increases. So, for example, algal abundances may be so high in eutrophic lakes that omnivores drive some consumer species at lower intermediate trophic levels extinct. This may be one contributor to why species richness often is lower in high-productivity ecosystems (Mittelbach et al. 2001). Such effects are only possible for prey on which the omnivore can support a population in isolation.

The priority effect that either N_1 or P_1 but not both will coexist with R_1 also depends on the specific pattern of support afforded to P_1 from feeding on the other two species (see fig. 5.5). The omnivore must be able to support a population on the resource, and unable to support a population on the consumer alone but still inflict enough mortality on the consumer to depress its population.

ADDING SPECIES TO THE OMNIVORE FOOD CHAIN

Once a single species at each position is present to form the basic intraguild predation module, we can then begin to evaluate the requirements needed for additional resource, consumer, and predator species to invade and coexist in the community. In this section, I follow the systematic evaluation established in chapters 3 and 4 to ask what is required for multiple species to invade and coexist at each trophic level in turn: multiple resources, then multiple consumers, and finally multiple predators. For each, the following set of equations describes multiple species at each of these three trophic levels where predators feed on both consumers and resources:

$$\frac{dP_k}{P_k dt} = \frac{\sum_{i=1}^{p} w_{ik} v_{ik} R_i + \sum_{j=1}^{q} n_{jk} m_{jk} N_j}{1 + \sum_{i=1}^{p} v_{ik} u_{ik} R_i + \sum_{j=1}^{q} m_{jk} l_{jk} N_j} - x_k - y_k P_k$$

$$\frac{dN_j}{N_j dt} = \frac{\sum_{i=1}^{p} b_{ij} a_{ij} R_i}{1 + \sum_{i=1}^{p} a_{ij} h_{ij} R_i} - \sum_{k=1}^{s} \frac{m_{jk} P_k}{1 + \sum_{i=1}^{p} v_{ik} u_{ik} R_i + \sum_{j=1}^{q} m_{jk} l_{jk} N_j} - f_j - g_j N_j. \quad (5.4)$$

$$\frac{dR_i}{R_i dt} = c_i - d_i R_i - \sum_{j=1}^{q} \frac{a_{ij} N_j}{1 + \sum_{i=1}^{p} a_{ij} h_{ij} R_i} - \sum_{k=1}^{s} \frac{v_{ik} P_k}{1 + \sum_{i=1}^{p} v_{ik} u_{ik} R_i + \sum_{j=1}^{q} m_{jk} l_{jk} N_j}$$

Again, this is a daunting set of equations, but interpreting their dynamics in terms of species invasibility and coexistence is simply a matter of continuing the isocline analyses we have been exploring. I focus primarily on the situations with linear functional responses.

Multiple Resources Invading an Omnivore Food Chain

Exact analytical conditions cannot be specified for the invasibility of the second resource (invasion 13 in fig. 5.1), but it can easily be understood by exploring the geometric relationships among the isoclines.

Adding R_2 changes this to a four-dimensional problem. The isocline system can still be visualized in the $P_1 - N_1 - R_1$ three-dimensional subspace once the abundance of R_2 is specified, but the positions of the R_2, N_1, and P_1 isoclines in the $P_1 - N_1 - R_1$ subspace change as R_2's abundance changes. When viewed in this subspace, the R_2 isocline is a plane that intersects the N_1 axis at $(c_2 - d_2 R_2)/a_{21}$, intersects the P_1 axis at $(c_2 - d_2 R_2)/v_{21}$, and runs parallel to the R_1 axis. R_2 can invade and increase if its isocline lies above the point of the three-species equilibrium $[R^*_{1(N_1, P_1)}, 0, N^*_{1(R_1, P_1)}, P^*_{1(R_1, N_1)}]$ when $R_2 = 0$ (fig. 5.8(a)–(b)). Because its isocline runs parallel to the R_1 axis, whether it is above or below this equilibrium can be evaluated by considering the isoclines and the equilibrium point in the $P_1 - N_1$ plane through the three-species equilibrium.

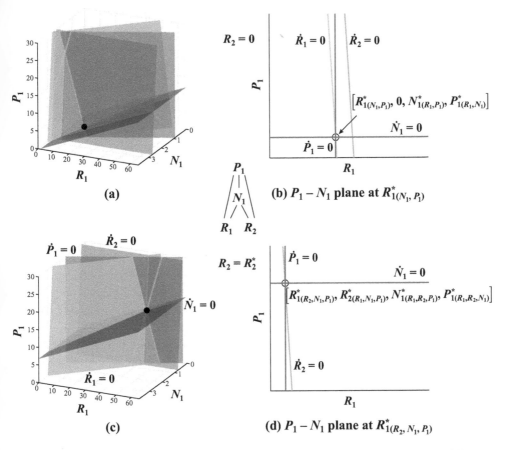

FIGURE 5.8. An isocline configuration for an intraguild predation/omnivory community module containing a consumer (N_1) that feeds on two resources (R_1 and R_2) and a predator (P_1) that feeds on both the consumer and the resources. Panels (a) and (c) show the three-dimensional $P_1 - N_1 - R_1$ subspaces of the full four-dimensional system. Because of this, the positions of the R_2, N_1, and P_1 isoclines change in this subspace as R_2's abundance changes. Panels (a) and (b) show the isocline configurations when $R_2 = 0$, with the full subspace representation in (a) and the plane through $R^*_{1(N_1,P_1)}$ in (b). Panels (c) and (d) show the isocline configurations for $R_2 = R^*_{2(R_1,N_1,P_1)}$. All isoclines and equilibria are identified as in figure 5.2, with the addition of the R_2 isocline identified accordingly. The parameter values are as follows: $c_1 = 2.25$, $c_2 = 2.0$, $d_1 = d_2 = 0.04$, $a_{11} = a_{21} = 0.75$, $b_{11} = b_{21} = 0.1$, $h_{11} = h_{21} = 0.0$, $f_1 = 0.1$, $g_1 = 0.0$, $m_{11} = 0.4$, $n_{11} = 0.1$, $l_{11} = 0.0$, $v_{11} = v_{21} = 0.01$, $w_{11} = w_{21} = 0.1$, $u_{11} = u_{21} = 0.0$, $x_1 = 0.1$, and $y_1 = 0.01$.

If the R_2 isocline passes above the three-species equilibrium when $R_2 = 0$, R_2's abundance will increase when it invades. As R_2 increases in abundance, when viewed in the $P_1 - N_1 - R_1$ subspace, the R_2 isocline moves toward the R_1 axis, the P_1 isocline moves toward the P_1 axis, and the N_1 isocline moves toward the N_1 axis (cf. 5.8(a)–(b) and 5.8(c)–(d)). Once the community reaches the new four-species equilibrium, P_1 has increased in abundance, but R_1 and N_1 may have increased or decreased as compared to their abundances at the three-species equilibrium lacking R_2, depending on exactly the abundance changes and the relative attack coefficients of the consumer and predator on the two resources (fig. 5.8(c)–(d)).

The IGP simulation module with two basal resources can be accessed at https://mechanismsofcoexistence.com/omnivory_3D_2R_1N_1P_IGP.html.

A third resource species (call it R_3) can then invade this community containing R_1 and R_2 if its isocline passes above the four-species equilibrium when $R_3 = 0$. Any potential invading resource species that can satisfy the comparable criterion for itself would be able to invade this community. Again, once it invades and the community comes to equilibrium, P_1 will increase and N_1 may increase or decrease. Additional resource species can continue to invade the community if they can satisfy the same criterion given the current predator and consumer abundances. Moreover, some resource species invasions may cause previously invading resources to go extinct because of apparent competitive effects mediated through both N_1 and P_1. The final membership of the community will be the same regardless of the sequence of invasions from the resource species pool. Those resources that are not coexisting in the final community are those whose isoclines pass below the abundances of the consumer and predator at this final equilibrium. Thus, multiple resource species can potentially invade and coexist with a single consumer and a single intraguild predator.

The logic of understanding which additional resource species can invade an omnivory food chain is simply the logic of the apparent competition criteria discussed in chapters 3 and 4, but now with both a consumer and a predator imposing mortality on the invading resource species. Any resource species that has adequate productivity in the ecosystem in which the community is developing (i.e., large enough c_i) relative to the combined mortality imposed by the consumer and the predator will be able to invade and coexist. As with analogous situations we have considered previously, saturating functional responses and consumer and predator self-limitation make the invasion criteria for resources less stringent because both will reduce the abundances and the feeding rates of coexisting consumer and predator (see chapters 3 and 4).

As in chapter 4, "Multiple Resources Invading a Three-Trophic-Level Food Chain," the addition of the predator also feeding on the invading resources offers only small but substantive differences in empirical approach than what is discussed in chapter 3, "Apparent Competitors." For the invading resources, the issue is simply whether each can support a population given the level of mortality now from multiple enemies above them in the food web.

The more substantive issues arise for the species above them in the food web. As noted in the previous section, where only one resource species was present, increasing the productivity of the basal resource can inflate the omnivore's abundance to a level that will drive the intermediate consumer extinct. If the invading resources are more beneficial to the omnivore than to the consumer, the consumer may not coexist if certain resource species are present. Such effects would manifest as patterns of species loss at intermediate trophic levels across communities with different compositions of species lower in the food web.

Multiple Consumers Invading an Omnivore Food Chain

A second consumer invading an omnivory food chain faces the same obstacles as a consumer invading a linear food chain with a predator at the top that only feeds at the consumer trophic level (invasion 14 in fig. 5.1). Namely, it must be able to support a population on the available resource level but with the added mortality inflicted by the predator. However, the omnivorous predator adds some additional dynamical considerations to the problem faced by the invading consumer.

Call the resident consumer N_1 and the invading consumer N_2. With all linear functional responses ($h_{1j} = u_{11} = l_{j1} = 0$) and no consumer or predator self-limitation ($g_j = y_1 = 0$), the consumers face exactly the same issue for coexistence as with no intraguild predation. When viewed in the $N_1 - N_2 - R_1$ subspace, each consumer isocline is parallel to both consumer axes and intersects the R_1 axis at $R_1 = f_j/(a_{1j}b_{1j}) + m_{j1}P_1$ (fig. 5.9(a)). Thus, each consumer isocline's minimum intersection point on the R_1 axis (i.e., when $P_1 = 0$) is its R-star value (chapter 3, "Resource Competitors"), and this intersection point increases at a rate equal to P_1's attack coefficient times the P_1 abundance (McPeek 2014). The four-species equilibrium,

$$[R^*_{1(N_1,N_2,P_1)}, N^*_{1(R_1,N_2,P_1)}, N^*_{2(R_1,N_1,P_1)}, P^*_{1(R_1,N_1,N_2)}],$$

if it exists, is defined by the R_1 abundance at which the two consumer isoclines are coincident for a given abundance of P_1. The consumer isoclines will be coincident at some positive resource abundance given a predator abundance if the consumers satisfy inequality (4.9), and this point defines $R^*_{1(N_1,N_2,P_1)}$ and $P^*_{1(R_1,N_1,N_2)}$ (McPeek 2014). Intraguild predation does not change the trade-off criterion that permits the two consumers to coexist at all: as with the simple diamond/keystone predation module, the trade-off permitting consumer coexistence requires that the intraguild predator impose disproportionate mortality on the consumer that is the better resource competitor (i.e., inequality 4.9).

The IGP simulation module with two consumers can be accessed at https://mechanismsofcoexistence.com/omnivory_3D_1R_2N_1P_IGP.html.

The additional considerations that intraguild predation introduces involve the R_1 and P_1 isoclines and thus the R_1 and P_1 abundances. In the $N_1 - N_2 - R_1$ subspace, the P_1 isocline now intersects all three axes instead of being parallel to the R_1 axis (cf. figs. 4.5(a) and 5.9(a)). Because P_1 is not self-limiting, the P_1 isocline intersection points with these axes do not change with P_1's abundance. In contrast, the R_1 isocline also intersects all three axes in this view (just as in the simple diamond/keystone predation module), but all three intersection points along the respective axes move toward the origin as P_1's abundance increases. Thus, the line of intersection between the R_1 and P_1 isoclines in this view moves to lower values of R_1 as P_1's abundance increases. The four-species equilibrium exists if the two consumer isoclines are coincident at a point on the R_1 axis that causes them to intersect the line of intersection between the R_1 and P_1 isoclines (fig. 5.9(a)). Assuming that the two consumers satisfy inequality (4.9) so that N_1 is a better resource competitor than N_2, the resulting four-species equilibrium is stable if

$$\frac{m_{11}n_{11}}{a_1} > \frac{x_1}{c_1 - d_1 R^*_{1(N_1,N_2,P_1)} - v_{11} P^*_{1(R_1,N_1,N_2)}} > \frac{m_{21}n_{21}}{a_2}, \qquad (5.5)$$

which is very similar to inequality (4.10): only the denominator of the center term in inequality (5.5) has an extra term for the predator eating the resource as compared to inequality (4.10). As with the diamond/keystone predation module lacking intraguild predation, inequality (5.5) identifies the trade-off between the predator and the resource in which the predator must be a better resource competitor for the consumer whose abundance is more inflated by the resource (McPeek 2014).

Inequality (5.5) also shows that the addition of intraguild predation alters the balance of predator death rate and resource net productivity that is needed for all four species to coexist. The same properties of the consumers' interactions with the resource and predator bracket this ratio, whether the predator is an omnivore or not (cf. inequalities (4.10) and (5.5)). Intraguild predation alters the ratio of predator intrinsic death rate to resource productivity that permits the two consumers to coexist. Consequently, the two consumers must be capable of coexisting at combinations of lower resource abundance and higher predator abundance (McPeek 2014).

If inequality (5.5) is not true because the inner ratio is larger than both outer ratios (i.e., P_1 has a large intrinsic death rate relative to the resource productivity), N_2 (the poorer resource competitor in this case) will be excluded, but whether N_1 can coexist with R_1 and P_1 will depend on the relative abilities of these three

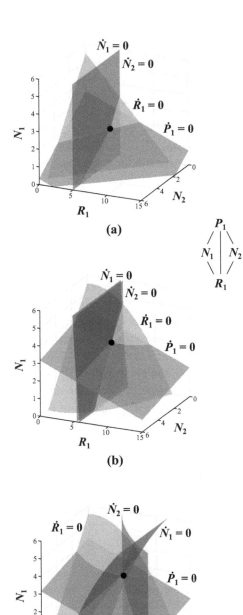

(a)

(b)

(c)

FIGURE 5.9. Various isocline configurations for an intraguild predation/omnivory community module containing two consumers (N_1 and N_2) that feed on a resource (R_1) and a predator (P_1) that feeds on both consumers and the resource, when the consumers and predator have (a) linear functional responses, (b) saturating functional responses, and (c) saturating functional responses with consumers that are directly self-limiting. The equilibria and isoclines are identified as in figure 5.2, with the addition of the N_2 and P_1 isoclines identified accordingly. The parameter values common to all three panels are as follows: $c_1 = 2.5$, $d_1 = 0.2$, $a_{11} = 0.5$, $a_{12} = 0.45$, $b_{11} = b_{12} = 0.1$, $f_1 = 0.1$, $f_2 = 0.15$, $m_{11} = 0.3$, $n_{11} = n_{21} = 0.1$, $v_{11} = 0.05$, $w_{11} = 0.1$, $x_1 = 0.2$, and $y_1 = 0.0$. Panel (a) has $g_1 = g_2 = 0.0$ and $h_{11} = h_{12} = l_{11} = l_{21} = u_{11} = 0.0$, (b) has $g_1 = g_2 = 0.0$ and $h_{11} = h_{12} = l_{11} = l_{21} = u_{11} = 0.2$, and (c) has $g_1 = g_2 = 0.025$ and $h_{11} = h_{12} = l_{11} = l_{21} = u_{11} = 0.2$.

species to satisfy inequalities (5.2) and (5.3). Biologically, this means that the predation pressure on the better competitor is not great enough to increase R_1's abundance (i.e., $R^*_{1(N_1,P_1)}$) enough to permit the poorer competitor to invade and coexist. In contrast, if this inner ratio of inequality (5.5) is smaller than both outer ratios, only the poorer resource competitor (i.e., N_2 in this case) can possibly coexist with R_1 and P_1. Biologically, this outcome means that the predator abundance when coexisting with only the poorer competitor (i.e., $P^*_{1(R_1,N_2)}$) or only the resource (i.e., $P^*_{1(R_1)}$) is high enough to prevent the better competitor from invading the three-species food chain and coexisting.

Also, as with the simple diamond/keystone predation module, if inequality (5.5) is reversed so that the consumer that is less susceptible to the intraguild predator is also the consumer for which the intraguild predator is a better resource competitor, the four-species equilibrium is unstable (McPeek 2014). This would again establish *priority effects* for the two alternative communities, depending on whether each consumer can coexist alone with R_1 and P_1. Neither consumer would be able to invade if the community were at or near the other dynamical state.

If the equilibrium for the four-species community is stable, generally no more than two consumers can coexist if only one resource and one predator are present. The only situation in which more than two consumers can coexist is if the isoclines of all three consumers intersect at precisely one point in the $P_1 - R_1$ face, which is extremely unlikely. If the three isoclines intersect at different points, then only two consumers can coexist. However, which two will coexist from the pool of available consumer species depends on the ecosystem properties that influence resource productivity (i.e., c_1) and predator mortality (i.e., f_1), because the consumers' isocline intersection point depends on the abundances of the resource and predator.

For example, figure 5.10 shows the communities that develop along a productivity gradient when the species pool consists of the three consumers (as in fig. 5.10(a)).

FIGURE 5.10. The equilibrium abundances and isocline configurations of a pool of three consumers (N_1, N_2, N_3) that feed on a resource (R_1) and a predator (P_1) that feeds on all consumers and the resource along a productivity gradient. Panel (a) shows the isocline configurations of these five species in the $P_1 - R_1$ face of the full five-dimensional representation. Only the position of the R_1 isocline changes as c_1 changes along the gradient. Panel (b) shows the equilibrium abundances of the species and the configurations that develop in communities along the gradient of c_1. Isoclines and abundances are identified as in figure 5.9, with the addition of N_3. The other parameter values are as follows: $d_1 = 0.2$, $a_{11} = a_{12} = a_{13} = 0.5$, $b_{11} = b_{12} = b_{13} = 0.1$, $h_{11} = h_{12} = h_{13} = 0.0$, $f_1 = 0.1$, $f_2 = 0.15$, $f_3 = 0.3$, $g_1 = g_2 = g_3 = 0.0$, $m_{11} = 0.3$, $m_{21} = 0.15$, $m_{31} = 0.1$, $n_{11} = n_{21} = n_{31} = 0.1$, $l_{11} = l_{21} = l_{31} = 0.0$, $v_{11} = 0.05$, $w_{11} = 0.1$, $u_{11} = 0.0$, $x_1 = 0.1$, and $y_1 = 0.0$.

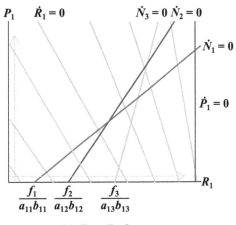

(a) $P_1 - R_1$ face

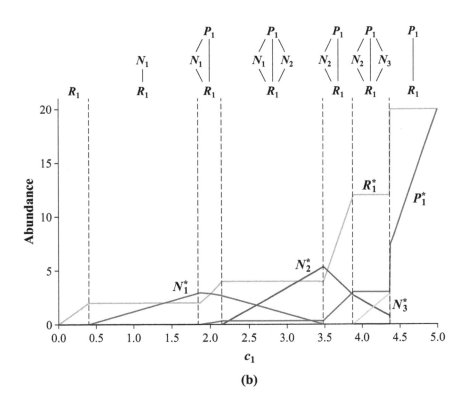

(b)

The intersection point of the R_1 isocline with each axis increases as c_1 increases, and so its position changes along the gradient. As c_1 increases from very low values, N_1 (the consumer that can support a population at the lowest R_1 abundance) can invade and coexist with R_1 (fig. 5.10(b)). At somewhat higher productivities, the predator can invade and coexist when the combined abundances of R_1 and N_1 can support its population. The next consumer to invade and coexist is the one with the lowest intersection point with the N_1 isocline in the $P_1 - R_1$ face, which in this case is N_2. In the range where N_1 and N_2 coexist, increasing c_1 causes the abundance of N_1 (the better resource competitor) to decrease and that of N_2 (the better apparent competitor) to increase. In locations with ever higher values of c_1, eventually N_1 is extirpated, and further increases allow N_3 to coexist with N_2. However, at very high productivity levels, none of the consumers are able to coexist with the predator. Therefore, communities drawn from the same species pool that develop at different points along a productivity gradient or a gradient that influences the intraguild predator's intrinsic death rate are predicted to have different coexisting consumers or no consumers at all.

Saturating functional responses ($h_{1j} > 0$, $u_{11} > 0$, $l_{j1} > 0$) greatly complicate the analytical expressions that summarize the criteria for coexistence (i.e., those analogous to inequalities (4.11) and (4.12)) with the added complication of intraguild predation, but the geometrical relationships among the isoclines needed to produce a stable four-species equilibrium in this case remain the same as above with linear functional responses (see fig. 5.9(b)). In addition, the constraint that only two consumers can coexist if the system comes to a stable point equilibrium remains the same, because the two consumer isoclines are parallel to all consumer axes and so must be coincident at the same R_1 and P_1 abundances. In short, all the same possible outcomes discussed above with linear functional responses result in the same relative geometries of the isoclines, but with the resource and consumer isoclines now being bowed.

Finally, the addition of consumer self-limitation ($g_j > 0$) again causes the consumer isoclines to bend away from their own axes, which causes them to intersect at points above the lowest possible resource abundances at which they can support populations. Thus, more than two consumers can coexist in this diamond/keystone predation module with intraguild predation. Figure 5.9(c) shows two consumers coexisting with all the same parameter values as figure 5.9(b), except that each consumer now also directly limits its own abundance. This causes R_1 to equilibrate at a higher abundance and P_1 to equilibrate at a lower abundance. Any invading consumer with an isocline that passes to the left of the four-species equilibrium will be able to coexist with these consumers, and additional consumers can invade and coexist if they too have isoclines that pass to the left of the new equilibria. Here again, direct self-limitation removes the constraint on the number of species at a given trophic level set by the number of other species at other trophic levels that limit their abundances.

The studies that would elucidate mechanisms in any specific community here are the same as those described in chapter 4, "Multiple Consumers Invading a Three-Trophic-Level Food Chain." Consequently, all of the following also applies when the top predator is or is not an omnivore.

The replacement of species at intermediate trophic levels along environmental gradients (e.g., fig. 4.10) offers great opportunities to perform strong tests of mechanisms. As noted above, a strong prediction made by models of omnivory is that the omnivore should drive species at intermediate trophic levels extinct at very high productivities (Diehl and Feißel 2000, Mylius et al. 2001, Amarasekare 2008a, Verdy and Amarasekare 2010, Gellner and McCann 2012). What the present analysis also shows is that a series of species replacements at intermediate trophic levels may also occur along such gradients. The order of consumer species replacement shows a strong and predictable pattern in the relative abilities to acquire and utilize resources and to avoid the omnivore. With little or no self-limitation, coexistence of the two consumers in any one community requires that one be better at utilizing the resource and the other be better at avoiding the omnivore (or predator in chapter 4, given that the criteria are qualitatively identical).

In systems where such species replacements along an environmental productivity gradient are apparent, the consumers that coexist in communities at the lower end of the gradient should be able to support populations at lower resource levels (i.e., the better competitors) but experience greater rates of per capita mortality from the predator (i.e., poorer at avoiding the predator). At the lower end of the gradient, the omnivore is at lower abundance, and so even though the consumers at this end of the gradient are individually more vulnerable, the total predation pressure on them is lower because of the lower omnivore abundance. In contrast, consumers in communities at the higher end of the gradient should be less able to utilize the resource at lower levels but much less vulnerable to the omnivore. These differences in species abilities are what determine the positions and shapes of the consumer isoclines in figure 5.10(a).

Consequently, observational and experimental studies that quantify the resource acquisition and utilization abilities and the predator avoidance abilities of the various consumers along the gradient will be invaluable in determining whether these species replacements are caused by the bottom-up effects of productivity across the communities. In general, consumer species should get better at predator avoidance and worse at utilizing resources in communities with higher basal productivities.

Multiple Intraguild Predators Invading an Omnivore Food Chain

Now consider the conditions required for one or more additional predator species to invade the basic omnivory food chain (invasion 15 in fig. 5.1). Again, initially assume that neither the consumer or the predators are self-limiting, and identify the already coexisting predator as P_1 and the invading predator as P_2. Because both predators are omnivores that can feed on both R_1 and N_1, two limiting factors impinge on the top trophic level. Thus, unlike with a simple three-trophic-level food chain (chapter 4, "Multiple Predators Invading a Three-Trophic-Level Food Chain"), having only one species at each trophic level can permit two coexisting intraguild predators at the top of the food web even with no predator self-limitation (i.e., $y_1 = y_2 = 0$).

The conditions for invasion and coexistence in this case are quite analogous to the conditions that must be satisfied for two consumers to coexist on two resources (chapter 3). Coexistence of these two predators requires that their isoclines intersect at a point giving positive abundances for both R_1 and N_1. Because their isoclines are parallel to both predator axes with both linear and saturating functional responses and with no self-limitation, this intersection can be evaluated in the $N_1 - R_1$ face. Coexistence in this case requires that each be the better competitor for one of the two prey species, just as with two consumers feeding on two resources. This requires that

$$\frac{v_{11}\left(w_{11} - x_1 u_{11}\right)}{v_{12}\left(w_{12} - x_2 u_{12}\right)} > \frac{x_1}{x_2} > \frac{m_{11}\left(n_{11} - x_1 l_{11}\right)}{m_{12}\left(n_{12} - x_2 l_{12}\right)}, \tag{5.6}$$

which results if P_1 is the better competitor for R_1 and P_2 is the better competitor for N_1 (fig. 5.11). This inequality is exactly analogous to inequality (3.20) for two consumers coexisting on two resources.

For the four-species equilibrium

$$[R^*_{1(N_1,P_1,P_2)}, N^*_{1(R_1,P_1,P_2)}, P^*_{1(R_1,N_1,P_2)}, P^*_{2(R_1,N_1,P_1)}]$$

to exist and be stable or to be a stable attractor of a limit cycle, the resource and consumer isoclines must also intersect at the four-species equilibrium, such that

$$\frac{v_{11}}{m_{11}} > \frac{c_1 - d_1 R^*_1 - \dfrac{a_{11} N^*_1}{1 + a_{11} h_{11} R^*_1}}{\dfrac{b_{11} a_{11} R^*_1}{1 + a_{11} h_{11} R^*_1} - f_1} > \frac{v_{12}}{m_{12}} \tag{5.7}$$

(fig. 5.11(b)). Inequality (5.7) again defines a trade-off whereby each prey species is the better apparent competitor against the enemy that is not its best resource

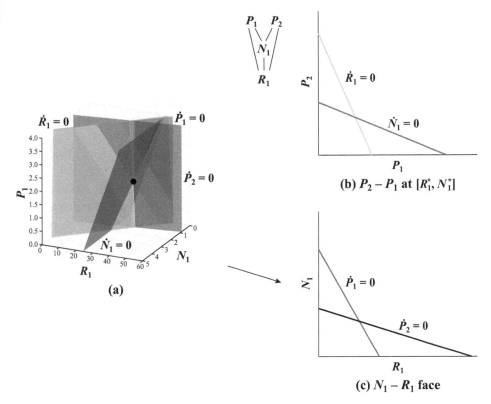

(a)

(b) $P_2 - P_1$ at $[R_1^*, N_1^*]$

(c) $N_1 - R_1$ face

FIGURE 5.11. An isocline configuration for an omnivory/intraguild predation community module containing a consumer (N_1) that feeds on a resource (R_1) and two predators (P_1 and P_2) that both feed on the consumer and the resource. Panel (a) shows the three-dimensional $P_1 - N_1 - R_1$ subspace of this four-dimensional system, (b) shows the $P_2 - P_1$ plane through the four-species equilibrium, and (c) shows the $N_1 - R_1$ face where $P_1 = P_2 = 0$. The equilibria and isoclines are identified as in figure 5.2, with the addition of the P_2 isocline identified accordingly. The parameter values are as follows: $c_1 = 2.0$, $d_1 = 0.04$, $a_{11} = 0.5$, $b_{11} = 0.1$, $h_{11} = 0.0$, $f_1 = 0.1$, $g_1 = 0.0$, $m_{11} = 0.27$, $m_{12} = 0.6$, $n_{11} = n_{12} = 0.1$, $l_{11} = l_{12} = 0.0$, $v_{11} = 0.02$, $v_{12} = 0.008$, $w_{11} = w_{12} = 0.1$, $u_{11} = u_{12} = 0.0$, $x_1 = x_2 = 0.1$, and $y_1 = y_2 = 0.01$.

competitor (cf. inequality (5.7) to inequality (3.19)). Thus, the criteria for two intraguild predators to coexist on one resource and one consumer is exactly analogous to the conditions required for two consumers to coexist on two resources. As with adding consumers to the basic omnivory/intraguild predation community module, at most two predators can coexist with one resource species and one consumer species at a stable point equilibrium if neither is self-limiting ($y_k = 0$). Moreover, each predator must be able to coexist with the resource and consumer on its own.

If the inequalities in (5.7) are reversed but inequality (5.6) is true so that each prey is the better apparent competitor against the predator that is its best resource competitor, the four-species equilibrium is unstable, but each three-species equilibrium with only one predator present may be locally stable (when evaluated on its own). In this case, neither predator would be able to invade the community with the other predator already present. This situation again results in priority effects for each predator excluding the other, but both being potentially able to coexist alone with R_1 and N_1.

The IGP simulation module with two intraguild predators can be accessed at https://mechanismsofcoexistence.com/omnivory_3D_1R_1N_2P_IGP.html.

As with the consumers, at most two omnivorous predator species can coexist at a stable point equilibrium with one resource and one consumer. If the isocline of any predator in the species pool passes wholly below the isoclines of the other predators already present in the area bounded by the R_1 and N_1 isoclines, that predator will drive the others extinct to coexist by itself with R_1 and N_1.

Alternatively, if three predator species' isoclines all intersect in state space where the R_1 and N_1 isoclines intersect to satisfy inequality (5.7), the two predators with the intersection point at the lowest abundance of N_1 that also satisfies all the above criteria will be the two that coexist. For the three predators depicted in figure 5.12, each can coexist with the two prey singly, and all pairwise combinations can also coexist. However, if all three invade the community, P_2 and P_3 will

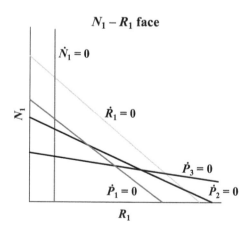

FIGURE 5.12. An isocline configuration showing only the $N_1 - R_1$ face for an omnivory/intraguild predation community module containing a consumer (N_1) that feeds on a resource (R_1) and three predators (P_1, P_2, P_3) that all feed on the consumer and the resource. Given this configuration of isoclines, only P_2 and P_3 can coexist because the intersection of their isoclines occurs at the lowest combination of R_1 and N_1.

eventually be the predators that coexist with R_1 and N_1. The productivity of the ecosystem in which the community develops will also influence which two predators can coexist in analogous ways to consumer coexistence (i.e., fig. 5.10(b)) by determining which combinations of predators can satisfy inequality (5.7) at their equilibrium.

Finally, just as with consumers feeding on multiple resources, more than three predators can possibly invade and coexist on only one resource and one consumer, if the predators are self-limiting to some degree ($y_k > 0$). Invasions would follow the same principles as with multiple self-limiting consumers: any predator with an isocline that passes below the current community equilibrium in state space would experience sufficient R_1 and N_1 abundances to have a positive population growth rate when rare.

Again, testing the mechanisms in this section empirically are very similar to those discussed in chapter 3, "Resource Competitors" and "Multiple Consumers Eating Multiple Resources."

Multiple Species at Each Trophic Level of an Omnivory Food Web

This chapter and the previous one provide a comprehensive picture of what is required for multiple species to be capable of invading and coexisting at a stable point equilibrium on a given trophic level. Levin (1970) provided a general proof for those constraints in his description of how the concept of a niche should be expanded. His argument identifies *limiting factors*—the number of independent factors that directly generate density dependence for the species. Density dependence is not limited to the species' own abundance, but rather includes the abundance of each species with which it directly interacts and the abundance of each depletable abiotic resource. For example, if only one resource species is present, only one consumer can coexist with it. This is because, at the consumer level, only one resource is available to directly interact with any consumers. The consumer's abundance depends directly on only the abundance of the resource—that is the limiting factor. Simply thinking about "intraspecific versus interspecific competition" among species at a given trophic level without any knowledge of the limiting factors that shape their indirect interaction is itself limiting to a true understanding about *why* species coexist.

One can identify the number of limiting factors impinging on a given species by simply counting the number of species that appear in its per capita population growth rate equation (Levin 1970). Those are the species and abiotic resources that directly influence the abundance of that species. Thus, in the basic omnivory food chain, two other species directly influence the consumer's abundance (ignoring direct intraspecific density dependence for the moment: $g_j = 0$), and two other species directly influence the predator's abundance (again ignoring direct intraspecific density dependence for the moment: $y_k = 0$). This is why at most two consumers or two predators can coexist at a stable equilibrium if only one resource and one species at the other trophic level are present.

Omnivory does not qualitatively change the invasibility criteria for species at the lower trophic levels, but it does make their criteria quantitatively more difficult (McPeek 2014). Species at the bottom position now suffer mortality from an additional species higher in the food web. However, the consumer in the middle is squeezed the most. Feeding by the omnivore on the basal species means that the intermediate consumer now has less food available for it and a more abundant predator to face (Polis et al. 1989, Polis and Holt 1992, Holt and Polis 1997). In addition, increasing productivity at the base of the food web now does not benefit the intermediate consumer (e.g., fig. 4.3), but rather favors the resource species to the detriment of the consumer (e.g., fig. 5.10: Diehl and Feißel 2000, Amarasekare 2008a, Verdy and Amarasekare 2010, Gellner and McCann 2012).

As should be expected by feeding on more than one species, omnivory has decided benefits for the omnivores. By feeding on more than one trophic level below it, the omnivore does not have to be able to support a population on any single prey species to coexist, even if only one intermediate consumer is present (see fig. 5.4). Moreover, omnivory increases the number of limiting factors influencing each species at the top of the food web, which means that more species can stably coexist at the top. Hence, omnivory increases the possible species richness of the food web.

The addition of multiple species at each level of the food web with omnivory makes no difference, except for the additional limiting factors that impinge on the omnivores. For example, once a second resource species has invaded and coexists in the simple omnivory food chain (see fig. 5.8), opportunities for more species at other trophic levels to invade have been created. The consumer in the new food web now feeds on two resource species and is fed upon by one omnivorous predator species. This means that three factors limit the consumer's abundance. Thus, three consumer species that all feed upon the two resources and are fed upon by the omnivore could potentially coexist with one another.

Figure 5.13 shows four different configurations of species properties that give coexistence of three such consumers with two resources and one predator. In the first, the three consumers have the same rank in competitive abilities on both resources, and the predator imposes the most mortality on the best competitor and the least mortality on the poorest competitor (fig. 5.13(a)). In the second, each of two consumers is the best competitor on one resource and the worst on the other, and the predator imposes greater mortality on these two than on the consumer that

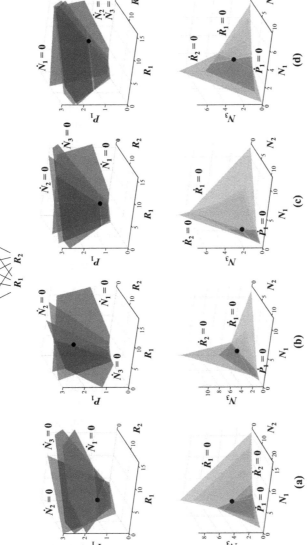

FIGURE 5.13. Isocline configurations for various parameter combinations of an omnivory/intraguild predation community module containing three consumers (N_1, N_2, N_3) and a predator (P_1) that feed on two resources (R_1 and R_2). Panels (a)–(d) each show a different parameter combination. The top graph in each panel shows the N_1, N_2, and N_3 isoclines in the $P_1 - R_2 - R_1$ subspace, and the bottom graph shows the P_1, R_1, and R_2 isoclines in the $N_3 - N_2 - N_1$ subspace. Equilibria and all isoclines are identified as in figures 5.8 and 5.9, with the addition of the N_3 isocline identified accordingly. Parameter values common to all four panels are as follows: $d_i = 0.2$, $b_{ij} = n_{jk} = w_{ik} = 0.1$, $h_{ij} = l_{jk} = u_{ik} = 0.0$, $f_j = 0.1$, $g_j = 0.0$, $x_i = 0.1$, and $y_1 = 0.0$. Panel (a) has $a_{11} = 0.8$, $a_{12} = 0.25$, $a_{13} = 0.8$, $a_{21} = 0.3$, $a_{22} = 0.16$, $a_{23} = 0.3$, $v_{11} = 0.01$, $v_{21} = 0.02$, $m_{11} = 0.6$, $m_{21} = 0.05$, and $m_{31} = 0.15$. Panel (b) has $a_{11} = 0.2$, $a_{12} = 0.3$, $a_{13} = 0.8$, $a_{21} = 0.8$, $a_{22} = 0.3$, $a_{23} = 0.2$, $v_{11} = 0.01$, $v_{21} = 0.02$, $m_{11} = 0.2$, $m_{21} = 0.1$, and $m_{31} = 0.2$. Panel (c) has $a_{11} = 0.2$, $a_{12} = 0.3$, $a_{13} = 0.4$, $a_{21} = 0.8$, $a_{22} = 0.3$, $a_{23} = 0.4$, $v_{11} = 0.01$, $v_{21} = 0.02$, $m_{11} = 0.2$, $m_{21} = 0.1$, and $m_{31} = 0.2$. Panel (d) has $a_{11} = 0.2$, $a_{12} = 0.3$, $a_{13} = 0.4$, $a_{21} = 0.8$, $a_{22} = 0.8$, $a_{23} = 0.3$, $a_{13} = 0.4$, $a_{21} = 0.8$, $a_{23} = 0.4$, $v_{11} = 0.01$, $v_{21} = 0.01$, $m_{11} = 0.15$, $m_{21} = 0.15$, and $m_{31} = 0.23$.

is intermediate in competitive ability on both resources (fig. 5.13(b)). In the final two, one consumer is the best competitor on one resource and the worst competitor on the other, and the other two consumers have the same rank competitive abilities on both resources relative to one another; but the predator imposes different patterns of relative mortality on the three consumers (fig. 5.13(c)–d)). As when three or more resources are present for consumers (i.e., fig. 3.13), multiple configurations of species abilities and performances permit coexistence. Dominant trade-offs only exist when species face only two limiting factors. With three or more limiting factors, the combinatorics of species properties that would permit coexistence also expands much faster than the number of limiting factors. This is true because each species does not need to be most limited by a different factor, as is implied when one is considering only two limiting factors (see fig. 3.13 and McPeek 2019b).

Thus, adding species at one trophic level increases the potential for increasing species richness at other trophic levels. To continue the example, if three consumers can coexist with two resources and one predator, this then opens the possibility of four more predators coexisting, because five limiting factors control the abundance of the one predator in all the situations illustrated in figure 5.13. Then, if any more predators from the pool of available colonists can invade and coexist, more consumers could potentially invade and coexist after them. Thus, *species diversity at one trophic level begets species diversity at adjacent trophic levels, both above and below.*

All of this only applies if the community will eventually approach a stable point equilibrium. The temporal variability generated by the community entering a stable limit cycle or chaotic orbit opens up additional possibilities for new species to coexist (Koch 1974b, Hsu et al. 1978b, 2015, Armstrong and McGehee 1980). In fact, in many situations, more species than limiting factors may coexist when the community displays a limit cycle or chaos (discussed in chapter 8). The same is true if species at higher trophic levels also directly limit their own abundances through various forms of interference.

With substantial species richness at all three trophic levels, identifying the specific factor or ability that permits a particular species into the community may be difficult. Remember that some may be present because those that are better at the key abilities needed at a trophic level also limit their own abundances to some degree (i.e., $g_j > 0$ and $y_k > 0$), and so permit others into the community that would not otherwise be able. However, for many, the combinations of abilities that make them successful may be readily apparent, based on the success of identifying such properties throughout the history of ecology.

However, in such diverse communities, the present analysis makes a very strong prediction about how functional diversity at one trophic level creates opportunities at other trophic levels. Moreover, adding a predator or omnivore at the top has the same effect as adding a resource species at the bottom: all species additions create the possibility of more additional species with the appropriate abilities to be able to coexist on other trophic levels.

Testing whether this is the case is then a simple proposition. In fact, these experimental tests mirror the environmental impacts of nutrient loading and overharvesting discussed in chapter 4. The critical experiment to test the circuitry of limiting factors in the food web is to inflate or depress one abiotic resource or one species and see which other species are more significantly impacted at other trophic levels. For example, imagine the discussion of the paradox of the plankton in this context. In this case, the limiting abiotic resources are at the base of our food web, with phytoplankton at the first intermediate level and zooplankton herbivores, omnivores, invertebrate predators, and fish above them. In an experiment that greatly increases the abundance of one of the abiotic nutrients, say, phosphorus, we would predict that some of the phytoplankton species may go extinct because of the apparent competitive effects on other limiting abiotic resources mediated by changes in the phytoplankton species abundances that are most limited by phosphorus (see chapter 3, "Resource Competitors"). Likewise, some of the phytoplankton species are present not because they are very successful resource competitors but because they are relatively good apparent competitors against some of the zooplankton species (e.g., Leibold 1989). Therefore, deleting specific zooplankton species should also cause extinctions among the phytoplankton species.

Exacerbating or removing potentially limiting factors removes the advantage that permits some species at other trophic levels to coexist. Each coexisting species is present because of the demographic balance it is able to strike among all the ecological interactions in which it must directly engage. However, the strengths of these direct interactions are influenced by the other species that also engage directly with these limiting factors. *This is why indirect interactions between species are so important.* Consider the consumer species in figure 5.13. In each of the scenarios illustrated there, deleting R_2 from the community would cause one or more of the three consumers to then go extinct, and those would be the consumers that most depend on R_2 for their presence (i.e., the consumers that are best at utilizing R_2). In contrast, experimentally inflate R_2's abundance, and the consumers that are more limited by R_1 will be more greatly impacted because of the apparent competitive effects mediated through the consumers that are more limited by R_2. Likewise, delete P_1 from the community, and the consumers that are best at avoiding predation from P_1 will go extinct because of changes in the abundances of the other consumers that will reduce resource levels. Or experimentally inflate P_1's abundance, and the set of consumers for which mortality from P_1 is more limiting to their coexistence will be most impacted.

Thus, sorting through the complexity of diverse food webs is far from hopeless. We simply need to think in terms of limiting factors. Moreover, understanding the causal structure of communities allows us to make a richer set of predictions about the consequences of species loss and gain. Such understanding is scientifically desirable and should have great practical importance as well in many areas of agriculture (e.g., pest management) and conservation (e.g., consequences of nutrient loading, species invasions, and predicting extinction debts due to species loss).

SUMMARY OF MAJOR INSIGHTS

- The basic omnivory/intraguild predation module provides many different configurations for the intraguild predator to coexist with a consumer and resource species because of the various resource competitive and apparent competitive relationships among the three species.
- Invasibility criteria for all species in an omnivory community module are similar to analogous modules without omnivory, except with the added complications of predation by the omnivore across trophic levels.
- Omnivory greatly increases the range of possible dynamical complexities (e.g., point equilibria, priority effects, limit cycles, chaos) in more areas of parameter space. For example, limit cycles and chaotic dynamics are possible even if all foragers have linear functional responses.
- Omnivory promotes greater diversity at higher trophic levels by increasing the number of limiting factors impinging on those upper trophic levels via their greater diet diversity. By the same token, omnivory will also promote greater diversity at lower trophic levels if it adds to the number of species that feed on that level.
- Species replacements should occur at intermediate trophic levels among communities arrayed along productivity gradients, with consumers that are better resource competitors at the lower end of the gradient changing to consumers that are better apparent competitors at the higher end of the gradient.

Mutualists, Symbionts, and Facilitators in a Food Web

Walk out onto the rolling landscape of the Konza tallgrass prairie in Kansas, USA, and you will be surrounded by mutualisms. In the summer, over 350 species of bees, flies, beetles, butterflies, and moths pollinate scores of plant species (Welti and Joern 2018). Many of the 90 grass and sedge species exchange nutrients and water with the 29 species of mycorrhizal fungi that can be found on this prairie (Hartnett and Wilson 1999, Smith et al. 1999). Approximately 40 legume species exchange nitrogen and carbon compounds with the rhizobial bacteria growing in their root nodules (Towne and Knapp 1996, Stanton-Geddes and Anderson 2011). Every animal inhabiting this landscape has an associated gut microbiome that aids in extracting nutrition from its food (Colman et al. 2012, Esposti and Romero 2017). Among other influences, these mutualistic interactions shift the competitive relationships among the hundreds of plant species and increase species abundances at higher trophic levels because of the increased plant biomass that is available to those higher trophic levels (Smith et al. 1999, Kula et al. 2004).

Swim out onto a Caribbean coral reef, and you will be surrounded by mutualisms that are even more fundamental to the survival of species in those communities. The corals that are the foundation for the reef have symbiotic relationships with various *Symbiodinium* species of dinoflagellates (zooxanthellae) (Muscatine and Porter 1977, Dubinsky and Jokiel 1994, Knowlton and Rohwer 2003), and the dissolution of this interaction can have catastrophic effects on the entire ecosystem (Glynn 1993). These dinoflagellates also have similar associations with sponges, gorgonians, clams, conchs, and benthic foraminifera (Rowan 1998, Knowlton and Rohwer 2003). Many arthropods, mollusks, chaetognaths, annelids, and fish also harbor bioluminescent dinoflagellates (Widder 2010, Sparks et al. 2014, Davis et al. 2016). Shrimp species set up stations where fish come to have ectoparasites cleaned from them (Losey et al. 1999, Côté 2000, Becker and Grutter 2004, Caves et al. 2018).

Symbiotic interactions are prominent components of all ecosystems. In fact, all eukaryotes base their metabolic existence on a fundamental mutualism that was established between an archaeal host and a bacterial endosymbiont (Margulis 1970, Gray et al. 1999).

Despite the importance and ubiquity of mutualistic interactions among species in all ecosystems, mutualisms are a neglected component of the study of interactions within communities and food webs (Thompson 1994, 2005, Palmer et al. 2015). Bronstein (2015b) provides a scholarly and comprehensive review of the history of studies into mutualisms, and this review should be read by anyone who wants to understand how a scientific discipline develops. The pity is that this "subdiscipline" within community ecology developed largely outside our larger discussion of the networks of species interactions.

Many different types of species interactions fall under the rubric of mutualism. The first of these categories are *symbiotic mutualisms*, which are defined as "those in which there is prolonged physical intimacy between partner species" (Bronstein 2015b, p. 7). Examples include associations between corals and their zooxanthellae; plants and rhizobial bacteria; mycorrhizal fungi and lichens; and animals and their gut microbiomes (Douglas 2015). Bronstein (2015a) reserves the descriptor of *mutualism* for the more general type of species interaction where long-term intimate contact does not occur but where both species receive reciprocal net benefits because of the interaction. Examples of these types of associations include plants and their pollinators and seed dispersers, sea anemones and their territorial clown fish defenders, ants protecting plants that provide them with extrafloral nectaries or domatia, and fish and their cleaner shrimp that remove ectoparasites. Bronstein (2015a, p. 7) associates the term *facilitation* with "an interaction in which the presence of one species alters the environment in a way that enhances growth, survival, or reproduction of a second, neighboring species (Callaway 2007)."

Mutualisms are often also categorized as either obligate or facultative. An *obligate mutualism* is one in which one species will not be present in the community unless its mutualist partner is present. In other words, the species cannot satisfy its invasibility criterion unless its mutualistic partner is already present in the community. A *facultative mutualism* is one in which the mutualist partners can invade and coexist in the community without the other, but both benefit from the other's presence. Because I am concerned with the consequences of mutualisms on the rest of the food web, I only consider facultative mutualisms here. However, little change is needed in the analyses presented here to turn those into obligate interactions.

Mutualisms can also be categorized based on the mechanisms of the interactions that generate benefits between the mutualist partners (Bronstein 2015b). In the first category are *transport mutualisms*, in which one partner receives a food benefit while transporting some individual life stage for its partner. Examples include bees moving pollen among individual plants or ants dispersing seeds away from parental plants. In the second category are *nutrition mutualisms*, in which the partners exchange nutrition with one another. Examples include gut microbiomes and their hosts, rhizobia and mycorrhizal associations with plants, and corals and their zooxanthellae. In the final category are *protection mutualisms*, such as acacias and their ant mutualists, sea anemones and their clown fish, and fish and their cleaner shrimp, in which one species attracts another species as a protector that deters or removes some herbivore, predator, or pathogen. I explore model representations of these three categories in this chapter.

MUTUALISMS AS CONSUMER-RESOURCE INTERACTIONS

Modeling mutualistic interactions among species has largely followed the same lines as other types of species interactions in community ecology, namely, mutualisms have been on two separate tracks. Typically, the same kinds of phenomenological models based on the generalized Volterra-Lotka-Gause competition model have been used: competition between two species is easily converted to a mutualism by simply changing the sign of the interspecific interaction coefficients from negative to positive. In fact, Gause and Witt (1935) did just that in their original paper showing that two competitors could coexist using the Gause model of competition: if the interspecific competition terms are positive, they showed how the model could also generate a mutualism. This parallel approach of using various forms of the Volterra-Lotka-Gause competition model with positive interspecific terms has formed many preconceptions about the consequences of mutualisms (Vandermeer and Boucher 1978, Goh 1979, Travis and Post 1979, Dean 1983, Wolin and Lawlor 1984, Boucher 1985, Wright 1989, Saavedra et al. 2016).

Right from the start, one limitation was apparent: what May (1981) called the *orgy of mutual benefaction*, in which mutualist partners would increase to infinite abundance if their mutual benefits were sufficiently high. The problem results from linear isoclines not intersecting (Vandermeer and Boucher 1978)—just as higher-order interactions with competition forms of these basic models were modified so that the isoclines were nonlinear but still devoid of any substantive mechanistic feature (Travis and Post 1979, Dean 1983, Wolin and Lawlor 1984, Wright 1989).

How to get mechanism into models of mutualism? While the thought of "mutualism" typically conjures ideas about altruism—the sacrifice of personal gain for the betterment of another—mutualistic interactions among species in nature typically are much more aligned with the underlying principles of a capitalist economy described by Adam Smith (1776) in *Wealth of Nations*. These interactions typically imply some form of trade between the species (Noë and Peter 1994), and are often modeled in an economic markets framework (e.g., Wyatt et al. 2014, Akçay 2015). Many mutualistic interactions do involve sacrifice on the part of one or both species, but the sacrifice is typically made to increase their own benefit in other ways. Individuals of a species give up something to individuals of the other species in order to potentially benefit in some other way, in the same way that money is exchanged for goods and services in a market economy.

In an ecological context, this trade typically involves one species consuming some biomolecular product (e.g., inorganic molecules, glucose and other sugars, amino acids and proteins, vitamins) or tissue (e.g., pollen, seeds, elaiosomes) of its mutualist partner. Almost all the interactions described in the first few paragraphs of this chapter involve one or both species consuming some material produced by its partner. As a result, many researchers have come to this realization and are now exploring the dynamics of mutualistic interactions among species as another form of consumer-resource interactions (Herre et al. 1999, Holland and DeAngelis 2009, 2010, Jones et al. 2012, Bronstein 2015b). The tissue or fluids that are consumed from one species constitute the cost incurred by that species in order to accrue some fitness benefit, be that greater seed production via pollination, the ability to consume alternative products from individuals of the other species (as with the plant-fungal associations), or the ability to recruit defenders against other more harmful consumers (as with ants foraging for herbivorous insects on the leaves and stems of plants with extrafloral nectaries).

That is not to say that all mutualisms are consumptive in nature. Others "trade" or "exploit" services (to continue the metaphor) that do not involve direct consumption. For example, foragers in mixed-species flocks of birds all benefit from mutual protection from predators via herd dynamics and mutual vigilance. Also, many facilitative effects of one species are the result of its altering local abiotic conditions in ways that boost the survival of other species (Callaway 1997, 2007, Stachowicz 2001, Bruno et al. 2003, Brooker et al. 2007, Bronstein 2009, Brooker and Callaway 2009).

In this section, I follow the framework set out by Holland and DeAngelis (2009, 2010) to develop more mechanistic models of the three mechanistic categories of mutualistic interactions. Their work develops models for two types of mutualisms involving consumer-resource interactions. They identify a *unidirectional consumer-resource interaction* as one in which one partner consumes tissue from the other, and the consumer inadvertently provides some service to the other species (e.g., pollination, seed dispersal) whose tissue it is consuming (fig. 6.1(a)). This is typically the mechanism of transport mutualisms. They identify a *bidirectional consumer-resource interaction* as one in which each mutualist partner consumes tissue from the other (i.e., nutritional mutualisms) and each may incur some benefit from this consumption (e.g., plants and mycorrhizal fungi, corals and their associated dinoflagellate zooxanthellae) (fig. 6.1(b)). I shorten these

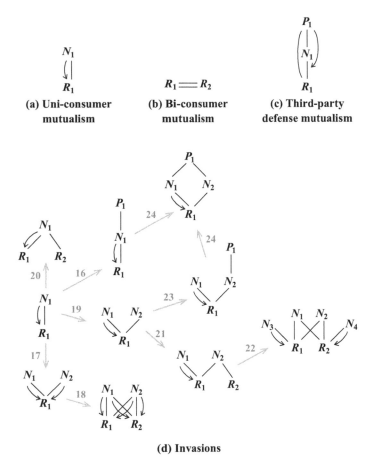

(a) Uni-consumer mutualism

(b) Bi-consumer mutualism

(c) Third-party defense mutualism

(d) Invasions

FIGURE 6.1. Module diagrams of the three basic types of mutualistic interactions. (a) A uni-consumer mutualism involves one species, the uni-consumer (N_1), which consumes some tissue or fluid of its resource species (R_1), but in the process of this consumption provides some fitness subsidy to the resource. The arrow pointing from the uni-consumer to the mutualist resource partner signifies the benefit accrued to the partner. (b) A bi-consumer mutualism involves two species that each consume material from the other, but the benefit of consuming the other species' material outweighs the cost of being consumed. The bi-consumer mutualism is signified by two consuming lines between them. (c) A third-party defense mutualism involves one species (R_1) sacrificing tissue to one consumer (P_1) to attract this species as protection to interfere with the feeding of another consumer (N_1). The arrow from P_1 to the interaction between R_1 and N_1 signifies the interaction modification that occurs because of the abundance of P_1. Panel (d) diagrams the species invasions considered in this chapter of a simple community already containing a uni-consumer and its mutualist resource partner. For each community module, R_i represents a resource species, N_j a consumer species, and P_k a predator species. A line between two species identifies a trophic link where the species above consumes the species below. Uni-consumers and their mutualist partners have an arrow between them signifying the benefit accrued to the partner. Arrows show transitions between modules that result from the invasion by different types of species. Numbers are associated with each arrow so that different types of transitions can be identified in the text.

TABLE 6.1. Summary descriptions of the additional parameters used in the models presented in this chapter

Parameter	Description
p_{ij}	Rate of interactions between species i and j that result in a fitness benefit accruing to species i in a uni-consumer mutualism
ϕ_i	Parameter defining asymptotic maximum increase in fitness for species i interacting with various uni-consumer mutualist partners (the asymptotic maximum is $1/\phi_i$)

monikers by calling them *uni-consumer* or *bi-consumer mutualisms*, respectively, to simply identify the number of species consuming tissue of a partner. I also subdivide the uni-consumer category into those in which the consumer species inadvertently provides some service directly to its partner (e.g., pollination, seed dispersal) and those in which the consumer species inadvertently provides defense against other consumer species (e.g., defense mutualisms, such as ants and acacia trees, ant-tended aphids, fish and cleaner shrimp). I call the latter *third-party defense mutualisms* to highlight the mechanistic features of the interactions (fig. 6.1(c)). Table 6.1 identifies the new parameters used in models in this chapter.

Uni-consumer Mutualism

Many bee species garner all or a substantial portion of their nutrition from foraging for nectar and pollen from flowers. In this interaction, the bee species is the consumer that is reducing a fitness component of the plant by feeding on these tissues and fluids, and its foraging greatly increases its own fitness. In addition, the foraging activity of the bees inadvertently provides a fitness benefit to the plant by increasing the fertilization rate of the plant's ovules so that the plant produces more offspring. Thus, this interaction falls squarely in the category of uni-consumer mutualism (fig. 6.1(a)).

Following the general framework of Holland and DeAngelis (2009, 2010), I describe the interaction between these two species with this basic consumer-resource model:

$$\frac{dN_1}{N_1 dt} = \frac{b_{11} a_{11} R_1}{1 + a_{11} h_{11} R_1 + z_1 N_1} - f_1 - g_1 N_1$$

$$\frac{dR_1}{R_1 dt} = c_1 - d_1 R_1 + \frac{p_{11} N_1}{1 + \phi_1 p_{11} N_1} - \frac{a_{11} N_1}{1 + a_{11} h_{11} R_1 + z_1 N_1}$$

(6.1)

First, note the general similarity between this model and the classic Rosenzweig and MacArthur (1963) model for a purely predator-prey interaction (equations (3.3) and (3.6)). The resource (e.g., plant) species R_1 follows logistic population growth in the absence of the consumer, with an intrinsic birth rate of c_1 and a density-dependent rate of fitness decrease of d_1. The consumer (e.g., bee) species N_1 has a saturating functional response for feeding on the resource, with an attack coefficient of a_{11} and handling time of h_{11}, and its numerical response resulting from this saturating functional response has a conversion efficiency of b_{11}. Revilla (2015) provides justifications for using these forms of functional responses in these types of interactions. The consumer also has an intrinsic death rate of f_1, just as in a model for a predator feeding on a prey. In the model I consider here, the consumer also may experience two forms of direct self-limitation: the density-dependent increase in its per capita death rate that occurs at rate g_1 and the feeding interference in the form of a Beddington-DeAngelis functional response with parameter z_1 (Beddington 1975, DeAngelis et al. 1975). As we will see, this direct consumer density dependence of some form is essential to this interaction being a mutualism if nothing else limits the consumer's abundance. All of these model features should be very familiar to the reader by now, because they all have identical effects in this mutualistic interaction as they did in the predator-prey forms of the interactions in chapters 3–5.

The single feature of this model that makes this species interaction potentially a mutualism is the third term of the resource equation (i.e., $\rho_{11}N_1/(1+\phi_1\rho_{11}N_1)$). This term of the model describes the resource's increase in fitness due to the interaction with the consumer and so represents the inadvertent fitness service (e.g., reproductive enhancement via pollination) provided by the consumer (Holland and DeAngelis 2009, 2010, Wang et al. 2011, 2012). This service saturates at a maximum per capita fitness benefit to the resource at high consumer abundance (Holland and DeAngelis 2009, 2010). Holland and DeAngelis used a Michaelis-Menten function to describe this fitness gain, but I use a functional form akin to Holling's (1966) type II functional response. The parameter ϕ_1 controls the level at which this per capita fitness benefit to the resource asymptotes, which is $1/\phi_1$. Each resource species is assumed to have a maximum per capita limit to this benefit. For example, in the absence of bees, some fraction of a plant's ovules will be fertilized, and the action of bees increases this fraction until all the ovules that a plant can produce are fertilized (i.e., the maximum fitness benefit). ρ_{11} is the rate at which consumer and resource interact to produce this benefit to the resource, which determines the rate at which the benefit approaches this asymptote with consumer abundance.

This fitness benefit term has important consequences for the resource isocline (fig. 6.2). The isocline functions are

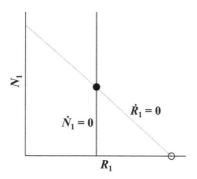

(a) $h_{11} = 0, \rho_{11} = 0, 1/\phi_{11} = 0, g_1 = 0, z_1 = 0$

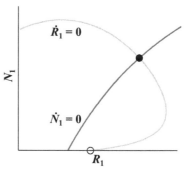

(d) $h_{11} > 0, \rho_{11} > 0, 1/\phi_{11} > 0, g_1 > 0, z_1 = 0$

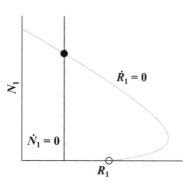

(b) $h_{11} = 0, \rho_{11} > 0, 1/\phi_{11} > 0, g_1 = 0, z_1 = 0$

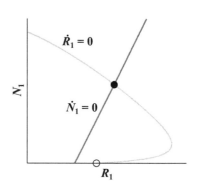

(e) $h_{11} > 0, \rho_{11} > 0, 1/\phi_{11} > 0, g_1 = 0, z_1 > 0$

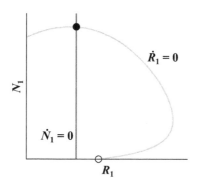

(c) $h_{11} > 0, \rho_{11} > 0, 1/\phi_{11} > 0, g_1 = 0, z_1 = 0$

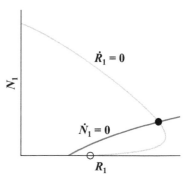

(f) $h_{11} > 0, \rho_{11} > 0, 1/\phi_{11} > 0, g_1 > 0, z_1 > 0$

$$\frac{dN_1}{dt} = 0 : R_1 = \frac{(f_1 + g_1 N_1)(1 + z_1 N_1)}{a_{11}(b_{11} - h_{11}(f_1 + g_1 N_1))}$$

$$\frac{dR_1}{dt} = 0 : -d_1 a_{11} h_{11} R_1^2 + \left(c_1 a_{11} h_{11} - d_1(1 + z_1 N_1) + \frac{a_{11} h_{11} \rho_{11} N_1}{1 + \phi_1 \rho_{11} N_1} \right) R_1$$

$$+ \left(c_1 + \frac{\rho_{11} N_1}{1 + \phi_1 \rho_{11} N_1} \right)(1 + z_1 N_1) - a_{11} N_1 = 0.$$

As in all the results explored in chapter 3, because R_1 experiences logistic population growth, its isocline intersects the R_1 axis at $R_{1(0)}^* = c_1 / d_1$, which is its equilibrium abundance in the absence of the uni-consumer. (This is, therefore, a facultative mutualism—it would be an obligate mutualism if $c_1/d_1 < 0$.) Figure 6.2(a) shows the isocline configuration for this model when N_1 has a linear functional response ($h_{11} = 0$) and provides no benefit to R_1 (i.e., $\phi_1 = \infty$, which in this case has a maximum at $1/\phi_1 = 1/\infty = 0$), which is the configuration for the basic consumer-resource interaction from chapter 3. If N_1 provides some level of benefit to R_1, the R_1 isocline remains anchored to the R_1 axis at c_1/d_1, but the fitness benefit accrual from N_1 causes the R_1 isocline to bulge into higher values of R_1 at a given level of N_1; at positive N_1 values, the fitness supplement to R_1 shifts the value of its isocline to higher values of R_1, and the point at which it intersects the N_1 axis shifts up (fig. 6.2(b)). As a result, much of the R_1 isocline has R_1 values above c_1/d_1. The portion of the R_1 isocline with R_1 values greater than c_1/d_1 is where the N_1 abundance provides greater fitness benefit through the incidental service than

FIGURE 6.2. Various isocline configurations for a uni-consumer (N_1) feeding on a single resource (R_1) to which it also provides some service that augments the resource's fitness. (a) The basic relationship in which the uni-consumer has a linear functional response does not limit its own abundance by any feeding interference or other form of direct intraspecific density dependence, and provides no fitness subsidy to the resource. In this configuration, the open circle defines the unstable equilibrium of R_1 by itself (i.e., $R_{1(0)}^*$), and the filled black circle identifies the stable two-species equilibrium (i.e., $[R_{1(N_1)}^*, N_{1(R_1)}^*]$). (b) The isocline system with all parameters the same as in (a), except that the uni-consumer now provides a fitness subsidy to the resource. (c) All parameters as in (b), except that the uni-consumer now has a saturating functional response. (d) All parameters as in (c), except that the uni-consumer now limits its own abundance via some form of direct intraspecific density dependence. (e) All parameters as in (c), except that the uni-consumer now limits its own abundance via feeding interference. (f) All parameters as in (c), except that the uni-consumer now limits its own abundance via both feeding interference and some other form of direct intraspecific density dependence. Note that this interaction can only be considered a mutualism for the parameter combinations in (d)–(f), because $R_{1(N_1)}^* > R_{1(0)}^*$ in these panels, but $R_{1(0)}^* > R_{1(N_1)}^*$ in panels (a)–(c). The following parameters are used, except as otherwise indicated: $c_1 = 1.0$, $d_1 = 0.02$, $a_{11} = 0.25$, $b_{11} = 0.1$, $h_{11} = 0.05$, $\rho_{11} = 1.5$, $\phi_1 = 0.5$, $f_1 = 0.6$, $g_1 = 0.05$, and $z_1 = 0.05$.

harm through consumption, and the portion with R_1 values less than c_1/d_1 have the harm from consumption greater than the benefit through the service. As in predator-prey forms of this model, the saturated feeding by the consumer ($h_{11} > 0$) causes the R_1 isocline to be concave up as well (fig. 6.2(c)–(f)).

The interaction is considered a mutualism if the equilibrium abundances of both species are higher in the presence of the other as compared to the other's absence (Holland and DeAngelis 2009, Bronstein 2015b). Obviously, this is always true for N_1 because it consumes R_1, but this is not necessarily true for R_1. As in all the predator-prey forms of these models, the N_1 isocline intersects the R_1 axis at $f_1/(a_{11} (b_{11} - f_1 h_{11}))$, and the invasibility criterion for N_1 still demands that the N_1 isocline intersect the R_1 axis below where the R_1 isocline intersects this axis (i.e., $f_1/(a_{11} (b_{11} - f_1 h_{11})) < c_1/d_1$) (chapter 3). If the N_1 isocline is a vertical line because only indirect resource limitation limits the consumer's abundance (i.e., $g_1 = 0$ and $z_1 = 0$), this interaction is not a mutualism, because R_1's equilibrium abundance will be lower than it would be in the absence of the consumer (fig. 6.2(b)–(c)). In this case, all the fitness benefit afforded to R_1 immediately rebounds to increasing N_1—an interesting indirect self-benefit! In other words, this community module would equilibrate at a consumer abundance at which the benefits to the resource of interacting with the consumer do not outweigh the costs of this interaction, when measured in the consequence to its abundance. Biologically, in this case, the consumer would simply be considered another herbivore or parasite feeding on a specialized tissue or fluid of the plant, even though it had some positive contribution to a fitness component of the plant.

The uni-consumer mutualism simulation module can be accessed at https://mechanismsofcoexistence.com/mutualistsUni_2D_1R_1N.html.

A locally stable equilibrium can exist where R_1's abundance is inflated by the presence of N_1 even if the N_1 isocline is a vertical line, but in this case N_1 cannot successfully invade the community when rare (i.e., the N_1 isocline intersects the R_1 axis above where the R_1 isocline does) (fig. 6.3). The community would only track to the two-species locally stable equilibrium if the consumer begins at a sufficiently high abundance. In this case, the alternative stable states consist of the two species present, or only the resource being present if the consumer's abundance falls below the separatrix dividing the domains of attraction for these two locally stable equilibria.

To be a mutualism, some other process must also contribute to limiting N_1's abundance in addition to the indirect resource limitation afforded by consuming the resource. If only these two species are present, this would mean that the consumer would have to directly limit its own abundance through feeding interference ($z_1 > 0$; fig. 6.2(e)) or some other form of direct self-limitation that either decreases its per capita birth rate or increases its per capita death rate with increases in its

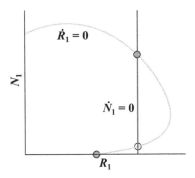

FIGURE 6.3. The isocline configuration for a uni-consumer (N_1) feeding on a single resource (R_1) to which it also provides some service that augments the resource's fitness. However, the N_1 isocline intersects the R_1 axis above R_1's equilibrium abundance in N_1's absence. This creates two locally stable equilibria (filled gray circles) and an unstable equilibrium (open circle) that lies on the separatrix boundary between the domains of attraction for these two equilibria. As a consequence, N_1 cannot invade a community containing only R_1. Parameters are as in figure 6.2, except $f_1 = 1.0$, $g_1 = 0$, and $z_1 = 0$.

own abundance ($g_1 > 0$; fig. 6.2(d)), or both (fig. 6.2(f)). These sources of direct self-limitation cause the N_1 isocline to bend away from the N_1 axis and intersect the R_1 axis so that it can invade, but also to intersect the R_1 isocline at a point that results in a stable equilibrium, where its benefit to the resource outweighs the harm due to feeding to permit the resource to equilibrate at an increased abundance (figs. 6.2(d)–(f)). Clearly, in each case, N_1's abundance is also higher because of the fitness supplement to R_1 (i.e., the shift in the R_1 isocline causes the isoclines' intersection point to move to the higher values for both species).

A specialist predator feeding on the consumer (chapter 4) or a pathogen of the consumer (chapter 7) will have a comparable effect to direct density dependence in causing the consumer's isocline to tilt in a way that potentially permits this interaction to be a mutualism (invasion 16 in fig. 6.1(d)). Thus, as in chapter 4, adding a predator that feeds only on the consumer to the uni-consumer model (equations (6.1)) results in the following model:

$$\frac{dP_1}{P_1 dt} = \frac{n_{11} m_{11} N_1}{1 + m_{11} l_{11} N_1} - x_1 - y_1 P_1$$

$$\frac{dN_1}{N_1 dt} = \frac{b_{11} a_{11} R_1}{1 + a_{11} h_{11} R_1 + z_1 N_1} - \frac{m_{11} P_1}{1 + m_{11} l_{11} N_1} - f_1 - g_1 N_1. \tag{6.2}$$

$$\frac{dR_1}{R_1 dt} = c_1 - d_1 R_1 + \frac{\rho_{11} N_1}{1 + \phi_1 \rho_{11} N_1} - \frac{a_{11} N_1}{1 + a_{11} h_{11} R_1 + z_1 N_1}$$

All the parameters are as defined in chapters 3 and 4.

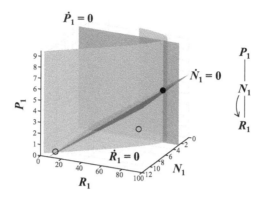

FIGURE 6.4. The isocline configuration for a uni-consumer (N_1) feeding on a single resource (R_1) to which it also provides some service that augments the resource's fitness, and the uni-consumer is itself fed upon by a predator (P_1). The R_1 isocline is identified as $\dot{R}_1 = 0$, the N_1 isocline as $\dot{N}_1 = 0$, and the P_1 isocline as $\dot{P}_1 = 0$. The parameters are as follows: $c_1 = 1.0$, $d_1 = 0.02$, $a_{11} = 0.25$, $b_{11} = 0.1$, $h_{11} = 0$, $\rho_{11} = 1.5$, $\phi_1 = 0.5$, $f_1 = 0.2$, $g_1 = 0$, $z_1 = 0$, $m_{11} = 0.4$, $n_{11} = 0.1$, $l_{11} = 0$, $x_1 = 0.1$, and $y_1 = 0$.

Compare this isocline system for the uni-consumer mutualism model (fig. 6.4) to that of a simple three-trophic-level food chain (see fig. 4.2(a)) to see the effects of adding the mutualism (i.e., $\rho_{11} > 0$ and $\phi_1 > 0$). As in the simple food chain, population regulation imposed by P_1 on N_1 causes the N_1 isocline to tilt away from the P_1 axis. The P_1 isocline intersects the N_1 axis (remember, this occurs at $x_1/(m_{11}$ $(n_{11} - x_1 l_{11}))$ and is parallel to the R_1 axis with $y_1 = 0$: if it intersects the R_1 isocline in the bulge that extends beyond c_1/d_1 (as in fig. 6.4), the three species will equilibrate such that R_1's abundance is above where it would be in the absence of N_1, making the interaction between R_1 and N_1 a mutualism. In this case, N_1's abundance does not increase as a consequence of the benefit, since its abundance is fixed by the P_1 isocline: P_1 is the beneficiary of N_1's fitness subsidy to R_1. Only if the enemy limiting N_1's abundance is itself limited by more than simply feeding on N_1 (e.g., P_1 limits its own abundance via $y_1 > 0$, experiences its own feeding interference, or has an enemy feeding on it) will N_1 increase as a consequence of the benefit it provides to R_1.

Demonstrating that this interaction is a mutualism requires more than simply identifying possible fitness benefits. For example, pollinators offer plants clear fitness benefits, but pollen may still be limiting on the stigma of flowers even though pollinators routinely visit (Burd 1994). Conversely, pollen limitation does not mean that pollinators are not inflating a plant's abundance by visiting. The fitness benefit term in the resource equation is the mechanistic representation of such fitness benefits. Consider a basic experiment done to quantify pollen limitation using three treatments: (1) pollinators are prevented from visiting flowers, (2) pollen is artificially added to other flowers in excess, and (3) control flowers are open to the natural pollinator community (e.g., Knight 2003, Urbanowicz et al. 2018). The level of seed fertilization in treatment (1) quantifies the baseline of this fitness component if the mutualist partner were absent, whereas the level of seed fertilization in treatment (2) quantifies the asymptotic fitness benefit (i.e., akin to $1/\phi_1$). The difference between treatments (2) and (3) quantifies the level of pollen limitation, which is how far this fitness benefit is from the asymptote, and the difference between treatments (1) and (3) quantifies the fitness benefit afforded to the plant by the current level of pollinators.

One must realize, though, that quantifying a substantial fitness benefit for the resource from interacting with the uni-consumer mutualist here does not establish that this interaction is a mutualism. That conclusion is only justified if the resource's equilibrium abundance is higher when the uni-consumer is present than when it is absent (e.g., fig. 6.2(a)–(c) versus 6.2(d)–(f)), which simply requires quantifying the abundance of the resource in the presence and absence of the uni-consumer.

A test of invasibility for the uni-consumer mutualist is an even more critical issue here, since a locally stable equilibrium for the uni-consumer can exist even if it cannot invade the community if only the resource in question were present (as in fig. 6.3). For example, an additional resource species may be present in the community that also inflates the uni-consumer's per capita growth rate when rare that permits its invasion, even though it could not support a population solely on its mutualist partner.

Finally, one of the critical predictions to emerge from this particular model is that some other process must limit the abundance of the uni-consumer in order for the resource abundance to be inflated by the fitness benefit. Otherwise, the uni-consumer abundance should increase to a level where the fitness cost to the resource of consumption outweighs the fitness benefit (e.g., fig. 6.2(a)–(c) versus 6.2(d)–(f)). Studies that identify those other limiting factors are therefore critical to demonstrating the complete mechanism of this interaction. Also note that these alternative limiting factors must be something that depresses the uni-consumer's abundance. For example, the uni-consumer having an alternative resource will not cause its isocline to bend away from its own axis (see fig. 3.5), and so would not contribute to its interaction with the first resource being a mutualism. The limiting factor must be either some form of direct self-limitation or some enemy (i.e., predator or disease).

Bi-consumer Mutualism

The second type of mutualism considered here involves interactions where each species is a consumer of its interaction partner. Most all of these fall squarely in Bronstein's (2015b) nutritional category. They include such classic examples as plants interacting with mycorrhizal fungi or nodule-forming rhizobial bacteria, corals and giant clams interacting with their dinoflagellate zooxanthellae, and our own interactions with our gut microbiome. Some would be considered mutualisms (e.g., plants and mycorrhizal fungi), while others are symbioses (e.g., corals and zooxanthellae). In all of these interactions, each interaction partner is consuming some material from its partner, typically in the form of small biochemical molecules. For example, rhizobial bacteria consume carbohydrates from their plant partners, and the plants consume the small nitrogen-containing molecules that the rhizobial bacteria produce from their fixing atmospheric molecular nitrogen (Poole et al. 2018). Therefore, no extra benefit or service must be hypothesized beyond the dynamics of consumption of these materials from the partner.

Following the model developed by Holland and DeAngelis (2009, 2010), consider two consumers R_1 and R_2 that each display logistic growth in the absence of the other, and each consumes parts of the other according to a standard saturating functional response. The model is then

$$\begin{aligned}
\frac{dR_1}{R_1 dt} &= c_1 - d_1 R_1 + \frac{b_{21} a_{21} R_2}{1 + a_{21} h_{21} R_2} - \frac{a_{12} R_2}{1 + a_{12} h_{12} R_1} \\
\frac{dR_2}{R_2 dt} &= c_2 - d_2 R_2 + \frac{b_{12} a_{12} R_1}{1 + a_{12} h_{12} R_1} - \frac{a_{21} R_1}{1 + a_{21} h_{21} R_2}
\end{aligned} \tag{6.3}$$

Holland and DeAngelis (2009, 2010) use Michaelis-Menten functions for the saturating functional responses, but I use Holling's formulation to allow comparison with the consumer-resource models of chapter 3 and to easily allow me to compare the effects of linear versus saturating functional responses in one set of equations (see also Revilla 2015). All the same parameter interpretations as in chapter 3 hold here. The functional and numerical responses in this case should be considered augmentations or deficits to the per capita fitness of each because of the interaction with the other. For brevity of presentation, I assume that both species can support populations in the absence of the other (i.e., both $c_i > 0$), which means they would be considered facultative mutualists. I leave it to the reader to explore what is possible if one or both species were obligate mutualists (i.e., $c_i < 0$).

Basic insights are gained by first considering this interaction when both functional responses are linear (i.e., $h_{12} = h_{21} = 0$). In this case, the isocline of each resource is a line that intersects its own axis at $R_{i(0)}^* = c_i/d_i$ (i.e., its equilibrium abundance in the absence of the other species) and intersects the other resource's axis at $c_i/(a_{ij} - b_{ji}a_{ji})$ (in this section, the resource in question is subscripted i and the other is subscripted j). Many possible relationships between these two linear isoclines exist, depending on the strengths of the interaction (a_{ij}) and the fitness benefits that each species (b_{ij}) garners from this interaction. I consider four here.

The bi-consumer mutualism simulation module can be accessed at https://mechanismsofcoexistence.com/mutualistsBi_2D_2R.html.

First, a feasible two-species equilibrium will exist if each species' isocline intersects the other resource's axis at a positive value that is below $R^*_{i(0)}$, the equilibrium value of the other resource in its absence (fig. 6.5(a)). This relationship between the isoclines is true if

$$a_{ji} > a_{ij}b_{ij} \text{ and } 0 < \frac{c_j}{a_{ji} - a_{ij}b_{ij}} < \frac{c_i}{d_i}$$

for both species. The first set of inequalities $(a_{ji} > a_{ij}b_{ij})$ ensures that the isocline of species j intersects the axis of species i at a positive value: this inequality implies that each species is harmed more by losing the material to the other species than it is benefited by consuming the material of the other species. In other words, the mutual consumer-resource interaction has a net fitness decrement for each species. The second set of inequalities implies that this fitness decrement is substantial relative to its intrinsic growth rate. This results in two locally stable equilibria with only one of the consumers present, and the feasible two-species equilibrium is unstable (fig. 6.5(a)). This situation is analogous to the unstable Volterra-Lotka-Gause competition outcome in which the strength of interspecific competition is greater than intraspecific competition for both species (inequality (2.10), and see Gause and Witt 1935, Slobodkin 1961). With these relationships, neither resource could invade a community in which the other were present at its equilibrium abundance—a priority effect.

If the benefit of feeding on the other is high enough to have the isocline of the other species intersect each resource's axis above $R^*_{i(0)}$ but still have the consumer interactions be a net fitness cost,

$$a_{ji} > a_{ij}b_{ij} \text{ and } 0 < \frac{c_i}{d_i} < \frac{c_j}{a_{ji} - a_{ij}b_{ij}},$$

each resource will be able to invade a community containing the other, and the two will coexist at a stable equilibrium (fig. 6.5(b)). This situation is analogous to the stable Volterra-Lotka-Gause competition outcome in which the strength of intraspecific competition is greater than interspecific competition for both species (inequality (2.11), and see Gause and Witt 1935, Slobodkin 1961). However, the abundance of each species when coexisting with the other is less than their

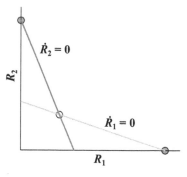

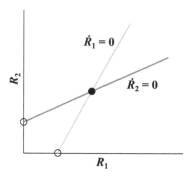

(a) $a_{12} = a_{21} = 0.3$, $b_{12} = b_{21} = 0.1$,
$h_{12} = h_{21} = 0$

(c) $a_{12} = a_{21} = 0.1$, $b_{12} = b_{21} = 1.5$,
$h_{12} = h_{21} = 0$

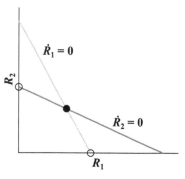

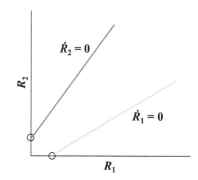

(b) $a_{12} = a_{21} = 0.1$, $b_{12} = b_{21} = 0.5$,
$h_{12} = h_{21} = 0$

(d) $a_{12} = a_{21} = 0.1$, $b_{12} = b_{21} = 2.5$,
$h_{12} = h_{21} = 0$

FIGURE 6.5. Various isocline configurations for the interaction between two bi-consumers (R_1 and R_2) with linear functional responses that feed on one another. (a) In this parameter configuration, the cost of being fed upon by the other consumer greatly outweighs the benefit of feeding on the other bi-consumer. This produces two locally stable equilibria ($[R^*_{1(0)}, 0]$ and $[0, R^*_{2(0)}]$) that are separated by the unstable two-species equilibrium (i.e., $[R^*_{1(R_2)}, R^*_{2(R_1)}]$) that lies on the separatrix that defines the domains of attraction to the two stable equilibria. (b) If the benefit of feeding on the other species is greater than the cost of being fed upon but not substantially so, the two bi-consumers coexist at a stable two-species equilibrium $[R^*_{1(R_2)}, R^*_{2(R_1)}]$. However, in this case, the two species would each be considered parasites of the other, since neither has $R^*_{i(R_j)} > R^*_{i(0)}$. (c) If the benefit of feeding on the other species is substantially larger than the cost of being fed upon, and the two isoclines still intersect at positive abundances for both species, this interaction will be a mutualism for one or both species. (d) If the benefit from feeding on the other species is too great relative to the cost, the orgy of mutual benefaction will result. The parameters $c_1 = 1.0$, $d_1 = 0.1$ are used in all panels; parameters that differ are given beneath each panel. Open circles identify unstable equilibria, filled gray circles identify the presence of multiple locally stable equilibria that create a priority effect based on initial conditions, and filled black circles identify globally stable equilibria.

abundances alone (i.e., $R^*_{i(0)} > R^*_{i(R_j)}$). Thus, this set of parameter values results in a reciprocally parasitic interaction between the two species (Holland and DeAngelis 2009).

For the species to coexist as mutualists, the benefit of feeding on the other must be higher than the cost of being fed upon by the other species for both species. The consequence of making the benefit greater than the cost is that each species' isocline now intersects the other species' axis at a negative value:

$$a_{ji} < a_{ij}b_{ij}, \text{ so } \frac{c_j}{a_{ji} - a_{ij}b_{ij}} < 0 < \frac{c_i}{d_i}.$$

If the isoclines still intersect at positive abundances to produce a feasible two-species equilibrium, both species can invade and coexist (fig. 6.5(c)). Moreover, their abundances at this equilibrium are now greater than when alone (i.e., $R^*_{i(R_j)} > R^*_{i(0)}$), showing that this interaction under these circumstances is a mutualism. If the benefits are substantially greater than the costs with linear functional responses, the orgy of mutual benefaction will occur when the slopes of the lines diverge enough that their isoclines no longer intersect at positive abundances (fig. 6.5(d)).

If the consumers have saturating functional responses (i.e., $h_{12} > 0$, $h_{21} > 0$), the same basic issues hold. With saturating functional responses, each isocline still intersects the species' own axis at $R^*_{i(0)} = c_i/d_i$, but the exact point where each isocline intersects the other species' axis is now a complicated equation of quadratic and quotient terms in both abundance variables, which is not very useful at generating analytical insights. However, changes in the benefits of feeding on the other consumer relative to the costs of being fed upon by the other consumer shifts these two points in similar ways to the above descriptions with linear functional responses. With the benefit of feeding on the other resource much lower than the cost of being fed upon, neither species can invade the community if the other is present. Thus, only one species will be present, depending on which species is first to invade (fig. 6.6(a))—again, a priority effect.

With the benefit lower than but near the cost for each species, the two isoclines may intersect at three different points of positive abundance for the two species, giving three locally stable and two unstable equilibria (fig. 6.6(b)) (Holland and DeAngelis 2009, 2010). One of these locally stable equilibria has both species present. However, neither of these species can invade the community when rare and the other is present because the equilibria with only one species present are themselves locally stable. In this case, where the community finally arrives will depend on which basin of attraction the community starts.

The two species will only coexist at a globally stable two-species equilibrium when the benefit for both species of feeding on the other species is greater than the

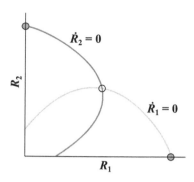

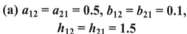

(a) $a_{12} = a_{21} = 0.5$, $b_{12} = b_{21} = 0.1$,
$h_{12} = h_{21} = 1.5$

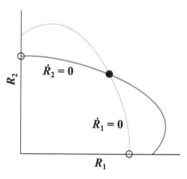

(c) $a_{12} = a_{21} = 0.1$, $b_{12} = b_{21} = 0.5$,
$h_{12} = h_{21} = 1.5$

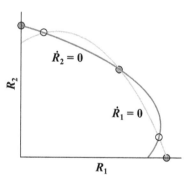

(b) $a_{12} = a_{21} = 0.12$, $b_{12} = b_{21} = 0.1$,
$h_{12} = h_{21} = 1.5$

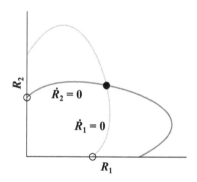

(d) $a_{12} = a_{21} = 0.1$, $b_{12} = b_{21} = 1.5$,
$h_{12} = h_{21} = 1.5$

FIGURE 6.6. Various isocline configurations for the interaction between two bi-consumers (R_1 and R_2) with saturating functional responses that feed on one another. (a) The isocline configuration that results when the cost of being fed upon is substantially greater than the benefits of feeding on the partner. As with linear functional responses (see fig. 6.5(a)), each consumer by itself is a locally stable equilibrium (filled gray circles) with an unstable two-species equilibrium (open circle) that lies on the separatrix between the two domains of attraction. (b) If the cost of being fed upon is only slightly higher than the benefits of feeding on the partner, three locally stable and two unstable equilibria occur. However, in this case, neither species can increase when rare. (c) If the benefits of feeding on the partner are somewhat greater than the costs of being fed upon, a stable two-species equilibrium emerges. However, the two bi-consumers are each parasites of the other, since neither has $R^{*}_{i(R_j)} > R^{*}_{i(0)}$. (d) Finally, when the benefits are substantially greater than the costs, the interaction can be a mutualism. The filled black circles in (c) and (d) identify stable equilibria. The parameters $c_1 = 1.0$, $d_1 = 0.1$ are used in all panels; parameters that differ are given beneath each panel.

cost of being fed upon by the other species (fig. 6.6(c)–(d)). Moreover, this inter-action is a reciprocal parasitism if the benefit is only somewhat larger than the cost for each (fig. 6.6(c)). The interaction only becomes a mutualism if the benefit is substantially larger than the cost for each (fig. 6.6(d)). Again, the benefits must outweigh the costs if the two species are to be able to jointly satisfy their invasibil-ity criteria and coexist, and these relative benefits must be substantial if they are to be mutualists.

The first issue to resolve for any type of bi-consumer mutualism is what mate-rials are traded between species. This information provides key insights about the critical variables to measure about this trade. For example, plants typically pro-vide carbohydrates to their mycorrhizal fungi partners in exchange for nitrate and phosphate compounds (e.g., Kiers et al. 2011). By knowing the commodities that are important in mediating this interaction, measures of these exchanges in other studies will verify that responses are influenced by this exchange.

As with the uni-directional mechanism, this interaction can shift between being a mutualism, in which both species benefit, and a parasitism, where one or more species are diminished depending on the environmental circumstances in which the interaction takes place (figs. 6.5–6.6) (Holland and DeAngelis 2010). Therefore, a critical issue to test (and not assume) is whether the interaction results in an increase in abundance for both species. This would be tested by a simple experiment in which each species is present in the community with and without its mutualist partner, and evaluating their abundances. Following on from this, studies that identify the environmental factors that shift the balance of trade from mutualism to parasitism will permit the prediction of subsequent community effects because of the change in this interaction.

Third-Party Defense Mutualism

The final type of mutualism to be considered here are species interactions in which individuals of one species recruit individuals of a predator to spend time near them in order to defend them from attacks by other consumers. Some of the most iconic mutualisms fall into this category. Three species of acacia trees recruit ants to live on them by secreting nectar from specialized structures on their stems, called extrafloral nectaries, and providing homes for the ants (i.e., domatia) (Palmer et al. 2008). These ants protect the trees by deterring the foraging activities of various herbivores (including elephants!) (Palmer et al. 2008). Plants with extrafloral nec-taries that presumably attract ant protectors are scattered across the angiosperm phylogeny and are found throughout the tropics (Weber and Keeler 2013).

Some of the more cryptic mutualisms also fall into this category. For example, many crayfish species have associated branchiobdellidan annelids (Brown et al. 2002, 2012, Skelton et al. 2013, Thomas et al. 2016). The branchiobdellidans forage on the bacteria, algae, and protozoans that grow on the exoskeleton and gills of the crayfish, and in fact will only reproduce while on a crayfish (Creed et al. 2015). Cleaning of these fouling organisms from the gills and exoskeleton typically increases the growth rate and survival of the crayfish hosts, but the annelids can generate fitness costs for crayfish when at high densities by consuming gill tissue (Brown et al. 2002, 2012, Thomas et al. 2016). Another surprising mutualism is found in the rocky intertidal of southern New England, where the red alga *Chondrus crispus* benefits from its association with two snail species (Stachowicz and Whitlatch 2005). In areas lacking both snails, *Chondrus* are overgrown by fouling organisms, such as ascidians and bryozoans, but grazing by the two snail species on the early life stages of these fouling species prevents overgrowth. The snails also benefit from this interaction because large growths of *Chondrus* form refuges from their crab predators (Stachowicz and Whitlatch 2005).

The basic structure of a third-party defense mutualism is the intraguild predation community module from chapter 5. Imagine a plant that is fed upon by some herbivore consumer, but the plant also produces nectar in extrafloral nectaries to attract ants. These ants consume the herbivorous insect (or elephants, if you would prefer to think about the ant-acacia system) and interfere with the feeding by the herbivorous insect because of their mere presence. In this model, the ant feeds on the herbivorous insect directly and consumes nectar from the plant, which makes this an intraguild predation module: the plant is the resource in the module, the herbivore is the intraguild consumer, and the ant is the intraguild predator.

The additional feature that enhances a defense mutualism is that the presence of the intraguild predator can interfere with the feeding rate of the herbivore on the plant. A simple way to model such interference is with the intraguild consumer having a Beddington-DeAngelis functional response (Beddington 1975, DeAngelis et al. 1975) that depends on the abundance of the intraguild predator instead of its own abundance:

$$\frac{dP_1}{P_1 dt} = \frac{w_{11}v_{11}R_1 + n_{11}m_{11}N_1}{1 + v_{11}u_{11}R_1 + m_{11}l_{11}N_1} - x_1 - y_1 P_1$$

$$\frac{dN_1}{N_1 dt} = \frac{b_{11}a_{11}R_1}{1 + a_{11}h_{11}R_1 + z_{11}P_1} - \frac{m_{11}P_1}{1 + v_{11}u_{11}R_1 + m_{11}l_{11}N_1} - f_1 - g_1 N_1. \qquad (6.4)$$

$$\frac{dR_1}{R_1 dt} = c_1 - d_1 R_1 - \frac{a_{11}N_1}{1 + a_{11}h_{11}R_1 + z_{11}P_1} - \frac{v_{11}P_1}{1 + v_{11}u_{11}R_1 + m_{11}l_{11}N_1}$$

This model of three interacting species is almost identical to the intraguild preda-tion model of equations (5.1) considered in chapter 5. The only difference is the $z_{11}P_1$ term in the denominator of the consumer's functional response: z_{11} has a double subscript in this case to identify this as the parameter associated with the effect of P_1 on N_1 and not N_1 on itself. This term models feeding interference in the consumer that reduces its feeding rate on the resource as the abundance of the predator increases: this is the feeding deterrence that benefits the plant by attracting the predator to forage for consumers on and near it. The functional response of the predator feeding on the resource describes its consumption of the attracting material (e.g., nectar from extrafloral nectaries).

Figure 6.7 illustrates the consequences of this interspecific feeding interfer-ence to the geometry of the isoclines for this community module and so how it changes the equilibrium abundances of these species. With linear functional responses (i.e., $h_{11} = u_{11} = l_{11} = 0$) and no feeding interference of the consumer by the predator (i.e., $z_{11} = 0$), this is a three-species intraguild predation module in which the isoclines of all species are planes (e.g., fig. 5.2(a) versus fig. 6.7(a)). For the parameter values used in figure 6.7(a), R_1 (the plant) equilibrates at an abun-dance of $R_1^* = 6.0$ in the presence of N_1 (the herbivore) but with no P_1 (i.e., no ants) (i.e., the point of intersection between the N_1 and R_1 isoclines in the $N_1 - R_1$ face identified with an open circle). In the presence of P_1 but still with no feeding inter-ference, R_1's abundance increases to $R_1^* = 13.7$ (i.e., the filled black circle). Thus, the presence of P_1 even with no added feeding interference between N_1 and P_1 has a net benefit on the plant because it decreases N_1's abundance.

Adding feeding interference to the interaction between N_1 and P_1 (i.e., $z_{11} > 0$ in equations (6.4)) alters the shapes of the R_1 and N_1 isoclines. First, remember that the R_1 isocline is a plane that intersects all three axes in the basic intraguild predation module with all linear functional responses. Adding consumer feeding interference caused by the predator's abundance does not change the points where the R_1 isocline intersects the three axes, but it does cause the R_1 isocline to gener-ate a large bow: think of the R_1 isocline as now being shaped like a triangular sail of a sailboat that is taut in its connections between the anchor points on the P_1 and R_1 axes, taut between the N_1 and R_1 axes, and bowing caused by the looseness of the sail in the $P_1 - N_1$ face (the right panel in fig. 6.7(b) shows the full extent of the R_1 isocline in the three-dimensional space, and the left panel shows only the very bottom of the space around the three-species equilibrium). As a consequence, at low abundances of P_1, the R_1 isocline bows away from the P_1 axis.

The third-party defense mutualism simulation module can be accessed at https://mechanismsofcoexistence.com/mutualists3PD_3D_1R_1N_1P.html.

Likewise, while the N_1 isocline already tilts away as a plane from the P_1 axis in the absence of such interference, $z_{11} > 0$ generates a curvature in the N_1 isocline

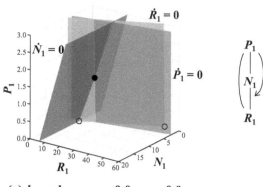

(a) $h_{11} = l_{11} = u_{11} = 0.0$, $z_{11} = 0.0$

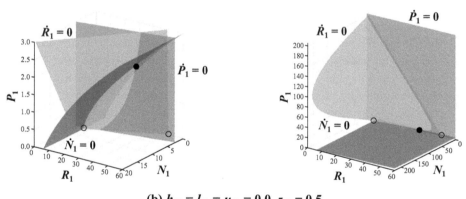

(b) $h_{11} = l_{11} = u_{11} = 0.0$, $z_{11} = 0.5$

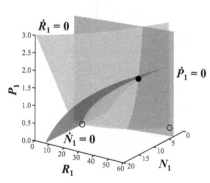

(c) $h_{11} = l_{11} = u_{11} = 0.1$, $z_{11} = 0.5$

that causes it to be even farther from the P_1 axis as P_1's abundance increases. As a result of these changes in isocline shapes, feeding interference causes R_1 to equilibrate at an even higher abundance ($R_1^* = 36.25$ in the example shown in fig. 6.7(b)). If the predator and consumer have saturating functional responses, the resource abundance will equilibrate at an even higher value for a given level of feeding interference both because the N_1 isocline intersects the R_1 axis at a higher value (see chapter 4 and 5) and because the curvature away from the P_1 axis is accentuated (fig. 6.7(c)).

Coexistence of these three species involves the same invasibility criteria as with simple intraguild predation (chapter 5). This is because the points of intersection among the three isoclines and the three abundance axes are not altered by the consumer's feeding interference in response to the intraguild predator. Thus, the addition of the mechanistic feature that permits the mutualism has no effect on whether these species can invade and coexist. Feeding interference only changes the location of their equilibrium abundances by altering the shapes of the resource and consumer isoclines away from the axes. This model of mutualism via feeding interference illustrates how third-party defense mutualisms embody the proverb that "the enemy of my enemy is my friend." For the plant that is attracting the protector, its abundance increases with higher levels of interspecific feeding interference imposed on its enemy by its enemy's enemy.

Many of the same types of studies for the basic intraguild predation module in chapter 5 are relevant here. The additional feature of the interaction that merits scrutiny here is the mechanism by which the benefiting species attracts the defense mutualist and how the defense mutualist interferes with the consumer. With a simple intraguild predation module, P_1 typically can be considered a mutualist with the resource given that P_1 must be the poorer resource competitor to coexist

---

FIGURE 6.7. Various isocline configurations for interactions among the species involved in a third-party defense mutualism. (a) The isocline relationship for a parameter combination producing a three-species stable equilibrium, but without any feeding interference in N_1 generated by the abundance of P_1. This is simply a basic intraguild predation module. (a) Two views on the isoclines when N_1 experiences feeding interference generated by the abundance of P_1, and both the consumer and the predator have linear functional responses. The panel on the right shows the full extent of the R_1 isocline, and the left panel zooms in to view the isocline relationships at the very bottom on the P_1 axis. (c) The situation using the same parameters as in (b), except that both the consumer and the predator have saturating functional responses. The isoclines are identified as in figure 6.4. The parameters are as follows, except as noted with each panel: $c_1 = 2.0$, $d_1 = 0.04$, $a_{11} = 0.25$, $b_{11} = 0.1$, $f_1 = 0.15$, $g_1 = 0$, $z_1 = 0$, $v_{11} = 0.01$, $w_{11} = 0.1$, $m_{11} = 0.15$, $n_{11} = 0.1$, $l_{11} = 0$, $x_1 = 0.1$, and $y_1 = 0$.

with N_1, and the presence of P_1 typically increases R_1's abundance because it decreases N_1. Therefore, to call such an interaction a third-party defense mutualism, R_1 individuals should in some way attract P_1 individuals to their vicinity, and those P_1 individuals must decrease feeding by N_1 on R_1 through either interference (e.g., Palmer et al. 2008) or direct killing (e.g., Brown et al. 2002, Stachowicz and Whitlatch 2005, Thomas et al. 2016).

MUTUALISTS IN FOOD WEBS

For whatever reason, the study of consumer-resource interactions to build food webs and the study of mutualistic interactions have generally operated on independent tracks. As I said in chapter 1, I believe this has occurred primarily because the focus on Volterra-Lotka-Gause competition as the framework for understanding coexistence orphans species interactions like mutualisms, symbioses, and facilitations outside the general theoretical landscape of communities as food webs. Even without the insight that most mutualisms involve the act of consuming material from another individual (Holland and DeAngelis 2009, 2010, Jones et al. 2012, Wang et al. 2012), mutualistic species interactions occur among species that are also involved in simple consumer-resource interactions with many other species. Thus, the effects of mutualisms also indirectly affect the outcomes of interactions in other parts of the food web.

In this section, I explore some of these indirect effects that flow from mutualistic interactions to the rest of the food web. To keep this topic manageable, I focus on uni-consumer mutualisms embedded in larger community modules. I hope this chapter will spark a greater exploration of mutualisms embedded in larger food webs and thereby remove this artificial divide between those who study mutualisms and those who study consumers and resources.

Multiple Uni-consumer Mutualists and Resources

A plant species is typically visited by more than one pollinator, and pollinator species typically forage on multiple plant species (Bascompte et al. 2003, 2006, Jordano et al. 2003). Given these feeding relationships, the natural question arises about how many pollinator species can be supported by their resources, just as was considered in chapter 3 for resource competitors (invasion 17 in fig. 6.1(d)). Ample evidence exists to infer competition among mutualists such as pollinators for access to the rewards supplied by their partners (Palmer et al. 2003). The augmentation of plant abundance by the actions of their mutualist partners also raises the possibility that the actions of one mutualist partner might facilitate the invasion and coexistence of other mutualist partners that would be unable to inhabit the community without the presence of other species (Johnson and Amarasekare 2013).

With the addition of multiple uni-consumers and multiple resources, equations (6.1) are expanded to

$$\frac{dN_j}{N_j dt} = \frac{\sum_{i=1}^{p} b_{ij} a_{ij} R_i}{1 + \sum_{i=1}^{p} a_{ij} h_{ij} R_i} - f_j - g_j N_j$$

$$\frac{dR_i}{R_i dt} = c_i - d_i R_i + \frac{\sum_{j=1}^{q} \rho_{ij} N_j}{1 + \phi_i \sum_{j=1}^{q} \rho_{ij} N_j} - \sum_{j=1}^{q} \frac{a_{ij} N_j}{1 + \sum_{i=1}^{p} a_{ij} h_{ij} R_i}$$

(6.5)

To simplify the analysis without any loss of generality, the feeding interference terms have not been included (i.e., $z_j = 0$ from equations (6.1)). Direct self-limitation in the uni-consumers will be generated only by having $g_j > 0$ here. This formulation of the model also assumes that the fitness subsidy to R_i has a maximum per capita value at $1/\phi_i$, and that the sum of all uni-consumers weighted by their interaction strengths contributes to approaching this maximum. For example, plants may have a maximum number of ovules to be fertilized, and more pollinator visits will increase the fraction of fertilized ovules. However, the number fertilized cannot exceed the maximum number of ovules an individual possesses, which is determined by factors independent of the number of pollinators.

First, consider the invasion of additional uni-consumers to coexist with only one resource species. The criteria for the invasibility of additional uni-consumers in this case are completely analogous to those of multiple consumers competing for one resource that was explored in chapter 3. However, the fitness subsidy augments the resource's abundance and so reduces the stringency of these invasibility criteria. Assume that N_1 is the best uni-consumer competitor for R_1: it has the lowest $f_j/(a_{1j}(b_{1j} - h_{1j}f_j))$ and so intersects the R_1 axis at the lowest value. As in chapter 3, any uni-consumer with an isocline that passes to the left of the two-species equilibrium $[R^*_{1(N_1)}, N^*_{1(R_1)}]$ in the $N_1 - R_1$ face can increase when rare and so invade (fig. 6.8(a)). With the fitness subsidy to R_1, even uni-consumers with $c_1/d_1 < f_j/(a_{1j}(b_{1j} - h_{1j}f_j))$ can potentially satisfy this criterion (fig. 6.8(c)). (Remember that in the pure competition case, coexisting separately with the resource was a necessary condition for potential coexisting consumers.) Thus, inferior uni-consumer competitors may be able to coexist with the superior uni-consumer competitors even

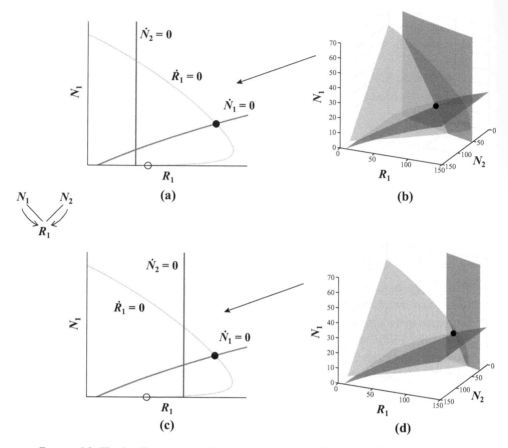

FIGURE 6.8. The isocline structure for two uni-consumers (N_1 and N_2) interacting with one resource (R_1) for two sets of parameters. Panels (a) and (b) show the isocline relationships in which N_1 can support a population at a lower R_1 abundance, and each uni-consumer can support populations on R_1 in the absence of the other. Panel (a) shows the isoclines in the $N_1 - R_1$ face, and (b) shows the full three-dimensional structure of the isoclines. Panels (c) and (d) show the same system, except that N_2 cannot support a population on R_1 in the absence of N_1. However, N_2 can invade and coexist when the other two species are present. The open circles in (a) and (c) identify $(R^*_{1(0)}, 0, 0)$ and the filled black circles identify $(R^*_{(N_1)}, N^*_{(R_1)}, 0)$. The R_1 isocline is identified as $\dot{R}_1 = 0$, the N_1 isocline is $\dot{N}_1 = 0$, and the N_2 isocline is $\dot{N}_2 = 0$. The parameters are as follows for both scenarios: $c_1 = 1.0$, $d_1 = 0.02$, $\phi_1 = 0.5$, $a_{11} = 0.05$, $b_{11} = 0.1$, $h_{11} = 0.05$, $\rho_{11} = 1.5$, $f_1 = 0.04$, $g_1 = 0.02$, $a_{12} = 0.025$, $b_{12} = 0.1$, $h_{12} = 0.05$, $\rho_{12} = 1.5$, and $g_2 = 0.005$. The difference between the two scenarios is that in (a) and (b) $f_2 = 0.1$, and in (c) and (d) $f_2 = 0.2$.

if they cannot coexist with R_1 by themselves. This facilitation of the presence of other species that could not coexist otherwise is a fundamental contribution that mutualisms make to the structure of food webs.

Once the second uni-consumer invades, the community will approach a new stable equilibrium (fig. 6.8(b) or 6.8(d)). R_1's equilibrium abundance may increase if the addition of the second uni-consumer increases its fitness subsidy relative to the fitness loss due to their feeding, but its abundance may also decrease because the fitness subsidy increase does not offset losses to consumption by the two uni-consumers. Analogous to chapter 3, additional uni-consumers may also invade and coexist if their isoclines pass to the left of the new equilibrium for previously invading uni-consumers and R_1 (i.e., their isoclines are between the origin and the equilibrium point for the community they are invading). Additional uni-consumers can continue to invade and coexist until no potential invaders can meet this criterion, and so would be excluded.

All the empirical issues discussed in testing multiple consumers feeding on a single resource (chapter 3, "Resource Competitors") apply here. However, a few interesting complications arise. First, note that not all mutualist partners need to be able to invade a community in which only the resource and no other mutualist partners are present. The fitness subsidy afforded to R_1 by the uni-consumers can facilitate inferior uni-consumer competitors (in the sense of having higher $f_j/(a_{1j} (b_{1j} - h_{1j}f_j))$ values) to be members of the community, even if those inferior uni-consumers could not coexist with R_1 on their own. This is an interesting example of *facilitation*. Therefore, in addition to evaluating whether each mutualist partner can satisfy its own invasibility criterion with all the other mutualist partners present, studies that determine whether each mutualist partner can invade when only the resource is present determine which mutualists are facilitated by the presence of other mutualists.

The other difference between pure resource competition and competition among mutualist partners for a resource is the difference in abundance responses that should occur to changes in mutualist partner abundances. With pure resource competition, experimentally increasing the abundance of one consumer should cause a decrease in the abundances of both the other consumer and the resource. However, with multiple uni-consumers interacting with one resource, experimentally increasing the abundance of one uni-consumer may increase or decrease the other species' abundances. Which result occurs will depend on where on the resource isocline the equilibrium lies. Remember that the resource isocline has the shape of a triangular boat sail filled with wind (see fig. 6.8). If the equilibrium falls on the upper part of the sail (i.e., the portion of the R_1 isocline where the derivative with respect to R_1 or N_2 is negative), the response to increasing the abundance of one uni-consumer mutualist will be the same as with resource

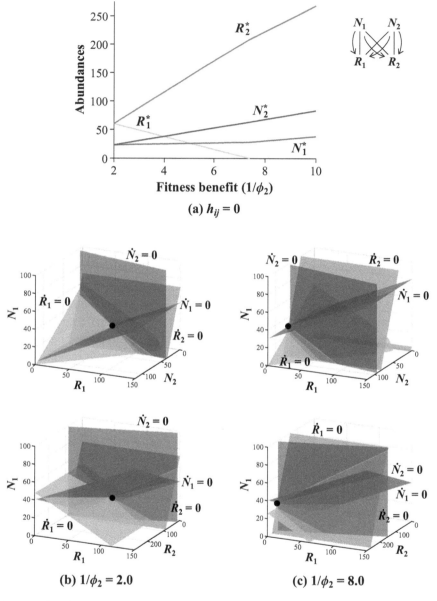

FIGURE 6.9. The effects of differences in the asymptotes of the fitness subsidies from two uni-consumers interacting with two resources. Panels (a)–(c) show results for uni-consumers with linear functional responses. Panel (a) presents the equilibrium abundances of the four species for different values of the maximum possible per capita fitness subsidy that R_2 can receive while holding the maximum fitness subsidy for R_1 constant. Panels (b) and (c) show the isocline structures for two points along the gradient. Panels (d)–(f) show the results for the same gradient but

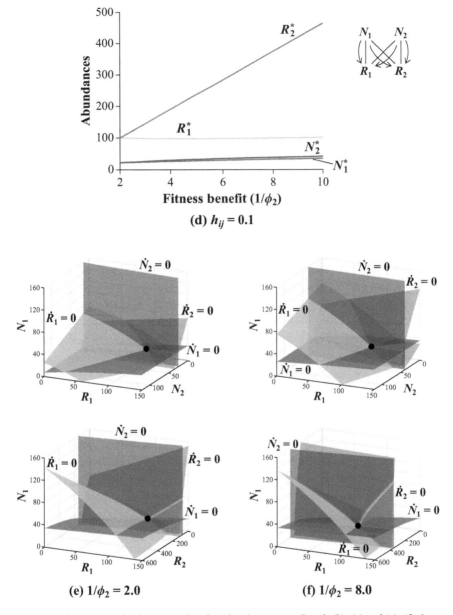

(d) $h_{ij} = 0.1$

(e) $1/\phi_2 = 2.0$

(f) $1/\phi_2 = 8.0$

with both uni-consumers having saturating functional responses. Panels (b)–(c) and (e)–(f) show different three-dimensional representations of this four-dimensional system. The R_1 isocline is identified as $\dot{R}_1 = 0$, the R_2 isocline is $\dot{R}_2 = 0$, the N_1 isocline is $\dot{N}_1 = 0$, and the N_2 isocline is $\dot{N}_2 = 0$. The parameter values are as follows, except as noted on each panel: $c_1 = 1.0$, $d_1 = 0.02$, $\phi_1 = 0.5$, $c_2 = 1.0$, $d_2 = 0.02$, $a_{11} = 0.05$, $b_{11} = 0.1$, $\rho_{11} = 1.5$, $f_1 = 0.1$, $g_1 = 0.015$, $a_{12} = 0.025$, $b_{12} = 0.1$, $\rho_{12} = 1.5$, $f_2 = 0.1$, and $g_2 = 0.015$.

competition, namely, the abundances of the resource and the other uni-consumer will both decrease. Here, the uni-consumers would be considered competitors because of this abundance result.

However, if the three- or four-species equilibrium falls in the lower part of the sail (i.e., the portion of the R_1 isocline where the derivative with respect to R_1 or N_2 is positive), increasing the abundance of one uni-consumer will cause both the resource and the other uni-consumer to increase in abundance. Both increase because the equilibrium falls on the portion of the R_1 isocline where the fitness benefit to the resource of having more uni-consumers still outweighs the costs. Here, the uni-consumers are also in effect mutualists with one another.

Likewise, if multiple resources are also present, the qualitative criteria for multiple uni-consumers and multiple resources that are mutualists to coexist are not fundamentally different from the situation with pure consumers (see chapter 3, "Multiple Consumers Eating Multiple Resources"). Because mutualism demands that the uni-consumers also either directly limit their own abundances or be limited by predators or diseases, (1) more uni-consumer species than resource species can coexist, and (2) the uni-consumers do not need to show any particular pattern of trade-offs in resource utilization (invasion 18 in fig. 6.1(d)). Hence, hundreds of pollinators can coexist with scores of plants, as on the Konza Prairie.

The fitness benefit garnered by consumers also influences community structure in the same fashion as the resources' intrinsic birth rates. For example, if the uni-consumers have linear functional responses, a difference in the asymptotic fitness benefit between resources causes apparent competition between them, and if the disparity is large enough, the resource with the lower fitness asymptote may not be able to coexist (fig. 6.9(a)–(c)). In contrast, if the uni-consumers have saturating functional responses, differences in the fitness benefit asymptotes among resources have very little effect on the abundances of the uni-consumers or those resources with lower asymptotes (fig. 6.9(d)–(f)). Thus, saturating functional responses can make the abundance of one uni-consumer almost indifferent to changes in the abundance of the other. Understanding the form of consumer functional responses in uni-directional mutualism is key to understanding how multiple mutualist partners may coexist in a community module.

Other Consumers and Resources

Uni-consumer mutualists not only have effects on the abundances of their mutualist partners and other uni-consumers but also indirectly influence other functional groups in the food web, because their actions inflate the abundances of their mutualist partners. In this section, I explore community modules in which only one pair of species are a uni-consumer mutualism and the remaining consumer and resource species in the community provide or receive no fitness benefits from the other species other than direct consumption.

First, consider another consumer feeding on the resource that has a uni-consumer mutualist partner (invasion 19 in fig. 6.1(d)). If the resource species is a plant, this second consumer may be a nectar robber, a seed predator, an herbivore of leaf tissue, or a galler consuming stem or leaf tissues (e.g., Bronstein et al. 2003, Morris et al. 2003, Wang et al. 2012). Equations (6.5) can describe the dynamics between one resource species R_1 and two consumers N_1 and N_2, where N_1 is a uni-consumer mutualist with R_1 (i.e., $\rho_{11} > 0$) and N_2 is simply a consumer of R_1 (i.e., $\rho_{12} = 0$). While the mutualistic interaction between R_1 and N_1 can confuse the issue, this is a competitive interaction between the uni-consumer mutualist and the regular consumer, and all the same conditions for coexistence of two consumers feeding on one resource apply (see chapter 3, "Resource Competitors").

If N_2, the pure consumer, is the better resource competitor (i.e., $f_2/(a_{12} (b_{12} - h_{12}f_2)) < f_1/(a_{11} (b_{11} - h_{11}f_1)))$ and does not also limit its own abundance via some form of direct self-limitation (i.e., $g_2 = 0$), N_2 will drive R_1's abundance to a level at which N_1 cannot support a population (fig 6. 10(a)). The uni-consumer and consumer would only coexist in this case if the consumer also limits its own abundance to the degree where its isocline bends away from the N_2 axis enough so that its point of intersection with the R_1 isocline in the $N_2 - R_1$ face places the N_1 isocline between that point and the origin (fig. 6. 10(d)). The new three-species equilibrium can easily have $R^*_{1(N_1,N_2)} > c_2/d_2 > c_1/d_1$ (fig. 6.10(c)), even though N_2 by itself would cause R_1 to equilibrate with $R^*_{1(N_2)} < c_2/d_2 < c_1/d_1$. Thus, uni-consumer mutualists can elevate resource abundance even with superior competitors also feeding on the resource.

If the uni-consumer mutualist is also the better resource competitor (i.e., $f_1/(a_{11} (b_{11} - h_{11}f_1)) < f_2/(a_{12} (b_{12} - h_{12}f_2)))$, it can facilitate the membership of inferior resource competitors in the community. The community module depicted in fig. 6.11 has the uni-consumer N_1 as the better resource competitor, and the pure consumer N_2 cannot support a population on the resource in the absence of the uni-consumer because it has $c_1/d_1 < f_2/(a_{12} (b_{12} - h_{12}f_2))$. N_2 can only invade and coexist in this community if N_1 is already present and coexisting with the resource. Additional consumers may also be facilitated if their isoclines pass below the new equilibrium.

This same facilitation of consumers that would be unable to coexist with the resource in the absence of the uni-consumer mutualist would also happen for subsequent consumers in the situation where the uni-consumer is the poorer competitor above if the multispecies equilibrium has $R^*_{1(N_1,N_2)} > c_1/d_1$. These two examples illustrate the critical role that uni-consumer mutualists, such as pollinators and seed

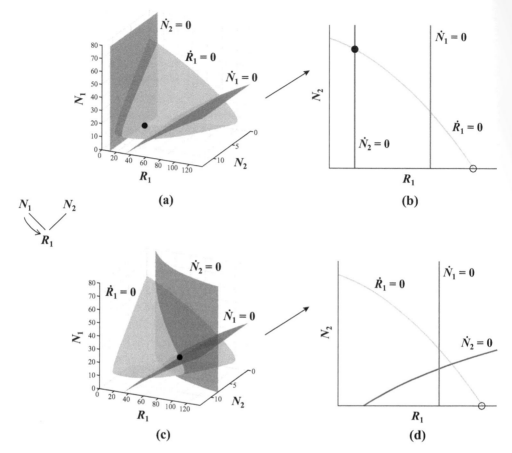

FIGURE 6.10. The isocline configurations for two situations of a uni-consumer N_1 interacting with one resource R_1 and another pure consumer N_2 that also feeds on the resource. Panel (a) shows the full three-dimensional system of isoclines for these three species when the pure consumer N_2 can support a population at a lower abundance of R_1 than can N_1. Panel (b) shows the $N_2 - R_1$ face of this system. Panels (c) and (d) show the same relationships as in (a) and (b), except that N_2 also experiences some level of direct intraspecific density dependence. The isoclines are identified as in figure 6.9. The parameter values are as follows: $c_1 = 1.0$, $d_1 = 0.02$, $\phi_1 = 0.5$, $c_2 = 1.0$, $d_2 = 0.02$, $a_{11} = 0.05$, $b_{11} = 0.1$, $h_{11} = 0.05$, $\rho_{11} = 1.5$, $f_1 = 0.15$, $g_1 = 0.01$, $a_{12} = 0.25$, $b_{12} = 0.1$, $h_{12} = 0.05$, and $f_2 = 0.25$. In (a) and (b), $g_2 = 0$, and in (c) and (d), $g_2 = 0.125$.

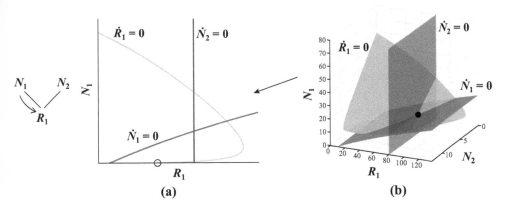

FIGURE 6.11. The isocline configuration for a uni-consumer N_1 interacting with one resource R_1 and another pure consumer N_2 that also feeds on the resource. In contrast to figure 6.10, in this figure the uni-consumer can support a population at a lower abundance of R_1 than can N_2. Panel (a) shows the $N_1 - R_1$ face, and (b) shows the full three-dimensional isocline system. The isoclines are identified as in figure 6.9, and the parameters are the same as in figure 6.10, with these exceptions: $f_1 = 0.05$, $g_1 = 0.02$, $f_2 = 1.0$, and $g_2 = 0$.

dispersers, may play in facilitating the presence of species at higher trophic levels in food webs because of how they augment resource abundances at the base of the web. Species that would otherwise be unable to invade and support populations in the community can do so because the actions of uni-consumer mutualists foster greater resource abundances. Analogous results are obtained for similar types of community modules with bi-consumer and third-party mutualisms as well.

This section illustrates the simplest examples of how mutualists can facilitate other functional groups to coexist in a community. In the absence of the uni-consumer mutualist, the resource abundance is not adequate to support some other pure consumer (fig. 6.11). If this other consumer can invade and coexist, its presence now provides opportunities for its predators to then potentially invade and coexist if this new consumer is abundant enough for its predators to invade and coexist (chapters 4 and 5). For example, the mycorrhizal associations of those 29 fungal species with the dozens of grasses and forbs on the Konza Prairie enhance the abundances of grasshoppers that feed on those plants (Kula et al. 2004), and predators of those grasshoppers are also presumably more abundant as well. These are the kinds of indirect effects on food web structure that are ignored when we focus our theoretical attention on only a general competition theory instead of a mechanistic theory built around the diversity of species interactions.

However, at this point, species deletion experiments would identify these additional consumers that are supported because of the actions of mutualists. So, for example, if the mutualists of some resource species are removed from the community, some pure consumers of that resource may be driven extinct because of the decrease in resource availability caused by that removal, and in turn their predators would be driven extinct as well.

Uni-consumers may also feed on other resource species besides the one to which they supply a fitness subsidy. For example, hummingbirds pollinate flowers they forage on for nectar, but they also feed on other resources (Lucas 1893, Henderson 1927, Young 1971, Ramsey 1988, Abrahamczyk and Renner 2015, Magalhães et al. 2018). Because the fitness subsidy to the resource mutualist indirectly inflates the abundance of the uni-consumer mutualist, the resource receiving the subsidy may also indirectly compete with other resource species on which the uni-consumer feeds via apparent competition. To explore this in equations (6.5), allow only one uni-consumer N_1 feeding on two resources R_1 and R_2. R_1 is the uni-consumer's mutualist partner (i.e., $\rho_{11} = 0$ in equations (6.5)), but R_2 receives no fitness subsidy from N_1 (i.e., $\rho_{21} = 0$ in equations (6.5)). This community module is identical to the simplest apparent competition module considered in chapter 3, "Apparent Competitors," but with the fitness subsidy to R_1 from the actions of N_1. Each resource influences the abundance of the other via their shared effects on N_1's abundance.

Begin with the uni-consumer N_1 coexisting with R_1. As with simple apparent competition modules, the second resource species R_2 can invade (invasion 20 in fig. 6.1(d)) if its isocline passes above the equilibrium abundance of the two other species $[R^*_{1(N_1)}, N^*_{1(R_1)}]$ in the $N_1 - R_1$ face (fig. 6.12(a)). If R_2 can pass this invasibility criterion, the abundances of the uni-consumer and R_2 will increase and R_1 will typically decrease (unless $[R^*_{1(N_1)}, N^*_{1(R_1)}]$ is located on the underside of the bulge of the R_1 isocline) to the new stable three-species equilibrium. Also, remember the reciprocal condition that the R_1 isocline must pass above the boundary equilibrium point $(R^*_{2(N_1)}, N^*_{1(R_2)})$ in the $N_1 - R_2$ face (fig. 6.12(b)) for R_1 to be able to invade and coexist. Invasion of subsequent resources, whether they receive a fitness subsidy from N_1 or not, must all be able to pass analogous invasibility criteria.

This community module also displays very interesting community structure patterns across ecological gradients, and the patterning depends qualitatively on whether the consumer has a linear or saturating functional response for feeding on the resources. With a linear functional response, a large fitness subsidy to R_1 (i.e., a smaller value of ϕ_1 and so a larger asymptote of $1/\phi_1$) can preclude R_2 from being a member of the community. Because the fitness subsidy to R_1 indirectly benefits N_1 by creating more resource on which N_1 feeds, N_1's abundance increases (fig. 6.13(a)). As a result, the per capita death rate of R_2 increases, and it equilibrates at lower abundances in communities where the other resource has a higher

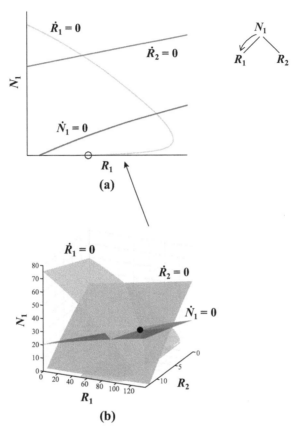

FIGURE 6.12. The isocline configuration for a uni-consumer N_1 interacting with one resource R_1 to which it provides a fitness subsidy and another resource R_2 to which it does not provide a fitness subsidy. Panel (a) shows the $N_1 - R_1$ face, and (b) shows the full three-dimensional system of isoclines. The isoclines are identified as in figure 6.9. The parameters are as follows: $c_1 = 1.0$, $d_1 = 0.02$, $\phi_1 = 0.5$, $c_2 = 2.0$, $d_2 = 0.02$, $a_{11} = 0.05$, $b_{11} = 0.1$, $h_{11} = 0.05$, $a_{21} = 0.05$, $b_{21} = 0.1$, $h_{21} = 0.05$, $\rho_{11} = 1.5$, $f_1 = 0.025$, and $g_1 = 0.02$.

maximum fitness subsidy. Moreover, if the fitness subsidy to R_1 is high enough, R_2 may be unable to invade and coexist. Also, as with apparent competition in pure consumer-resource interactions (chapter 3, "Apparent Competitors"), increasing the intrinsic birth rate c_i of one resource decreases the abundance of the other resource, and either resource will not be able to coexist if the other resource has a substantially higher c_i (fig. 6.13(c), (e)).

In contrast, if the uni-consumer has a saturating functional response for feeding on the two resources, the resources are themselves mutualists over much of the parameter space. If the asymptotic fitness subsidy to R_1 is very small, an

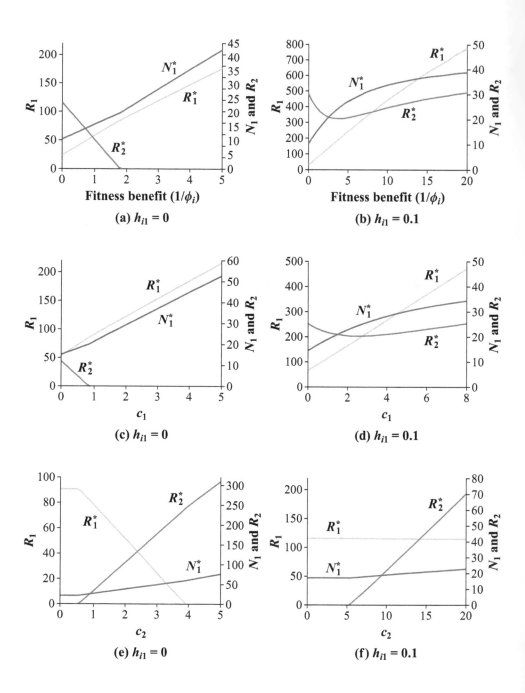

(a) $h_{i1} = 0$

(b) $h_{i1} = 0.1$

(c) $h_{i1} = 0$

(d) $h_{i1} = 0.1$

(e) $h_{i1} = 0$

(f) $h_{i1} = 0.1$

increase in the subsidy causes R_2's abundance to decrease. However, if the asymptotic fitness subsidy is moderate to high, R_2's abundance increases if the subsidy is increased (fig. 6.13(b)). The same is true for differences in the per capita intrinsic birth rate of R_1, making the resources apparent mutualists in communities with high values of c_1 (fig. 6.13(d), (f)). Increasing either the asymptotic fitness subsidy or the productivity of R_1 increases R_1's abundance and so saturates the feeding capabilities of N_1, which in turn reduces its feeding impact on R_2. Alternatively, the abundances of R_1 and N_1 are relatively insensitive to a change in the productivity of R_2 via changes in c_2 (fig. 6.13(f)).

These effects on community structure also apply with longer chains of species interaction networks. Consider an interaction chain having at one end a uni-consumer N_1 that feeds on its mutualist partner R_1. R_1 is also fed upon by a pure consumer N_2 that provides it no fitness benefit (e.g., N_2 is an herbivore), and N_2 feeds on a second resource species R_2 to which it also provides no fitness benefit (invasion 21 in fig. 6.1(d)). As in the two examples above, increasing the asymptotic fitness subsidy for R_1 causes a decrease in R_2's abundance until it cannot maintain a population if the consumers have linear functional responses (fig. 6.14(a)). In contrast, if both consumers have saturating functional responses, R_1 and R_2 are apparent competitors at low asymptotic subsidy levels but apparent mutualists at high asymptotic subsidy levels to R_1 (fig. 6.14(b)).

These results show that the form of the functional response of a uni-consumer can cause dramatically different responses to changes in the environment (e.g., changes in the relative productivities of the two resources) or differences in the properties of the interacting species (e.g., the potential maximum fitness subsidy that can be garnered by a resource interacting with the uni-consumer). All the empirical issues discussed in chapter 3, "Apparent Competitors," apply here as well.

FIGURE 6.13. For the community module pictured in figure 6.12 of a uni-consumer interacting with two resources, one of which it provides a fitness subsidy, the equilibrium abundances of these three species are shown at different points—along a gradient of the maximum fitness benefit for R_1 when the consumer has (a) a linear functional response and (b) a saturating functional response; along a gradient of the intrinsic per capita birth rate for R_1 when the consumer has (c) a linear functional response and (d) a saturating functional response; and along a gradient of the intrinsic per capita birth rate for R_2 when the consumer has (e) a linear functional response and (f) a saturating functional response. The symbols and lines are identified as in figure 6.9. The parameters are as follows, unless otherwise specified: $c_1 = 1.0$, $d_1 = 0.02$, $\phi_1 = 0.5$, $c_2 = 1.0$, $d_2 = 0.02$, $a_{11} = 0.05$, $b_{11} = 0.1$, $a_{21} = 0.05$, $b_{21} = 0.1$, $\rho_{11} = 1.5$, $f_1 = 0.025$, and $g_1 = 0.02$.

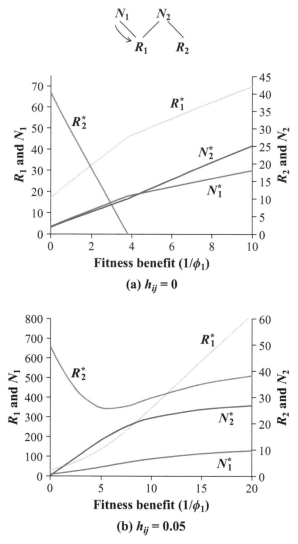

FIGURE 6.14. The equilibrium abundances of four interacting species along a gradient of the maximum fitness benefit to a resource from a uni-consumer. The community contains the uni-consumer N_1 that provides a fitness subsidy to a resource R_1. This resource and another, R_2, are also fed upon by a pure consumer N_2. The changes in the equilibrium abundances of the four species are shown along the gradient when the consumers have linear functional responses (a), and when the consumers have saturating functional responses (b). The symbols and lines are identified as in figure 6.9. The parameters are as follows, unless otherwise specified: $c_1 = 1.0$, $d_1 = 0.02$, $\phi_1 = 0.5$, $c_2 = 1.0$, $d_2 = 0.02$, $a_{11} = 0.05$, $b_{11} = 0.1$, $p_{11} = 1.5$, $f_1 = 0.05$, $g_1 = 0.01$, $a_{12} = 0.25$, $b_{12} = 0.1$, $a_{22} = 0.1$, $b_{22} = 0.1$, $f_2 = 0.05$, and $g_2 = 0.75$.

Uni-consumer mutualisms will also influence the structure of competition between consumers for multiple resources. Consider the case of two pure consumers that both feed on two resources. This is the classic module that MacArthur and colleagues used as the foundational basis for the development of competition theory and that Tilman and others used to elaborate MacArthur's work into a mechanistic understanding of the interaction of resource and apparent competition (MacArthur and Levins 1967, MacArthur 1969, 1970, 1972, Hsu et al. 1978a, 1978b, Tilman 1980, 1982, Kleinhesselink and Adler 2015, Letten et al. 2017, McPeek 2019b). Two pure consumers N_1 and N_2 each feed on two resources R_1 and R_2: $\rho_{11} = \rho_{12} = \rho_{21} = \rho_{22} = 0$ in equations (6.5), giving the community module considered in chapter 3, "Multiple Consumers Eating Multiple Resources." Add to this module a specialist uni-consumer for each resource, with N_3 interacting with R_1 (i.e., $\rho_{13} > 0$) and N_4 interacting with R_2 (i.e., $\rho_{24} > 0$) (invasion 22 in fig. 6.1(d); see also Lee and Inouye 2010).

The addition of the uni-consumer mutualists to this community module does not alter the fundamental relationships between the two pure consumers and the resources. In fact, the positions of the pure consumer isoclines are unchanged by the addition of the uni-consumer mutualists, and the consumer isoclines are independent of all axes except those for the two resources; so their isoclines still must satisfy inequality (3.19) in order to intersect (fig. 6.15(a)). As intuition would suggest, the addition of the uni-consumer mutualists alters the positions of the resource isoclines. As in chapter 3, the critical issue for coexistence of N_1 and N_2 is the relative point of intersection of the R_1 and R_2 isoclines with the N_1 and N_2 axes (fig. 6.15(b)–(d)). The fitness subsidies provided by the uni-consumer mutualists cause the R_1 and R_2 isoclines to intersect the pure consumer axes at points that are the ratio of the net per capita fitness subsidy divided by the attack coefficient for each consumer for feeding on the respective resources. If the R_1 and R_2 isoclines still intersect at positive pure consumer abundances, the two pure consumers will invade and coexist in the community. This is true when

$$\frac{a_{11}}{a_{21}} > \frac{c_1 - d_1 R^*_{1(R_2,N_1,N_2,N_3,N_4)} + \dfrac{\rho_{13} N^*_{3(R_1,R_2,N_1,N_2,N_4)}}{1+\phi_1\rho_{13}N^*_{3(R_1,R_2,N_1,N_2,N_4)}}}{c_2 - d_2 R^*_{2(R_1,N_1,N_2,N_3,N_4)} + \dfrac{\rho_{24} N^*_{4(R_1,R_2,N_1,N_2,N_3)}}{1+\phi_1\rho_{24}N^*_{4(R_1,R_2,N_1,N_2,N_3)}}} > \frac{a_{12}}{a_{22}}. \tag{6.6}$$

This criterion is exactly analogous to inequality (3.19) derived for the pure consumer form of this module in chapter 3, except that the resource productivities include the net fitness subsidies from their uni-consumer mutualist partners. This inequality ensures that all four consumer isoclines are coincident at abundances where the resource isoclines intersect (fig. 6.15(d), left) and is analogous to the

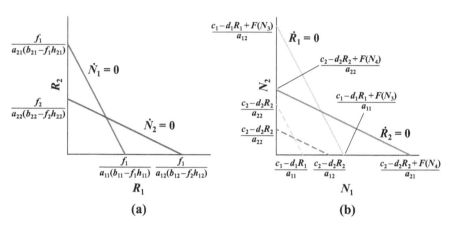

(a)

(b)

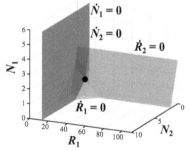

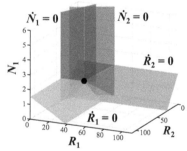

(c) N_3 and N_4 absent

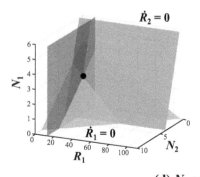

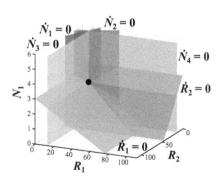

(d) N_3 and N_4 present

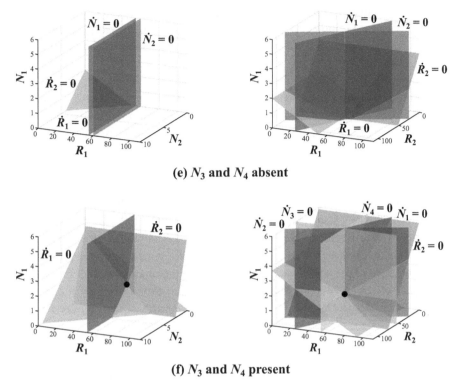

(e) N_3 and N_4 absent

(f) N_3 and N_4 present

FIGURE 6.15. The isocline configurations for a community module of two pure consumers (N_1 and N_2) that are competing for two resources (R_1 and R_2), and each resource has a specialist uni-consumer (N_3 for R_1 and N_4 for R_2) that provides a fitness subsidy to it. (a) Coexistence of the pure consumers requires that their isoclines intersect at positive resource abundances in the $R_2 - R_1$ face, if they do not limit their own abundances via either feeding interference or another mechanism of direct intraspecific density dependence. (b) For the pure consumers to coexist, their isoclines must intersect at positive resource abundances when the resources are at their equilibrial abundances. The dashed lines show the positions of the resource isoclines with no fitness subsidies from their uni-consumer specialist partners (i.e., when $N_3 = N_4 = 0$), and the corresponding solid isoclines are their positions with the fitness subsidy at equilibrium. Thus, the fitness subsidies to the resource translate to higher abundances of the pure consumers. Panels (c) and (d) show the isocline structures for a parameter combination where the pure consumers can coexist with the resources in the absence of the uni-consumer specialists when the uni-consumer specialists are (c) absent and (d) present. Panels (e) and (f) show a comparable situation, but the two pure consumers cannot coexist with the resources unless the uni-consumer specialists are both present. The isoclines are identified as in figures 6.7 and 6.9, with the addition of two uni-consumer isoclines labeled $\dot{N}_3 = 0$ and $\dot{N}_4 = 0$ in (d) and (f). The parameters are as follows: $c_1 = 1.0$, $d_1 = 0.02$, $\phi_1 = 0.5$, $c_2 = 1.0$, $d_2 = 0.02$, $\phi_2 = 0.5$, $b_{ij} = 0.1$, $h_{ij} = 0$, $a_{11} = 0.5$, $a_{21} = 0.25$, $g_1 = 0$, $a_{12} = 0.25$, $a_{22} = 0.5$, $g_2 = 0$, $a_{13} = 0.05$, $\rho_{13} = 1.5$, $f_3 = 0.01$, $g_3 = 0.025$, $a_{24} = 0.05$, $\rho_{24} = 1.5$, $f_4 = 0.01$, and $g_2 = 0.025$; in (a)–(d), $f_1 = 1.0$ and $f_2 = 1.0$; in (e)–(f), $f_1 = 4.0$ and $f_2 = 4.0$.

condition requiring that the resource supply point (which now includes the resource fitness subsidies) be within the wedge formed by the pure consumer consumption vectors in Tilman's (1980, 1982) formulation of the problem for pure consumers.

Obviously, if the fitness subsidies to the two resources displace their isoclines so that they no longer intersect at a feasible equilibrium for the two pure consumers, the pure consumer whose isocline is completely below the other's will be unable to coexist because of apparent competition between the two resources. Not only are the basic productivities of the resources important to the coexistence of their consumers, the actions of their mutualist partners also influence whether resource competitors that are pure consumers of those resources can coexist.

The actions of the uni-consumer mutualists may also facilitate consumers that would be unable to invade without heightened abundances of the resources because of their fitness subsidies. Figure 6.15(e)–(f) shows the isocline systems for this community with higher intrinsic death rates for the two pure consumers. In the absence of the uni-consumer mutualists, neither pure consumer can coexist with the resources, because the combined resource abundances are too low for either N_1 or N_2 to support a population. However, with N_3 and N_4 present, the resource isoclines are heightened enough to permit N_1 and N_2 to both invade and coexist. This example illustrates that the consequences of mutualistic interactions can facilitate greater species richness and diversity at other functional positions in a food web. The converse of this insight is that the loss of mutualists, such as plant pollinators and seed dispersers, can cause a cascading loss of other species from the food web. For example, the expulsion of zooxanthellae from corals to cause their bleaching has dire consequences throughout the reef food web (e.g., Hempson et al. 2017). Bi-consumer and third-party mutualists have similar effects on species at other positions in the food web. This example should also serve as a powerful reminder of the critical role that the indirect effects of species interactions play in shaping overall food web structure, and this critical role cannot be ignored when we consider what mechanisms shape species coexistence and foster richness and diversity.

This community module offers a very interesting extension on the standard issues involved in competition for multiple resources (Tilman 1982, Grover 1997, Letten et al. 2017, McPeek 2019b). In the parlance of Tilman's competition framework, the uni-consumer mutualists alter the position of the supply point for the resources to the consumers (i.e., inequality (6.6)). Or in the present context, the uni-consumers alter the apparent competitive abilities of the two resources.

Here again is another example of why studying the competition between the consumers while ignoring the indirect interactions that form the mechanism of that interaction can be wildly misleading. For example, the mycorrhizal fungal associations shift the competitive relationships among C3 and C4 plants on the Konza Prairie (Hartnett and Wilson 1999, Smith et al. 1999). Experiments that either include or omit the mutualist partners of the resources in their execution could lead to strong differences in the conclusions that would be drawn about the abilities of each to coexist in a community with the other consumer. Moreover, apparent differences in their competitive abilities would result from experiments that differed in the factors influencing the demographic performances of the uni-consumer mutualists. In fact, both of the pure consumers may only be able to coexist because of the presence of the uni-consumer mutualists inflating the resource abundances to adequate levels.

More Trophic Levels

These effects of mutualisms on the abundances of species at the same trophic levels as the mutualism partners will obviously also propagate to the species at other trophic levels both above and below. Adding predators of the consumers to the general community model specified by equations (6.5) gives

$$
\frac{dP_k}{P_k\,dt} = \frac{\displaystyle\sum_{j=1}^{q} n_{jk} m_{jk} N_j}{1 + \displaystyle\sum_{j=1}^{q} m_{jk} l_{jk} N_j} - x_k - y_k P_k
$$

$$
\frac{dN_j}{N_j\,dt} = \frac{\displaystyle\sum_{i=1}^{p} b_{ij} a_{ij} R_i}{1 + \displaystyle\sum_{i=1}^{p} a_{ij} h_{ij} R_i} - \sum_{k=1}^{s} \frac{m_{jk} P_k}{1 + \displaystyle\sum_{j=1}^{q} m_{jk} l_{jk} N_j} - f_j - g_j N_j. \tag{6.7}
$$

$$
\frac{dR_i}{R_i\,dt} = c_i - d_i R_i + \frac{\displaystyle\sum_{j=1}^{q} \rho_{ij} N_j}{1 + \phi_i \displaystyle\sum_{j=1}^{q} \rho_{ij} N_j} - \sum_{j=1}^{q} \frac{a_{ij} N_j}{1 + \displaystyle\sum_{i=1}^{p} a_{ij} h_{ij} R_i}
$$

We have already explored the requirements for invasion and the resulting community changes of adding a predator that can feed on a single uni-consumer that itself feeds on one resource to which it also supplies a fitness subsidy (see fig. 6.7). We have also seen that this fitness subsidy inflates the resource abundance and increases the abundances of other consumers that feed on it and may even facilitate the presence of some consumers that would otherwise not be able to invade and coexist (e.g., fig. 6.11).

What is then required for a predator of these consumers to invade and coexist (invasion 23 in fig. 6.1(d))? Begin with a community of one uni-consumer N_1, one resource R_1 that receives a fitness subsidy from N_1, and a pure consumer N_2 that feeds on but does not subsidize R_1 (as in fig. 6.11). What is required for a predator P_1 that feeds only on N_2 (i.e., $m_{11} = 0$, $m_{21} > 0$) to be able to invade and coexist with these three species? Again, the answer is no different than what was identified in chapters 4 and 5. In the absence of P_1, this community has a stable equilibrium at $[R^*_{1(N_1,N_2)}, N^*_{1(R_1,N_2)}, N^*_{2(R_1,N_1)}]$ (see fig. 6.11). The predator can invade and coexist if the abundance of its prey gives it a positive per capita growth rate when it is rare, which means that its isocline must pass below this equilibrium point. Since this predator's isocline is parallel to the R_1 and N_1 axes, this means that the P_1 isocline must intersect the N_2 axis below this equilibrium value, which requires that

$$N^*_{2(R_1,N_1)} > \frac{x_1}{m_{21}\left(n_{21} - x_1 l_{21}\right)}$$

(fig. 6.16). Because N_2's equilibrium abundance is inflated by the fitness subsidy provided by N_1 to R_1 (e.g., fig. 6.11), this fitness subsidy makes the invasibility criterion for P_1 much less stringent. Also, P_1's presence is facilitated by the mutualism subsidy if N_2's abundance is not adequate for P_1 to invade when N_1 is absent and so cannot subsidize R_1's fitness to thereby inflate its abundance. If P_1 can invade and coexist, the community comes to a new equilibrium, with fewer N_2, more R_1, and more N_1 (fig. 6.16).

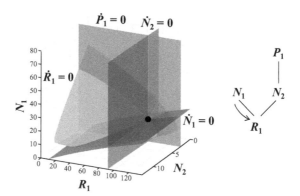

FIGURE 6.16. The isocline configuration for a resource that interacts with a uni-consumer providing it a fitness subsidy and a pure consumer, and the pure consumer has a predator feeding on it. The filled black circle is the four-species equilibrium. The isoclines are identified as in figure 5.7. The parameters are as follows: $c_1 = 1.0$, $d_1 = 0.02$, $\phi_1 = 0.5$, $a_{11} = 0.05$, $b_{11} = 0.1$, $h_{11} = 0.05$, $\rho_{11} = 1.5$, $f_1 = 0.05$, $g_1 = 0.02$, $a_{21} = 0.25$, $b_{12} = 0.1$, $h_{12} = 0.05$, $f_2 = 1.0$, $g_2 = 0$, $m_{21} = 0.4$, $n_{21} = 0.1$, $l_{21} = 0.05$, $x_1 = 0.1$, and $y_1 = 0$.

If the predator feeds on both the uni-consumer and the pure consumer (i.e., $m_{11} > 0$, $m_{21} > 0$), this set of species interactions becomes a diamond/keystone predation module (invasion 24 in fig. 6.1(d)). For example, this model would describe an arthropod predator, such as a spider species, that feeds on both the bee pollinator of a plant species as well as an insect herbivore that feeds on the plant. For simplicity, I only consider here cases in which the uni-consumer is a better resource competitor than the pure consumer (again, meaning simply that its isocline intersects the R_1 axis at a lower point than does N_2's isocline), and the uni-consumer limits its own abundance to some degree via some form of direct self-limitation. These assumptions ensure that the two consumer isoclines intersect at positive abundances.

Coexistence and the resulting community structure depend on the relative feeding abilities of the consumers and predator. If all species have linear functional responses, the isoclines of the consumers intersect at positive abundances of all species when two inequalities are satisfied:

$$\frac{a_{11}b_{11}}{a_{12}b_{12}} > \frac{f_1 + g_1 N^*_{1(R_1,N_2,P_1)}}{f_2 + g_2 N^*_{2(R_1,N_1,P_1)}} \quad \text{and} \quad \frac{m_{11}}{m_{21}} > \frac{a_{11}b_{11} - g_1 N^*_{1(R_1,N_2,P_1)}}{a_{12}b_{12} - g_2 N^*_{2(R_1,N_1,P_1)}}. \tag{6.8}$$

This criterion is analogous to inequality (4.9) for the diamond/keystone module with only pure consumers in chapter 4, "Diamond/Keystone Predation." The first of the inequalities in (6.8) is again simply the statement that the uni-consumer is the better resource competitor. The second demands that the ratio of the mortalities inflicted by the predator on the two consumers be greater than the ratio of the responses of the two consumers feeding on the resource discounted by the strength of their direct density-dependent effects on themselves. Because foraging by the uni-consumer will have much less impact on the resource than that of the pure consumer, this latter inequality will typically also be satisfied. With saturating functional responses, similar relationships among the parameters (and so species abilities) must hold, but the equations describing the criteria are extremely complicated and so provide little practical insight.

In addition, the bulge in the R_1 isocline caused by the fitness subsidy from N_1's actions greatly relaxes the criteria for the R_1 and P_1 isoclines to intersect, as compared to when both consumers are merely consumers (chapter 4, "Diamond/Keystone Predation"). If the isoclines intersect to form a feasible four-species equilibrium, this equilibrium is stable if the fitness benefit is not extremely large; the species coexist in a limit cycle if the fitness benefit is extremely large.

If the predator inflicts greater per capita mortality on the pure consumer than on the uni-consumer (i.e., $m_{21} > m_{11}$), community structure changes with the maximum fitness benefit just as the pure consumer diamond/keystone module changes with changes in the productivity of the resource. For example, consider a

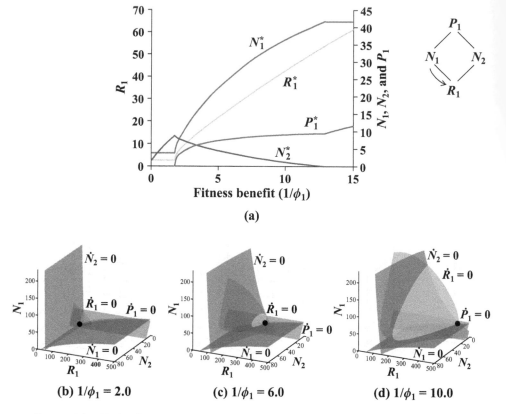

FIGURE 6.17. The isocline configurations for a diamond/keystone predation module in which one of the consumers is a uni-consumer that provides a fitness subsidy to the resource. In the example illustrated here, the predator imposes greater mortality on the pure consumer. Panel (a) shows the changes in species abundances for different values of the maximum fitness benefit, and (b)–(d) show the isocline configurations for the different subsidy levels along this gradient. The isoclines are identified as in figures 6.7 and 6.9. The parameters are as follows: $c_1 = 1.0$, $d_1 = 0.02$, $a_{11} = 0.05$, $b_{11} = 0.1$, $h_{11} = 0.05$, $\rho_{11} = 1.5$, $f_1 = 0.05$, $g_1 = 0.02$, $a_{21} = 0.25$, $b_{12} = 0.1$, $h_{12} = 0.05$, $f_2 = 0.5$, $g_2 = 0$, $m_{11} = 0.03$, $n_{11} = 0.1$, $l_{11} = 0.05$, $m_{21} = 0.14$, $n_{21} = 0.1$, $l_{21} = 0.05$, $x_1 = 0.125$, and $y_1 = 0$.

community with the parameter combination depicted in figure 6.17. When the maximum fitness subsidy is small ($1/\phi_1 < 1.6$), the consumer abundances are not sufficient to support the predator, but above that the predator can invade and coexist in the diamond module (fig. 6.17(a)). If the level of maximum fitness subsidy is increased above this level, R_1 and P_1 increase, the uni-consumer N_1 increases, but the pure consumer N_2 decreases. Moreover, if the maximum fitness subsidy is high enough, N_2 cannot coexist because P_1 is too abundant. Figure 6.17(b)–(d) shows how the isocline shapes and positions change along this gradient.

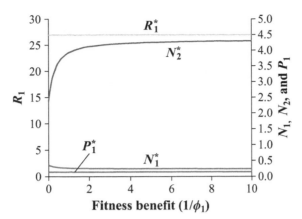

FIGURE 6.18. The effects of different maximum fitness subsidies for a diamond/keystone predation module in which one consumer is a uni-consumer and the predator imposes greater mortality on the uni-consumer (cf. fig. 6.17(a)). The isoclines are identified as in figures 6.7 and 6.9.
The parameters are as in figure 6.17, except as follows: $m_{11} = 0.5$, $m_{21} = 0.03$, and $x_1 = 0.025$.

In contrast, if the predator imposes more mortality on the uni-consumer than
on the pure consumer and all four species coexist, a higher level of the maximum
fitness subsidy only substantially alters the pure consumer's abundance (fig. 6.18).
Here, the pure consumer is the only beneficiary of the mutualistic interaction
between the uni-consumer and the resource.

These examples also highlight the fact that mutualist partners of one species
can facilitate the coexistence of species at other functional groups and trophic
levels. Empirical examinations of food webs should focus much greater attention
on these important drivers of community structure. The actions of mutualists
mimic the actions of greater productivity for basal resources. Thus, we could predict that changes in species abundances across the food web would change along
an environmental gradient that altered the abundance of a uni-consumer mutualist
that benefited the basal resource, just as that food web would change along a productivity gradient for the basal resource (e.g., figs. 4.3(a)–(b), 4.6(a)–(b)). Alternatively, experimentally altering the abundance of the uni-consumer mutualist of
a basal resource would have comparable effects for the rest of the food web as
fertilizing the basal resource.

In addition to providing greater opportunities for more species to invade and
coexist at other food web positions, the presence of symbiotic interactions embedded in food webs may also enhance the overall stability of communities. For
example, Qian and Akçay (2020) performed simulation studies of community
assembly by serially adding randomly selected species of various interaction
types to communities and studying the stability properties of the resulting

assemblages. They found that communities with a higher proportion of mutualists increased the diversity of species that coexisted, the stability of the community overall, and the persistence of species over the long term.

I could continue with exploring community modules with more richness and more functional diversity. However, I believe this is a good point to stop, given that this survey covers almost all the simple modules (as chapters 4 and 5 showed, the results for food chains and diamond/keystone modules do not qualitatively change with the addition of omnivory/intraguild predation, and this is also true with these mutualisms). Hopefully, this very brief tour of food webs where some species interactions are uni-consumer mutualisms will inspire others to explore these richer community structures with uni-consumer mutualisms and the other types of mutualisms.

SUMMARY OF MAJOR INSIGHTS

- Some degree of direct self-limitation may be needed to foster the invasibility of mutualist partners and prevent overexploitation in some forms of mutualistic interactions.
- In third-party defense mutualisms, interspecific feeding interference may be the most crucial component making the interaction a mutualism.
- Interactions shift between being mutualisms and exploitation/parasitism based on how environmental contexts alter the costs and benefits of the interaction.
- Competition among mutualists for access to their partners follows the same rules as competition for limiting resources.
- Mutualists can shift the competitive relationships among their partners and can shift competitive relationships among other consumers through shifting the apparent competitive abilities of their partners.
- Mutualisms at the base of the food web provide greater opportunities for invasion and coexistence at higher trophic levels by increasing species abundances.

CHAPTER 7

Pathogens in a Food Web

Walk out into the Carpinteria Salt Marsh, and you will be surrounded by a diverse food web both in and out of the Pacific waters of coastal California, USA. Pickleweed (*Salicornia virginica*) is the primary plant, with open mudflats and creek channels cutting through the marsh. Epiphytes grow on the pickleweed and the other plants on the marsh, and phytoplankton wash in with the tidal waters. Many aquatic and terrestrial grazers and filterers consume those algae and plants, including annelid worms, isopods, amphipods, copepods, ostracods, nemerteans, crabs, shrimp, snails, dipterans, clams, and mussels. Eight fish species feed on those plants, grazers, and filterers, and thirty-six bird species feed there as well (Lafferty, Hechinger, et al. 2006). What is not readily apparent is that each of those species is a host to an array of pathogens and parasites, including viruses, bacteria, and a diverse metazoan contingent (Lafferty, Hechinger et al. 2006, Kuris et al. 2008, Lafferty et al. 2008). Those pathogens alter the abilities of infected hosts to engage in the interactions with their food resource species and their predators and so decrease their survival and fecundity rates.

Likewise, all the species in the pelagic zones of freshwater lakes are also hosts to diverse pathogen and parasite assemblages. For example, *Daphnia dentifera* and *D. pulicaria*, the cladoceran algal grazers in the lakes of southwestern Michigan, USA, are infected with six to eight bacterial and fungal pathogens (Duffy et al. 2010). These infections reduce their feeding and fecundities, and make them more conspicuous and so more susceptible to fish predators (Ebert et al. 2000, Decaestecker et al. 2003, Duffy et al. 2005, Searle et al. 2016).

All species have pathogens and parasites that infect them, and each typically supports diverse pathogen assemblages. Humans, certainly the best studied host, support roughly 1,415 different pathogen and parasite species, including viruses, bacteria, protozoa, fungi, and helminths (Cleaveland et al. 2001). Many of those pathogens and parasites infect many different host species in addition to humans. Moreover, interactions with pathogens are not separated from the rest of the ecological environment in which the host is embedded. For example, *Vibrio cholera*, the bacterium that causes cholera, is a naturally occurring and abundant community member in lakes, estuaries, and oceans around the world (Cottingham and Butzler 2006).

Although they are diverse and abundant in every ecosystem, parasites and pathogens are typically community members that are sorely neglected (Collinge and Ray 2006, Lafferty, Dobson, et al. 2006). Because they alter the demographic rates of their hosts, they can influence the coexistence of their hosts in the community and in turn influence the effects of their hosts on the coexistence of other species in the community. Host-pathogen dynamics are a rich area of conceptual and theoretical exploration (e.g., Anderson and May 1991, Keeling and Rohani 2008). However, modeling the effects of pathogens on the interactions among host species is a less explored set of interactions (Holt and Dobson 2006).

In this chapter, I offer a brief exposition on the possible effects that pathogens may have on the coexistence of their hosts and the species with which their hosts interact. My goal here is simply to highlight the major ways pathogens may influence the coexistence of their host species in a community and the species with which the host interacts. I hope that this brief exposition, taken along with other recent approaches (e.g., Hall et al. 2005, Borer et al. 2007, Holt and Roy 2007, Bonsall and Holt 2010), will spark further interest in how pathogens embedded in food webs might influence coexistence of their host species in those food webs.

BASIC MODELS OF PATHOGEN-HOST INTERACTIONS

Pathogenic organisms have a fantastic diversity of life cycles. Some pathogens cannot survive for more than a few seconds outside of a host, while others can persist in the environment without a host for decades or centuries (Caraco and Wang 2008, Roche et al. 2011). Some infect a single host, and others must pass through a series of hosts to complete their entire life cycle (Gandon 2004, Buhnerkempe et al. 2015). Some pathogens only reduce the reproductive capacities of an individual for a short time (e.g., by reducing the motivation for feeding), while others are lethal (Anderson and May 1978a, 1978b). Consequently, a chapter is too short to treat the full panoply of possible effects that pathogens may have on the coexistence of their hosts in a community. However, here I try to highlight some of the major consequences that pathogens may have on host coexistence and present some of the approaches to modeling these interactions.

As briefly discussed in chapter 2, some of the very first theoretical models of species interactions considered the consequences of a pathogen on host dynamics (Ross 1911, Lotka 1912). Because pathogens detrimentally impact the demographic rates of their hosts, their consequence is typically a reduction in the population size of their host. In fact, Anderson and May (1980) showed that an interaction with a pathogen can generate population regulation in an otherwise exponentially growing host population.

One important feature of the host-pathogen interaction is the subdivision of the host population into various classes of individuals, based on their status relative to the pathogen. Some individuals are *susceptible* to being infected but are not currently harboring pathogens. While individuals may vary in their susceptibility to being infected in real populations, for present purposes, I assume that all uninfected individuals are equally susceptible to being infected. Some individuals are *infected* with the pathogen and dealing with the consequences of that infection. Likewise, I assume that all infected individuals suffer the same fitness consequences. Other individuals may have been previously infected and now are recovered from the infection: if these individuals are Bacteria and Archaea, which have CRISPR, or vertebrates, which have either variable lymphocyte receptors or immunoglobin antibodies and thus acquire immunity to future infections, this constitutes a separate class of *recovered* individuals. However, if recovered individuals do not acquire any added protection from having been previously infected with the pathogen, they simply return to the susceptible class, which I assume throughout here.

Distinguishing these classes of individuals in the host population is important because their infection status will influence their demographic rates and their abilities to interact with other species (Anderson and May 1991, Keeling and Rohani 2008). Obviously, pathogens may increase the per capita death rates or decrease the per capita birth rates of infected individuals. Pathogens may also alter the feeding rates of consumers or the susceptibility and nutritional quality of hosts when they are the prey of some predator. The typical approach to modeling diseases is therefore to have equations describing the abundances of each class of individuals. Thus, the models used here are very familiar in their basic structure, but include scaling parameters that relate the demographic parameters among classes and include terms that describe the mechanisms and rates of transition of host individuals among these classes. For simplicity, I focus on modeling situations without a "recovered" class, and so only include susceptible and infected hosts as with nonvertebrate eukaryotic species.

Those readers who are already familiar with the theoretical disease dynamics literature will be familiar with the "SIR"—susceptible-infected-recovered— formulation of such models. However, the typical SIR formulation tracks the relative frequencies of the host classes and not their absolute abundances (Anderson and May 1991, Keeling and Rohani 2008). Abundances are critical for our purposes, and so the reader should keep in mind when comparing the results below to other works of the potential difference in variables being modeled (see Keeling and Rohani 2008 for an excellent exposition on the differences).

Another important consideration for host-pathogen models is whether pathogen individuals can survive for any amount of time outside a host in the external environment. If pathogens can survive outside their hosts, they will form an environmental reservoir population that will constitute an alternative route for

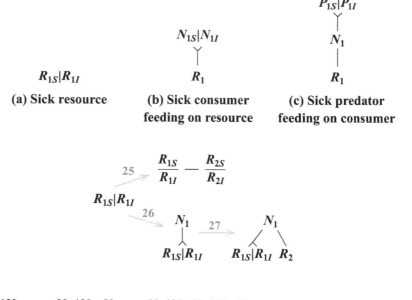

(a) Sick resource

(b) Sick consumer feeding on resource

(c) Sick predator feeding on consumer

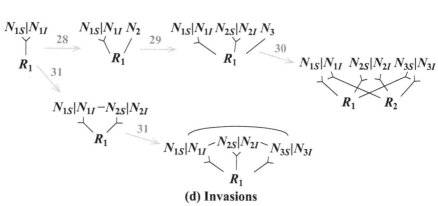

(d) Invasions

FIGURE 7.1. Module diagrams of some basic pathogen-infected species and community modules in which they are embedded. (a) A single resource species that is infected with a pathogen consists of two classes of individuals—those that are infected (R_{1I}) and those that are uninfected but susceptible to being infected (R_{1S}). (b) A food chain consisting of a consumer that is infected with a pathogen and so contains susceptible (N_{1S}) and infected (N_{1I}) individuals, and both consumer classes feed on a single resource species. (c) A food chain consisting of a predator that is infected with a pathogen, in which both predator classes of individuals feed on a single consumer species, which in turn feeds on a single resource species. Panel (d) provides a road map of the species invasions that are considered in this chapter into a simple community already containing either an infected resource or infected consumer. The narrow line between two infected species implies that infected individuals of each species can cross-infect the other.

susceptible hosts to become infected (Caraco and Wang 2008, Roche et al. 2011). Therefore, the size of this free-living pathogen population must also be modeled. In contrast, if pathogens can only survive in hosts, the models simply need to keep track of infected host individuals, since infected hosts are synonymous with the presence and abundance of pathogens. The main qualitative difference between these two types of pathogens is that the presence of an environmental reservoir of free-living pathogens makes extirpation of the pathogen population more diffi-cult. Because my main focus is simply the major consequences of pathogens being present, for simplicity, I focus on pathogens that cannot live outside their hosts and leave it to others to explore what differences may result in community dynamics for pathogens with different life cycles and capabilities.

Sick Resource Species

The first model to consider is simply a resource species that is infected with a patho-gen that cannot live outside its host (fig. 7.1(a)). This model is structured according to the principles used by Anderson and May (1980, 1981), but I have modified the specifics of their model to align it with the model of resource species R_1 used in previous chapters. Imagine a single resource species with logistic population growth, just as has been considered throughout this book (i.e., equation (3.1)). Some number of individuals are infected with the pathogen, R_{1I}, and the remaining individuals are uninfected but susceptible to becoming infected, R_{1S}: thus, $R_1 = R_{1S} + R_{1I}$. Infections occur when susceptible and infected individuals come into contact, and so the rate of infection is modeled as simple mass action at rate β_{11} (Anderson and May 1980, 1981, 1991, Keeling and Rohani 2008). (Many other forms of transition dynamics may occur (McCallum et al. 2001), but for simplicity I focus on only mass action here.) Individuals recover from the pathogen infection at rate v_1. Finally, fitness dif-ferences between infected and susceptible individuals are modeled by the classes having different intrinsic rates of increase and different strengths of density depen-dence. These assumptions imply the following model:

$$\frac{dR_{1S}}{dt} = c_1 R_{1S} + \alpha_1 c_1 R_{1I} - d_1 R_{1S} \left(R_{1S} + R_{1I} \right) - \beta_{11} R_{1S} R_{1I} + v_1 R_{1I}$$
$$\frac{dR_{1I}}{dt} = -\gamma_1 d_1 R_{1I} \left(R_{1S} + R_{1I} \right) + \beta_{11} R_{1S} R_{1I} - v_1 R_{1I}$$

(7.1)

The last two terms in each equation describe the mass action of susceptible indi-viduals becoming infected and so transitioning to the infected pool at rate β_{11} and the recovery of infected individuals transitioning back to the susceptible pool at rate v_1 (Anderson and May 1981). The parameter α_1 scales the infected class's

TABLE 7.1. Summary descriptions of the additional parameters used in the models presented in this chapter

Parameter	Description
α_i	Scaling parameter for the decrement of the intrinsic rate of increase for infected individuals of resource i
β_{ii}'	Infection rate of susceptible individuals of species i' when they come in contact with infected individuals of species i
v_i	Recovery rate of infected individuals of species i
δ_i	Scaling parameter for the increase in susceptibility to predation in prey species i
θ_j	Scaling parameter for the increase in the intrinsic death rate of consumer or predator j
ψ_{jk}	Scaling parameter for the increase in the attack coefficient for predator k feeding on consumer j

intrinsic rate of increase to be a product of that of the susceptible class, with the assumption that infected individuals' intrinsic rate of increase may be less than that of susceptible/uninfected individuals (i.e., $\alpha_1 \le 1$). The parameter γ_1 scales the infected class's density dependence, with the assumption that infected individuals may experience a greater consequence of density dependence than that of susceptible individuals (i.e., $\gamma_1 \ge 1$). Either of these may be because of differences in the per capita birth or death rates of susceptible and infected individuals (e.g., see chapter 2). Table 7.1 lists new parameters for models presented in this chapter.

The one-resource infected simulation module can be accessed at https://mechanismsofcoexistence.com/pathogen_2D_1R.html.

Both classes of individuals will be present in the population if the isoclines for equations (7.1) intersect at a feasible equilibrium for both classes in the $R_{1S} - R_{1I}$ space (fig. 7.2(a)). The equations for the isoclines are

$$\frac{dR_{1S}}{dt} = 0 : R_{1I} = \frac{c_1 R_{1S} - d_1 R_{1S}^2}{d_1 R_{1S} + \beta_{11} R_{1S} - v_1 - c_1 \alpha_1}$$

$$\frac{dR_{1I}}{dt} = 0 : R_{1I} = \frac{(\beta_{11} - d_1 \gamma_1) R_{1S} - v_1}{d_1 \gamma_1}.$$

The R_{1S} isocline for susceptible individuals intersects the R_{1S} axis at $R_{1(0)}^* = c_1 / d_1$, its equilibrium abundance in the absence of the pathogen, but does not intersect the R_{1I} axis. Instead, the R_{1S} isocline asymptote occurs at

$$R_{1S} = \frac{c_1 \alpha_1 + v_1}{d_1 + \beta_{11}}.$$

This asymptote at low R_{1S} abundance places a minimum on the frequency of susceptible individuals if the population is viable (the vertical dashed black line in fig. 7.2(a)). The R_{1I} isocline for infected individuals is a line that intersects the R_{1I} axis at $-v_1/(d_1\gamma_1)$ and the R_{1S} axis at $v_1/(\beta_{11} -d_1\gamma_1)$, and has a slope of $(\beta_{11} - d_1\gamma_1)/(d_1\gamma_1)$. If these isoclines intersect at feasible abundances for the two classes, it is stable (Anderson and May 1981).

The change in total population abundance because of the infectious disease is caused by the demographic differences between susceptible and infected individuals. If the intrinsic rates of increase and the strengths of density dependence for the two are identical (i.e., $\alpha_1 = 1$ and $\gamma_1 = 1$), the total population size is the same as for a completely uninfected population, namely, $R_{1(0)}^* = c_1 / d_1$, and the relative frequencies of the two classes are determined by the infection (β_{11}) and recovery (v_1) dynamics. For the infectious pathogen to maintain itself in the resource population (i.e., $R_{1I}^* > 0$), the inequalities

$$\frac{v_1}{\beta_{11} - d_1\gamma_1} < \frac{c_1}{d_1} \text{ and } \beta_{11} > d_1\gamma_1 \qquad (7.2)$$

must be satisfied. These are true when the R_{1I} isocline intersects the R_{1S} axis at a positive value below $R_{1(0)}^* = c_1 / d_1$; in other words, the ratio of the rate of recovery of infected individuals to the net addition of infected individuals must be less than $R_{1(0)}^*$ but must intersect the R_{1S} axis so as to intersect the R_{1S} isocline at a feasible equilibrium value. Thus, if the infection rate is not sufficiently high relative to the recovery rate of hosts and loss of individuals from the population due to death, the pathogen cannot coexist with the host.

When the strength of density dependence is higher for infected individuals than that for susceptible individuals (i.e., $\gamma_1 > 1$) (e.g., the per capita death rate of infected individuals increases with total abundance or the per capita birth rate of infected individuals decreases faster than that for susceptible individuals), the abundance of infected individuals declines and the abundance of susceptible individuals increases. For example, figure 7.2(b) shows the equilibrium abundances of the two individual classes for pathogens that cause different strengths of density dependence for infected individuals. Pathogens that generate stronger density dependence in infected individuals cause the R_{1I} isocline to have a shallower slope and to intersect the R_{1I} axis at less negative values, which increases R_{1S}'s abundance and decreases R_{1I}'s abundance (figs. 7.2(c)–(e)). If the fitness decrement of being infected is large enough, the isoclines will no longer intersect at a feasible equilibrium (i.e., inequalities (7.2) are not satisfied), and only susceptible individuals will be present (i.e., $\gamma_1 > 12.4$ giving $d_1/\gamma_1 > 0.248$ in fig. 7.2). In other words, the fitness consequences of the pathogen are so great to the host that the

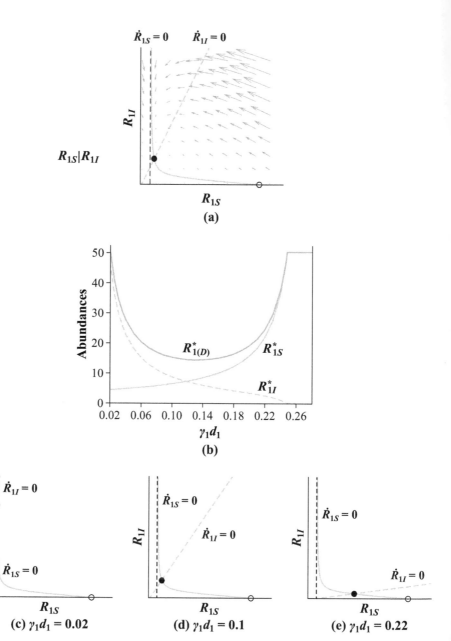

$\dot{R}_{1S} = 0$ $\dot{R}_{1I} = 0$

R_{1I}

$R_{1S}|R_{1I}$

R_{1S}

(a)

Abundances

$R^{*}_{1(D)}$ R^{*}_{1S}

R^{*}_{1I}

0.02 0.06 0.10 0.14 0.18 0.22 0.26

$\gamma_1 d_1$

(b)

R_{1I}

$\dot{R}_{1I} = 0$

$\dot{R}_{1S} = 0$

R_{1S}

(c) $\gamma_1 d_1 = 0.02$

R_{1I}

$\dot{R}_{1S} = 0$

$\dot{R}_{1I} = 0$

R_{1S}

(d) $\gamma_1 d_1 = 0.1$

R_{1I}

$\dot{R}_{1S} = 0$

$\dot{R}_{1I} = 0$

R_{1S}

(e) $\gamma_1 d_1 = 0.22$

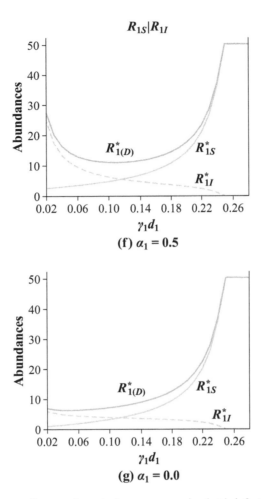

FIGURE 7.2. State space diagrams for a single resource species that is infected with a pathogen that cannot live outside its host. In each panel, features associated with the susceptible individuals that are not currently infected with the pathogen (R_{1S}) are shown as solid lines, and features associated with currently infected individuals (R_{1I}) are dashed lines. Panels (a) and (c)–(e) show the isoclines for these two individual classes for various parameter combinations. Unstable equilibria are identified by open circles, and stable equilibria are identified by filled black circles. The grid of arrows in (a) shows how the abundances of each individual class change away from the isoclines, which identify the general trajectories of abundance change. Panels (b), (f), and (g) show equilibrium abundances for the susceptible and infected individuals and the total number of individuals ($R_{1(D)}$) for various combinations of density dependence experienced by infected individuals ($\gamma_1 d_1$). In all panels, $c_1 = 1.0$, $d_1 = 0.02$, $\beta_{11} = 0.25$, and $v_1 = 0.1$. Additional parameters are as follows: (a) $\alpha_1 = 1$, $\gamma_1 = 4.0$; (c) $\alpha_1 = 1$, $\gamma_1 = 1.0$; (d) $\alpha_1 = 1$, $\gamma_1 = 5.0$; (e) $\alpha_1 = 1$, $\gamma_1 = 11.0$; (f) $\alpha_1 = 0.5$, $\gamma_1 = 1.0$; and (g) $\alpha_1 = 0.0$, $\gamma_1 = 1.0$.

pathogen cannot maintain a viable population on the host and so will go extinct. Consequently, the host population will then equilibrate at $R_{1(0)}^* = c_1/d_1$ with all susceptible but uninfected individuals. As a result, total population size initially decreases and then increases with an increasing density-dependent fitness difference between susceptible and infected individuals (fig. 7.2(b)).

In contrast, if only the intrinsic rate of increase for infected individuals is less than that of susceptible individuals (i.e., $\alpha_1 < 1$), the total abundance of both individual classes is reduced, and pathogens that cause a greater difference decrease the abundances of both susceptible and infected individuals (fig. 7.2(b) and (f)–(g)). This is because the asymptote of the R_{1S} isocline decreases with decreasing values of α_1 but does not affect the R_{1I} isocline. However, these effects are only significantly manifested if susceptible and infected hosts have similar strengths of density dependence.

Thus, as one might imagine, the expected consequence of a disease afflicting a host population is to decrease total host population size (i.e., $R_{1(D)}^* < R_{1(0)}^*$). If the fitness consequence for being infected is small, infected individuals will be common and overall population size will be depressed only somewhat. If the fitness consequences are very large, infected individuals will be rare and, consequently, overall population size will be depressed only somewhat. The greatest depression in overall population size occurs when the fitness decrement is moderate but infected individuals are still relatively common at equilibrium.

This model offers a set of very straightforward predictions to test. For example, the severity of the fitness consequences of infections should be directly proportional to the relative frequency of infected individuals in the resource population. Comparative analyses of different pathogens infecting host populations would immediately test this seemingly intuitive prediction. Moreover, populations that are infected with a pathogen should have a lower abundance than a population that is not infected. This latter prediction suggests the obvious experiment of treatments with and without various pathogens infecting a resource species and allowing each replicate to approach its equilibrium abundance.

The degree of abundance depression because of a pathogen infection would be a more difficult prediction to test. For example, imagine a series of pathogens that all infect a single resource species, and these pathogens cause different per capita death rates in the host. These pathogens can be ordered from most severe to weakest effect along the x axis in figure 7.2(b). If replicate populations of the resource were infected with one of these different pathogens, the abundances of infected and susceptible individuals and total population size should follow the same qualitative pattern as figure 7.2(c)–(e), namely, a qualitative U-shaped pattern in total abundance, with susceptibles decreasing and infected individuals increasing when moving from more to less severe pathogens.

Sick Consumer Species

A pathogen may afflict not only basal resource species in models of food webs but also species at higher trophic levels. Therefore, consider a resource that is fed upon by a single consumer that is itself afflicted with a pathogen (see fig. 7.1(b)). In this basic model, pathogens are assumed to have two possible effects on infected consumer individuals: increased per capita death rate and decreased feeding on their own prey. These features can be easily added to the Rosenzweig and MacArthur (1963) model of a consumer species feeding on a resource species (i.e., equations (3.3)) to give

$$\frac{dN_{1S}}{dt} = \frac{b_{11}a_{11}R_1N_{1S}}{1+a_{11}h_{11}R_1} + \frac{b_{11}\delta_1a_{11}R_1N_{1I}}{1+\delta_1a_{11}h_{11}R_1} - f_1N_{1S} - \beta_{11}N_{1S}N_{1I} -$$

$$\frac{dN_{1I}}{dt} = -\theta_1f_1N_{1I} + \beta_{11}N_{1S}N_{1I} - v_1N_{1I}. \tag{7.3}$$

$$\frac{dR_1}{dt} = R_1\left(c_1 - d_1R_1 - \frac{a_{11}N_{1S}}{1+a_{11}h_{11}R_1} - \frac{\delta_1a_{11}N_{1I}}{1+\delta_1a_{11}h_{11}R_1}\right)$$

N_{1S} is the abundance of susceptible consumers, and N_{1I} is the abundance of infected consumers whose pathogen cannot live outside a host. Most of the parameters in equations (7.3) were introduced in chapter 3. Again, susceptible individuals are infected according to mass action at a rate of β_{11} and recover from the infection at rate v_1, just as in equations (7.1) for an infected resource. The two new parameters in this model are δ_1, which scales the attack rate of infected consumers relative to that of susceptible consumers ($0 \leq \delta_1 \leq 1$), and θ_1, which scales the intrinsic death rate of infected consumers relative to that of susceptible consumers ($\theta_1 \geq 1$). For simplicity, the consumer does not experience any form of direct self-limitation in the model considered here.

The isoclines from these equations are

$$\frac{dN_{1S}}{dt} = 0 : N_{1S} = \frac{v_1N_{1I} + \dfrac{b_{11}\delta_1a_{11}R_1N_{1I}}{1+\delta_1a_{11}h_{11}R_1}}{f_1 + \beta_{11}N_{1I} - \dfrac{b_{11}a_{11}R_1}{1+a_{11}h_{11}R_1}}$$

$$\frac{dN_{1I}}{dt} = 0 : N_{1S} = \frac{v_1 + \theta_1f_1}{\beta_{11}}.$$

$$\frac{dR_1}{dt} = 0 : N_{1S} = \left(c_1 - d_1R_1 - \frac{\delta_1a_{11}N_{1I}}{1+\delta_1a_{11}h_{11}R_1}\right)\frac{1+a_{11}h_{11}R_1}{a_{11}}$$

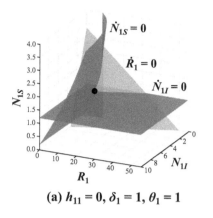

(a) $h_{11} = 0, \delta_1 = 1, \theta_1 = 1$

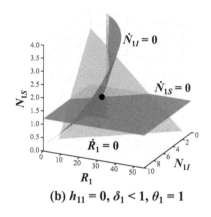

(b) $h_{11} = 0, \delta_1 < 1, \theta_1 = 1$

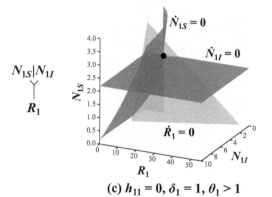

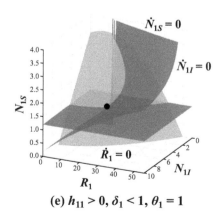

(c) $h_{11} = 0, \delta_1 = 1, \theta_1 > 1$

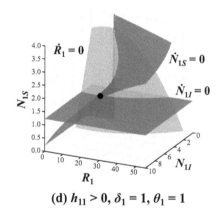

(d) $h_{11} > 0, \delta_1 = 1, \theta_1 = 1$

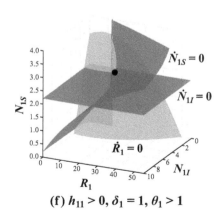

(e) $h_{11} > 0, \delta_1 < 1, \theta_1 = 1$

(f) $h_{11} > 0, \delta_1 = 1, \theta_1 > 1$

The resource isocline is largely unchanged because of the infected consumer. The R_1 isocline still intersects its own axis at $R^*_{1(0)} = c_1/d_1$ and the N_{1S} axis at c_1/a_{11}, but it also intersects the N_{1I} axis at $c_1/(\delta_1 a_{11})$. If the consumer has a linear functional response ($h_{11} = 0$), the R_1 isocline is a plane; with a saturating functional response ($h_{11} > 0$), the R_1 isocline again bows away from the origin but is a straight line between its points of intersection with the N_{1S} and N_{1I} axes in the $N_{1S} - N_{1I}$ face (fig. 7.3).

The consumer infected simulation module can be accessed at https://mechanismsofcoexistence.com/pathogen_3D_1R_1N.html.

The consumer is now the species with two isoclines in this three-dimensional state space. Because both susceptible and infected consumers contribute to the birth of susceptible consumers, the N_{1S} isocline now has a much more complicated shape. It is still a vertical line in the $N_{1S} - R_1$ face that intersects the R_1 axis at $f_1/(a_{11}(b_{11} - f_1 h_{11}))$, but it bows away from the $N_{1S} - N_{1I}$ face; and this bowing becomes more pronounced with higher values for handling time (cf. fig. 7.3(a)–(c) with 7.3(d)–(f)) and greater levels of feeding depression of infected individuals (i.e., larger δ_1) (cf. fig. 7.3(a) and (d) with 7.3(b) and (f)). The N_{1I} isocline is a plane that intersects the N_{1S} axis at $(v_1 + f_1 \theta_1)/\beta_{11}$ and is parallel to both the R_1 and N_{1I} axes. Consequently, if the pathogen can persist in the consumer population, the equilibrium abundance of susceptible consumers is defined by the point of intersection of the N_{1I} isocline with the N_{1S} axis (fig. 7.3).

The pathogen can coexist in this community if these three isoclines intersect at a feasible point with the resource and both susceptible and infected consumers having positive abundances at this point equilibrium. Thus, for the pathogen to be able to invade and coexist in the community, the N_{1I} isocline must intersect the N_{1S} axis below the equilibrial abundance of the consumer in the pathogen's absence (i.e., the point where the R_1 and N_{1S} isoclines intersect in the $N_{1S} - R_1$ face); that is,

$$\frac{v_1 + f_1 \theta_1}{\beta_{11}} < N^*_{1(R_1)}. \tag{7.4}$$

This inequality identifies the demographic constraints on pathogen coexistence: namely, the pathogen will not be able to invade if the infection rate β_{11} is too

---

FIGURE 7.3. State space diagrams for a consumer infected with a pathogen (N_{1S} and N_{1I}) feeding on a resource (R_1). The isocline for R_1 is identified as $\dot{R}_1 = 0$, and the isoclines for the two consumers are identified for susceptible ($\dot{N}_{1S} = 0$) and infected ($\dot{N}_{1I} = 0$) individuals. Equilibria are identified as in figure 7.2. Unless otherwise stated, parameter values in all panels are as follows: $c_1 = 1.0$, $d_1 = 0.02$, $a_{11} = 0.3$, $b_{11} = 0.1$, $f_1 = 0.25$, $\beta_{11} = 0.25$, and $v_1 = 0.05$. In (a)–(c), $h_{11} = 0.0$; in (d)–(f), $h_{11} = 0.1$. In (a), (c)–(d), and (f), $\delta_1 = 1.0$; in (b) and (e), $\delta_1 = 0.5$. In (a)–(b) and (d)–(e), $\theta_1 = 1.0$; in (c) and (f), $\theta_1 = 2.0$.

low relative to the recovery rate v_1 or if the increase in the consumer per capita mortality due to the infection θ_1 is too substantial. Also, note the similarity between this criterion and the analogous criteria from the previous model, namely, inequalities (7.2).

Because the position of the N_{1I} isocline is influenced by the infection and recovery rates and the proportional increase in intrinsic death rate of infected individuals, decreased feeding by infected individuals (i.e., $\delta_1 < 1$) has no effect on the equilibrium abundance of susceptible consumers. However, decreased feeding by infected individuals causes the resource abundance to increase, which in turn typically also causes an increase in the infected individual abundance (fig. 7.4). This occurs because a lower feeding level by infected individuals (i.e., $\delta_1 < 1$) causes the resource isocline to intersect the N_{1I} axis at higher values and causes the N_{1S} isocline to bow farther away from the $N_{1S} - N_{1I}$ face (see fig. 7.3). In fact, for all parameter combinations I have examined, N_{1I}^* with no feeding diminution ($\delta_1 = 1$) was less than N_{1I}^* when infected individuals did not feed at all ($\delta_1 = 0$). Consequently, a diminution of feeding by infected individuals on its own will not cause the pathogen to go extinct (i.e., $N_{1I}^* = 0$).

In contrast, pathogens that increase the intrinsic death rate of infected consumers (i.e., $\theta_1 > 1$) decrease the relative abundance of infected consumers and increase that of susceptible consumers (fig. 7.4). With a high enough death rate due to infection, inequality (7.4) will not be satisfied, and the pathogen will not be able to coexist in the community. Thus, infections by pathogens causing such high death rates will cause short-term perturbations to the demography of the consumer but eventually will go extinct and so not cause long-term chronic depression of consumer abundance.

At the community level, the most important consequence is the indirect effect the pathogen has on resource abundance. If the pathogen has absolutely no fitness effects on consumers (i.e., $\delta_1 = \theta_1 = 1$), the equilibrium resource abundance is the same as when the pathogen is absent because the total consumer abundance is the same in these two situations. With diminution of feeding (i.e., $\delta_1 < 1$) or increasing intrinsic death rate (i.e., $\theta_1 > 1$) because of infection, even though the total number of consumers may be greater, the equilibrium resource abundance is always higher than when the pathogen is absent (fig. 7.4). This is an example of a *hydra effect*, in which a process that increases the consumer's abundance also increases the abundances of the resources on which the consumer feeds (Abrams 2009b, Sieber and Hilker 2012, Cortez and Abrams 2016). One example may be the association between *Metschnikowia* infections and increased abundances of both their *Daphnia* hosts and the phytoplankton on which the *Daphnia* feed (Dr. Meghan Duffy, personal communications). Thus, pathogens that affect the demographic rates of their consumer hosts may increase resource abundance.

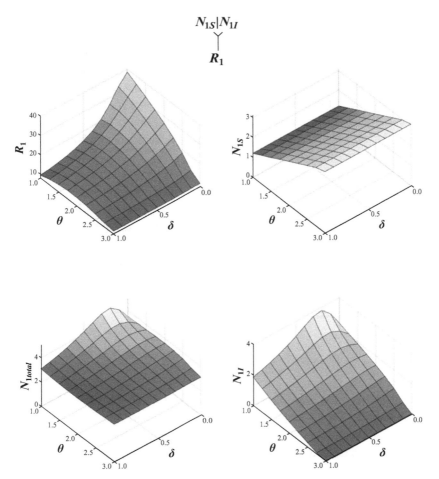

FIGURE 7.4. Surfaces illustrating the equilibrium abundances of a consumer that is infected with a pathogen feeding on a resource. The panels show the equilibrium abundances of the resource (R_1), the susceptible consumers (N_{1S}), the infected consumers (N_{1I}), and the total number of consumers (N_{1total}) along gradients of the proportional decline in the attack coefficient of infected consumers (δ_1) and the increase in the intrinsic death rate of infected individuals (θ_1). Parameters are as specified in figure 7.3(d), except $h_{11} = 0.05$, and changes in δ_1 and θ_1 are as specified on the axes.

A major issue this model raises that deserves serious empirical scrutiny is the prediction that the abundances of both the consumer and the resource should be higher if the consumer is infected with a pathogen that diminishes its ability to feed on the resource. All the studies described in chapter 3 for this system should be completed, but the presence of a pathogen in the consumer population also warrants this prediction to be tested.

A number of comparative observational studies would test the validity of the predictions. For example, comparisons of the same consumer-resource pair across sites where the consumer is infected with various pathogens that cause different levels of death and feeding diminution and sites where the consumer is not infected would provide data to test this prediction. Obviously, the sites would vary in many abiotic factors that would cause many parameters of the model to be different. However, if enough sites for each pathogen were included in the data set, statistical methods would parse differences among sites from pathogen effects to test the prediction.

However, the most definitive test would be an experiment that manipulated the presence and absence of various pathogens infecting the consumer population. Pathogens that mainly kill the consumer should cause increased resource abundance but have differing consequences on the relative frequencies of infected and susceptible consumers, depending on the severity of the fitness consequences. In contrast, pathogens that primarily reduce the short-term feeding rates of the consumer without killing them (i.e., that make infected individuals sick but not dead), should have higher resource and consumer abundances with the pathogen as compared to when the pathogen is absent.

Sick Predator Species

To confirm that these results are not restricted to a two-trophic-level food chain, consider a three-trophic-level food chain in which the predator is infected with a pathogen that cannot live outside its host (see fig. 7.1(c)). The following model is an analogous extension of equations (4.1), but neither N_1 nor P_1 experience any direct self-limitation:

$$\frac{dP_{1S}}{dt} = \frac{n_{11}m_{11}P_{1S}N_1}{1+m_{11}l_{11}N_1} + \frac{n_{11}m_{11}\delta_1 P_{1I}N_1}{1+m_{11}l_{11}\delta_1 N_1} - x_1 P_{1I} - \beta_1 P_{1S}P_{1I} + v_1 P_{1I}$$

$$\frac{dP_{1I}}{dt} = -\theta_1 x_1 P_{1I} + \beta_1 P_{1S}P_{1I} - v_1 P_{1I}$$

$$\frac{dN_1}{dt} = N_1\left(\frac{b_{11}a_{11}R_1}{1+a_{11}h_{11}R_1} - \frac{m_{11}P_{1S}}{1+m_{11}l_{11}N_1} - \frac{m_{11}\delta_1 P_{1I}}{1+m_{11}l_{11}\delta_1 N_1} - f_1\right).$$

$$\frac{dR_1}{dt} = R_1\left(c_1 - d_1 R_1 - \frac{a_{11}N_1}{1+a_{11}h_{11}R_1}\right)$$

(7.5)

The isocline system for this model with saturating functional responses for both N_1 and P_1 is illustrated in figure 7.5: compare this system with that shown in

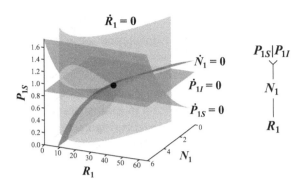

FIGURE 7.5. State space diagram for a resource (R_1) being fed upon by a consumer (N_1), which in turn is fed upon by a predator that is infected by a pathogen (susceptible P_{1S} and infected P_{1I}) that cannot live outside a host. The resource and consumer isoclines are identified as in figure 7.3, and the P_{1S} and P_{1I} isoclines are identified in the figure as $\dot{P}_{1S} = 0$ and $\dot{P}_{1I} = 0$, respectively. Equilibria are identified as in figure 7.2. The parameter values are as follows: $c_1 = 2.0$, $d_1 = 0.04$, $a_{11} = 0.5$, $b_{11} = 0.1$, $h_{11} = 0.1$, $f_1 = 0.1$, $m_{11} = 0.4$, $n_{11} = 0.1$, $l_{11} = 0.1$, $x_1 = 0.1$, $\delta_1 = 0.70$, $\theta_1 = 1.2$, $\beta_{11} = 0.25$, and $v_1 = 0.1$.

figure 4.2(b) for an uninfected P_1. Again, δ_1 scales the attack rate of infected predators relative to that of susceptible predators ($0 \le \delta_1 \le 1$), and θ_1 scales the intrinsic death rate of infected predators relative to that of susceptible predators ($\theta_1 \ge 1$).

A pathogen infecting P_1 causes analogous changes at lower trophic levels, as seen when the pathogen infects N_1 in the absence of P_1 (cf. figs. 7.4 and 7.6). The fitness costs of infection may either decrease or increase P_1's abundance, depending on the exact combination of diminution in attack coefficient and increase in intrinsic death rate for infected predators. However, consistent with the results for the consumer (as in fig. 7.4), an infected top predator causes an increase in N_1's abundance and in turn a decrease in the basal resource's abundance (fig. 7.6). An infected predator causes a *trophic cascade* down the food chain. Thus, these results confirm that a pathogen infecting the species at the top of a food web will have similar consequences to species at lower trophic levels that propagate down the food web.

All the same issues and studies described in the previous section apply here. The most interesting question is whether a pathogen infecting a top predator can change its abundance and feeding performance enough to generate a trophic cascade. Thus, experiments or observational studies comparing food chains in which the top predator is and is not infected with a costly but coexisting pathogen would be crucial.

$$\frac{dR_{1S}}{dt} = c_1 R_{1S} + \alpha_1 c_1 R_{1I} - d_1 R_{1S}\left(R_{1S} + R_{1I}\right) - \beta_{11} R_{1S} R_{1I} - \beta_{12} R_{1S} R_{2I} + v_1 R_{1I}$$

$$\frac{dR_{1I}}{dt} = -\gamma_1 d_1 R_{1I}\left(R_{1S} + R_{1I}\right) + \beta_{11} R_{1S} R_{1I} + \beta_{12} R_{1S} R_{2I} - v_1 R_{1I}$$

$$\frac{dR_{2S}}{dt} = c_2 R_{2S} + \alpha_2 c_2 R_{2I} - d_1 R_{2S}\left(R_{2S} + R_{2I}\right) - \beta_{21} R_{2S} RI - \beta_{22} R_{2S} R_{2I} + v_2 R_{2I}$$

$$\frac{dR_{2I}}{dt} = -\gamma_2 d_2 R_{2I}\left(R_{2S} + R_{2I}\right) + \beta_{21} R_{2S} R_{1I} + \beta_{22} R_{2S} R_{2I} - v_2 R_{2I}$$

$$. \quad (7.6)$$

The only additional feature here is that susceptible individuals in each species can now obtain the pathogen from infected individuals of both species. The parameters β_{ii}' define the transmission rate from infected individuals of species i' to species i. All other parameters are as defined for equations (7.1) with appropriate subscripting to identify species. Holt and Pickering (1985) analyzed a related model (hosts had exponential, instead of logistic, population growth in the absence of the pathogen) to explore when a shared pathogen may prevent coexistence of the two host species. Begon et al. (1992) analyzed a model more closely allied to equations (7.6), where both hosts show logistic population growth but the demographic transitions between various classes are not as explicitly identified; the results presented here reinforce their results.

I forego a visual presentation of the four isoclines here. They follow the shapes presented in figure 7.2(a) projected into three and four dimensions, but they angle toward or away from the two axes of the infected resources because of the dependencies on those individual classes in the model.

The first result that is apparent for having multiple infected species is to increase the relative frequency of infected individuals in each species (see also Begon et al. 1992). For example, figure 7.7 presents the abundances of two resource species that share a pathogen and cross-infect one another, and both have a significant fitness decrement in both the intrinsic growth rate ($\alpha_i < 1$) and the strength of density dependence ($\gamma_i > 1$) for infected individuals. With an intrinsic growth rate of $c_i = 1.0$ and only one resource species present, each species would have equilibrium abundances of the two individual classes of $[R_{iS(0)}^*, R_{iI(0)}^*] = 7.775, 3.750]$, which gives a total abundance of $R_{i(D)}^* = 11.525$. However, with both species present, these abundances become $[R_{iS(R_{i'})}^*, R_{iI(R_{i'})}^*] = [2.62, 3.31]$ and so a total population size for each is $R_{i(R_{i'},D)}^* = 5.93$. Having another species that shares the pathogen substantially decreases the total abundance of each species. Thus, having an environmental reservoir of pathogens (i.e., infected heterospecifics) that are continually infecting a species does have serious detrimental consequences.

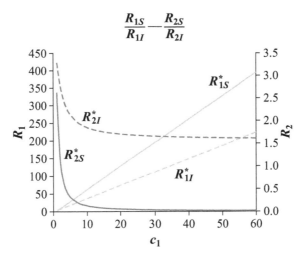

FIGURE 7.7. Equilibrium abundances of susceptible (R_{iS}) and infected (R_{iI}) individuals of two resource species that are both infected by the same pathogen that cannot survive outside a host along a gradient of the intrinsic rate of increase for R_1. This interaction is an example of apparent competition mediated by a pathogen. Individual classes for the two species are identified as in figure 7.2. Parameters are as specified in figure 7.5, except as follows: $c_2 = 1.0$, $d_1 = d_2 = 0.02$, $\alpha_1 = \alpha_2 = 0.25$, $\gamma_1 = \gamma_2 = 8.0$, $\beta_{11} = \beta_{22} = 0.25$, $\beta_{12} = \beta_{21} = 0.05$, and $v_1 = v_2 = 0.05$.

Given the linkage of abundances created by a shared pathogen, environmental differences that influence the equilibrium abundance of one resource species may indirectly influence the other—namely, the resources may be apparent competitors (Begon et al. 1992, Bowers and Turner 1997, Bonsall and Holt 2010). Shared pathogens can also be very detrimental to rarer host species because of the apparent competitive interactions between the hosts that are mediated by the pathogen. Figure 7.7 shows the differences in equilibrium abundances from equations (7.6) across a gradient of intrinsic growth rates for R_1 but where R_2 has the same intrinsic growth rate over the entire gradient. Imagine communities of R_1 and R_2 that develop at different sites, and the sites differ in the values of c_1, so the equilibrium abundance of R_1 would be different across sites were it the only species present. However, R_2 would equilibrate at the same abundance at every site, were it the only species present. At sites where c_1 is substantially greater than c_2, nearly all R_2 individuals are infected, and their abundances decrease with increasing c_1. However, in this model, R_2 can coexist no matter the productivity of R_1. Other model formulations can result in the elimination of one host species because of apparent competition via the shared pathogen (Begon et al. 1992, Bowers and Turner 1997, Bonsall and Holt 2010).

These results confirm that a pathogen will act as an apparent competitive mediating agent between the species that it infects (Holt and Pickering 1985, Begon et al. 1992, Bonsall and Holt 2010). Obviously, the strengths of the fitness decrements that the pathogen causes in the various hosts and the transmission rates between host species will all shape the strength and importance of this indirect effect among the species. Rare species may be driven extinct at a lower abundance of the common species if the pathogen causes a disproportionate fitness decrement to the rare species. Conversely, if the fitness decrement is much more severe in a species that would be more common in the pathogen's absence, rare species would have a disproportionate indirect effect on the common species' abundance because it serves as a pathogen reservoir.

All the empirical tests suggested in chapter 3 for the apparent competition module are appropriate here, with the pathogen substituted for the consumer species. Important metrics to quantify in these types of systems will be to measure the fitness decrements, reflected in the difference in equilibrium abundances of each species in populations that do and do not harbor the pathogen. These studies compare $R_{i(0)}^*$ to $R_{iS(0)}^* + R_{iI(0)}^*$, which is the demographic cost to the population of the fitness cost of the infection. These measures will be critical for predicting the relative apparent competitive effects of species that are mediated by the shared pathogen and for determining the drivers of the various species' abundances in any particular community. For example, is the rarer species in this interaction at a disadvantage in a particular community because its fitness costs from the pathogen are substantially larger or because the abiotic environment and other conditions at that site are poorer (e.g., smaller c_i)?

Consumer Feeding on a Sick Resource Species

In the basic model section above, I explored a pathogen-infected resource species living by itself. Now imagine that a consumer of that resource invades the community, resulting in a pathogen-free consumer feeding on a pathogen-infected resource (see invasion 26 in fig. 7.1(d)) (see also Packer et al. 2003 and Hall et al. 2005 for versions of this model in which the consumer abundance is held constant). In this scenario, further assume that infected resource individuals potentially suffer three fitness decrements relative to susceptible individuals: again, their intrinsic rates of increase may be lower than susceptible individuals and infected individuals may experience stronger density dependence; and, additionally, infected individuals may be more vulnerable to the consumer. The dynamics of these two species are governed by the following equations:

$$\frac{dN_1}{dt} = \frac{b_{11}a_{11}R_{1S}N_1 + b_{11}\psi_{11}a_{11}R_{1I}N_1}{1 + a_{11}h_{11}(R_{1S} + \psi_{11}R_{1I})} - f_1 N_1$$

$$\frac{dR_{1S}}{dt} = c_1 R_{1S} + \alpha_1 c_1 R_{1I} - dR_{1S}(R_{1S} + R_{1I}) - \frac{a_{11}R_{1S}N_1}{1 + a_{11}h_{11}(R_{1S} + \psi_{11}R_{1I})} - \beta_{11}R_{1S}R_{1I} + v_1 R_{1I}, \quad (7.7)$$

$$\frac{dR_{1I}}{dt} = -\gamma_1 dR_{1I}(R_{1S} + R_{1I}) - \frac{\psi_{11}a_{11}R_{1I}N_1}{1 + a_{11}h_{11}(R_{1S} + \psi_{11}R_{1I})} + \beta_{11}R_{1S}R_{1I} - v_1 R_{1I}$$

where α_1 scales the decrease in infected individuals' intrinsic rate of increase relative to that of susceptible individuals ($\alpha_1 \leq 1$), γ_1 scales the strength of density dependence for infected individuals relative to that of susceptible individuals ($\gamma_1 \geq 1$), ψ_{11} scales the increase in the attack coefficient of the consumers on infected individuals relative to that of susceptible individuals ($\psi_{11} \geq 1$), and all other parameters here are as defined for equations (7.1) and (7.3).

The isoclines for this community module should be relatively familiar by now. The R_{1I} isocline has the same positioning in the $R_{1S} - R_{1I}$ face and angles away from the N_1 axis. The R_{1S} isocline is a much more complex surface: it has the same shape in the $R_{1S} - R_{1I}$ face as in the consumer's absence, and it warps to intersect the N_1 axis at all points at and above c_1/a_{11}. The N_1 isocline is a plane that intersects the R_{1S} axis at $f_1/(a_{11}(b_{11} - f_1 h_{11}))$, the R_{1I} axis at $f_1/(\psi_{11}a_{11}(b_{11} - f_1 h_{11}))$, and runs parallel to the N_1 axis (fig. 7.8).

The three fitness costs for infected resource individuals influence the isoline shapes differently, and so affect the equilibrium abundances of the three types of individuals differently. Decreasing the intrinsic rate of increase ($\alpha_1 < 1$) of infected resource individuals decreases the asymptote of the R_{1S} isocline, but leaves the other two isoclines unchanged (fig. 7.8(b)). As a result, pathogens that cause lower α_1 values cause the consumer to equilibrate at a lower abundance, and if the fitness cost to the resource's intrinsic rate of increase is great enough, the consumer may not be able to coexist with the resource harboring a pathogen (e.g., $\alpha_1 < 0.255$ in fig. 7.9(a)). If N_1 can maintain a population on the diseased resource, the total resource abundance is constant, but infected individuals are at higher relative abundance for a greater intrinsic rate of increased cost.

Increasing the strength of density dependence ($\gamma_1 > 1$) for infected resource individuals decreases the slope of the R_{1I} isocline in the $R_{1S} - R_{1I}$ face but leaves the other two isoclines unchanged (fig. 7.8(c)). Consequently, pathogens that cause a greater strength of density dependence also cause the consumer to equilibrate at a lower abundance, and can cause the consumer to be unable to coexist with a diseased resource if the strength is too high (e.g., $\gamma_1 > 4.15$ in fig. 7.9(b)). Again, if the consumer coexists with the diseased resource, the total resource abundance is constant across density dependence strengths, but here the relative

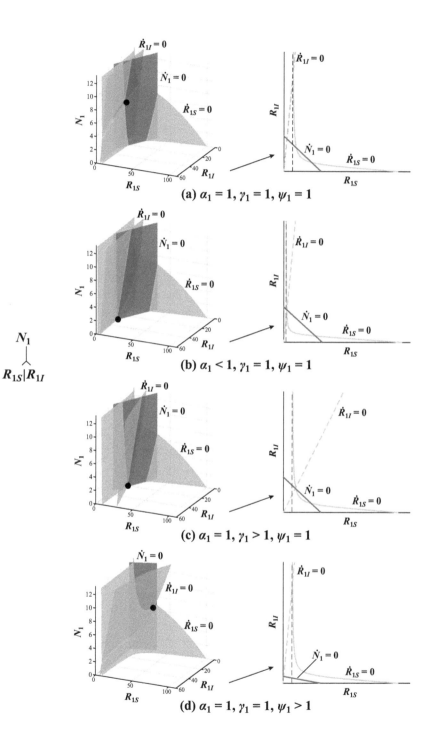

(a) $\alpha_1 = 1$, $\gamma_1 = 1$, $\psi_1 = 1$

(b) $\alpha_1 < 1$, $\gamma_1 = 1$, $\psi_1 = 1$

(c) $\alpha_1 = 1$, $\gamma_1 > 1$, $\psi_1 = 1$

(d) $\alpha_1 = 1$, $\gamma_1 = 1$, $\psi_1 > 1$

abundance of infected resource individuals decreases with increases in this fitness cost.

Finally, increasing the vulnerability of infected resources to consumer attacks
($\psi_{11} > 1$) causes the R_{1I} isocline to tilt more steeply away from the N_1 axis and
causes the N_1 isocline to intersect the R_{1I} axis closer to the origin (fig. 7.8(d)).
Unlike the other two fitness costs, making infected resource individuals more
susceptible to being killed and eaten by the consumer causes a shift in the relative
abundances of susceptible and infected resources. Total resource abundance is
different for different levels of susceptibility, and the pathogen cannot invade and
coexist if it makes infected resource individuals too susceptible to the consumer
(i.e., $\psi_{11} > 5.45$ in fig. 7.9(c)).

Thus, pathogen fitness costs may decrease the resources abundance to levels at
which the consumer cannot invade and coexist. However, these must be intermediate levels of cost to the resource: pathogens that cause too great a fitness
cost will be unable to invade and support a sustaining population on the
resource host.

One important feature for the rest of the community is that any of the fitness
costs to the resource from harboring the pathogen causes the consumer to equilibrate at a lower abundance. Consequently, other resources that might have been
excluded because of mortality imposed on them by the consumer may now be
able to invade and coexist in the community (see invasion 27 in fig. 7.1(d)). This
represents a simple extension of the issues of apparent competition considered in
chapter 3, "Apparent Competitors." Figure 7.10 shows the equilibrium abundances of two resource species and one consumer, where R_1 is infected with a
pathogen, R_2 is not, and N_1 feeds on both (dynamical equations not shown). For
the parameter combinations shown, in the absence of the pathogen (or with no
fitness costs to infected individuals), R_2 is an inferior apparent competitor that

---

FIGURE 7.8. State space diagram for a consumer (N_1) feeding on a resource that is infected by a
pathogen that cannot live outside a host. Infected (R_{1I}) resource individuals may suffer a fitness
cost as a diminished intrinsic rate of increase (scaled by α_1) and proportionally stronger density
dependence (scaled by γ_1), or be more susceptible to predation (scaled by ψ_1) relative to susceptible (R_{1S}) individuals. Graphs on the left in each panel show the full three-dimensional state
space of their isoclines, and those on the right show the respective $R_{1I} - R_{1S}$ faces. In (a), infected
individuals suffer no fitness costs of infection; in (b), infected resource individuals have a lower
intrinsic rate of increase ($\alpha_1 = 0.25$); in (c), infected individuals experience stronger density
dependence ($\gamma_1 = 4.0$); and in (d), infected resource individuals are more susceptible to predation
by the consumers ($\psi_1 = 5.0$). The resource and consumer isoclines are identified as in figure 7.3,
and equilibria are identified as in figure 7.2. Additional parameter values for (a)–(c) are as fol-
lows: $c_1 = 2.0$, $d_1 = 0.02$, $a_{11} = 0.25$, $b_{11} = 0.1$, $h_{11} = 0.03$, $f_1 = 0.65$, $\beta_{11} = 0.25$, and $v_1 = 0.05$.

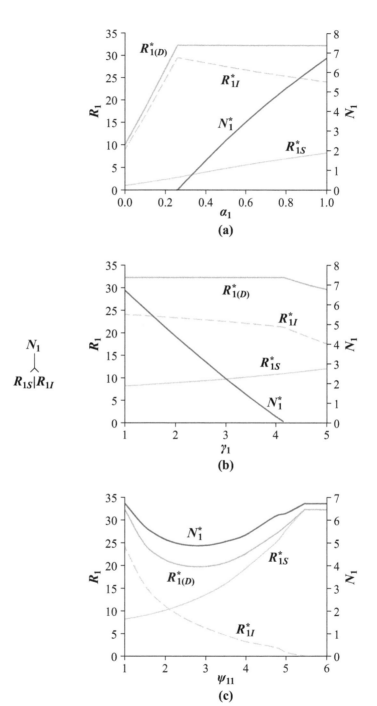

N_1
$\perp$
$R_{1S}|R_{1I}$

cannot coexist with N_1 in the presence of R_1. However, if the pathogen causes a sufficiently severe fitness decrement in R_1, all three species coexist at a stable equilibrium (fig. 7.10 shows various combinations for decreased intrinsic rate of increase ($\alpha_1 < 1$) and increased density dependence ($\gamma_1 > 1$), but increased feeding susceptibility ($\psi_{11} > 1$) also can permit R_2 to coexist).

Observational studies comparing sites with and without the pathogens in the resources will provide a trove of natural history and quantitative information on how consumer populations respond to pathogens in their resources. For example, are there resource pathogens that can drive a consumer extinct by lowering the resource's abundance below levels at which the consumer can maintain a population? Are most resource pathogens that coexist with a consumer relatively benign? Here again, experimental studies comparing the abundances of the resources with and without various pathogens are an indicator of their demographic costs, and this information paired with observational studies of their prevalence in the field with various types of consumers feeding on them will be invaluable as a starting point for evaluating these types of models.

A definitive test would, however, involve experimental studies of one consumer-resource system with different resource pathogens having a great range of fitness costs. As in the section "A Sick Consumer Species," pathogens with different fitness costs are points along the x axes in the various panels of figure 7.9, and the equilibrium abundances of the consumer and resource should follow the qualitative pattern that this figure illustrates.

The indirect consequences of a pathogen infecting one resource to the abundances of other species in the food web are also important issues of inquiry. As figure 7.10 illustrates, pathogens can influence the abilities of other resources to invade and coexist in the community because of their effects on the abundances of consumers that feed on the infected resources. Similar effects will propagate to other trophic positions in the food web. For example, a predator may not be able to invade and coexist by feeding on N_1 if N_1's abundance is reduced by R_1 being infected with a particular pathogen.

FIGURE 7.9. The equilibrium abundances for a consumer (N_1) feeding on a resource that is infected by a pathogen that cannot live outside a host when the pathogen inflicts fitness costs of various magnitudes through (a) diminishing the intrinsic rate of increase (greater cost means a smaller value of α_1), (b) increasing the strength of density dependence (greater cost means a larger value of γ_1), and (c) increasing the susceptibility of the resource to be fed upon by the consumer (greater cost means a larger value of ψ_1). The other parameters are as specified in figure 7.8.

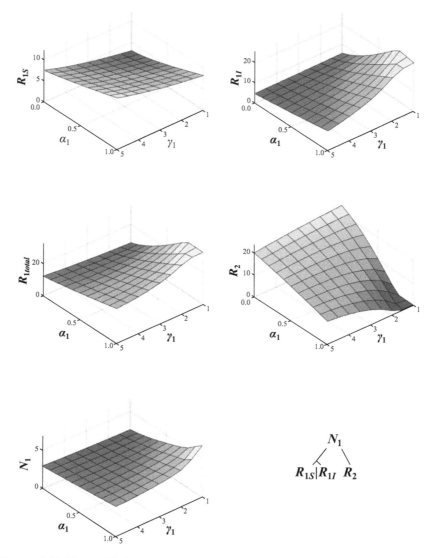

FIGURE 7.10. The equilibrium abundances for a consumer (N_1) feeding on a resource that is infected by a pathogen that cannot live outside a host when the pathogen inflicts fitness costs of both diminishing the intrinsic rate of increase (greater cost means a smaller value of α_1) and increasing the strength of density dependence (greater cost means a larger value of γ_1) to varying degrees. The other parameters are as specified in figure 7.8.

Pathogen-Mediated Resource Competition

In this section, I examine whether species-specific pathogens as enemies can permit multiple consumer species to coexist while feeding on a single resource. As in chapters 3 and 4, the addition of resources only accentuates the number of consumers that can coexist, and so I focus here on a single resource and leave it to the reader's imagination and future work to extend this explicitly to adding resources and even more consumer species.

To explore this, begin with a single resource species being fed upon by a single consumer, which may or may not be infected with a pathogen that cannot live outside the host—in other words, equations (7.3). In the absence of the pathogen, the consumer and resource come to a stable equilibrium at $[R^*_{1(N_{1S})}, N^*_{1S(R_1)}, 0]$, which is equivalent to the equilibrium discussed in chapter 3, "The First Consumer Invades." If the pathogen can invade and coexist, the community will move to a new equilibrium at $[R^*_{1(N_{1S}, N_{1I})}, N^*_{1S(R_1, N_{1I})}, N^*_{1I(R_1, N_{1S})}]$. From the "Sick Consumer Species" section above, we know the resource abundance is higher if the consumer population harbors a pathogen that causes some fitness decrement in the consumer, meaning that $R^*_{1(N_{1S})} < R^*_{1(N_{1S}, N_{1I})}$ (see fig. 7.4).

The basic premise considered here, which is akin to the Janzen-Connell hypothesis discussed in chapter 9, is that a species-specific pathogen will permit additional resource competitors into the community that could not coexist without the pathogen infecting N_1. This can be modeled directly by simply adding a second consumer to equations (7.3) (see invasion 28 in fig. 7.1(d)):

$$\frac{dN_{1S}}{dt} = \frac{b_{11}a_{11}R_1N_{1S}}{1+a_{11}h_{11}R_1} + \frac{b_{11}\delta_1 a_{11}R_1N_{1I}}{1+\delta_1 a_{11}h_{11}R_1} - f_1N_{1S} - \beta_{11}N_{1S}N_{1I} + v_1 N_{1I}$$

$$\frac{dN_{1I}}{dt} = -\theta_1 f_1 N_{1I} + \beta_{11}N_{1S}N_{1I} - v_1 N_{1I}$$

$$\frac{dN_2}{dt} = \frac{b_{12}a_{12}R_1N_2}{1+a_{12}h_{12}R_1} - f_2 N_2 \qquad (7.8)$$

$$\frac{dR_1}{dt} = R_1\left(c_1 - d_1 R_1 - \frac{a_{11}N_{1S}}{1+a_{11}h_{11}R_1} - \frac{\delta_1 a_{11}N_{1I}}{1+\delta_1 a_{11}h_{11}R_1} - \frac{a_{12}N_2}{1+a_{12}h_{12}R_1} \right)$$

Following the naming conventions established in chapter 3 (i.e., inequalities (3.13)), N_2 has an isocline that intersects the R_1 axis at $f_2/(a_{12}(b_{12}-f_2 h_{12}))$, which is above $R^*_{1(N_{1S})} = f_1/(a_{11}(b_{11}-f_1 h_{11}))$ (fig. 7.11). Thus, this second consumer can invade and coexist if

$$\frac{f_2}{a_{12}(b_{12}-f_2 h_{12})} < R^*_{1(N_{1S}, N_{1I})}, \qquad (7.9)$$

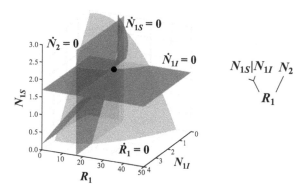

FIGURE 7.11. State space diagram for two consumers (N_1 and N_2) feeding on a single resource (R_1), where N_1 is infected with a pathogen that cannot live outside the host and so contains both susceptible (N_{1S}) and infected (N_{1I}) individuals. N_2 is not infected with a pathogen, and is a poorer resource competitor for R_1 than N_1. The resource and consumer isoclines are identified as in figure 7.3, and equilibria are identified as in figure 7.2. The parameter values are as follows: $c_1 = 1.0$, $d_1 = 0.02$, $a_{11} = 0.25$, $a_{12} = 0.2$, $b_{11} = b_{12} = 0.1$, $h_{11} = h_{12} = 0.1$, $f_1 = f_2 = 0.25$, $\delta_1 = 0.25$, $\theta_1 = 1.5$, $\beta_{11} = 0.25$, and $v_1 = 0.05$.

which again is simply the statement that N_2 has a positive per capita population growth rate when rare at this resource level. Thus, a pathogen infecting the better competitor can generate opportunities for poorer resource competitors to invade and coexist, and so enemies such as pathogens can promote consumer species richness just as Janzen and Connell hypothesized.

In fact, pathogens permit additional consumers to coexist on a single resource in the exact fashion of a specialized predator feeding on the better competitor (equations (4.6) in chapter 4, "Multiple Consumers Invading a Three-Trophic-Level Food Chain") by preventing the better competitor from depressing resource availability to its minimum possible. However, it cannot be emphasized enough that this does not alleviate resource competition between the two consumers. What permits the second consumer to invade and coexist is the additional limiting factor that depresses the better competitor's abundance below its maximum and so increases the resource abundance to a level at which the poorer competitor has a positive per capita growth rate when rare.

Whether the second consumer can invade, therefore, depends on the fitness costs incurred by the better competitor because of the pathogen. If the fitness cost is not severe enough, the resource abundance may not increase to a level that permits the poorer competitor to invade (e.g., $\delta_1 > 0.725$ for the diminution in the consumer's attack coefficient in infected individuals in fig. 7.12(a) or $\theta_1 < 1.4$ for the increase in the intrinsic death rate of infected consumers in fig. 7.12(b)). Likewise, if the fitness

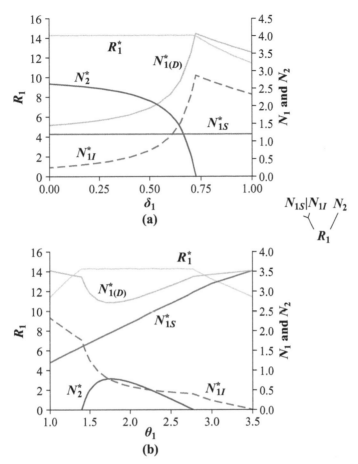

FIGURE 7.12. Equilibrium abundances for the resource and consumer species depicted in figure 7.11, where N_1 is infected with various pathogens along gradients of the fitness costs in (a) the proportional decrease in attack coefficient (δ_1) and (b) the increase in the intrinsic death rate (θ_1) for infected N_1 individuals (i.e., N_{1I}). All other parameters are as in figure 7.11.

cost is too severe, infected individuals will be too rare in the consumer population to adequately increase resource abundance to permit coexistence by the second consumer (e.g., $\theta_1 > 2.77$ in fig. 7.12(b)). Thus, pathogens that generate intermediate levels of fitness costs in infected hosts are expected to have the greatest indirect impact on community species richness by permitting inferior resource competitors to invade and coexist. Interestingly, if N_2 can coexist, the abundance of N_{1I} generally decreases below levels that would be expected in N_2's absence (fig. 7.12).

Obviously, a third consumer could invade and coexist with N_1 and N_2 if N_2 was also infected with a pathogen that reduced its fitness and thus reduced its impact on the resource, and if this third consumer could satisfy its invasibility criterion at this resource abundance (see invasion 29 in fig. 7.1(d)). As in chapter 3, this would continue until all the potential invading consumers that could satisfy their invasibility criteria had invaded. All consumers except for the poorest competitor would have to harbor a fitness-reducing pathogen, because this is what permits additional consumers to invade the community. This also extends to multiple consumers feeding on multiple resources, as in chapter 3 (see invasion 30 in fig. 7.1(d)). In this case, a fitness-reducing pathogen works in analogous fashion to direct self-limitation (i.e., $g_j > 0$ in chapter 3) or a specialized predator (chapter 4, equations (4.6); and see Grover (1994)) in permitting the number of coexisting consumers to exceed the number of limiting resources. Here, each pathogen is an additional limiting factor that permits additional consumers to potentially invade and coexist.

Various observational and experimental studies would definitively test the prediction that species-specific pathogens can permit more consumer species to coexist. Any study of these issues should fulfill the analyses outlined in chapter 3 (e.g., What are the limiting resources and how many of them are there? Do resource and consumer abundances covary as expected across environmental gradients and in manipulative experiments?). Here, the issue is the presence of the pathogens, the fitness costs incurred by hosts because of their pathogens, and the indirect effects of these fitness costs on the availability of these limiting resources. If the pathogen is missing from the better competitor, either absent from an area in an observational study or removed in an experimental manipulation, do other consumers decrease in abundance or go extinct because of a change in the available resources? For example, for the two consumers depicted in figure 7.12 with a pathogen that fosters their coexistence, if the pathogen is removed from N_1, does N_2 then go extinct? An observational study of their distributions should find infected N_1 individuals in local communities where N_2 is present, and no infected N_1 individuals or individuals infected with very benign or very severe pathogens in areas where N_2 is absent. Likewise, in a community with infected N_1 individuals and N_2 present, experimentally removing the pathogens from N_1 individuals should cause N_2 to go locally extinct.

Pathogen-Mediated Keystone Predation

If pathogens can foster the coexistence of resource competitors, perhaps they can have similar effects on coexistence as top predators do in other community modules, such as keystone or intraguild predation modules. Borer et al. (2007) addressed how interactions between pathogens or parasitoids might be structured in such ways. However, in this section I address a different issue, namely, whether a single pathogen can act as a keystone or intraguild predator on two hosts (see also Bonsall and Hassell 1998, Tompkins et al. 2000).

To consider this issue, start with a single consumer that is infected with a pathogen and that is feeding on a single host. Then allow a second consumer that feeds on the same resource and that can be a host to the same pathogen to invade (see invasion 31 in fig. 7.1(d)):

$$\frac{dN_{1S}}{dt} = \frac{b_{11}a_{11}R_1N_{1S}}{1+a_{11}h_{11}R_1} + \frac{b_{11}\delta_1a_{11}R_1N_{1I}}{1+\delta_1a_{11}h_{11}R_1} - f_1N_{1S} - \beta_{11}N_{1S}N_{1I} - \beta_{12}N_{1S}N_{2I} + v_1N_{1I}$$

$$\frac{dN_{1I}}{dt} = -\theta_1f_1N_{1I} + \beta_{11}N_{1S}N_{1I} + \beta_{12}N_{1S}N_{2I} - v_1N_{1I}$$

$$\frac{dN_{2S}}{dt} = \frac{b_{12}a_{12}R_1N_{2S}}{1+a_{12}h_{12}R_1} + \frac{b_{12}\delta_2a_{12}R_1N_{2I}}{1+\delta_2a_{12}h_{12}R_1} - f_2N_{2S} - \beta_{21}N_{2S}N_{1I} - \beta_{22}N_{2S}N_{2I} + v_2N_{2I}. \quad (7.10)$$

$$\frac{dN_{2I}}{dt} = -\theta_2f_2N_{2I} + \beta_{21}N_{2S}N_{1I} + \beta_{22}N_{2S}N_{2I} - v_2N_{2I}$$

$$\frac{dR_1}{dt} = R_1\left(c_1 - d_1R_1 - \frac{a_{11}N_{1S}}{1+a_{11}h_{11}R_1} - \frac{\delta_1a_{11}N_{1I}}{1+\delta_1a_{11}h_{11}R_1} - \frac{a_{12}N_{2S}}{1+a_{12}h_{12}R_1} - \frac{\delta_2a_{12}N_{2I}}{1+\delta_2a_{12}h_{12}R_1}\right)$$

Assuming that N_2 is the poorer resource competitor, it can invade the community only under certain combinations of fitness costs incurred by the two consumers from this shared pathogen.

The fitness costs of the pathogen in the two consumers may be quite different. For example, the HIV-1 virus is much less detrimental to the fitness of chimpanzees than to humans (Ehret et al. 1996, Davis et al. 1998, de Groot and Bontrop 2013). For simplicity here, assume that this pathogen does not affect the feeding rate of either consumer ($\delta_1 = \delta_2 = 1$) but does inflate the intrinsic death rate of one or both ($\theta_1 > 1$ or $\theta_2 > 1$ or both). Figure 7.13 shows an example parameter combination set for two consumers over a range of fitness costs for each. As when the pathogen only infects N_1 (e.g., fig 7.12(b)), the cost incurred by the superior resource competitor (N_1) must be greater than some minimum to permit resource abundance to increase to a level at which the inferior resource competitor (N_2) can invade and coexist. Pathogens that cause greater cost in N_1 cause the relative frequency of infected N_1 individuals to decrease, and again the fitness cost may be so large that the pathogen cannot sustain a population in N_1 (e.g., $\theta_1 > 5$ in fig. 7.13).

When the fitness cost to the inferior resource competitor (N_2) is very low and the cost to the superior competitor (N_1) is relatively high, N_2 can invade and coexist with the pathogen, but N_1 cannot coexist if N_2 is present. With a pathogen that causes this set of fitness costs on the two consumers, N_1 is excluded because the resource abundance is not adequate to offset the added mortality due to the pathogen being maintained at high levels in N_2 (fig. 7.13).

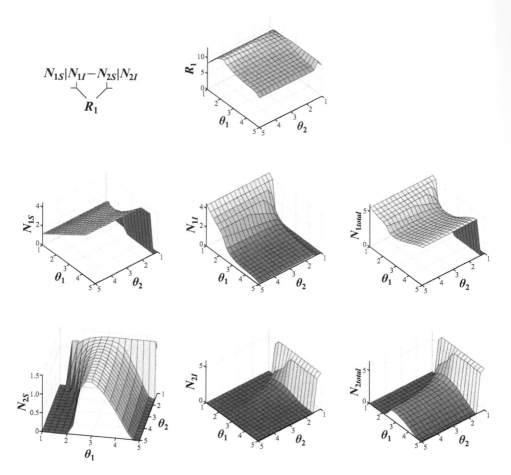

FIGURE 7.13. Equilibrium abundances for two consumers that are both infected by a pathogen that cannot live outside the host and that is passed between the two consumers. Both consumers feed on a single resource species. The panels show the equilibrium abundances when the pathogen causes different levels of increase for the intrinsic death rate of infected individuals in each consumer. Infected and susceptible individuals of both consumer species are identified via subscripts, as in figure 7.11. All parameter values are as follows: $c_1 = 2.0$, $d_1 = 0.04$, $a_{11} = 0.3$, $a_{12} = 0.25$, $b_{11} = b_{12} = 0.1$, $h_{11} = h_{12} = 0.0$, $f_1 = f_2 = 0.25$, $\delta_1 = \delta_2 = 1.0$, $\beta_{11} = \beta_{22} = 0.25$, $\beta_{12} = \beta_{21} = 0.05$, and $v_1 = v_2 = 0.05$.

In contrast, the inferior resource competitor's abundance (N_2) is relatively insensitive to the fitness cost of the pathogen infection to itself if N_1 is present, but its abundance depends critically on the fitness cost to the other consumer (fig. 7.13). N_2's abundance is highest for pathogens with intermediate fitness costs to N_1, and this is because they have the highest abundance of susceptible N_2 individuals. If pathogen cost is relatively low in N_1, it is more limiting to N_2's abundance; but if N_1's pathogen cost is relatively high, N_2's abundance is low primarily because of low resource availability. Thus, pathogen fitness cost in N_1 determines the relative importance of various limiting factors to the equilibrium abundance of N_2.

Additional consumers that also become infected with the same shared pathogen can invade and coexist in this model (results not shown). However, this may be a function of the model's structure and not a general feature of pathogens. Unlike with a keystone predator, this pathogen model permits more than two consumers to coexist at a stable point equilibrium with only one resource species and the single shared pathogen—that is, two limiting factors. The structure of this model assumes that the pathogen population is effectively infinitely large and so not limiting the infection rate. Therefore, the limiting factors associated with infections in this model derive from the fitness differences among the susceptible and infected individuals. If the pathogen population is modeled separately and is not assumed to be infinitely large and therefore influences the infection rate, it would represent a separate limiting factor and then limit the number of host species that may coexist.

To make mechanistic progress in understanding this interaction, the relative competitive hierarchy of the consumers must be known, since which consumer is the superior resource competitor in the absence of the pathogen is critical to interpretation. Therefore, empirical studies of competitive abilities in the absence of the pathogen—as described in chapter 3—are the antecedents of understanding the pathogen effects on these interacting species.

One of the most intriguing issues suggested by this model is the dependence of fitness costs across species on the coexistence of each consumer and the indirect interactions among them. For multiple consumers to coexist on a single limiting resource, the better resource competitor must experience a significant fitness decrement when infected. However, the fitness decrement cannot so severe that it permits a substantial fraction of infected hosts to be the equilibrium state for the species. If this is true, the inferior competitor can sustain a similar or even greater level of fitness decrement and still coexist. Thus, to determine whether the pathogen mediates coexistence between the consumers, experiments that compare the outcome of resource competition among the consumers in the presence and absence of the pathogen are essential.

SUMMARY OF MAJOR INSIGHTS

- Pathogens must have a rate of new infections greater than the demographic detriments of being infected and recovery rate to be maintained as a chronic infection in its host. Only chronic pathogen infections will have persistent effects on species interactions and community structure.

- Pathogens have their effects on species interactions via reduced host abundance and shifting performance characteristics (e.g., feeding rates and susceptibility to predators of infected individuals). Pathogens that have moderate fitness effects have the greatest impact on species interactions because they will have the highest frequency of infected hosts in the population.
- Pathogens thus act as limiting factors on their hosts, and so can increase the number of coexisting species in a community via affecting their host's ability to depress resource abundances or inflate predator abundances.
- Pathogens can shift resource competitive and apparent competitive relationships among their hosts, foster coexistence of resource competitors on a single resource, drive apparent competitors extinct, and cause trophic cascades, among other effects.
- Pathogens that infect multiple species link those hosts in apparent competitive, diamond/keystone predation, or omnivory/intraguild predation modules.

Temporal Variability

Float in the open waters of Trout Lake among all those phytoplankton, zooplankton, and fish species, and you will notice that the environmental conditions you experience change almost continuously. During the daylight hours, the temperature increases and then decreases, and the light shining on the water's surface changes angle and intensity. At night, the light is extremely diminished, which eliminates the photosynthetic activity of autotrophs and changes the ability of visual predators to find their prey. The next day is cloudy, cooler, and rainy. In the spring, the water is cold and murky because of the flush of nutrients and sediment from spring runoff as the snow melts. In the summer, the water is clear and warm. In the winter, the surface water freezes to cut the water off from the atmosphere. Averages over a season also vary from year to year. Such temporal variation in environmental conditions occurs in all ecosystems around the globe.

One of Hutchinson's (1961) main explanations for why so many phytoplankton species can potentially coexist in the pelagic zones of lakes like Trout Lake was this temporal environmental variation. As I have argued in previous chapters, the great diversity of species at higher trophic levels offers an adequate set of limiting factors to explain much of the phytoplankton diversity. However, that is not to say that temporal environmental variation does not also contribute to allowing some of these species to coexist.

Chesson (1978) provides one of the best reviews of the general sources and levels of variability in demographic rates and population dynamics, and I summarize his classification here. The lowest level of such variation results from the differences in the phenotypic traits among individuals within a population. The models explored in this book describe the dynamics of the average fitness of individuals in the population (Lande 1982, 2007, Charlesworth 1994, Lande et al. 2009), and so this source of variation is absent from our considerations. The second source of variation Chesson identifies is within-individual variation, given that survival and fecundity are probabilistic. Thus, in a finite population, some level of *demographic stochasticity* will determine the exact growth rate a population experiences. Given that demographic stochasticity largely does not contribute to promoting coexistence (Anderies and Beisner 2000), I do not consider this further. Because the population growth rate when rare is the central concept of coexistence, demographic stochasticity will be important whenever a species invades a new community. However, for our purposes, the critical issue to evaluate is the *expected* per capita growth rate when rare, and not any single instantiation. Variation in environmental conditions through time (i.e., on hourly, daily, seasonal, and yearly time scales) can influence whether a given species can coexist in a community. Chesson also discusses within-patch and between-patch scales of spatial variation in ecological conditions; the influences of spatial variation on coexistence are considered in the next chapter. One final source of temporal variation that can influence coexistence, but which was not discussed by Chesson, is the endogenous cycling—namely, limit cycles and chaotic dynamics—that can occur in a temporally constant environment.

Given temporal variation in demographic rates, the criterion for invasibility is not deterministic, but rather becomes a stochastic criterion. Turelli and Gillespie (1980) described a stochastic version of the invasibility criterion for population genetic models and a simple model of two-species competition. They showed that two competing species will have a stationary positive density distribution if the two boundary equilibria with only one species present are both unstable. In other words, each species has an expected per capita growth rate when each is rare and the other is at their long-term demographic steady state. Chesson (1982) described a similar concept that he called *stochastic boundedness* for a lottery model of competition (discussed later in this chapter). Chesson and Ellner (1989) extended these ideas to show that *mutual invasibility* in a stochastic environment implies stochastic boundedness (see also Turelli 1978a, 1981, Ellner 1989, Schreiber et al. 2011).

In this chapter I explore how both endogenously generated population cycling and temporal variation in environmental conditions can influence coexistence of species in communities at different functional positions. Temporal variation in some environmental features has no effect on coexistence whatsoever, although such temporal variation may prolong the persistence of species that will ultimately go extinct (Huston 1979, Chesson and Huntly 1997, Fox 2013, Hastings et al. 2018). Other sources and forms of temporal variation can make coexistence more difficult for some types of species, but still other forms and sources may foster coexistence.

ENDOGENOUSLY GENERATED POPULATION CYCLING

In chapters 3–7, we explored in great depth and diversity the various conditions required for the mutual invasibility and coexistence of species in various food webs in areas of parameter space that result in locally or globally stable equilibria.

In this section, we explore how these coexistence criteria may be different in areas of parameter space where *endogenously driven limit cycles or chaotic dynamics* result (Turchin 2003). In a limit cycle, the populations follow an identical trajectory through repeating cycles of increasing and decreasing over time (Strogatz 2015). This oscillation of abundances is caused by the relative balance between mechanistic features of the interactions that promote stability and mechanistic features that promote instability (May 1972). Chaotic dynamics are similar to limit cycles in that the populations' cycles are caused by overresponding to changes in the abundances of community members; unlike limit cycles, chaotic dynamics follow a general trajectory path, but passes around this cycling trajectory are not identical across cycles (Hirsch et al. 2012, Strogatz 2015).

The cycling trajectory of a limit cycle or chaos orbits an abundance equilibrium. This equilibrium represents an attractor because it pulls community abundance dynamics that begin outside the orbit but within the basin of attraction for the equilibrium to enter the orbit around it (May 1972, Hirsch et al. 2012, Strogatz 2015). However, the equilibrium is also unstable in the sense that the community abundances will also move away from this equilibrium to enter the cycling trajectory when the community begins near the equilibrium (May 1972, Hirsch et al. 2012, Strogatz 2015). Thus, one can speak of the limit cycle as being a "stable" limit cycle (Hirsch et al. 2012), or of the volume encompassing the chaotic dynamical trajectory as a "strange attractor" (Ruelle and Takens 1971). The parameter areas where these oscillatory attractors occur are difficult if not impossible to identify analytically in most cases (Strogatz 2015). Thus, numerical simulations are typically employed to identify these areas and the mechanistic properties that these parameter areas represent.

The basic Rosenzweig-MacArthur model (i.e., equations (3.3), assuming all $g_j = 0$)—the base model for this entire analysis—has cycling dynamics in some areas of parameter space (Rosenzweig and MacArthur 1963, Rosenzweig 1969, 1971, Gilpin and Rosenzweig 1972). With only one consumer feeding on one resource, the two species will oscillate in abundance if the resource isocline has a "hump" (i.e., if the maximum value of the R_1 isocline in the N_1 dimension is greater than c_1/a_1) (Rosenzweig 1969), which occurs when

$$\frac{c_1}{d_1} > \frac{1}{a_{11}h_{11}}, \tag{8.1}$$

and if the consumer isocline intersects the resource isocline at the apex or to the left of the resource isocline's peak, which occurs when

$$\frac{f_1}{a_{11}(b_{11} - f_1h_{11})} < \frac{c_1a_{11}h_{11} - d_1}{2a_{11}d_1h_{11}}. \tag{8.2}$$

The left sides of these two inequalities should look familiar: the left-hand side of inequality (8.1) is $R^*_{1(0)}$, and the left-hand side of inequality (8.2) is $R^*_{1(N_1)}$. The right-hand side of inequality (8.1) is the resource abundance at which the functional response of the consumer is at half saturation, a quantity that will become important in a coexistence context below. The right-hand side of inequality (8.2) is the resource abundance for the point on the R_1 isocline that has the largest N_1 value (i.e., the apex of the "hump"). If both of these inequalities are true, the consumer and resource coexist in a limit cycle (fig. 8.1).

More generally, for a stable limit cycle to occur with a single consumer and resource in this model, two things must be true. First, the consumer must have a saturating functional response that causes the resource's isocline to have a peak at a nonzero resource abundance. Second, the consumer's isocline must intersect the resource isocline at a resource abundance at or below the peak in the resource isocline. These two criteria are needed for limit cycles or chaotic dynamics to occur regardless of the mathematical function that is used to model a saturating functional response (Rosenzweig 1971, Hesaaraki and Moghadas 2001, Fussmann and Blasius 2005, Seo and Wolkowicz 2018).

To empirically test the first of these requirements, the functional response of the consumer must be quantified. This is done by measuring the per capita feeding

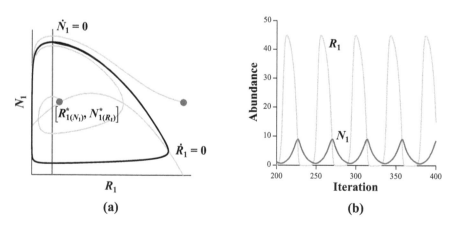

Figure 8.1. An example of a stable limit cycle in the interaction between a resource R_1 and consumer N_1. Panel (a) shows the isocline system for the two species, with the R_1 isocline identified as $\dot{R}_1 = 0$, and the N_1 isocline as $\dot{N}_1 = 0$. The filled gray circles identify two different starting positions; the gray lines are the transient trajectories for each of these two initial conditions as they approach the stable limit cycle, which is given in black. Panel (b) shows the abundance trajectories for the two species over time as they circle the limit cycle. The parameter values are as follows: $c_1 = 1.0$, $d_1 = 0.02$, $a_{11} = 0.3$, $b_{11} = 0.1$, $h_{11} = 0.3$, $f_1 = 0.125$, and $g_1 = 0.0$.

rate of individual consumers on a range of resource abundances over long enough time periods to incorporate search, capture success, handling, processing, satiation, and digestive pauses to affect the rate measured (Jeschke et al. 2002). The resource abundance gradient should be extended to high enough values to permit estimation of both the rate of deceleration in increase of the feeding rate and the asymptotic feeding rate. These two quantities will be important in estimating the resource abundance at which the consumer's feeding rate is at half saturation (inequality (8.1)). They will be important in the next section as well.

A formal empirical test would also estimate the shapes of the isoclines for both species to see if they have the form in figure 8.1. Remember that an isocline maps the abundance at which the species in question has a zero total population growth rate, given the abundances of all the other species in the community. Therefore, to empirically map the R_1 isocline, the following experiment would be done. Establish an abundance gradient of the consumer from near zero to a high value in replicated treatments, and hold the consumer's abundance constant for the entire course of the experiment. In each replicate, initiate a population of the resource and allow the population to equilibrate to the local consumer abundance. If the R_1 isocline has a hump, at high treatment values of the consumer, different replicates within a given consumer abundance level would need to be initiated at very low resource abundances and others at higher resource abundances. This is because the R_1 isocline has low and high resource abundance points for a given consumer abundance above c_1/a_1.

Mapping the consumer's isocline would involve the analogous experiment in which the resource's abundance is held constant at a range of values and the consumer population in each replicate is allowed to change and potentially equilibrate in response. If the N_1 isocline is a vertical line (fig. 8.1(a)), consumer populations at lower resource abundances will go extinct, but above a given resource abundance (i.e., $R_1 = f_1(a_{11} (b_{11} - f_1 h_{11}))$) they will increase. If the consumer has some other limiting factors that cause its isocline not to be a vertical line, its population should still go extinct below a given resource abundance but then increase and equilibrate at resource abundances above that point (chapter 3).

With these estimates of the isoclines, one can evaluate whether their relative positions and shapes are consistent with the dynamics of a limit cycle.

Resource Competition

This endogenously driven cycling of a consumer and a resource creates opportunities for additional consumers to coexist on the single resource (Koch 1974a, 1974b, Armstrong and McGehee 1976, 1980, Hsu et al. 1977, 1978b, McGehee and Armstrong 1977). These opportunities arise when the consumers differ in the degree of nonlinearity in their functional responses. Armstrong and McGehee (1980) illustrated situations where two or more consumers could coexist if one had a saturating functional response and would coexist alone with the resource in a limit cycle while the other consumer had a linear functional response. Hsu et al. (1978b) provided a more general analysis to show that both consumers could have saturating functional responses and coexist in a limit cycle on one resource. Figure 8.2(a) illustrates one such example where two consumers coexist on one resource because of endogenous cycling of their abundances. Two consumers that coexist in such a limit cycle will satisfy the following inequalities (Hsu et al. 1977, 1978b):

$$\frac{f_1}{a_{11}(b_{11} - f_1 h_{11})} < \frac{f_2}{a_{12}(b_{12} - f_2 h_{12})} \text{ and } \frac{1}{a_{11} h_{11}} < \frac{1}{a_{12} h_{12}}. \tag{8.3}$$

(The identities of the two consumers as 1 and 2 are arbitrary.) In words, these inequalities state that coexistence requires that the better resource competitor (the left inequality in (8.3)) must also have a functional response with a lower half-saturation level (the right inequality in (8.3)) (Hsu et al. 1978b). Taken together, this means that the functional response of the better competitor satiates at a lower level than does that of the poorer competitor. The first inequality in (8.3) is simply the typical statement about which consumer can support a population at the lower resource abundance. The second states a difference in the *relative nonlinearity* of their responses to resource abundance needed for their coexistence. If one of these inequalities is reversed, the consumer with the lower $R^*_{1(N_j)}$ but also the higher half-saturation constant will drive the other consumer extinct.

Inequalities (8.3) are only necessary conditions for coexistence. To coexist in a limit cycle, the average abundance of the resource over the limit cycle must be higher than the $R^*_{1(N_j)}$ value for the consumer with the higher $R^*_{1(N_j)}$ (i.e., the poorer resource competitor) (Hsu et al. 1977, 1978b, Armstrong and McGehee 1980). However, even this does not ensure coexistence. Two consumers will coexist only in tiny slivers of parameter space where other balances are struck between the two species, and the conditions for coexistence seem to change in different areas of adjacent parameter space (Hsu et al. 1978b, Abrams and Holt 2002, Xiao and Fussmann 2013). For example, figure 8.2(b) shows an area of parameter space where two consumers coexist in a limit cycle while feeding on one resource, but each drives the other extinct over the majority of the parameter space considered. For this example, the parameters for N_1 are constant, and the attack coefficient and intrinsic death rate of N_2 are varied. In the area where only N_1 coexists with R_1, N_1 is a much better competitor for the resource than N_2 (i.e., this means that the N_1

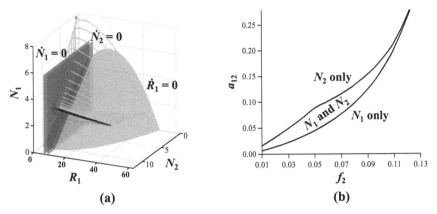

FIGURE 8.2. An example of two consumers (N_1 and N_2) coexisting in a stable limit cycle on one resource (R_1). Panel (a) shows the isoclines of the three species (identified as in figure 3.8) and the transient trajectory to the stable limit cycle (identified as in figure 8.1). Panel (b) shows the areas of parameter space where only one consumer coexists with the resource, and where the two consumer species both coexist with the resource for combinations of the attack coefficient a_{12} and intrinsic death rate f_2 of N_2. The parameter values in (a) are as follows: $c_1 = 1.0$, $d_1 = 0.02$, $a_{11} = 0.3$, $b_{11} = 0.1$, $h_{11} = 0.3$, $f_1 = 0.125$, $g_1 = 0.0$, $a_{12} = 0.1$, $b_{12} = 0.1$, $h_{12} = 0.3$, $f_2 = 0.07$, and $g_2 = 0.0$. These same values are used to map out the parameter space in (b), except for varying a_{12} and f_2.

isocline intersects the R_1 axis at a much lower point than does the N_2 isocline, or in this case mathematically $R_{1(N_1)}^* \ll R_{1(N_2)}^*$ since $g_1 = g_2 = 0$). In the area where the two consumers coexist, N_1 is a better competitor for the resource than N_2, but their isoclines intersect the R_1 axis close together (as in fig. 8.2(a)). Similarly narrow parameter bands permitting coexistence also occur along gradients of consumer attack coefficients and intrinsic death rates (Abrams and Holt 2002, Xiao and Fussmann 2013).

The examples in figure 8.2(b) also identify areas where coexistence in limit cycles does not conform to the rules for coexistence at point equilibria. For example, the boundary between the area of coexistence and the area where only N_2 coexists with R_1 (i.e., the upper boundary of the "N_1 and N_2" area) at values with $f_2 < 0.05$ is defined when the parameter combinations result in $R_{1(N_1)}^* = R_{1(N_2)}^*$. Just across this boundary in this range, N_2 now has the lower $R_{1(N_j)}^*$ but a higher half-saturation point for its functional response, and so it drives N_1 extinct because of its lower $R_{1(N_j)}^*$—just as would be expected at a point equilibrium for the same community module. However, at this boundary for values of $0.05 < f_2 < 0.125$, $R_{1(N_1)}^*$ is still less than $R_{1(N_2)}^*$, and the right inequality in (8.3) is also satisfied. Thus, combinations of parameters for the consumer species above but near this boundary for $0.05 < f_2 < 0.125$ satisfy both inequalities in (8.3) and have the average resource abundance well above the $R^*_{1(N_j)}$'s for both species, but the consumer with the higher $R^*_{1(N_j)}$ and with the higher half-saturating functional response drives the other consumer extinct.

Breaking the rules outlined in chapter 3 for this community module is further illustrated by considering three potential consumers feeding on one resource. Figure 8.3 shows the average abundances of the three consumers and the resource along a gradient for the intrinsic death rate for N_2. All other parameter values are fixed across all simulations such that $R^*_{1(N_1)} < R^*_{1(N_3)}$ and $a_{13} < a_{12} < a_{11}$. With $f_2 < 0.056$, N_2 has the lowest $R^*_{1(N_j)}$ and so drives the other two consumers extinct (below the leftmost vertical dashed line). However, with $0.056 < f_2 < 0.059$, both inequalities in (8.3) are satisfied such that $R^*_{1(N_1)} < R^*_{1(N_2)}$, but N_2 drives N_1 extinct (i.e., between the two leftmost vertical dashed lines). In the range $0.059 < f_2 < 0.077$, N_1 and N_2 coexist, with their relative abundances shifting with values of f_2. In the very narrow window of $0.07705 < f_2 < 0.0771$, all three consumers coexist in a limit cycle. Most surprisingly, above this range, only N_1 and N_3 coexist, even though $R^*_{1(N_1)} < R^*_{1(N_2)} < R^*_{1(N_3)}$.

Which consumers will coexist because of limit cycles also depends on the environmental conditions in which their interaction takes place. For example, Rosenzweig (1971) defined the *paradox of enrichment* based on the observation that cycling in this model is more likely at higher values of the resource's equilibrium abundance in the absence of any consumers (i.e., $R^*_{1(0)}$). Which consumers will coexist via cycling also depends on such environmental effects. As an

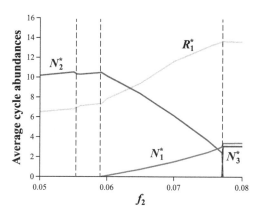

FIGURE 8.3. The average abundances of three consumers (N_1, N_2, N_3) that coexist in varying combinations with one resource (R_1) in limit cycles along a gradient of the intrinsic death rate f_2 for N_2. The parameter values are as follows: $c_1 = 1.0$, $d_1 = 0.02$, $a_{11} = 0.3$, $b_{11} = 0.1$, $h_{11} = 0.3$, $f_1 = 0.125$, $g_1 = 0.0$, $a_{12} = 0.1$, $b_{12} = 0.1$, $h_{12} = 0.3$, $g_2 = 0.0$, $a_{13} = 0.05$, $b_{13} = 0.1$, $h_{13} = 0.3$, $f_3 = 0.05$, and $g_3 = 0.0$.

example, figure 8.4 shows the coexistence pattern of three consumers along a gradient of $R^*_{1(0)}$. In these simulations, all parameters are the same across all simulations except d_1. When d_1 is large, $R^*_{1(0)} = c_1 / d_1$ is small, and when d_1 is small, $R^*_{1(0)} = c_1 / d_1$ is large. The three consumers have $R^*_{1(N_1)} < R^*_{1(N_2)} < R^*_{1(N_3)}$, caused by $a_{13} < a_{12} < a_{11}$ and $f_3 < f_2 < f_1$. Again, a gradient of $R^*_{1(0)}$ represents a gradient of different basal productivities for ecosystems in which the communities develop.

When $R^*_{1(0)}$ is very low (i.e., $R^*_{1(0)} < 35.7$ in fig. 8.4(a)), only N_1, the consumer with the lowest $R^*_{1(N_j)}$, coexists with R_1, even though R_1's average abundance is greater than $R^*_{1(N_1)}$ and $R^*_{1(N_2)}$. At somewhat higher levels of ecosystem productivity ($35.7 < R^*_{(0)} < 64.5$), N_1 and N_2 coexist in a limit cycle. Here again, note that for the coexisting pair, the consumer with the lower $R^*_{1(N_j)}$ also has the lower half-saturation constant for its functional response. In this range, R_1's average abundance is still less than $R^*_{1(N_3)}$. In the next higher range of productivities ($64.5 < R^*_{1(0)} < 100.0$), only N_2 coexists with R_1, even though R_1's average abundance is greater than $R^*_{1(N_j)}$ for all three consumers. As productivity is increased further ($100.0 < R^*_{1(0)} < 144.9$), N_2 and N_3 coexist; and in ecosystems with the highest productivities considered ($144.9 < R^*_{1(0)}$), only N_3 persists with R_1.

Thus, the basic consumer-resource model predicts that consumers should replace one another along a productivity gradient for the single resource species (i.e., for increasing values of $R^*_{1(0)}$) when the community displays limit cycles or chaos. Consequently, the R-star rule (Tilman 1982) does not strictly apply when consumers display limit cycles or chaotic dynamics. These consumers replace one another along the $R^*_{1(0)}$ gradient because of their differences in base demographic parameters. As $R^*_{1(0)}$ is increased, the resource abundance cycle increases in both amplitude and period, which means that the resource abundance spikes to higher values but also remains at low values for longer periods (fig. 8.4(b)–(d)). At the low end of the productivity gradient, the consumer with the higher attack rate that can more effectively harvest the resource is favored, because the resource spends only short periods at low abundance (fig. 8.4(b)). However, at the high end of the productivity gradient, a consumer with a small intrinsic death rate is favored so as to persist through the prolonged periods of low resource abundance (fig. 8.4(d)).

With multiple resource species (> 3), oscillations may be more pronounced and occur over a greater range of parameter space and so foster many consumers in the system by this mechanism (Huisman and Weissing 1999, 2001a, 2001b, 2002, Beninca et al. 2008, 2015). In fact, more resource species present may expand the areas of parameter space where oscillations occur and thus increase the likelihood of consumer coexistence via these mechanisms (Huisman and Weissing 2001a, 2001b, 2002). Moreover, which consumers come to dominate and which are driven extinct may be largely unpredictable because of the sensitivity of chaotic dynamics to initial conditions (Huisman and Weissing 2001b).

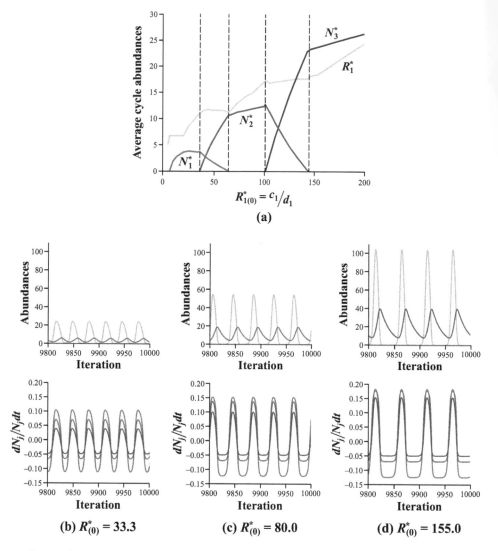

FIGURE 8.4. The average abundances of three consumers (N_1, N_2, N_3) that coexist in varying combinations with one resource (R_1) in limit cycles along a productivity gradient defined by the equilibrium resource abundance $R_{1(0)}^*$ in the absence of all consumers. This gradient is generated by varying the strength of density dependence d_1 for R_1. Panel (a) shows the average abundances for communities along this gradient. Panels (b)–(d) show representative examples for three different points on the gradient for (top) abundance trajectories over time and (bottom) the change in per capita population growth rate over time. The parameter values are as given in figure 8.3, with one addition: $f_2 = 0.07$.

One of the important conclusions to emerge from analyses of coexistence by limit cycles and chaotic dynamics for species that would be excluded otherwise is that these species must be ecologically very similar to the species that are already present in the community. Thus, the results of experiments quantifying the positions of the consumer isoclines that were described in the previous section should place their isoclines very close to one another, if they coexist in a stable limit cycle or chaotic dynamics. The experiments quantifying their functional responses should also produce empirical results that satisfy inequalities (8.3).

Another prediction to emerge here is that the amplitude of population cycles should be larger in communities occurring in more productive ecosystems, and different species should coexist at different points along such ecological gradients. Observational studies that document such patterns and correlate the demographic rates of the various species along the gradient could be used to test what favored species persist in the various communities. The model results suggest that the demographic ability to withstand conditions of low resource availability (fig. 8.4) may be key.

Apparent Competition

Cycling in community dynamics does not alter the fundamental criteria for coexistence of additional resource species in apparent competition modules. Each resource species must be able to increase when rare, given the consumer's abundance in its absence, which means that the resource's isocline must pass above the equilibrium point for the community it is invading (chapter 3). This is true whether the equilibrium is a point equilibrium or an attractor for a limit cycle.

Cycling caused by a consumer's saturating functional response can cause unexpected responses to species additions and losses in this module. Typically, the invasion of a new resource species to an apparent competition module is expected to increase the consumer's abundance, and thereby decrease the abundances of the other resource species that are already present because of the increased predation pressure on them (Holt 1977). A strongly saturating functional response causes the per capita predation pressure imposed by the consumer to diminish as prey abundances increase. With more resources, the consumer's feeding rate can reach a tipping point at which the addition of more resource species will cause a decrease in the per capita feeding rate of the consumer (Abrams and Matsuda 1996). Beyond this point, adding more resource species will cause the abundances of the other resource species to increase, either at stable point equilibria (Abrams and Matsuda 1996) or in average abundances in limit cycles (Abrams et al. 1998). In effect, the resource species respond to perturbations in one another's abundances as mutualists (Abrams and Matsuda 1996, Abrams et al. 1998).

Food Chain, Keystone, and Intraguild Predation

Three-trophic-level food chains can also display limit cycles and chaos (Hastings and Powell 1991, Abrams and Roth 1994, McCann and Yodzis 1994), and so community modules formed by adding species to a simple food chain may also be possible when cycling dynamics occur. With a diamond/keystone module, limit cycles and chaos occur in some areas of parameter space, and Abrams (1999) showed that coexistence of the two intermediate consumers was more difficult with cycling dynamics. My own numerical explorations of parameter space also suggest that at most a third consumer can be added to a cycling diamond module.

Interestingly, the basic intraguild predation module (i.e., a three-trophic-level food chain in which the predator also feeds on the resource) will display limit cycles and chaotic dynamics even when both consumer and predator have linear functional responses (Holt and Polis 1997, Tanabe and Namba 2005, Hsu et al. 2015). Such cycling with linear functional responses occurs when the predator has a large attack coefficient but a very small conversion efficiency for feeding on the resource. When such cycles occur, multiple consumers can be added to the community (McPeek 2014). When the consumer and predator have saturating functional responses and cycling dynamics occur, again numerical explorations suggest that at most three consumers can coexist.

EXOGENOUSLY GENERATED TEMPORAL VARIABILITY

Endogenously generated temporal variation in abundances is caused by the dynamical feedbacks between the interacting species. These feedbacks cause overcompensations in responses to one another. These overcompensations in response are what generate the limit cycles and chaotic dynamics. The abiotic background in which the interactions occur do not change. Such temporal dynamics present the opportunities for coexistence of more species that could be supported in the community, given the number of limiting factors.

The abundances of species may also vary through time because of a completely different set of mechanisms, namely, *exogenously generated temporal variability*. The abiotic conditions of the ecosystem in which a community develops are not static. Every time it rains, many abiotic factors—temperature, relative humidity, soil moisture—change. These abiotic changes may change the demographic properties of the species inhabiting that ecosystem. In the models considered in this book, those changes would be reflected as changes in the parameters of the equations or perturbations of the abundances of limiting factors, such as water and mineral nutrients. One could add descriptive relationships to these equations to directly link abiotic factors to the parameter values (e.g., Litchman and Klausmeier 2001, Litchman et al. 2004, Klausmeier 2010). However, that is beyond the scope of what I want to accomplish here. I simply generate variation in the parameters directly and so only imply the underlying causes of that temporal variation in parameter values.

The temporal variability in parameters may themselves change smoothly through repeating cycles. The simple daily cycle of light availability creates a smooth change in the demographic parameters for algae that utilize that light to produce sugars through photosynthesis (Litchman and Klausmeier 2001, Litchman et al. 2004). Yearly seasonal cycling in temperature and precipitation also generate temporally repeated cycles that directly affect the demographic rates of populations. Many species evolve life history or migratory adaptations (e.g., the winter diapause of many insects, hibernation in vertebrates, winter migrations of birds) to respond to these environmental cycles. Other longer cycles in climatic variation, such as the El Niño–La Niña cycles in the Pacific Ocean, also drive demographic variation on continental and global scales. With such variability, the parameters become smooth functions of time (e.g., sine waves).

Exogenously driven temporal variability in parameters may also change abruptly in a much more stochastic and unpredictable fashion. Short-term weather patterns may be quite unpredictable. Or the differences between the conditions in consecutive years will not be identical. Such variability may be modeled as sampling parameter values from a specified probability distribution where the moments of the distribution are specified. For example, the intrinsic rate of increase for a resource species may be modeled as a random variable sampled from a probability distribution with a specified mean and variance.

Both forms of exogenously generated temporal variability will drive temporal variation in species abundances. This variation in species abundances is driven by the continual change in the positions of the species' isoclines in abundance space, because of changes in parameter values. As a result, the position of the community equilibrium varies continually, because of change in the balance of demographic forces acting on species with variable parameters. This temporal demographic variability within species with variable parameters causes their abundances to change through time. In some species, it will propagate to other community members whose demography is influenced by those species. For example, if the abundance of a resource fluctuates because of temporal variation in its intrinsic growth rate, the abundance of a consumer that feeds on it will also vary temporally because of the change in resource availability through time.

The important feature this generates in the dynamics of the community that can permit additional species into the food web is not typically the creation of variation in abundances per se; rather, the important feature is the pattern of covariation that these interactions create (Levins 1979). Covariances are created both within and between species. First, temporal variation in some parameter will create a covariance with its own abundance. For example, temporal variance in the attack coefficient of a consumer will typically generate a positive covariance with its own abundance: larger values of its attack coefficient generally cause larger values in its own abundance. This variability can also generate covariation between the parameters of one species and the abundances of others in the community: the attack coefficient of a consumer typically has a negative covariance with the abundances of its prey. The abundances of various species may also covary.

Only certain forms of variation and covariation permit new species to invade and coexist that could not in a static environment (Levins 1979, Fox 2013). As with endogenously generated temporal variability, the additional species that can be added to the community must also be similar to those that can coexist in a temporally constant environment.

One Resource and One Consumer

Begin with the Rosenzweig-MacArthur (1963) model for a single consumer feeding on a single resource. All the parameters in this model are as defined in chapter 3 and are static, except for the attack coefficient, which varies with time—a_1 (t). The model is then

$$\frac{dN_1}{N_1 dt} = \frac{b_{11} a_{11}(t) R_1}{1 + a_{11}(t) h_{11} R_1} - f_1$$

$$\frac{dR_1}{R_1 dt} = c_1 - d_1 R_1 - \frac{a_{11}(t) N_1}{1 + a_{11}(t) h_{11} R_1}$$

(8.4)

The temporal variation in the attack coefficient (symbolized as $Var(a_{11})$) generates variation in both species abundances (i.e., $Var(R_1)$ and $Var(N_1)$) and covariation among the consumer's attack coefficient and their abundances (i.e., $Cov(a_{11}, R_1)$, $Cov(a_{11}, N_1)$, and $Cov(R_1, N_1)$). These variances and covariances change the long-term per capita population growth rates of these species, which in turn cause their long-term average abundances to be different from the equilibrium abundances that would result in a static environment if the attack coefficient were held constant at its average value.

To see this, the Taylor expansions of the per capita growth rates of the two species are calculated to the quadratic order at the long-term average per capita growth rates (Lewontin and Cohen 1969, Levins 1979, Chesson 1994). This gives the following:

$$
\frac{dN_1}{N_1 dt} = \frac{b_{11}\overline{a}_{11}\overline{R}_1}{1+\overline{a}_{11}h_{11}\overline{R}_1} - f_1 + \frac{b_{11}(1-\overline{a}_{11}h_{11}\overline{R}_1)}{\left(1+\overline{a}_{11}h_{11}\overline{R}_1\right)^3}Cov(a_{11}, R_1)
$$

$$
-\frac{b_{11}h_{11}\overline{R}_1^2}{\left(1+\overline{a}_{11}h_{11}\overline{R}_1\right)^3}Var(a_{11}) - \frac{\overline{a}_{11}^2 b_{11}h_{11}}{\left(1+\overline{a}_{11}h_{11}\overline{R}_1\right)^3}Var(R_1),
$$

$$
\frac{dR_1}{R_1 dt} = c_1 - d_1\overline{R}_1 - \frac{\overline{a}_{11}\overline{N}_1}{1+\overline{a}_{11}h_{11}\overline{R}_1} - \frac{h_{11}\overline{R}_1\overline{N}_1}{\left(1+\overline{a}_{11}h_{11}\overline{R}_1\right)^3}Var(a_{11}) \tag{8.5}
$$

$$
-\frac{2\overline{a}_{11}h_{11}\overline{N}_1}{\left(1+\overline{a}_{11}h_{11}\overline{R}_1\right)^3}Cov(a_{11}, R_1) - \frac{1}{\left(1+\overline{a}_{11}h_{11}\overline{R}_1\right)^2}Cov(a_{11}, N_1)
$$

$$
+\frac{\overline{a}_{11}^3 h_{11}^2 \overline{N}_1}{\left(1+\overline{a}_{11}h_{11}\overline{R}_1\right)^3}Var(R_1) - \frac{\overline{a}_{11}^2 h_{11}}{\left(1+\overline{a}_{11}h_{11}\overline{R}_1\right)^2}Cov(R_1, N_1)
$$

where an overbar signifies the long-term average of that quantity. The variances and covariances in equations (8.5) are functions of variation in $a_{11}(t)$ and how this drives variation in the abundances of the two species. Consequently, it is impossible to specify exact values for them. In addition, in this more complicated model with the consumer having a saturating functional response, the long-term average abundances are extremely tedious to calculate and are not informative. However, simulations of this model provide insights into how temporal variation in the attack coefficient drives these quantities.

Figure 8.5 presents results of simulations of equations (8.4) in which the attack coefficient is modeled as a random variable drawn from a normal distribution with specified mean and standard deviation. The covariation of the attack coefficient with the resource's abundance is generally negative (i.e., $Cov(a_{11}, R_1) < 0$) and with the consumer's abundance is generally positive (i.e., $Cov(a_{11}, N_1) > 0$). In addition, the covariance between the abundances of the consumer and resource is typically negative (i.e., $Cov(R_1, N_1) < 0$), and so these two covariances tend to increase the resource's average long-term per capita growth rate. $Cov(a_{11}, R_1)$ tends to decrease the consumer's long-term per capita growth rate. Variation in both the attack coefficient and R_1's abundance also directly depress N_1's long-term per capita growth rate. Consequently, greater variation in $a_1(t)$ causes an increase in the long-term average resource abundance above what its abundance would be in a static environment (i.e., $\overline{R}_{1(N_1)} > R^*_{1(N_1)}$) (fig. 8.5(a)); conversely, the consumer's long-term abundance is depressed below what its abundance would be in a

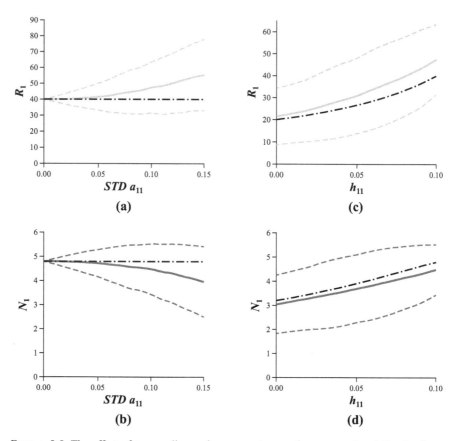

FIGURE 8.5. The effect of temporally varying parameters on the mean and variation in abundances of a consumer N_1 feeding on one resource R_1. Panels (a)–(b) show the effects of a temporally varying attack coefficient. Simulations were run for 25,000 iterations, and a random value was drawn on each iteration from a normal distribution with a specified mean and standard deviation to determine the attack coefficient value for that iteration. In panels (c)–(d), the same procedures were performed with the standard deviation in the attack coefficient constant across a gradient of the handling time of the consumer. In each panel, the mean (solid line) ± one standard deviation (dashed lines) over iterations 2,500–25,000 for the abundance of that species is illustrated for different standard deviations of each parameter. The dot-dashed line shows the expected equilibrium abundance in the absence of stochastic variation. The parameter values are as follows: $c_1 = 1.0$, $d_1 = 0.01$, $\bar{a}_{11} = 0.25$, $b_{11} = 0.1$, $h_{11} = 0.1$, $f_1 = 0.5$, and $g_1 = 0.0$. In (c)–(d), the standard deviation of the attack coefficient is $\sigma_{a_{11}} = 0.1$.

static environment (i.e., $\overline{N_{1(R_1)}} < N^*_{1(R_1)}$) (fig. 8.5(b)). The values of the static parameters also influence the magnitude of these differences. For example, greater values of the consumer's handling time cause these differences in long-term abundances to be larger (fig. 8.5(c)–(d)).

The stochastic Rosenzweig-MacArthur model simulation module can be accessed at https://mechanismsofcoexistence.com/variation_2D_1R_1N.html.

Extensive simulations indicate that variation in parameters has similar effects on community dynamics when the consumers have linear or saturating functional responses. Moreover, for this model, we can specify the long-term average abundances of these two species with a linear functional response exactly. With $h_{11} = 0$, equations (8.5) simplify to

$$
\frac{\overline{dN_1}}{N_1 dt} = b_{11}\overline{a}_{11}\overline{R}_1 - f_1 + b_{11}Cov(a_{11}, R_1)
$$
$$
\frac{\overline{dR_1}}{R_1 dt} = c_1 - d_1\overline{R}_1 - \overline{a}_{11}\overline{N}_1 - Cov(a_{11}, N_1)
$$

(8.6)

Because the long-term average per capita growth rates of both species must still equal zero at their demographic steady state, their long-term average abundances are found by setting these two equations to zero and simplifying (Levins 1979). This gives

$$
\overline{N_{1(R_1)}} = \frac{c_1\overline{a}_{11}b_{11} - d_1 f_1}{\overline{a}_{11}^2 b_{11}} + \frac{d_1 Cov(a_{11}, R_1)}{\overline{a}_{11}^2} - \frac{Cov(a_{11}, N_1)}{\overline{a}_{11}}
$$
$$
\overline{R_{1(N_1)}} = \frac{f_1}{\overline{a}_{11}b_{11}} - \frac{Cov(a_{11}, R_1)}{\overline{a}_{11}}
$$

(8.7)

The first term for each is exactly the equilibrium abundance of that species in a static environment. Their long-term average abundances are deviations from these values caused by the covariation between the attack coefficient and their abundances. Because $Cov(a_{11}, R_1)$ is generally negative, R_1's abundance is inflated by temporal variation in a_1 (t) (fig. 8.5). $Cov(a_{11}, N_1)$ is generally positive, and so covariation of the attack coefficient with both species' abundances contributes to depressing N_1's long-term average abundance. These changes in long-term average abundances are what will influence the entry of other species into this food web that would not be able to invade and coexist if the present species were in a temporally constant environment. Because some analytical insights can be easily developed and because of the similarity of general results between linear and saturating functional responses, I focus in the following sections on consumers with linear functional responses.

Resource Competition

First, consider the simple resource competition module in which a second consumer species N_2 is invading a simple food chain consisting of one resource R_1 and one consumer N_1. For illustrative purposes, assume that the resource has logistic population growth, both consumers have linear functional responses for feeding on the resource, neither consumer is self-limiting, and N_1 is a better competitor for the resource than N_2. In the environmental context of temporal constancy of parameters, the second consumer could not invade and coexist in this community (chapter 3, "Resource Competitors"). In this section, I explore what types of exogenously driven temporal variability will permit this second consumer to invade and coexist with the first. The model of this situation, where various parameters may fluctuate over time, is

$$\frac{dN_1}{N_1 dt} = b_{11} a_{11}(t) R_1 - f_1(t)$$

$$\frac{dN_2}{N_2 dt} = b_{12} a_{12}(t) R_1 - f_2(t). \tag{8.8}$$

$$\frac{dR_1}{R_1 dt} = c_1(t) - d_1(t) R_1 - a_{11}(t) N_1 - a_{12}(t) N_2$$

Consider the effect of temporal variability in each of these parameters in turn.

The first parameter to consider is the intrinsic growth rate of the resource c_1. Imagine that the intrinsic growth rate of the resource varies with time (i.e., $c_1(t)$) in a stochastic fashion with a defined mean $\overline{c_1}$ and variance $\sigma_{c_1}^2$, but all other model parameters are constant. This variability would be driven by variation in the abiotic environment that affects the basal productivity of the ecosystem. In this model, the abundances of the consumers will vary through time because of the stochastic variability in $c_1(t)$ that causes abundance variation in R_1. However, this temporal variability will not allow the second consumer to invade and coexist.

To understand why this is true, consider how the isoclines of the three species vary with temporal variation in $c_1(t)$. Temporal variability in $c_1(t)$ causes the resource isocline to vary in its positions where it intersects the three axes (i.e., $c_1(t)/d_1$ on the R_1 axis, $c_1(t)/a_{11}$ on the N_1 axis, and $c_1(t)/a_{12}$ on the N_2 axis), and so the R_1 isocline varies in position up and down as $c_1(t)$ varies. However, variation in $c_1(t)$ has no effect on the positions of the N_1 and N_2 isoclines. Regardless of the value of $c_1(t)$, N_1 is the better resource competitor, and so under no circumstances will N_2 have a positive average population growth rate when rare. The same insight can be derived by calculating the long-term expected value of the per capita population growth rates of the two consumers from equations (8.8), which gives

$$\frac{\overline{dN_j}}{N_j dt} = b_{1j} a_{1j} \overline{R_1} - f_j. \tag{8.9}$$

Thus, environmental variation that only causes variation in the demographic rates of the resource, and thus only the position of the R_1 isocline (i.e., c_1 (t), or d_1 (t)), does nothing to permit N_2 into the community because such variation does not alter the long-term average per capita growth rate of either consumer (Levins 1979).

Variation in the intrinsic death rates of the consumers f_j (t), with all other parameters held constant, also does not facilitate the invasion of the second consumer. In this case, each consumer's long-term average per capita growth rate is

$$\frac{\overline{dN_j}}{N_j dt} = b_{1j} a_{1j} \overline{R_1} - \overline{f_j}. \tag{8.10}$$

Variation in f_j (t) causes the intersection points of the consumer isoclines on the R_1 axis to vary through time (i.e., f_j $(t)/(a_{1j}b_{1j})$). Each consumer will have a mean and variance for this variation, which will generate an average $R^*_{1(N_j)} = \overline{f_j} / (a_{1j}b_{1j})$ for each consumer. However, the R_1 isocline does not change positions with variation in f_j (t). If these average values are close enough on the R_1 axis relative to the variances of each, the relative competitive abilities of the two consumers will reverse over time because of their overlapping $R^*_{1(N_j)}$ distributions. In addition, R_1's abundance will vary through time as the consumer abundances vary. However, this periodic reversal of resource competitive abilities would only slow the rate at which the consumer with the higher average $R^*_{1(N_j)}$ would be driven extinct (Chesson and Huntly 1997). Such a periodic reversal does nothing to change the long-term average per capita growth rates of the consumers. Thus, temporal variance in the consumers' intrinsic death rates also does not promote their coexistence.

Temporal variation in the consumers' attack coefficients a_{1j} (t) do alter the long-term per capita growth rates of the consumers and can, under certain circumstances, allow consumers that are poorer resource competitors to invade and coexist in communities where they would be excluded in a temporally constant environment (Levins 1979, Abrams 1984, Chesson 1994, Li and Chesson 2016). With all other parameters constant in equations (8.8), the average per capita population growth rates of the three species are

$$\frac{\overline{dN_1}}{N_1 dt} = b_{11}\overline{a}_{11}\overline{R}_1 + b_{11}Cov(a_{11}, R_1) - f_1$$

$$\frac{\overline{dN_2}}{N_2 dt} = b_{12}\overline{a}_{12}\overline{R}_1 + b_{12}Cov(a_{12}, R_1) - f_2. \qquad (8.11)$$

$$\frac{\overline{dR_1}}{R_1 dt} = c_1 - d_1\overline{R}_1 - \sum_{j=1}^{2}\overline{a}_{1j}\overline{N}_1 - \sum_{j=1}^{2}Cov(a_{1j}, N_j)$$

Temporal variation in a consumer's attack coefficient generates covariation for each consumer between the attack coefficient—and thus indirectly the competitive ability of the consumer (i.e., its $R^*_{1(N_j)} = f_1 / (a_{1j}(t)b_{1j})$ value)—and the resource's abundance $R_1(t)$, and for the resource between each consumer's attack coefficient and the consumer's abundance $N_j(t)$. This covariation is generated because the resource and consumer isoclines covary in position with variation in $a_{1j}(t)$. The resulting covariances can enhance or depress the average per capita growth rates of these various species, which is what can permit the inferior competitor to potentially invade and coexist.

These are terms that correspond to what Chesson (1994) identified as the interaction between "competition and the environment" in his characterization of the roles of temporal variability affecting coexistence (see also Chesson 2003, 2018, Li and Chesson 2016, Barabás et al. 2018). Following the spirit of his approach to evaluating this problem, the issues are whether these covariances change the various species' per capita growth rates enough to permit each to increase when it is rare and the other species are at their long-term demographic steady states (Turelli 1981, Chesson and Ellner 1989, Chesson 1994). From equations (8.11), the average steady state resource abundances when coexisting with only one of the consumers are

$$\overline{R}_{1(N_1)} = \frac{f_1 - b_{11}Cov(a_{11}, R_1)}{b_{11}\overline{a}_{11}} \quad \text{and} \quad \overline{R}_{1(N_2)} = \frac{f_2 - b_{12}Cov(a_{12}, R_1)}{b_{12}\overline{a}_{12}}, \qquad (8.12)$$

respectively, for the two consumers. Thus, coexistence requires that each consumer be able to have a positive population growth rate when the other is coexisting alone with the resource at the respective long-term resource abundances given in equations (8.12).

Given that the species are identified such that $f_1 / (b_{11}\overline{a}_{11}) < f_2 / (b_{12}\overline{a}_{12})$, N_1 will always be present if it can invade. The question then is whether N_2 has a positive per capita growth rate when it invades the community of N_1 and R_1 at their long-term demographic steady states—what Chesson (1994, 2000b, 2003, 2018) called the "long-term low density growth rate" being greater than zero. This requires that

$$\frac{\overline{dN_2}}{N_2 dt} = b_{12}\overline{a}_{12}\overline{R_{1(N_1)}} + b_{12}Cov(a_{12}, R_1) - f_2 > 0.$$

Substituting $\overline{R_{1(N_1)}}$ from equations (8.12) and rearranging gives the invasibility criterion for N_2 as

$$\frac{Cov(a_{12}, R_1)}{\overline{a}_{12}} - \frac{Cov(a_{11}, R_1)}{\overline{a}_{11}} > \frac{f_2}{\overline{a}_{12}b_{12}} - \frac{f_1}{\overline{a}_{11}b_{11}}. \qquad (8.13)$$

The right-hand side of this inequality is what Chesson (2000b) dubbed the "equalizing" component of this interaction, namely, the difference in the long-term average "fitnesses" between these two species. (This is not the definition of fitness that is associated with individual fitnesses that drive evolution, but rather a component of the per capita population growth rate of the species in this community context (Chesson 1994, 2000b, 2003, Li and Chesson 2016). Because I want to minimize confusion between ecological and evolutionary mechanisms, I continue to refer to these as components of per capita population growth rate instead of fitness.) The left-hand side of inequality (8.13) is Chesson's (2000b) "stabilizing" component of this indirect interaction—in this case, the difference between the two consumer species in the covariances of the attack coefficient and resource abundance standardized by their respective attack coefficient averages (Levins 1979, Li and Chesson 2016). Thus, when the "stabilizing" component of the interaction is greater than the "equalizing" component, N_2 will be able to invade and coexist with N_1 and R_1. Or to paraphrase Levins (1979), N_1 is consuming the mean of R_1's abundance and N_2 is consuming the (co)variance.

The results of extensive computer simulations indicate that the sign and magnitudes of these covariances change with the abundances of the consumers, and whether $a_{11}(t)$ and $a_{12}(t)$ covary themselves. Typically, $Cov(a_{1j}, R_1)$ is negative and is larger in magnitude when the consumer's long-term average abundance is higher. If $a_{11}(t)$ and $a_{12}(t)$ are themselves uncorrelated, $Cov(a_{11}, R_1)$ is typically more negative than $Cov(a_{12}, R_1)$ when N_2 is invading and even when the two consumers are coexisting, because N_1 will have a higher abundance resulting from $f_1/\overline{a}_{11}b_{11} < f_2/\overline{a}_{12}b_{12}$. Thus, if the difference between their standardized covariances is larger than the difference in their long-term average competitive abilities (i.e., inequality (8.13) is satisfied), N_2 can invade and coexist in the community with R_1 and N_1 (fig. 8.6).

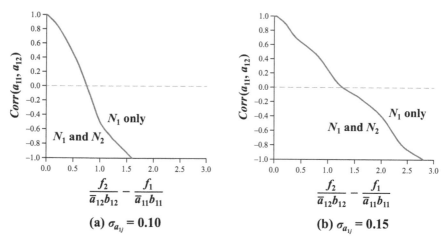

$$\textbf{(a)}\ \sigma_{a_{ij}} = \textbf{0.10}$$

$$\textbf{(b)}\ \sigma_{a_{ij}} = \textbf{0.15}$$

FIGURE 8.6. Regions of parameter space where two consumers (N_1 and N_2) can and cannot coexist with a resource (R_1) when the attack coefficients of both consumers vary temporally and may be interspecifically correlated. The x axis in (a) and (b) gives the difference in average competitive abilities between the two consumers, and the y axis gives the correlations between their attack coefficients. Simulations for these parameter combinations were run for 25,000 iterations to determine whether the two consumers would coexist. Differences in average competitive ability were generated by changing the intrinsic death rate for consumer 2 (i.e., f_2). Two random values were drawn on each iteration from a bivariate normal distribution with a specified mean, standard deviation, and correlation to determine the attack coefficient values for that iteration. Panel (a) shows the results for simulations in which the standard deviation of each consumer's attack coefficients was 0.10, and (b) shows the results when this standard deviation was 0.15. The other parameter values are as follows: $c_1 = 1.0$, $d_1 = 0.02$, $a_{11} = a_{12} = 0.25$, $b_{11} = b_{12} = 0.1$, and $f_1 = 0.1$.

One can imagine that $a_{11}(t)$ and $a_{12}(t)$ may themselves be correlated with one another. In fact, Chesson emphasized the importance of such interspecific covariation in species responses to environmental fluctuations as a central tenant of such covariation promoting coexistence (Chesson 1978, 1994, 2000b, Chesson and Warner 1981). This is because covariance between $a_{11}(t)$ and $a_{12}(t)$ alters that difference in the standardized covariances in inequality (8.13)—namely, enhancing or diminishing the "stabilizing" components of the interaction—and so alters the ability of N_2 to invade and coexist with R_1 and N_1. Larger positive covariation between $a_{11}(t)$ and $a_{12}(t)$ causes the standardized covariances in inequality (8.13) to be more similar, thus reducing the ability of N_2 to successfully invade (fig. 8.6). Biologically, this means that when environmental conditions are favorable for the invader, they are also good for the resident, and so they must be more similar in their "equalizing" components of per capita growth rate for N_2 to invade.

Conversely, negative covariation between $a_{11}(t)$ and $a_{12}(t)$ causes a greater difference between the standardized covariances in inequality (8.13): the environmental conditions that cause one consumer to have a large attack coefficient cause the other to have a small attack coefficient (Levins 1979, Abrams 1984, Li and Chesson 2016). Simulation results indicate that negative covariation between their attack coefficients causes the resident to have a more negative covariance $Cov(a_{11}, R_1)$; but the covariance for the invader $Cov(a_{12}, R_1)$ becomes positive, which makes the left-hand side of inequality (8.13) more positive (fig. 8.6). Therefore, the "equalizing" components can be more different between resident and invader and still have the invader be successful and coexist if their attack coefficients show strong negative covariation. Additionally, greater variation in the $a_{1j}(t)$'s increases the magnitude of these two covariances with species abundances, and so also increases the scope for N_2 to successfully invade (cf. fig. 8.6(a)–(b)).

The resource competition stochastic simulation module can be accessed at https://mechanismsofcoexistence.com/variation_3D_1R_2N.html.

Because such variation and covariation would be generated by changes in various environmental factors, what patterns of covariation across species might be expected in nature? Many of the environmental features that would influence the attack coefficients of various consumers should generate positive covariation because of shared effects on all consumers. For example, if the turbidity of a lake's water varies through time, such variation should affect all consumers similarly and so generate positive covariation between their attack coefficients. The same is true for variation in structural complexity. Variation in these factors should only foster the inclusion of new consumers that have very similar competitive abilities to consumers already present in the community (i.e., small differences in the equalizing effects in inequality (8.13)). One source of variation that would generate a negative covariation between the attack coefficients of two consumers would be diurnal variation between consumers that forage in the dark versus the light (e.g., Kohl et al. 2018). This suggests that consumers that forage for the same resources but are best at different times of the day may be more different in their average per capita growth rates and still coexist.

Other forms of temporal variation can also promote coexistence of resource competitors under specific circumstances. For example, Klausmeier (2010) extended the Rosenzweig-MacArthur model to study how long seasonal oscillations between good and bad periods may influence the coexistence of a consumer and resource and promote the coexistence of two consumers feeding on one resource. During the good season (which was meant to represent summertime growth), species dynamics followed the standard Rosenzweig-MacArthur consumer-resource model (a version of equations (3.3)), but during the extended bad season (meant to represent winter), populations of all species declined at constant per capita rates. The relative lengths of the two "seasons" influences which of fast-growing and slow-growing consumers could coexist with the resource. The fast-growing consumer dominated if the growing season was short, the slow-growing consumer dominated if the growing season was long, and the two consumers coexisted in a narrow intermediate range of relative growing season lengths (Klausmeier 2010). The two consumers also coexisted on one resource in narrow ranges of parameter space if, instead of alternating good and bad seasons, the two consumers alternated between being the superior and the inferior competitor with changing seasons (Klausmeier 2010). Similar results were found in models meant to simulate fluctuating light conditions for phytoplankton growth (Koch 1974a, Hsu 1980, Litchman and Klausmeier 2001, Litchman et al. 2004).

Of the mechanisms he enumerated, Hutchinson (1961) suggested that a reversal in competitive dominance was a likely explanation for why many phytoplankton can seemingly coexist in a lake with only a handful of limiting resources (and leaving aside for the moment that enemies are also limiting factors). This has always been a preferred explanation for the paradox of the plankton (e.g., Hutchinson 1961, Huisman and Weissing 1999, 2001a, 2002, Li and Chesson 2016). The mechanisms explored in this section deal with how those reversals in competitive dominance would need to occur if they promoted coexistence of some plankton species.

As in previous chapters, quantification of such metrics as the minimum resource abundance at which the consumer can maintain a population is essential to evaluating the mechanisms here (e.g., inequality (8.13)). Much of the rest of the analyses then involve estimating the covariation between fluctuating environmental features that drive variation in demographic rates and the abundances of the interacting species—what Chesson (2003, 2008) called the "covariance between environment and competition."

Apparent Competition

Because temporal variation in the resource's intrinsic growth rate c_1 (t) and the consumer's intrinsic death rate f_1 (t) does not alter the long-term average per capita growth rates of these species, variation in these parameters also does not change the criteria for new resources to invade. However, analogous to the last section, temporal variation in the attack coefficients a_{i1} (t) also fosters a greater range of resource species that are able to invade and coexist.

Again, start with a community in which R_1 and N_1 in equations (8.6) are coexisting at their long-term demographic steady state (i.e., the quantities given in equations (8.7)), with all parameters constant except a_{11} (t), which varies tempo-rally. Further assume that R_1 is the best apparent competitor among the pool of resource species that could invade this community (chapter 3, "Apparent Com-petitors"). Then allow another resource species R_2 to invade. R_2's long-term aver-age per capita growth rate at invasion is

$$\frac{\overline{dR_2}}{R_2\,dt} = c_2 - \overline{a}_{21}\,\overline{N_{1(R_1)}} - Cov(a_{21}, N_1) \qquad (8.14)$$

since $R_2 \approx 0$ when it invades. $\overline{N_{1(R_1)}}$ is as given in equations (8.7).

The apparent competition stochastic simulation module can be accessed at https://mechanismsofcoexistence.com/variation_3D_2R_1N.html.

Making this substitution, and querying when equation (8.14) is greater than zero gives the invasibility criterion for another apparent competitor invading and coexisting with the resource species already present. The resulting invasibility criterion is

$$\frac{c_2}{\overline{a}_{21}b_{12}} - \frac{c_1\overline{a}_{11}b_{11}}{\overline{a}_{21}^2 b_{11}} \frac{d_1 f_1}{} > \frac{d_1 Cov(a_{12}, R_1)}{\overline{a}_{11}^2} - \frac{Cov(a_{11}, N_1)}{\overline{a}_{11}} + \frac{Cov(a_{21}, N_1)}{\overline{a}_{21}}. \qquad (8.15)$$

The right-hand side of inequality (8.15) is generally less than zero, because the covariances between the attack coefficients with the consumer abundances are both generally positive and the covariance with the attack coefficient on the resi-dent will be larger at invasion, and because the covariance with R_1's abundance will generally be negative and the largest in magnitude. The left-hand side of inequality (8.15) is the same as in a temporally constant environment, which in that case simply has to be greater than zero for R_2 to invade. Because the right-hand side is generally negative, the covariation that depresses N_1's abundance now permits apparent competitors to invade and coexist that would not have been able to do so in a temporally constant environment.

The two-resource competition stochastic simulation module can be accessed at https://mechanismsofcoexistence.com/variation_3D_2R_2N.html.

Again, if the attack coefficients on the two resources themselves covary, this will alter the covariances between each and the species abundances, and so change the magnitude of the right-hand side of inequality (8.15). Simulation studies indi-cate that, as with resource competition, negative covariation between the attack coefficients increases the average per capita growth rate deficit for the inferior apparent competitor (i.e., the left-hand side of inequality (8.15) is more negative) that can invade and coexist for a given level of temporal variation. Simulation studies also show that consumers with greater handling times also permit apparent competitors with greater per capita growth rate deficits because of satiation by the resident consumers.

THE STORAGE EFFECT

Three species of fishes in the family Pomacentridae are found at nearly every site that is suitable for them in the Great Barrier Reef, with a much larger diversity of species being found regionally (Sale 1975). Individuals of these species defend territories in the crevices of rubble patches on the reef and graze the algae that grows around their territories. An individual can only recruit to the breeding population by acquiring a territory, and this occurs primarily when territorial individuals die (Sale 1975, 1977). Individuals seem to acquire new territories at random across species, and the species show little to no ecological differences (Sale 1975, 1977). In fact, Sale (1977) described these species as being "adapted to this unpredictable supply of spaces in ways which make interspecific competition for space a lottery in which no species can consistently win." He went on to propose that the high diversity of fishes overall on the reef was maintained because of the unpredictable nature of their interactions that "prevents development of an equilibrium community." Thus, Sale proposed that these species are not coexisting with one another at all, but rather are more akin to Hubbell's (2001) description of "neutral" species.

In response to Sale's (1977, 1978) descriptions of this assemblage of fishes, Chesson and Warner (1981) proposed a model to show that, although these species may obtain territories at random—in effect acting as the random winners of a lottery—the species may in fact be coexisting with one another. This was the first statement of the *lottery model* of competitive coexistence. Chesson (1982) presented a formal analysis of the lottery model for two competing species. He has since extended it to many competing species (Chesson 1994), and the model has been used as a basis of analyzing coexistence in other types of empirical systems besides reef fishes (e.g., Cáceres 1997, Sears and Chesson 2007, Angert et al. 2009).

Because one of the critical features of the model is overlapping generations, the lottery model is derived using difference equations instead of differential equations, and I use that formulation here (Chesson 1982, 1994). The development of the model here also closely follows the original development of Chesson (1982). Imagine two fish species that obtain territories on a reef at random as they become available because of territory holders dying. The model tracks the proportions of the two species through discrete time intervals $P_1(t)$ and $P_2(t)$. Since $P_1(t) + P_2(t) = 1$, the population dynamics model for one species also completely defines that for the other species.

Individuals of each species have a probability of dying in a time interval of $\delta_i(t)$, which may vary over time. Therefore, the total proportion of new territories that become available for colonization at time t is $\delta_1(t)P_1(t) + \delta_2(t)P_2(t)$. Each individual also produces some number of offspring (e.g., larval fish) that are available to colonize newly vacant territories. This birth rate $\beta_i(t)$ differs between the species and through time within each species. If newly created offspring do not secure a territory, they are lost from the system (e.g., currents carry them away). Each species is assumed to have a fixed ability to "compete" for newly open territories, specified by the parameter c_i.

Using these assumptions, the proportion of the sites occupied by species 1 in the next time interval, given the proportions of the two species in this time interval, is

$$P_1(t+1) = (1-\delta_1(t))P_1(t) + (\delta_1(t)P_1(t) + \delta_2(t)P_2(t))\frac{c_1\beta_1(t)P_1(t)}{c_1\beta_1(t)P_1(t) + c_2\beta_2(t)P_2(t)} \tag{8.16}$$

(Chesson 1982). The first term of this equation is the carryover of adult territory holders from the previous time interval (i.e., those that did not die). Because only a proportion of the population may die at each time interval, the population has overlapping generations if $\delta_1(t) < 1$. The second term is the proportion of new sites that become available during the time interval times the proportion of sites won by species 1 in the lottery competition for those newly available sites. Temporal variation influences the demography of both species because both their probabilities of dying $\delta_i(t)$ and their birth rates $\beta_i(t)$ vary with time.

One primary requirement for coexistence in the lottery model of equation (8.16) is that the species have overlapping generations. To see this, consider the counterfactual where all territory holders die at the end of each time interval so that all territories are vacated to start each new time interval. This means $\delta_1(t) = \delta_2(t) = 1$, and equation (8.16) reduces to

$$P_1(t+1) = \frac{c_1\beta_1(t)P_1(t)}{c_1\beta_1(t)P_1(t) + c_2\beta_2(t)P_2(t)}.$$

As a result, the ratio of the proportions of the two species changes according to

$$\frac{P_1(t+1)}{P_2(t+1)} = \frac{c_1\beta_1(t)P_1(t)}{c_2\beta_2(t)P_2(t)}.$$

Thus, the species with what Chesson (1982) called the larger long-term average "birth-competition" parameter $\overline{c_i\beta_i(t)}$ will drive the other species extinct in this lottery competition: if $\overline{c_1\beta_1(t)} > \overline{c_2\beta_2(t)}$, species 1 will eventually win; and if $\overline{c_1\beta_1(t)} < \overline{c_2\beta_2(t)}$, species 2 will eventually win.

To coexist, the two species must have overlapping generations (i.e., $0 < \delta_i(t) < 1$). Moreover, to coexist, each species must be able to invade the reef if the other is already present and currently occupying all the available territories. This is what Chesson and Ellner (1989) called stochastic boundedness away from zero abundance. When species 1 invades, the expected value of its per capita growth rate is

$$\overline{\ln(P_1(t+1)) - \ln(P_1(t))} = E\left(\ln\left(1 + \overline{\delta_1}\left(\overline{\frac{c_2\beta_2\delta_1}{c_1\beta_1\delta_2}} - 1\right)\right)\right), \qquad (8.17)$$

and a comparable equation can be written for species 2 (Chesson 1982). This represents the species' geometric mean of per capita growth rates over a long time period.

The important feature that overlapping generations cause in the long-term per capita growth rate is to create a strong nonlinearity with the ratio of the birth-competition parameter (fig. 8.7). As a result, the geometric mean per capita growth rate from equation (8.17) can be positive, even if the population is declining across many time intervals (Chesson 1986). Coexistence requires that the ratio of the birth-competition parameter in equation (8.17) vary around 1.0 (i.e., some periods above 1.0 and some years below 1.0) for both species, and this variation

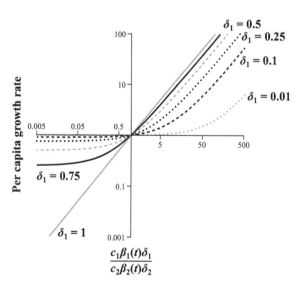

FIGURE 8.7. Relationships between the birth-competition parameter (a metric of the relative competitive abilities of two species) and the geometric mean population growth rate for various values of the probability of an individual dying (δ_1) in a given time interval.

for both species must be sufficiently large for both of their geometric mean per capita growth rates when rare to be greater than 1.0. Moreover, variation in offspring numbers promotes coexistence; variation in adult survival rate alone is not sufficient to foster coexistence (Chesson 1982, 1994).

The nonlinearity in per capita population growth rate (fig. 8.7) caused by overlapping generations is what permits species coexistence in the lottery model. Having adult individuals persist over many breeding periods (i.e., iteroparity) means that the population can persist with only slow decline over periods in which reproduction is low. Even in periods of complete reproductive failure, the rate of population decline is buffered by adult survival. Conversely, offspring from one very favorable breeding period will persist as a dominant cohort in the population across many subsequent breeding seasons. In effect, individuals from good breeding times can be "stored" in the population across many poor breeding cycles if adults have high survival. This is the *storage effect* (Chesson 1982, 1986, 1994).

Any life history feature that produces high survival of individuals across periods of low recruitment to the population can potentially generate the storage effect. The egg banks of many crustaceans, the seed and seedling banks of many plants, and long-lived adult stages in many different groups all can foster the storage effect under the right environmental conditions. As with the stochastic mechanisms considered in the previous section, the likelihood of coexistence among a group of species by this mechanism is greatly enhanced if the environmental conditions that promote high recruitment differ among the species (Chesson 1994, 2003, 2008, 2018, Barabás et al. 2018). This generates negative covariation across the species in how they respond to environmental changes, which can accentuate their responses to a good breeding period when they are rare, thus promoting invasibility (Chesson 2003, 2008).

Chesson has elaborated a series of methods to quantify the storage effect (Chesson 1994, 2003, 2008, 2018, Barabás et al. 2018), which have been applied in a number of types of communities (e.g., Cáceres 1997, Angert et al. 2009, Adler et al. 2010). A major problem with quantifying the magnitude of the storage effect is that a model must be built and analyzed for each situation, which makes empirical quantification difficult (Barabás et al. 2018). However, new simulation techniques have been developed that may alleviate the need to define a specific model for each application (Ellner et al. 2016).

DISTURBANCES

Disturbances come in many forms. Many disturbances come in the form of very short-term but dramatic changes in the demographic rates of species, usually

short-term spikes in mortality due to environmental causes. Fires, floods, hurri-
canes, logs slamming into the rocky shore all can create short-term spikes in mortality (e.g., Dayton 1971, Sousa 1979, Moritz 1997, Vandermeer et al. 2000,
Thonicke et al. 2001, Platt et al. 2002). Other types of disturbances come in the
form of habitat alterations (e.g., badger mounds that some plants favor) and are
probably better understood in the context of species interactions (e.g., Platt 1975).

Such rapid spikes in mortality cause reductions in abundances of many species
in the community experiencing the disturbance (Platt and Connell 2003). The
consequences of these abundance reductions from the community perspective are
generally interpreted to "prevent competitive exclusion" by continually perturb-
ing species below their equilibrium abundances at which inferior competitors
should be driven extinct (Grime 1977, Connell 1978, Huston 1979). Recovery of
the community from a major disturbance is typically a successional process with
directed species replacements as species colonize and interact (Platt and Connell
2003). In effect, repeated disturbances are thought to maintain the community in
a perpetual state of recovery, in which the superior competitors are below their
equilibrium abundances so that the process of competitive exclusion does not
happen for many species.

Connell (1978) elaborated these ideas in what he called the *intermediate disturbance hypothesis* based on data from tropical forest and coral reef diversities.
The hypothesis conjectures that species richness in a local community will be
highest at intermediate levels of both frequency and intensity of disturbances.
Species richness is predicted to be lower both at low disturbance frequencies and
intensities because competitive exclusion is completed, and at high disturbance
frequencies and intensities because fewer species will be able to withstand such
demographic perturbations.

The intermediate disturbance hypothesis has been criticized because disturbances per se do not prevent competitive exclusion (Chesson and Huntly 1997,
Fox 2013). A mechanism that prevents competitive exclusion implies that it promotes coexistence. As the previous sections illustrate, temporal variation in
demographic rates is not a panacea for generating coexistence. Only certain forms
of temporal variation will permit more species to coexist than in constant environmental conditions (Chesson and Huntly 1997, Fox 2013). For example, temporal
variation in the intrinsic death rate of consumers does not promote the coexistence
of more consumers, but temporal variation in their attack coefficients on prey
does. Also, many variance and covariance structures of temporal variation can
exclude species that would coexist in a temporally constant environment.

These critiques that many forms of disturbance do not promote coexistence of
additional species are certainly true. However, unless one assumes that all species
found together in a community are coexisting, I think the intermediate disturbance

hypothesis still has validity. Community dynamics after a disturbance event follow a successional pattern to approach a long-term demographic steady state (Connell and Slayter 1977, Willig and Walker 1999, Platt and Connell 2003). Disturbance may permit more species to colonize and persist for some time at a local site because many of the species in the local community are not at their long-term demographic steady states. Consequently, many species may be eventually driven extinct by species interactions as the community approaches that long-term demographic steady state (i.e., walking dead species). The issue to test for the intermediate disturbance hypothesis is what fraction of species along the disturbance gradient are (1) coexisting because of their general ecological superiority to deal with limiting factors (what Chesson (1994) called their "fluctuation-independent" superiority); (2) coexisting because of fluctuation-dependent mechanisms (i.e., relative nonlinearities, covariation among environmental features and abundances, or storage effect); and (3) long-term transients that will eventually go extinct as the community approaches its long-term demographic steady state.

SUMMARY OF MAJOR INSIGHTS

- Invasibility and coexistence criteria for a static environment sets the minimum bounds for coexistence in a temporally variable environment.
- Cycling and chaotic dynamics permit some species already close to their static coexistence boundary to cross over the boundary to coexist because of relative nonlinearities.
- Also, cycling and chaotic dynamics cause some coexistence rules for a static environment not to apply locally in some situations.
- Exogenously generated environmental variation offers greater scope for additional coexisting species to enter a community, via relative nonlinearities, covariances between this environmental variation and species abundances, and life stages with high survival through inhospitable periods (i.e., the temporal storage effect).
- Negative covariation between the responses of species to environmental variation can greatly facilitate coexistence of both species in many conditions.
- Greater environmental covariation creates greater opportunities for more marginal species to coexist.
- Environmental covariation does not per se increase a species' opportunities for coexistence. Some forms of covariation can exclude a species that would otherwise coexist in a community were the environment temporally static.

Spatial Variability on Local and Regional Scales

If you had followed E. Lucy Braun through the woods as she quantified the vegetation found on Black Mountain in eastern Kentucky—100 miles from where I grew up—you would have moved through a spatially variable landscape (Braun 1935, 1940). Across much of the mountain, you would have seen chestnut trees (*Castanea dentata*) dominating the canopy. However, as you move to different topographic areas of the mountain, the composition of tree and shrub species would shift dramatically. On the sunlit ridges at the top of the mountain, chestnut oak (*Quercus montana*), beech (*Fagus grandifolia*), scarlet and white oak (*Q. coccinea* and *Q. alba*), and hickory (*Carya* spp.), along with chestnuts, dominate the canopy layer. To the north of the ridge where it is slightly cooler and wetter, sugar maple (*Acer saccharum*) increases in frequency in the canopy and rhododendron (*Rhododendron calendulaceum*) dominates the understory; but to the south of the ridge where it is hotter and dryer, mountain laurel (*Kalmia latifolia*) dominates the understory. Down in the cool, shaded hollows surrounding the creeks, sugar and red maple (*A. saccharum* and *A. rubrum*), basswood (*Tilia heterophylla*), hemlock (*Tsuga canadensis*), beech, buckeye (*Aesculus octandra*), red oak (*Q. borealis*), and tulip trees (*Liriodendron tulipifera*) dominate the canopy with chestnuts.

Presumably, these changes in species' relative abundances across the mountain are the result of changes in the abiotic conditions across the landscape that influence the demographic rates of these plant species. For example, chestnut, beech, red and sugar maple, and various oaks are found in almost all areas on the mountain, but their local relative abundances shift greatly across the landscape (Braun 1935, 1940). Because the conspecific individuals in all these areas on the mountain can potentially breed with one another, the dynamics of each population of all these species on this single mountain are governed by some averaging of the local demographic rates that individuals experience in different areas of the mountain.

In the decades since Braun's work, many others have documented similar patterns of spatial heterogeneity in relative abundances in a number of different communities, including other forest trees (e.g., Whittaker 1956, Waring and Major 1964, Whittaker and Niering 1965, 1975, Beals 1969, Auerbach and Shmida 1993, Hemp 2005), birds occupying those forests (e.g., Miller 1951, Bond 1957, Beals 1960, James 1971, Terborgh 1971), zooplankton in the ocean (e.g., Fager and McGowan 1963), and all types of organisms on the rocks in the intertidal (e.g., Stephenson and Stephenson 1949, Southward 1958, Connell 1961, Paine 1966, 1974, Menge 1976, 1995). If all individuals of a species embedded in a community do not experience the same demographic rates, how will this spatial variation in demographic rates influence the invasibility criterion of each species? This is one question to be addressed in this chapter.

This is not the only scale at which spatial variation is important in influencing community structure and species coexistence. Individuals of different interacting species frequently operate on very different spatial scales. The individuals of some species may be completely sedentary in many life stages of their existence (e.g., terrestrial plants, epiphytic algae, barnacles, hydras), others may move only a few square meters for much of their lives (e.g., salamanders, mice, cleaner shrimp, territorial reef fishes), and still others may roam across tens to thousands of square kilometers in their lifetime (e.g., birds, bears, wolves, wolverines, tuna, sharks, whales). Because of these differences in spatial scales over which individuals operate, the movement of some species will link the demographics of essentially separate communities. For example, kingfishers and herons fly among multiple marshes, ponds, lakes, and rivers to forage for fish, and in so doing they will link the dynamics of fish populations across those water bodies (e.g., Dowd and Flake 1985, Kelly et al. 2008). How do the possible foraging decisions by these roaming species influence the spatial patterning of the communities they link together? This is another question to be addressed in this chapter.

Much of the patterning of species compositions across the landscape is caused by the fact that local ecological conditions are favorable for a species in one community but are hostile to that species in another. Many of our most famous studies in community ecology are studies of why species segregate to different habitat types. Zooplankton species composition is very different among ponds and lakes with different ecological conditions (e.g., Whittaker and Fairbanks 1958, Brooks and Dodson 1965, Elser and Carpenter 1988, Carpenter and Kitchell 1993). For example, comparing lakes with and without alewives in Connecticut, Brooks and Dodson (1965) documented that large-bodied cladoceran and copepod species occur in lakes lacking alewives but that small-bodied cladocerans and copepods dominate lakes with alewives. They attributed this difference to size-selective predation by the alewives. Likewise, size-selective predation by salamanders in small ponds in the Rocky Mountains causes differences in zooplankton composition between ponds where salamanders are and are not the top predators (Dodson

1970, Sprules 1972). In fact, many of the iconic studies of community ecology are studies of "competitive exclusion," which examine why one set of species is excluded and others can coexist in a particular community (e.g., Tansley 1917, Gause 1934, Crombie 1944, 1945, Connell 1961, DeBach and Sundby 1963, Paine 1966, 1969, 1974, Jaeger 1971, Tilman 1977, Hairston 1980, 1986).

While some species are segregated among communities in different habitats, others are common across these same communities. For example, in the high-altitude Colorado ponds, where many species are segregated among ponds dominated by *Ambystoma tigrinum* salamanders and ponds dominated by *Chaoborus americanus*, the copepod *Diaptomus coloradensis* is "ubiquitous" in both pond types (Dodson 1970, Sprules 1972). Likewise, *Enallagma* and *Lestes* damselfly species segregate among ponds and lakes in which fish versus large dragonflies are the top predators across eastern North America, because these species are differentially susceptible to these two predator types (McPeek 1990 a, 1990b, Stoks and McPeek 2003a, 2003b). In contrast to *Enallagma* and *Lestes*, *Ischnura* species are common in both fish and dragonfly lakes (McPeek 1998).

These species that are common across different community types may be present for two reasons. They may be coexisting in each of these distinct community types, which implies that they occupy similar functional roles across communities that force others to segregate. In contrast, they may be coexisting in only a subset of these communities and maintained in the others by continual immigration: in other words, their populations in some are sources and their populations in others are sinks (Shmida and Ellner 1984, Pulliam 1988). Exploring the dynamics resulting from such permanent movement of individuals from one local community to another (i.e., dispersal) is a third topic to be considered in this chapter. If individuals are not moving among local communities, each can be considered and studied in isolation, because what is found in one does not influence what is found in another. However, if individuals move between local communities, that dispersal may distort local community structures away from what local ecological conditions favor (Holyoak et al. 2005, Leibold and Chase 2017).

CONTINUOUS SPATIAL VARIATION: THE SPATIAL STORAGE EFFECT

In the discussion in chapter 8 of how temporal ecological variability can foster or retard the invasion and coexistence of species that would be excluded from a community occupying a temporally static environment, three mechanisms were identified that can foster the coexistence of these additional species: relative non-linearities in species responses to environmental conditions, covariances among environmental properties and species abundances, and the temporal storage effect. Mechanisms analogous to two of these, plus one other, can emerge from various patterns of spatial ecological variation that may foster or retard the coexistence of additional species in a community (Chesson 2000a, Snyder and Chesson 2004).

This section considers interactions among species in a single community. Spatial ecological variation implies that individuals at different positions in the area or volume occupied by the community have different demographic rates because the abiotic conditions and the abundances of the various community members vary in space. One example of such ecological variation is a meadow where the soil properties (e.g., texture, mineral availabilities, water availability) that influence plant demographic rates vary on short spatial scales. Because plant individuals also interact only with other individuals in their local neighborhood and their offspring (e.g., seeds) are dispersed around them with some spatial patterning as well, the spatial patterning of the underlying ecological variation can alter the per capita population growth rates of each species away from what would be expected in an environment that was simply the spatial average ecological condition. Likewise, thermal, light, and nutrient gradients in a lake can generate the same types of spatial ecological variation that may foster the coexistence of additional species that would not be present in the average environment.

One of the most enduring hypotheses about the factors that promote high species richness in tropical forest trees is the Janzen-Connell hypothesis (Janzen 1970, Connell 1971). In independent papers, Janzen and Connell both made the argument that species-specific enemies (e.g., herbivores and pathogens) may be important in reducing the competitive interactions among tree species by reducing competitor abundances. Janzen's (1970) version also contained a very important spatial component to his proposed mechanism. Namely, seedlings that are farther away from their parent may have higher survival. While his argument focused on spatial variability in sources of mortality for individual species, it falls within a broader category of spatially varying demographic rates within species and the potential for this variation to differ among species.

Because of the potential complexity and variety of ways that important processes may vary spatially in relation to one another, it is very difficult to enumerate general conclusions about how spatial variation may influence coexistence in the types of models that have been considered in this book to now. However, Chesson (2000a) developed a generalized framework of spatially and mechanistically implicit models that can be used to infer some level of general principles, and this general framework can be applied to specific interaction types on a case-by-case basis. In this section, I present this generalized framework primarily following the presentation of Snyder and Chesson (2004), which to me is the most lucid and general exposition of the issues.

In abstracting species interactions, imagine a guild of species that all occupy similar functional positions within the community of interest. For example, these species may all be consumers that feed on the same set of resources and that are fed upon by the same set of predators. Assume that the indirect interactions among these guild members all have negative interspecific consequences. Chesson (2000a) referred to these indirect interspecific interactions simply as "competition," be they resource or apparent competition. Likewise, space is not modeled explicitly but rather implicitly by inferring various relationships for spatial variances and covariances (Chesson and Huntly 1997, Chesson 2000a, Snyder and Chesson 2003, 2004, Snyder 2007, 2008) that can be derived in more explicit models (e.g., Bolker and Pacala 1999, Muko and Iwasa 2000, Bolker 2003, Bolker et al. 2003, Murrell and Law 2003).

As with the temporal storage effect, the model is developed in discrete time, so that the average abundance of N_j across all sites at time $t + 1$ is the product of the per capita population growth rate $\tilde{\lambda}_j$ averaged over all individuals in the population at time t times the average abundance of N_j at time t:

$$\overline{N}_j(t+1) = \tilde{\lambda}_j(t)\overline{N}_j(t)$$

(see equation 2 in Snyder and Chesson 2004). From this, the average per capita population growth rate is then

$$\tilde{\lambda}_j(t) = \frac{\overline{N}_j(t+1)}{\overline{N}_j(t)}.$$

Because of spatial ecological variation, each individual at each location in the population may have a different per capita growth rate $\lambda_{jx}(t)$, and so $\tilde{\lambda}_j(t)$ is the average across all individuals in the population (Chesson 2000a). If the relative local population density is defined as $v_j(x, t) = N_j(x, t)/\overline{N}_j(t)$, the average per capita population growth rate can be expressed as $\tilde{\lambda}_j(t) = \overline{\lambda_j(t)v_j(t)}$; that is, the local per capita population growth rates times their relative local population abundances averaged across all spatial locations.

The average of this product can then be expressed as

$$\tilde{\lambda}_j(t) = \overline{\lambda_j(t)v_j(t)} = \overline{\lambda}_j(t) + Cov(\lambda_j, v_j)(t), \tag{9.1}$$

where $\overline{\lambda}_j(t)$ is the spatial average of the local per capita population growth rates ($\lambda_j(x,t)$) weighted by local population abundance ($N_j(x,t)$), the covariance of $\lambda_j(x,t)$ and $v_j(x,t)$ is taken across space, and $\overline{v}_j(t) = 1$ (Chesson 2000a). This representation of the average per capita population growth rate can be further partitioned into four separate components (Chesson 2000a, Snyder and Chesson 2004):

$$\tilde{\lambda}_j(t) \approx \tilde{\lambda}_j'(t) + \Delta I - \Delta N + \Delta \kappa. \tag{9.2}$$

Here, $\tilde{\lambda}_j'(t)$ is the per capita population growth rate that the species would experience at the spatial average ecological condition, ΔI quantifies the *spatial storage effect*, ΔN quantifies the relative nonlinearity of species interactions, and $\Delta \kappa$ is the covariance between relative density and the local per capita growth rate. The first three components are directly analogous to comparable consequences of temporal ecological variation, whereas $\Delta \kappa$ is a unique consequence of spatial ecological variation. As with temporal variability, each of these terms may foster or retard coexistence of each species in the guild under consideration. To assess their effects on invasibility, this per capita population growth rate must as always be evaluated when species j is rare and the other species in the community are at their demographic steady states across the landscape (Chesson 2000a, Snyder and Chesson 2004).

The first term in equation (9.2), $\tilde{\lambda}_j'(t)$, is the per capita population growth rate that an invader would achieve in the absence of spatial variation and so "measures the effects of variation-independent coexistence mechanisms" (Chesson 2000a, p. 224). As with temporal ecological variation, this is then the per capita population growth rate that the invading species would achieve in the average environment. If the guild members were competing for resources, $\tilde{\lambda}_j'(t)$ would be the per capita population growth rate of the invading species at the average resource abundance; or if the guild members were apparent competitors, $\tilde{\lambda}_j'(t)$ would be the per capita population growth rate of the invading species at the average predator abundance. Remember that averages in this case are taken across spatial locations and weighted by local population abundance, because what is being averaged is the per capita growth rate contribution of each individual.

If spatial ecological variation had no discernible effect on coexistence of any species, all the relationships discussed in chapters 3–7 could be applied based on the demographics of the average environment. However, the remaining three terms in equation (9.2) cause deviations of the average per capita population growth rate away from what would be obtained in the average spatial environment. Thus, many species in the community are coexisting because their spatial average per capita growth rate is sufficient. Others for which this growth rate in the average spatial environment is insufficient are coexisting because these remaining three terms adequately boost their per capita growth rates when they are rare. And still others are excluded because these additional terms either are not sufficient to boost their per capita growth rates or actually diminish their per capita growth rates from what they would experience in the spatial average environment to prevent their membership in the community.

Just as nonlinearities in species responses to temporal variation can permit some inferior species to coexist (see chapter 8, "Endogenously Generated

Population Cycling"), nonlinearity in the relationship between local per capita population growth rates and local ecological conditions can also boost the average per capita growth rate of the entire population. The magnitude of this relative non-linearity effect is measured by ΔN in equation (9.2). For concrete examples, consider the relationships depicted in figure 9.1. The local per capita growth rate increases with the local environmental condition as an accelerating convex function (the curve in the center of each panel). For evaluating invasibility, the local environmental condition could be the local resource abundance set by the other consumer species present when the species of interest invades. The consequences of relative nonlinearities on the average per capita growth rate result from Jensen's inequality. If the relationship between local per capita growth rates and local environmental conditions is nonlinear, the average per capita population growth rate across space will not be the same as the per capita growth rate at the average environmental condition. In figure 9.1(a), the curve along the x axis shows the distribution of local spatial environmental conditions, which are determined by both the local abiotic environment and the other species that are present at their local demographic steady states that the invading species experiences (here again, remember that the various spatial conditions are weighted by the proportion of the population that experiences them); and the curve along the y axis shows the corresponding distribution of local per capita population growth rates that result because of the relationship between them (i.e., the center curve). Because the relationship is convex, the average per capita growth rate (i.e., at the dashed line reflected from the x axis) across all sites is higher than the per capita growth rate in the average location (i.e., the dashed line from the center curve only). Thus, the relative nonlinearity of the per capita growth rate across space increases the invader's per capita growth rate when invading. ΔN is measured as the per capita growth rate for the average environmental condition minus the average per capita growth rate across space, which is negative and so increases $\tilde{\lambda}_j(t)$ in equation (9.2).

Relative nonlinearity can also hinder invasibility. Instead of an accelerating relationship, figure 9.1(b) shows an asymptotic concave relationship between local per capita growth rate and local environmental condition. In this case, the per capita growth rate at the average environmental condition is greater than the average per capita growth rate across space, and so the relative nonlinearity for this invading species decreases its per capita growth rate at invasion (i.e., $\Delta N > 0$ and so decreases $\tilde{\lambda}_j(t)$ in equation (9.2)).

The term ΔI in equation (9.2) quantifies the magnitude of the spatial storage effect. Unlike the temporal storage effect, the spatial storage effect requires no specific life history feature to be manifested. The spatial storage effect results from covariation between environmental conditions and the strength of "competition," with competition being again a general catchall for negative indirect effects

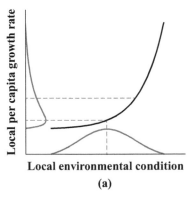
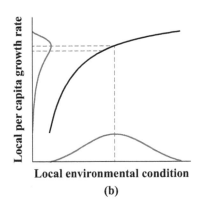

(a)   (b)

FIGURE 9.1. Two examples of how relative nonlinearity in per capita growth rate to spatially varying environmental conditions among individuals in a population can alter the overall average per capita growth rate of the entire population. In each panel, the center curve shows the relationship between some environmental condition (e.g., resource availability, predator abundance) and the per capita growth rate of a population of individuals that experience that local condition. The curve along the x axis ("Local environmental condition") is a hypothetical distribution of individuals within the population that experiences each of these environmental conditions that varies across space, and the curve along the y axis ("Local per capita growth rate") is the resulting distribution of local per capita population growth rates. The dashed lines that reflect from one axis to the other through the center curve identify the average local environmental condition and thus the per capita growth rate at that average environmental condition. The dashed line running only from the center curve to the y axis identifies the average per capita population growth rate. Panel (a) shows that a convex fitness relationship can increase the average per capita growth rate above what would be expected in the average environment, and (b) shows that a concave relationship can decrease the average per capita population growth rate below what would be expected at the spatial average.

among guild members mediated through resources or predators (Chesson 2000a), and so it's akin to the covariances among the environmental properties and species abundances generated by temporal variation. In his derivations, Chesson divided the various determinants of per capita population growth rate into effects of the environment (designated by E) and "competition" (designated by C), although the designation of any particular effect as one or the other can be quite arbitrary (Ellner et al. 2016, Barabás et al. 2018) and sometimes impossible to separate (Chesson 2000a). Here, competition is defined very broadly as any negative-negative direct or indirect interaction among species, including various forms of direct interference, indirect resource competition, and indirect apparent competition mediated by shared predators or pathogens (Chesson 2000a). The covariation between these two determinants for each guild member is its "spatial storage effect." (In the future, I hope that intrepid theoreticians will add positive interactions among mutualist partners to this construct so that a mix of species interaction types can be evaluated simultaneously.)

Specifically, the spatial storage effect is determined by the degree to which an invading guild member is able to exploit favorable environmental conditions without experiencing increased competition (again in the general sense of that word). For a spatial version of the lottery model analyzed by Snyder and Chesson (2004), the spatial storage effect is quantified as

$$\Delta I = F_r Cov(E_r, C_r) - F_i Cov(E_i, C_i), \tag{9.3}$$

where r identifies the resident species that is at its demographic steady state in the absence of the invading species, the invading species is designated by i, and F_j is the fecundity of the respective species (equation 10 in Snyder and Chesson 2004). The covariances in equation (9.3) depend on the distribution of environmental conditions in space, the degree of similarity in the spatial distribution of environmental factors that more limit each species in the guild, and the spatial distribution of the resident species, which determines strength of local competition for both the invader and the resident (see Snyder and Chesson 2004 for mathematical details; see also Murrell and Law 2003, Snyder and Chesson 2003, Snyder 2007, 2008).

The importance of the spatial storage effect in inflating the invader per capita population growth rate is increased by any mechanism that causes the abundances of the various species to be focused in different environmental conditions across space—partial spatial segregation—so as to accentuate the difference in the covariances in equation (9.3) (Bolker and Pacala 1997, 1999, Chesson 2000a, Snyder and Chesson 2004). For example, a common species may experience strong intraspecific competition (large C) in favorable conditions (large E), but only weak intraspecific competition (small C) in poorer areas (small E), which would result in a large positive value for $Cov(E_r, C_r)$. This resident covariance is increased if offspring remain near their parents in space (e.g., short seed dispersal kernels around adult plants). In contrast, an invading species will experience no competition initially with itself (very low contribution to its overall C at any location), and if the invader and resident species find different environmental conditions more favorable to their respective per capita growth rates, the invading species may actually have a negative $Cov(E_i, C_i)$ term because competitive effects of the resident species are focused on environmental conditions that are poor for the invader's growth (Snyder and Chesson 2004). This spatial difference in good and bad areas for each species reduces "competitive" effects between the invader and resident species and thus may boost the invader's per capita growth rate.

Finally, the term $\Delta\kappa$ specifically measures $Cov(\lambda_j, v_j)(t)$, the covariance between local per capita population growth rate and local relative population size in equation (9.1). Snyder and Chesson (2004) offered a number of scenarios that could increase or decrease this term as well.

As one can see from the derivation of the spatial storage effect, two sets of variables are critical for a robust examination of how spatial variation may enhance or retard coexistence of particular species in a community; namely, the spatial patterning of environmental factors that influence the per capita growth rate and the spatial patterning of limiting factors (e.g., resource species, predators, pathogens, mutualists) that are the consequence of that environmental patterning across a community. These are the underlying metrics that are needed to evaluate the various terms in equation (9.2).

As with all other types of analyses to now, the major ecological processes that influence the demography of the species of interest must first be known. Furthermore, to understand spatial effects, the spatial variation and covariation of these factors must be quantified across the community. For example, if one is examining the coexistence of the plant assemblage of a mountain meadow, quantifying the spatial variation and covariation of abiotic factors, limiting mineral nutrients, litter and water in the soil, and their herbivores and pathogens is required to associate spatial variation in these environmental drivers with spatial variation in local per capita population growth rates and the strengths of interspecific interactions. Information about how these environmental factors covary across space will also be critical for exploring how and why species may partially segregate along various ecological gradients. For example, Seabloom et al. (2005) provide an exemplar for how to quantify such spatial variation and covariation in species responses across space.

Chesson (2008) advocated calculating what he called the *community average* approach, which involves calculating the average of these various effects over all the species in the guild of study, and recent studies have reported only these average values (e.g., Angert et al. 2009). However, as he noted (Chesson 2008, pp. 151–152):

> Not so obvious is the fact that each of the quantities, ΔI, ΔN, and $\Delta \kappa$, in general has a particular value for each particular species. Thus, each of them really should be given a subscript i to indicate its species dependence. That a given mechanism may affect different species to different extents, or even in different ways, highlights the fact that these mechanisms need not be purely stabilizing in action, but may modify average fitness differences too.

Consequently, I see no value in evaluating the average of each term across species to the exclusion of evaluating the contributions for individual species. Here again, invasibility and thus coexistence is an issue about individual species. Each of these components may increase or decrease per capita growth rates of a species, and they may be different in both magnitude and sign across species. Thus, the averages across species are fairly meaningless. Does the spatial storage effect boost the per capita population growth rate of a species with a relatively low average growth rate enough to permit it to invade when rare? Or does a concave average nonlinearity or strong covariation with "competitor" abundances depress the per capita population growth rate of a species with relatively high average growth enough to prevent its coexistence? These are the questions that address coexistence of guild members, and not what the average of these various effects is across all species.

DISPERSAL LINKING LOCAL COMMUNITIES

One long-standing approach to spatial variation in community structure assumes that the assemblage of species in question occupies small habitat patches with largely inhospitable territory between. For example, the Glanville fritillary butterfly (*Melitaea cinxia* (L.)) is a rather sedentary butterfly that has two host plants, the ribwort plantain (*Plantago lanceolate* L.) and the spiced speedwell (*Vernoica spicata* L.), that grow in small, dry meadows on the Åland Islands in the Baltic Sea. In addition, a number of parasitoids attack the larvae of the Glanville fritillary, and a number of hyperparasitoids attack these parasitoids (Hanski 1999). Each meadow thus represents a community, and these local communities are linked by dispersal.

Patch Dynamics Ignoring Local Population Sizes

One approach to modeling such a system of multiple communities linked by dispersal is to treat each local community as a habitat patch and simply follow the number of patches that are occupied by various species in the community. These patch occupancy models all derive from Levins's (1969) first statement of a model of a *metapopulation*. The original metapopulation model analyzed by Levins followed the number of patches occupied by a single species. Shortly thereafter, Levins and Culver (1971) transformed this original model so that the proportion of patches occupied was the variable being tracked, and this is the metapopulation model that is most used today. Let p_1 be the fraction of habitat patches occupied by species 1 among all the possible patches it could colonize. Individuals of the species colonize new patches from old patches at a rate of m_1, and the populations in occupied patches go extinct at random and at a rate of x_1. (I follow Levins and Culver's original notation here, and so these are different definitions of p, m, and *x* than have been used in the rest of this book.) From these simple assumptions, the rate of change in the proportion of occupied patches is

$$\frac{dp_1}{dt} = m_1 p_1 (1 - p_1) - x_1 p_1 \tag{9.4}$$

(Levins and Culver 1971). This model has a stable equilibrium at

$$p_1^* = 1 - \frac{x_1}{m_1}, \tag{9.5}$$

which is feasible when $m_1 > x_1$ or, in words, when the colonization rate of new patches is greater than the extinction rate of occupied patches.

This equilibrium identifies that the proportion of patches occupied by species 1 is defined by the balance struck between extinction and colonization of patches by this species. However, if p_1^* patches are occupied, $1 - p_1^*$ are unoccupied. Perhaps some other species may be able to invade this system of communities—this *metacommunity* (Holyoak et al. 2005, Leibold and Chase 2017)—and displace species 1, or another species may be able to invade and exploit the unutilized patches. A number of analyses have been performed to address such questions while making various assumptions about the additional species and about the habitat patches (e.g., Cohen 1970, Levins and Culver 1971, Horn and MacArthur 1972, Slatkin 1974, Hastings 1980, Hanski 1981, 1983, Nee and May 1992).

Tilman (1994) analyzed a variant of these models to ask how many species that compete for patch occupancy could coexist. Assume all the patches to be ecologically identical and that species 1 will displace all other species if it occupies a patch. Now introduce a second species that cannot coexist with species 1 in any habitat patch but may be able to exploit the patches not occupied by species 1. In other words, if species 1 invades a patch occupied by species 2, species 2 will immediately go locally extinct. Given these assumptions, the dynamics of species 1 are still given by equation (9.4), and so its equilibrium patch occupancy proportion is still given by equation (9.5), but the dynamics of patch occupancy by species 2 are given by

$$\frac{dp_2}{dt} = m_2 p_2 (1 - p_1 - p_2) - x_2 p_2 - m_1 p_1 p_2 \tag{9.6}$$

(Tilman 1994). In this equation, the first term is the colonization of unoccupied patches by species 2, the second term is the extinction rate of patches occupied by species 2, and the third term is the displacement rate by species 1.

The poorer competitor can invade and coexist with the better competitor if

$$m_2 > m_1 \left(\frac{m_1 + x_2 - x_1}{x_1} \right)$$

(Tilman 1994). If the patch extinction rate for species 2 is greater than or equal to that for species 1 ($x_2 \geq x_1$), the poorer competitor must have a higher colonization rate than species 1 ($m_2 > m_1$) to invade and coexist (Hastings 1980, Tilman 1994). However, if $x_2 < x_1$, the poorer competitor can invade and coexist even if it has a lower patch colonization rate ($m_2 < m_1$) (Nee and May 1992).

Assume now that a pool of competitor species is available to colonize the system of patches, and species are identified by their rank competitive abilities (species 1 > species 2 > species 3 > . . . > species n). Tilman (1994) showed that if all the species have the same patch extinction rate, $x_1 = x_2 = . . . = x_j = x$, additional species can be added to the community that have

$$m_j > \frac{x}{\hat{s}_{j-1}^2},$$

where $\hat{s}_{j-1}$ is the equilibrium proportion of empty habitat patches with species 1, 2, . . .,j–1 already present at equilibrium. Therefore, an inferior competitor can invade and coexist if its patch colonization rate is greater than its patch extinction rate divided by the square of the proportion of empty patches. Because the fraction of empty patches decreases with every successfully invading species, this result implies a *competition-colonization trade-off* for species to coexist: poorer competitors must have an ever-increasing colonization advantage to exploit the ever-shrinking proportion of patches not already occupied by better competitors (see also Horn and MacArthur 1972, Hastings 1980, Yu and Wilson 2001, Calcagno et al. 2006).

In these models where local coexistence is assumed not to be possible, poorer competitors can only live off the patches left over by their superiors. In effect, this is quite analogous to the models of competition among consumers for a single resource that we explored in chapters 3–7. In those models, an inferior resource competitor could invade only if the superior competitor left some amount of the resource unexploited because of some other limitation on its abundance (e.g., consumer direct self-limitation, or a predator or pathogen also feeding on the consumer). Here, the poorer competitor is also relegated to what the superior competitor is not able to exploit because of its own extinction rate (note that $p_1^* = 1$ if $x_1 = 0$ in equation (9.5)).

Similar models have been constructed for predators and their prey that interact in such a patchy environment (Hastings 1977, 1978, Caswell 1978, Crowley 1979, Hanski 1981). These models also generally imply similar predictions to those from a single community. For example, a predator can mediate the coexistence of a superior and inferior competitor (here competition is for patch occupancy) (Caswell 1978, Hastings 1978, Crowley 1979, Hanski 1981).

Typically, these patch occupancy models are interpreted as each patch harboring small local populations of multiple individuals with continuous dispersal across the landscape. The ecological interactions that cause extinction or displacement are not specified but are assumed to occur because of the mechanisms (e.g., competitive exclusion, local disturbance, predator overexploitation) addressed in previous chapters; the mechanisms operating within patches are assumed but unseen. However, these models can also be interpreted as exploring competition *for* space among individuals. In this interpretation, each "patch" is simply a location that can support a single individual, and the model tracks the proportion of locations occupied by individuals of each species.

Metacommunities Incorporating Local Population Abundances

The previous sections briefly explored how spatial variation across the area or volume occupied by one community can influence the coexistence of species in that community. This spatial variation implies that different individuals in each population may experience different demographic rates because of their spatial location within this area or volume. In this section, I turn to spatial ecological variation that exists at the scale of communities found at widely separated locations that are connected in a large network of species interactions by movement of individuals between them. Imagine some collection of ponds in a local area. Each pond may have a unique set of local ecological conditions that makes each somewhat different from all the others: perhaps the ponds differ in the total availability of the mineral nutrients for the basal resources in the system, or they differ in the abilities of particular species to colonize them (e.g., fish species can only move among ponds and lakes that are connected by creeks and streams). If each of these ponds were completely isolated from all the others, they might each come to different steady states (i.e., point equilibria or limit cycles), given their local conditions. In the following, I specify what these differences are, in terms of parameters and species composition among localities.

However, dispersal of individuals of the various community members among the local communities may perturb each local community away from the abundances and community structure that the local conditions would favor (Leibold and Chase 2017). The space between local communities is unsuitable for the species under consideration to maintain its population, and so individuals that emigrate from one local community must move through this surrounding matrix to eventually immigrate into another local community or die. With such dispersal, local community structure then becomes a balance between what the local ecological conditions favor and the net flux of individuals among these local communities.

The recognition that dispersal among local communities can alter local community structure spawned the study of such metacommunities. Leibold and Chase (2017) presents a robust and in-depth analysis of the issues caused by dispersal. In this section, I analyze a few specific examples to illustrate major effects of dispersal and to highlight approaches for study.

Individuals of different species may use myriad strategies to decide when to leave a community and when to enter a new community (Clobert et al. 2001, 2009). The simplest dispersal strategy is that each individual simply has some fixed propensity to passively emigrate from their local community and settle in the first new suitable location that they encounter (e.g., Hastings 1982, Holt 1985). Many other strategies for leaving and entering communities may also be employed, which are variously based on cues about the perceived fitness returns the individual may receive from staying in its present community or entering another (e.g., Křivan and Sirot 2002, Poethke and Hovestadt 2002, Persson and De Roos 2003, Amarasekare 2004a, Matthysen 2005, Ronce 2007, Armsworth and Roughgarden 2008, Křivan et al. 2008, Fowler 2009, Cantrell et al. 2013). The leaving and entering strategies employed by individuals of a particular species will determine the exact quantitative effects of dispersal on community structure. However, most dispersal strategies cause local populations to deviate from what local ecological conditions favor in qualitatively similar ways (Amarasekare 2004a). Because of this, I focus on what are essentially the two endpoints in terms of deviations from local ecological expectations (Amarasekare 2004a): *passive dispersal*, in which a constant fraction of the current population emigrates from each population and disperses equally with some cost to all other available populations; and directed *habitat selection*, in which individuals only leave their current population and move to a new population if their fitness would be higher. For simplicity, I also focus on metacommunities that contain only two or three local communities linked by dispersers.

A SINGLE RESOURCE SPECIES

To build intuition about the consequences of dispersal to local equilibrium abundances, first imagine a simple metacommunity with two available locations and a single resource species that can potentially occupy both and that passively disperses between the two locations. The abundances of each population are given by $R_1^{(1)}$ and $R_1^{(2)}$, where the parenthetical superscript identifies the population's location. Each local population displays logistic population growth, but the intrinsic rates of increase and strengths of density dependence can differ between the

TABLE 9.1. Summary descriptions of the additional parameters used in the models
presented in this chapter

Parameter	Description
ω_i	Passive dispersal rate of species i leaving natal patch and moving to other patches
ξ_i	Proportion of individuals that die while moving from one patch to another

two locations. Finally, individuals emigrate from each location at a constant rate ω, and suffer mortality en route to the other location such that only a fraction ξ of them survive the journey ($0 \le \xi \le 1$) to immigrate into the other population. These specifications imply the following model for this resource species in the two locations:

$$\frac{dR_1^{(1)}}{dt} = c_1^{(1)} R_1^{(1)} - d_1^{(1)} (R_1^{(1)})^2 - \omega R_1^{(1)} + \xi \omega R_1^{(2)}$$
$$\frac{dR_1^{(2)}}{dt} = c_1^{(2)} R_1^{(2)} - d_1^{(2)} (R_1^{(2)})^2 - \omega R_1^{(2)} + \xi \omega R_1^{(1)}$$
(9.7)

Holt (1985) analyzed variations of this model with $\xi = 1.0$. If the two locations are identical in their ecological conditions, $c_1^{(1)} = c_1^{(2)}$ and $d_1^{(1)} = d_1^{(2)}$, passive dispersal has no effect on the local populations, because at equilibrium identical numbers of individuals leave, are killed in transit, and then enter both populations at each instant in time (McPeek and Holt 1992). Table 9.1 lists the new parameters used in models of this chapter.

Passive dispersal between the populations will distort local abundances away from what is locally favored if the locations differ in ecological conditions, and particularly if those differences regulate the populations to different equilibrium abundances (Holt 1985, McPeek and Holt 1992). If the locations are regulated to different abundances and a constant fraction of the population leaves in each unit of time, more individuals leave locations regulated to higher abundances. Consequently, more individuals immigrate than emigrate in locations with lower abundances, and fewer individuals immigrate than emigrate in locations with higher abundances. These differences in the net flux of individuals in and out of locations cause sites where local ecological conditions favor a higher abundance to have fewer individuals than the local ecological conditions will support, and locations where ecological conditions favor a lower abundance to have more individuals than the local conditions will support (fig. 9.2). A higher dispersal propensity causes the abundances in different locations to become more similar and so causes these deviations to be more pronounced. Mortality during dispersal reduces the

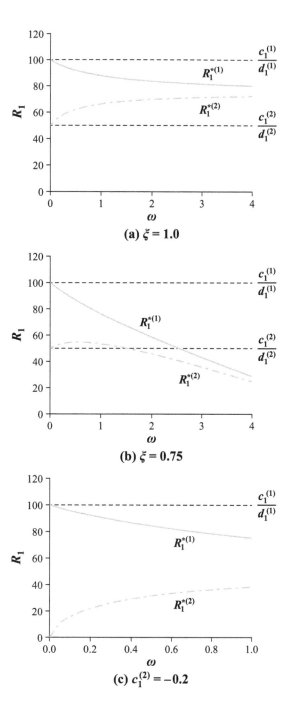

(a) $\xi = 1.0$

(b) $\xi = 0.75$

(c) $c_1^{(2)} = -0.2$

total number of individuals in all locations and can cause both populations to have a deficit of individuals if mortality during dispersal and dispersal propensity are both high (fig. 9.2(b)).

Sink populations occur in locations where individuals of the species cannot invade and coexist (Pulliam 1988). In equations (9.7), such a location would have $c_1^{(i)} < 0$, which would result in a negative and thus infeasible equilibrium abundance. For the sink species to exist in this location, other locations would have to support *source populations* with $c_1^{(i)} > 0$ to generate a continual supply of dispersers. The sink population is at a stable equilibrium abundance where the loss of individuals due to local population regulation is balanced by the gain via immigration from source populations. A higher passive dispersal propensity increases the equilibrium sink abundance and decreases the equilibrium source abundance (fig. 9.2(c)).

This simple example illustrates two important points about the distributions of species. First, the simple fact that a species is present at a location does not mean that it is coexisting there. A species in a sink location does not satisfy its local invasibility criterion, but rather is maintained at that location by continual immigration from other sites where it does satisfy its invasibility criterion. The second point is that the sink species has the same impacts on its local ecological environment as a source species would have. Sink species will depress resource abundances and inflate consumer, mutualist, and pathogen abundances, as we see later in this chapter. As a result, the presence of sink species will influence the invasibility criteria of other species that may try to become local community members.

At the other end of the dispersal strategy spectrum are those that only generate movement of individuals from locations where they receive lower fitness to locations where they receive higher fitness. Such strategies result in individuals distributed across populations according to an *ideal free distribution* (Fretwell and Lucas 1969). Many types of dispersal strategies can result in an ideal free

---

FIGURE 9.2. Equilibrium abundances for a single resource species that has populations in two locations and passively disperses between them. In each panel, the equilibrium abundances for different values of the propensity to passively disperse (ω) are shown. In (a) and (b), the local ecological conditions would regulate the population to $R_1^{*(1)} = 100$ in location 1 and to $R_1^{*(1)} = 50$ in location 2 in the absence of passive dispersal. Panel (a) shows results when all dispersers survive ($\xi = 1.0$), and (b) shows results when 75% of dispersers survive ($\xi = 0.75$). Panel (c) illustrates the equilibrium abundances when location 2 is a sink with $c_1^{(2)} = -0.2$ and all dispersers survive. The solid line is the equilibrium abundance in location 1 and the dot-dashed line is location 2. The thin dashed lines show the abundances to which populations in the two locations would be regulated in the absence of passive dispersal. The parameters are as follows: $c_1^{(1)} = 2.0$, $d_1^{(1)} = d_1^{(2)} = 0.02$.

distribution of individuals. It is achieved most rapidly if individuals can compare potential fitness returns among locations before they select which location to occupy (Fretwell and Lucas 1969). However, even some relatively passive dispersal strategies (e.g., when individuals emigrate from locations in inverse proportion to local abundances) can eventually result in an ideal free distribution of individuals across locations (McPeek and Holt 1992, Doebeli and Ruxton 1997, Doncaster et al. 1997, Cantrell et al. 2010).

Dispersal strategies that result in an ideal free distribution have an important consequence for spatial variation among communities, namely, that local abundances are exactly what local ecological conditions would favor (Holt 1984, Křivan and Sirot 2002, Cressman et al. 2004, Cressman and Křivan 2006, 2010, Křivan et al. 2008, Cantrell et al. 2010). For these strategies, once the metacommunity reaches an ideal free distribution, the number of emigrants and immigrants balance in each population such that no population is distorted in abundance away from what local population regulation mechanisms favor (Křivan et al. 2008). Moreover, most dispersal strategies that would be favored by natural selection but that cannot achieve an ideal free distribution will result in a qualitative pattern of abundance distortions that are consistent with the passive dispersal results above, where locations favoring higher abundances have deficits and locations favoring lower abundances have excesses of individuals (Amarasekare 2004a).

As with any mechanistic inquiry, the first essential piece of information needed to justify considering a mechanism is positive evidence that the mechanism could potentially influence the metrics under study, which in this case is positive evidence that individuals do disperse among the local communities under consideration. However, quantifying dispersal can be very difficult in practice, because distinguishing recent immigrants from residents in a population is usually impossible. Marking individuals in local populations and following exchange between local populations over time is the most direct metric of dispersal, but these methods provide more qualitative support for the operation of dispersal than quantitative estimates of dispersal rates, given that dispersal to populations outside the study area cannot be distinguished from mortality within the population. Artificial habitats for individuals to colonize also offer positive evidence of dispersal (e.g., McCauley 2006). In similar fashion, in my own work, I identified damselfly species that had high and low dispersal rates by quantifying their arrival as adults—the dispersal stage of their life history—at ponds and lakes where their larvae could not survive but that were near water bodies where they did support thriving populations (McPeek 1989).

Passive dispersal or dispersal syndromes that do not achieve an ideal free distribution should cause local abundances to be more similar to one another than expected simply based on local ecological conditions (fig. 9.2). An important metric that identifies such disparities is that the average absolute fitnesses of individuals in populations should show a strong pattern: individuals in high abundance populations should display positive average absolute fitnesses because of the net efflux of migrants that maintains local abundances below what local conditions could support, and individuals in low abundance populations should have negative average absolute fitnesses because of the net influx of migrants that inflates their abundances above what local conditions will support (Runge et al. 2006). To test this, replicate density manipulation experiments that bracket the natural abundances of a species could be performed in populations that span a large gradient of natural population abundances. The response variables in each population would be the estimates of the average absolute fitness at the natural abundance. If dispersal distorts local abundances away from what local conditions favor, a regression of average absolute fitness at the natural abundance against natural abundance should have a positive slope. If the cost of dispersing is not high, this regression should shift from less than replacement (e.g., $\ln(\bar{W}) < 0$) to more than replacement (e.g., $\ln(\bar{W}) > 0$) along this regression (fig. 9.2(a)). However, if dispersal is costly but still prevalent, all populations may be below expected abundances, but the positive relationship between average fitness and local abundance should hold ($\omega > 1.5$ in fig. 9.2(b)).

Throughout this book, tests for invasibility have been the standard for the coexistence of a species in a community. Fundamentally, testing invasibility is testing whether a species is a source or sink population in a community (Runge et al. 2006): each species that passes its invasibility criterion is a source, and each species that fails its invasibility criterion is a sink or walking dead. Thus, we have been testing this issue all along. Why a species is a sink species is again the mechanistic *why* question, and these questions lend much greater understanding than simply the yes/no source/sink issue.

However, failing an invasibility test does not distinguish a sink from a walking dead species. To do that, one must explore the broader source-sink structure of the potential metacommunity in which the local community is embedded (Leibold and Chase 2017). Studies to determine whether source populations are in nearby communities, whether individuals disperse from the sources to the sink, and whether dispersal into the local community could maintain the species locally all provide mechanistic explanations for the existence of a metacommunity structure. These are also the critical additional mechanistic features of a metacommunity that would discriminate a sink species (i.e., a mix of source and sink populations) from a walking dead species (i.e., all sink populations).

The other solid prediction that emerges from considering the effects of various dispersal strategies on metacommunity structure is that if individuals only move to locations that give them higher fitness so that an ideal free distribution results, populations within local communities will not be displaced away from abundances that local ecological conditions favor (Holt 1984, Křivan and Sirot 2002, Cressman and Křivan 2006, 2010, Křivan et al. 2008). If this were true, the above studies would show that individuals do move between communities regularly, but the regression of local per capita growth rate on local abundance should have a slope and intercept not different from zero. The first of these predictions of significant dispersal is key, because local population dynamics under an ideal free distribution are indistinguishable from the expectations with no dispersal at all (Holt 1984, Křivan and Sirot 2002, Cressman and Křivan 2006, Křivan et al. 2008). If an ideal free distribution is suggested by these results, studies that explore the cues used by individuals to make the necessary immigration decisions and then the habitat selection decisions of where to settle would be warranted (e.g., Resetarits 2001, Benard and McCauley 2008, Binckley and Resetarits 2008, Dupuch et al. 2009, Pintar et al. 2018).

LINK SPECIES AT THE TOP OF A FOOD CHAIN

Not all interacting species have population dynamics that operate on comparable spatial scales. In particular, consumers and predators higher in the food web sometimes tend to have larger bodies and thus range for food over much larger areas than many of their prey. Therefore, consumers may forage in what are effectively independent prey populations. For example, imagine a population of kingfishers consisting of individuals that forage on the fish in several ponds in a local area. The fish and the underlying food web that supports them in each pond constitute a local community. The kingfisher links the population dynamics of these separate communities together and so represents a type of *link species*. Holt (1984) analyzed a simple version of this scenario for a single consumer population feeding on two resource species that are spatially segregated. Oksanen et al. (1995) extended Holt's analysis to include saturating functional responses and consumer feeding interference, and then applied the predictions to data on armored catfish foraging on algae across multiple stream pools. This section follows their analyses closely.

Consider the simple model in which a consumer population at any instant in time is split between two locations, with τ proportion of the population in location 1 and $1 - \tau$ proportion of the population in location 2. The dispersal strategy employed by the consumer individuals determines the value of τ. Each location also potentially supports a resource population on which the consumer feeds, but the resource individuals do not disperse between the locations. Different resource species may be present at each location, or the same resource species may simply have populations in both locations with no dispersal between them. These assumptions imply the following set of equations:

$$\frac{dN_1}{N_1 dt} = \tau \left[\frac{b_{11}^{(1)} a_{11}^{(1)} R_1^{(1)} N_1}{1 + a_{11}^{(1)} h_{11}^{(1)} R_1^{(1)}} - f_1^{(1)} \right] + (1-\tau) \left[\frac{b_{11}^{(2)} a_{11}^{(2)} R_1^{(2)} N_1}{1 + a_{11}^{(2)} h_{11}^{(2)} R_1^{(2)}} - f_1^{(2)} \right]$$

$$\frac{dR_1^{(1)}}{R_1^{(1)} dt} = c_1^{(1)} - d_1^{(1)} R_1^{(1)} - \frac{b_{11}^{(1)} a_{11}^{(1)} \tau N_1}{1 + a_{11}^{(1)} h_{11}^{(1)} R_1^{(1)}}, \tag{9.8}$$

$$\frac{dR_1^{(2)}}{R_1^{(2)} dt} = c_1^{(2)} - d_1^{(2)} R_1^{(2)} - \frac{b_{11}^{(2)} a_{11}^{(2)} (1-\tau) N_1}{1 + a_{11}^{(2)} h_{11}^{(2)} R_1^{(2)}}$$

where all the parameters of this consumer-resource model are as defined in chapter 3.

Making the assumption that the resources equilibrate very rapidly to the number of consumers in their location (MacArthur 1969, Holt 1984, Oksanen et al. 1995), the foraging return that consumer individuals receive in each patch is given by the equation components in the respective square brackets of the consumer equation in (9.8). Figure 9.3 illustrates these functions for a saturating functional response.

If individuals move between the two locations at random and without cost, a constant proportion of the consumer population will occupy each location so that τ is a constant. Without loss of generality, assume that half the consumer population is found in each location ($\tau = 0.5$). When the consumer population invades

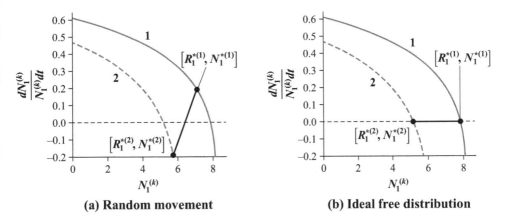

(a) Random movement   (b) Ideal free distribution

FIGURE 9.3. Relationships between per capita population growth rates for a consumer that feeds on a resource in two locations that differ in the productivity for the resource. The solid curve is for location 1 and the dashed curve is for location 2. The filled black circles identify the per capita growth rate in each location at equilibrium. Panel (a) shows the consumer's per capita growth rate in each location at equilibrium when the consumer population moves randomly between the two locations with time, and (b) shows the equilibrium when the consumer population moves according to a strategy that results in an ideal free distribution. The parameters are as follows: $c_1^{(1)} = 2.0$, $c_1^{(2)} = 1.5$, $d_1^{(1)} = d_1^{(2)} = 0.02$,, $a_{11}^{(1)} = a_{11}^{(2)} = 0.25$, $b_{11}^{(1)} = b_{11}^{(2)} = 0.1$, $h_{11}^{(1)} = h_{11}^{(2)} = 0.05$, and $\xi = 1.0$.

and begins to increase, equal numbers of consumers will be present in the two locations (fig. 9.3(a)). The consumer population will come to equilibrium when the net foraging returns balance. However, here again, the foraging returns (i.e., the average absolute fitness) to individuals in the two locations will be different, with one being positive and one being negative (Holt 1984, Oksanen et al. 1995). The location with a negative foraging return is a *pseudosink* (Watkinson and Sutherland 1995) in this case.

Thus, a link species that moves randomly among locations will have the same qualitative effect on the local resource abundances as when it passively disperses among communities. As a result, the more productive location has fewer consumers than the location can support, and the less productive location has more consumers than the location can support. Consequently, the more productive location has a higher resource abundance and the less productive location has a lower resource abundance than is expected. If the productivities between the locations are sufficiently disparate, the consumer can drive the resource in the lower productivity location extinct, which is a clear example of sink species affecting local community structure (Holt 1984).

In contrast, now assume that the consumer individuals apportion themselves between the two locations so that they all have the same fitnesses and are distributed according to an ideal free distribution. When this consumer invades, all individuals will be found in the location giving the higher foraging return (location 1 in fig. 9.3(b)). As the consumer population increases, individuals will remain in that location until its foraging return is depressed to the point equal to the maximum foraging return in the other location. As the consumer's abundance increases beyond this point, individuals will apportion themselves between the two locations so that their foraging returns are equal. The consumer population will increase until its total population growth rate is zero. This equilibrium has exactly the same numbers of resource and consumer individuals as the equilibrium that would result if the consumers did not move between locations such that the locations were independent communities (Holt 1984, Oksanen et al. 1995, Křivan and Sirot 2002, Cressman and Křivan 2006, Křivan et al. 2008).

All the same types of testing that were described in the previous section apply here as well. However, link species should show much greater movement rates between locations than dispersers, including individuals continually moving between locations.

SPECIALIST AND GENERALIST CONSUMERS ACROSS DIFFERENT HABITATS

In most types of ecosystems, species are confronted with different habitats scattered across the landscape or seascape. The ponds and lakes that are scattered around a region differ in the top predators that dominate them. These differences in top predators cause their prey to segregate among them so that distinct assemblages of species are found in different water bodies based on the identity of the top predator (Brooks and Dodson 1965, Dodson 1970, Zaret 1980, McPeek 1990b, Arnott and Vanni 1993, Wellborn et al. 1996, Stoks and McPeek 2003b). Other ecosystems can have other ecological factors that may drive patterns of community membership among local communities. For example, the prevalence of fires can shape the identity of the dominant plant assemblages (e.g., grassland, savanna, forest), which will in turn shape the assemblages of herbivores at the next trophic level. This spatial variability among local communities that are potentially linked by dispersal can determine spatial patterning across the food web. Moreover, some functional positions in the food web may be strongly influenced by these spatial differences and thus require species with special phenotypic properties to fulfill those functional roles, but other functional positions may exist in all local food webs.

Species that fill these locally specific roles are identified here as *habitat specialists*, because they are coexisting in only a specific subset of the available local communities; and species that fill the roles that are common to all food webs are identified as *habitat generalists*, because they coexist in many different community types that harbor specialists. In this section, I explore the ecological properties that define which species will be habitat specialists and generalists at the intermediate trophic level of a diamond/keystone predation food web (chapter 4) that develops in two locations that differ in their ecosystem properties.

First, imagine a metacommunity consisting of two locations where communities develop, and further imagine that the two locations support two different top predators that do not move between the locations. For example, in freshwater ponds and lakes, fish will be present if some route of colonization exists for them (e.g., stream connections), but other ponds and lakes that lack colonization routes for fish will be fish-free. As a result, different predators dominate the fish-free ponds and lakes (typically, large midges and copepods in the limnetic zone, and large invertebrate predators, such as dragonflies, in the littoral and benthic zones) (Dodson 1970, Sprules 1972, Crowder and Cooper 1982, Werner and McPeek 1994). The same resource species inhabits both locations, and a guild of multiple consumers are available to colonize both and passively disperse between the locations. For simplicity, I also assume that the resource does not disperse between the two locations, but this makes no difference to the outcomes described below. These assumptions lead to the following set of equations describing the species in the two locations:

$$\frac{dP_1^{(1)}}{dt} = \sum_{j=1}^{q} \frac{m_{j1}^{(1)} N_j^{(1)} P_1^{(1)}}{1+\sum_{k=1}^{q} m_{k1}^{(1)} l_{k1}^{(1)} N_k^{(1)}} - x_1^{(1)} P_1^{(1)}$$

$$\frac{dP_2^{(2)}}{dt} = \sum_{j=1}^{q} \frac{m_{j2}^{(2)} N_j^{(2)} P_2^{(2)}}{1+\sum_{k=1}^{q} m_{k2}^{(2)} l_{k2}^{(2)} N_k^{(2)}} - x_2^{(2)} P_2^{(2)}$$

$$\frac{dN_j^{(1)}}{dt} = \frac{b_{1j}^{(1)} a_{1j}^{(1)} R_1^{(1)} N_j^{(1)}}{1+a_{1j}^{(1)} h_{1j}^{(1)} R_1^{(1)}} - \frac{n_{j1}^{(1)} m_{j1}^{(1)} N_j^{(1)} P_1^{(1)}}{1+\sum_{k=1}^{q} m_{k1}^{(1)} l_{k1}^{(1)} N_k^{(1)}} - f_j^{(1)} N_j^{(1)} - \omega N_j^{(1)} + \xi \omega N_j^{(2)}$$

$$\frac{dN_j^{(2)}}{dt} = \frac{b_{1j}^{(2)} a_{1j}^{(2)} R_1^{(2)} N_j^{(2)}}{1+a_{1j}^{(2)} h_{1j}^{(2)} R_1^{(2)}} - \frac{n_{j2}^{(2)} m_{j2}^{(2)} N_j^{(2)} P_2^{(2)}}{1+\sum_{k=1}^{q} m_{k2}^{(2)} l_{k2}^{(2)} N_k^{(2)}} - f_j^{(2)} N_j^{(2)} - \omega N_j^{(2)} + \xi \omega N_j^{(1)}$$

$$\frac{dR_1^{(1)}}{R_1^{(1)} dt} = c_1^{(1)} - d_1^{(1)} R_1^{(1)} - \sum_{j=1}^{q} \frac{b_{1j}^{(1)} a_{1j}^{(1)} \tau N_j^{(1)}}{1+a_{1j}^{(1)} h_{1j}^{(1)} R_1^{(1)}}$$

$$\frac{dR_1^{(2)}}{R_1^{(2)} dt} = c_1^{(2)} - d_1^{(2)} R_1^{(2)} - \sum_{j=1}^{q} \frac{b_{1j}^{(2)} a_{1j}^{(2)} N_j^{(2)}}{1+a_{1j}^{(2)} h_{1j}^{(2)} R_1^{(2)}}$$

$$(9.9)$$

in which all the parameters are as defined in chapters 3 and 4. For simplicity, I also assume that all the consumers have the same passive dispersal rate between locations ω and the same survival rate during dispersal ξ. For this analysis, I also assume that none of the consumers or predators directly self-limit their own abundances, and the parameters are species-specific but do not differ among the locations.

As in chapter 4, only two consumers can coexist with the predator and resource in each location, but additional consumer species may be present as sink species in one location if they coexist in the other location. Only one consumer will be coexisting at a location if that guild member is superior at avoiding the local predator and exploiting the local resource population. Coexistence of two consumers requires that one be better at avoiding the local predator and the other be better at exploiting the resource (i.e., satisfying inequalities (4.9)) (Holt et al. 1994, Leibold 1996). Furthermore, if the predators in the two locations require different antipredator adaptations for a consumer to successfully reduce mortality from them, a trade-off will exist between the functional positions where predator avoidance is key. If this is true, different specialist consumers will coexist with the two different predators, but the same consumer that is better at exploiting the resource that is common to both sites will be a generalist coexisting in the communities at both locations (McPeek 1996).

Because the same functional position exists in both communities for the consumer that is better at exploiting the resource, dispersal between locations has no inherent limitation other than mortality during dispersal. In contrast, dispersal is

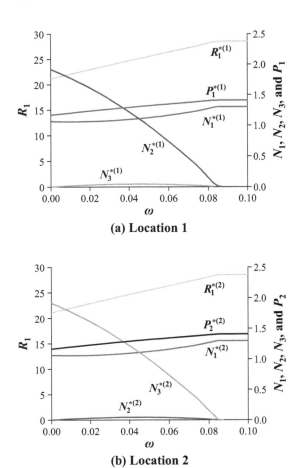

FIGURE 9.4. Relationships for how equilibrium abundances change with dispersal propensity for a pool of three consumers (N_1, N_2, N_3) that feed on a resource (R_1) in a metacommunity of two locations, where different intraguild predators are restricted to the two locations (P_1 in location 1 and P_2 in location 2). Consumer 1 is the best resource competitor, and consumers 2 and 3 are both equally poorer competitors, but consumer 2 is best at avoiding P_1 and consumer 3 is best at avoiding P_2. The horizontal axis of the two panels gives the dispersal rate of the three consumers between the two locations. The parameters are as follows: $c_1^{(1)} = c_1^{(2)} = 1.75$, $d_1^{(1)} = d_1^{(2)} = 0.04$, $a_{11}^{(1)} = a_{11}^{(2)} = 1.0$, $a_{12}^{(1)} = a_{12}^{(2)} = 0.2$, $a_{13}^{(1)} = a_{13}^{(2)} = 0.2$, $b_{1j}^{(i)} = n_{jk}^{(i)} = 0.1$, $h_{1j}^{(i)} = l_{jk}^{(i)} = 0.04$, $f_j^{(i)} = 0.25$, $m_{11}^{(1)} = m_{12}^{(2)} = 0.8$, $m_{21}^{(1)} = m_{32}^{(2)} = 0.1$, $m_{31}^{(1)} = m_{22}^{(2)} = 1.0$, $x_k^{(i)} = 0.1$, and $\zeta = 1.0$.

very costly for the consumers that are specialized on predator avoidance (fig. 9.4). At extremely low levels of passive dispersal (i.e., ω < 0.083), the predator-specializing consumers support very small sink populations in the other location. However, relatively low passive dispersal rates lead to the extinction of the specialist consumer that is better at avoiding the predator, even with no mortality of dispersers en route to the other location ($\xi = 1.0$). Thus, the ecological impact of some sink types can have profound consequences for the source populations of those species in the metacommunity.

Now imagine a metacommunity of two locations in which the same predator is found in both places, but a different resource species is found in each. This model is fashioned after many different terrestrial ecosystems where different plant species dominate different locations, but the same predators forage across these locations. Again for simplicity, I assume that all parameters are species-specific but do not differ between the two locations. I also assume here without loss of generality that the resource species do not disperse between locations but the consumers and predator do:

$$\frac{dP_1^{(1)}}{dt} = \sum_{j=1}^{q} \frac{m_{j1}^{(1)} N_j^{(1)} P_1^{(1)}}{1 + \sum_{k=1}^{q} m_{k1}^{(1)} l_{k1}^{(1)} N_k^{(1)}} - x_1^{(1)} P_1^{(1)} - \omega P_1^{(1)} + \omega \xi P_1^{(2)}$$

$$\frac{dP_1^{(2)}}{dt} = \sum_{j=1}^{q} \frac{m_{j1}^{(2)} N_j^{(2)} P_1^{(2)}}{1 + \sum_{k=1}^{q} m_{k1}^{(2)} l_{k1}^{(2)} N_k^{(2)}} - x_1^{(2)} P_1^{(2)} - \omega P_1^{(2)} + \omega \xi P_1^{(1)}$$

$$\frac{dN_j^{(1)}}{dt} = \frac{b_{1j}^{(1)} a_{1j}^{(1)} R_1^{(1)} N_j^{(1)}}{1 + a_{1j}^{(1)} h_{1j}^{(1)} R_1^{(1)}} - \frac{n_{j1}^{(1)} m_{j1}^{(1)} N_j^{(1)} P_1^{(1)}}{1 + \sum_{k=1}^{q} m_{k1}^{(1)} l_{k1}^{(1)} N_k^{(1)}} - f_j^{(1)} N_j^{(1)} - \omega N_j^{(1)} + \xi \omega N_j^{(2)} \qquad (9.10)$$

$$\frac{dN_j^{(2)}}{dt} = \frac{b_{2j}^{(2)} a_{2j}^{(2)} R_2^{(2)} N_j^{(2)}}{1 + a_{2j}^{(2)} h_{2j}^{(2)} R_2^{(2)}} - \frac{n_{j1}^{(2)} m_{j1}^{(2)} N_j^{(2)} P_1^{(2)}}{1 + \sum_{k=1}^{q} m_{k1}^{(2)} l_{k1}^{(2)} N_k^{(2)}} - f_j^{(2)} N_j^{(2)} - \omega N_j^{(2)} + \xi \omega N_j^{(1)}$$

$$\frac{dR_1^{(1)}}{R_1^{(1)} dt} = c^{(1)} - d_1^{(1)} R_1^{(1)} - \sum_{j=1}^{q} \frac{b_{1j}^{(1)} a_{1j}^{(1)} N_j^{(1)}}{1 + a_{1j}^{(1)} h_{1j}^{(1)} R_1^{(1)}}$$

$$\frac{dR_2^{(2)}}{R_2^{(2)} dt} = c^{(2)} - d_2^{(2)} R_2^{(2)} - \sum_{j=1}^{q} \frac{b_{2j}^{(2)} a_{2j}^{(2)} N_j^{(2)}}{1 + a_{2j}^{(2)} h_{2j}^{(2)} R_2^{(2)}}$$

With this model, where the resource base varies spatially, a greater range of possible outcomes occurs depending on the cost of dispersal. In this case, if two consumers coexist at each location, one is a specialist on the local resource at the expense of avoiding the predator and feeding on the resource at the other location, and the generalist is good at defending against the predator but poor at feeding on the resources in both locations (fig. 9.5). If dispersal is cost-free ($\xi = 1.0$), increased passive dispersal causes the generalist's abundance to increase in both locations and each specialist to maintain a significant sink population in the other location. In this case, passive dispersal generates two source consumers and one sink consumer in each location.

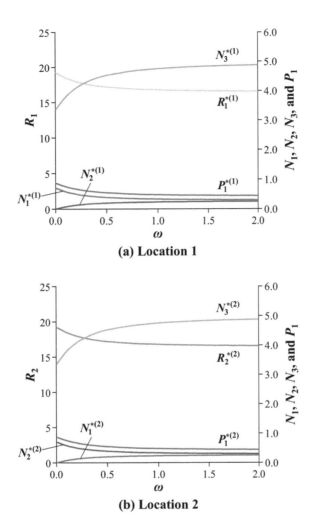

(a) Location 1

(b) Location 2

FIGURE 9.5. Relationships for how equilibrium abundances change with dispersal propensity for a pool of three consumers (N_1, N_2, N_3) that are fed upon by one predator (P_1) in two locations where different resource species are found (R_1 in location 1 and R_2 in location 2). Consumer 1 is the best competitor for resource 1, consumer 2 is the best competitor for resource 2, and consumer 3 is equally poor at competing for both resources but is the best at avoiding the predator. In this example, all dispersing consumers survive to reach the other locations ($\xi = 1.0$). The horizontal axis of the two panels gives the dispersal rate of the three consumers between the two locations. The parameters are the same as in figure 9.4, except for the following: $a_{11}^{(1)} = a_{22}^{(2)} = 1.0$, $a_{12}^{(1)} = a_{13}^{(1)} = a_{21}^{(2)} = a_{23}^{(2)} = 0.2$, $m_{11}^{(i)} = m_{21}^{(i)} = 1.0$, and $m_{31}^{(i)} = 0.1$.

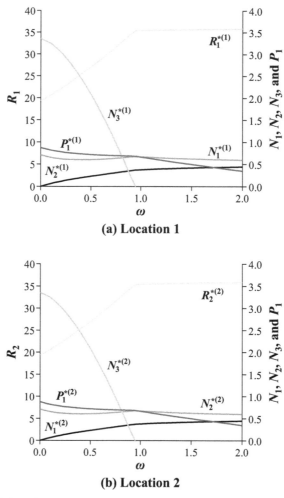

FIGURE 9.6. Relationships for how equilibrium abundances change with dispersal propensity for a pool of three consumers (N_1, N_2, N_3) that are fed upon by one predator (P_1) in two locations where different resource species are found (R_1 in location 1 and R_2 in location 2). This example is identical to that illustrated in figure 9.5, except that only 75% of dispersing consumers survive the journey to other locations ($\xi = 0.75$).

In contrast, if dispersers pay a significant mortality cost ($\xi < 1.0$), the generalist is now the consumer to be driven extinct at low dispersal rates because of the combined effects of high predation rates in both locations plus mortality due to dispersal (fig. 9.6). The specialists create source-sink relationships among the two locations at low to intermediate rates of dispersal, but can survive at much higher levels of dispersal because of their lower predator mortality in both locations.

These models demonstrate that the nature of species that are interpreted as habitat specialists and generalists depends critically on the nature of spatial variation in the food web. Although no systematic review has been performed to my knowledge, aquatic ecosystems—particularly standing bodies of water—differ spatially in top predators, from vernal ponds with no predators, to fishless ponds where dragonflies and salamanders dominate, to large lakes where fish dominate (Zaret 1980, Wellborn et al. 1996). In contrast, terrestrial ecosystems are typically defined by the dominant plant types—desert, prairie, tropical rain forest, temperate deciduous forest, boreal forest, tundra, taiga, riparian floodplain, cove versus ridgeline forest, and so on. Thus, the differences in the predictions from models (9.9) and (9.10) may offer important expectations to be tested for aquatic versus terrestrial food web patterning across spatial locations.

These models also make strong predictions about what processes should be more important in determining local and regional community structure and the properties of habitat specialists and generalists given the features that vary spatially among communities. The analyses of models (9.9) and (9.10) both imply that processes that influence mortality (i.e., predators and dispersal-associated mortality) will more greatly influence responses to spatial variation across the metacommunity than processes such as resource limitation. In the case where the communities differ in the top predator, only very small levels of cost-free passive dispersal can cause consumers that are specialized against the predators to be driven extinct (see fig. 9.4). Thus, predator specialists should show very low dispersal rates among ponds and lakes. This is in fact exactly what we have found in our studies of damselflies that specialize on living with different top predators in different lakes. For example, *Enallagma* and *Lestes* species segregate between lakes where fish are the top predators and lakes where dragonflies are the top predators (McPeek 1990b, 1998, Stoks and McPeek 2003b). Adults of *Enallagma* species show extremely low dispersal rates away from their natal lakes. In one study, we found that they showed almost no movement to ponds that were only a few meters from their natal pond (McPeek 1989). In contrast, I have observed adults of *Ischnura verticalis*, which are common in both fish and dragonfly lakes, that are kilometers from the nearest water body.

Likewise, in model (9.10), where the resource differs among locations, some amount of dispersal mortality combined with predator mortality could drive the species better adapted to the resources—and thus the species that experiences greater predator mortality—to extinction. Understanding the sources of population regulation for various community members is critical for understanding why species may or may not be successful in a variegated landscape.

These contrasting models of how food web functional groups vary among locations across the landscape also make strong predictions about the properties of species that will be habitat specialists and generalists. Not surprisingly, habitat specialists are those species that are better adapted to interacting with functional groups that are limited to only a subset of the available habitats. Again, *Enallagma* and *Lestes* damselfly species have adapted to live with the predator in only one pond type, and so are restricted to that specific pond type because they experience high mortality from the other predators (McPeek 1990b, 1998, Stoks and McPeek 2003b). Likewise, in high altitude ponds in the Rocky Mountains of Colorado, some cladoceran and copepod species are specialists in ponds that contain salamanders or *Chaoborus* midges as top predators because of their differential susceptibility to these predators (Dodson 1970, Sprules 1972).

What is often not appreciated is that other types of functional groups in the food web are common across many different community types. For example, in the spatially segregated predator system, a species that is better at utilizing the resource but experiences high predator mortality can coexist in both communities (see fig. 9.4). The advantage that this species enjoys in both communities is being the better resource competitor, but it can be lousy at avoiding any predator. In the littoral zones of ponds and lakes of North America, this functional group is filled by *Ischnura* species (McPeek 1998). Likewise, in the ponds of Colorado, *Diaptomus coloradensis* is common to both salamander and *Chaoborus* ponds, although the relative abilities of *D. coloradensis* at predator avoidance and resource competition are not known to my knowledge (Dodson 1970, Sprules 1972).

SOURCE-SINK DYNAMICS ACROSS A PRODUCTIVITY GRADIENT

In previous chapters, I have explored the abundance patterns of food web members in communities allayed along various environmental gradients (e.g., figs. 3.4, 4.3, 6.9). In these chapters, each community was assumed to exist in isolation from all others on the landscape with no dispersal of individuals among them. If individuals make choices to move among locations based on fitness differences (or cues for fitness differences) that cause them to primarily move to locations giving them higher fitness, these gradient abundance patterns illustrated in previous chapters would be adequate. However, as the past few sections have shown, passive dispersal among locations can generate distortions away from what local ecological conditions favor. An important study by Amarasekare (1998, 2004a, 2004b, 2007, 2008a, 2008b, 2008c, Amarasekare et al. 2004) of how various dispersal strategies influence abundance patterns across environmental gradients for various community modules serves as my model in this section.

Consider the pool of species containing one resource, three consumers, and one intraguild predator species, previously considered in figure 5.10. This pool of species occupies a metacommunity consisting of three locations that differ in productivity ($c_1^{(i)}$, where (i) identifies location) for the resource, but the parameters for the other species are identical in all the locations. (I forego presenting the full set of equations here, which is identical to equations (5.4) with the addition of terms for passive dispersal of all species among the three locations, as in (9.10).) Imagine that these three locations are located at different positions on the gradient depicted in figure 5.10 such that, in the absence of dispersal, R_1 and P_1 will coexist at all locations, N_1 and N_2 will coexist with them at location 1, only N_2 will coexist with them at location 2, and N_2 and N_3 will coexist with them at location 3. For this example, I assume all dispersers survive movement among locations.

At an extremely low level of passive dispersal for all species, all three consumers can be found in the communities at all three locations. Here, N_1 and N_3 have a source-sink structure across the communities. However, at dispersal rates that are somewhat higher but still extremely low, N_1 and N_3 are driven extinct in the metacommunity because of the loss of individuals from their source-to-sink populations and the distortions in the local abundance of N_2 across locations with consequent changes in the resource and intraguild predator abundances (fig. 9.7, and see Amarasekare 2008a, 2008b, 2008c). Abundance distortions because of dispersal strategies that do not result in ideal free distributions of species can drive others in the community extinct.

Other studies have explored more complicated dispersal patterns in larger metacommunities (e.g., Hauzy et al. 2010, Ristl et al. 2014, Melián et al. 2015, Thiel and Drossel 2018). The variety of possibilities is endless. These effects are probably the hardest to detect and explore empirically, because they require piecing together how the movement of various species distorts abundance patterns across the metacommunity. For example, consider how N_2's abundance changes with higher levels of passive dispersal when the other two consumers have gone extinct. Instead of increasing in some locations and decreasing in others (e.g., fig. 9.2), N_2's abundance either stays the same (location 1) or increases locally (locations 2 and 3) as dispersal rate is increased (fig. 9.7). This pattern is because of the indirect responses of R_1 and P_1.

SPATIOTEMPORAL VARIATION AND AUTOCORRELATION

In the last chapter and this, I have considered spatial and temporal variation in ecological conditions that affect community dynamics largely as separate entities. Certainly, many forms of ecological variation come as primarily one or the other. Daily and seasonal variations in temperature and sunlight intensity change consistently through time over local and regional areas of the landscape, and so represent synchronous temporal variation across the landscape; the temperature and sunlight intensity rise and fall in all the local communities together. Likewise, the

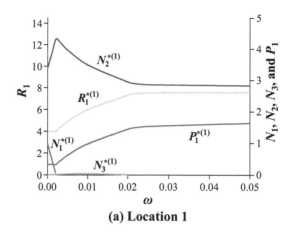

(a) Location 1

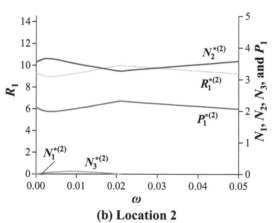

(b) Location 2

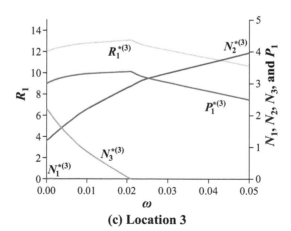

(c) Location 3

gradients of ecological conditions along an altitudinal transect up a mountain or among ponds and lakes that are found in areas with different bedrock will show consistent spatial differences that remain largely consistent through time; the soils on the ridgeline of the mountain will always be dryer than those at the base of the mountain. Also, the variation in and the covariation between ecological conditions and demographic outcomes for species have been the key features for understanding coexistence. However, the patterning of ecological variation itself through time and space in both the magnitudes of the variation and the spatial and temporal autocorrelation at various scales will also drive how species and entire communities are shaped by environmental variation.

The multispecies metapopulation models considered earlier in this chapter (both patch occupancy models and models that track local abundances) represent one extreme. These models assume a high degree of ecological variation locally, such that local demographic conditions swing between highly favorable to so bad that the population goes extinct. These models also assume that populations in patches go extinct at random, either because of changes in local abiotic conditions or because of the random colonization of another species. Therefore, the ecological conditions across patches show no spatial and no temporal autocorrelation. This means that knowing the ecological conditions in a patch now provides no information about the ecological conditions at any time in the future in that same patch or in a patch at any distance from the current one.

Introducing spatial or temporal autocorrelation to these metapopulation and metacommunity models can change the likelihood of persistence and the average per capita growth rates and abundances of species. For example, if disturbances are spatially positively autocorrelated such that many contiguous habitat patches all go extinct simultaneously, the metapopulation has a much lower probability of long-term persistence (Kallimanis et al. 2005). Alternatively, positive temporal autocorrelation can greatly increase the persistence of a metapopulation and inflate the abundances of the local populations, even when all the local populations are sinks (Roy et al. 2005).

FIGURE 9.7. Relationships for how equilibrium abundances change with dispersal propensity for a pool of three consumers (N_1, N_2, N_3) that are fed upon by one intraguild predator (P_1) and that all feed on one resource (R_1) in a metacommunity of three locations. The locations differ in their productivities for the basal resource. The horizontal axis in each panel gives the dispersal rate of the species among the locations; all species have the same dispersal rate and all survive dispersal. The parameters are as given in figure 5.10, with the following stipulations: $c_1^{(1)} = 3.0$, $c_1^{(2)} = 3.75$, $c_1^{(3)} = 4.25$, and $\xi = 1.0$.

Autocorrelation patterning on short and long scales for both spatial and temporal dimensions can also shape species responses. The most extreme examples are species that seasonally migrate in response to changing environmental conditions: birds and whales that annually migrate on a global scale between breeding and "wintering" grounds, large mammalian herbivores that migrate in response to seasonal weather patterns. Such migration is only evolutionarily feasible if ecological conditions are positively autocorrelated on short space and time scales but negatively autocorrelated on long space and time scales. These differences in autocorrelation on short and long scales provide the needed predictability but large magnitude of changes in conditions over time and through space that foster the evolution of migration patterns (Clobert et al. 2001, Ronce 2007, Blanquart and Gandon 2011, Drown et al. 2013).

These more complicated patterns combining spatial and temporal ecological variation have been explored primarily in the context of the fitness effects on individual species and largely not in the context of multiple interacting species. Moreover, not all species in a community will respond identically to spatiotemporal variation: the ducks and geese leave their breeding lake and fly south when the temperatures drop in the fall, but the fish cannot exercise that option and so must deal with winter conditions in that lake. Models exploring the community consequences of the combined spatial and temporal environmental variation will be a rewarding direction to extend the work summarized in these two chapters (e.g., Schmidt et al. 2000).

SUMMARY OF MAJOR INSIGHTS

- Spatial environmental variation across a single community influences the per capita population growth rates of member species via the effects of relative nonlinearities, the covariation between environmental conditions and demographic performance in species interactions (i.e., the spatial storage effect), and the covariation between local population growth rates and the local relative abundance.
- Multiple species can coexist in metapopulation models of patch occupancy as fugitive species utilize leftover patches as unused resources.
- Sink species do not satisfy their own invasibility criteria, but they will alter the invasibility criteria of other species through their interactions with other species in the community.
- Dispersal strategies that are not based on local fitness or abundance criteria distort local abundances away from what local ecological conditions favor. In general, communities in locations that support large populations will

have a net deficit of individuals below what local ecological conditions can support, and so species should in general experience net positive fitnesses. In contrast, communities in locations that support smaller populations will have a net excess of individuals above what local ecological conditions can support, and so species should in general experience net negative fitnesses.

- Sources of mortality are more important to spatial habitat patterning of species distributions across various community types than processes of resource partitioning.

CHAPTER 10

Ecologically Equivalent and Neutral Species Embedded in a Food Web

I chose to do my PhD dissertation on *Enallagma* damselflies for two reasons. First, *Enallagma* species segregate between two different habitat types—ponds and lakes in which fish are the top predators, and ponds and lakes where fish are absent (Johnson and Crowley 1980). Second, only one to three *Enallagma* species are found in fishless ponds and lakes, but ponds and lakes with fish have up to a dozen *Enallagma* species. Earl Werner and Don Hall, my PhD advisers, initially pressed me forcefully on whether this would turn out to be just a repetition of Paine's studies of keystone predation, where a top predator excludes the dominant competitor and so allows in many inferior competitors to coexist with it, but the dominant competitor excludes all other species in areas where the top predator is absent (Paine 1966, 1969, 1974). However, I convinced them that my natural history intuition about alternative predators in the two lake types might be correct. Luckily, Earl and Don had faith in my natural history intuition.

To work on this group, I first had to learn how to identify the 17 *Enallagma* species found in the ponds and lakes around the Kellogg Biological Station in southwestern Michigan, where I was doing my fieldwork. I went to the entomology collection on the top floor of the Natural Sciences Building on the Michigan State University campus and asked the faculty member in charge if I could see all the *Enallagma* larvae and adults, because I wanted to work on their ecological patterns across ponds and lakes. He sarcastically said "good luck" and took me to the appropriate cabinet. What I discovered that day—and the source of his sarcasm—was that if you are going to become good at telling these insect species apart, you are going to become an expert on insect reproductive structures (Eberhard 1985). About half of the local *Enallagma* in southwestern Michigan are nearly identical in adult color pattern and in both larval and adult morphology. Their only differences are in one set of minute structures on males and females: the claspers that males use to grasp females and the thoracic plates on females where these male claspers contact the female (Westfall and May 2006, McPeek et al. 2008, 2011, Siepielski et al. 2018). Little did I know in 1984 that this would lead me down a path toward considering neutral species in food webs (Siepielski et al. 2010).

Enallagma are not unique. Many clades contain species that are nearly impossible to discriminate unless one compares phenotypic traits involved in species mate recognition and mating or sequences their DNA. Like *Enallagma*, many of these cryptic species are members of the same local communities. Eighteen cryptic lacewing species in the *Chrysoperla carnea* group are found in North America and Eurasia. Species are only distinguishable phenotypically by their mating songs, and up to three species can be found foraging in the same bushes (Henry 1985, Henry et al. 1999). Many copepod species are only distinguishable by the morphologies of male antennae and fifth legs that the males use to grasp females during mating (Park 2000, Ferrari and Ueda 2005). Short stretches of small creeks and large rivers in both tropical and temperate regions can harbor more than 100 chironomid midge species that only a few taxonomic experts can identify to species after much painstaking work (Ruse 1995, Coffman and de la Rosa 1998, Langton and Casas 1999, Casas and Langton 2008). The prevalence of cryptic species diversity is not limited to the arthropods. For example, frog clades in both South America and southeast Asia have numerous sympatric cryptic species (Stuart et al. 2006, Funk et al. 2011). In fact, cryptic species are common in all major clades of animals, plants, and fungi (Bickford et al. 2007, Pfenninger and Schwenk 2007, Struck et al. 2018). Thus, we may be greatly underrepresenting the true number of species in many communities.

We know very little about the degree of ecological differentiation among these cryptic species complexes. Some are surely ecologically differentiated, like the three cryptic species in the skipper butterfly *Udranomia kikkawai* found in Costa Rica (Janzen et al. 2017). To now, we have considered species that are ecologically distinguishable, and so cryptic species that are ecologically differentiated pose no new conceptual challenges, but they may require new concepts to account for even more diverse assemblages of species to be able to coexist than we have discussed thus far.

In contrast, what are the ecological consequences of having multiple ecologically equivalent species in a community, such as *Enallagma* (Siepielski et al. 2010)? Other examples of ecologically equivalent and apparently neutral species have also been identified. For example, what was previously described as the ubiquitous amphipod *Hyallela azteca* is in fact a cryptic species complex with multiple cryptic species living together in ponds and lakes across North America (Witt and Hebert 2000, Witt et al. 2006). Experimental studies have also shown that some of these cryptic *Hyallela* species seem to be ecologically equivalent insofar as they show no difference in per capita population growth rates when rare and the others are at natural abundances (i.e., they have no demographic advantage or disadvantage when rare) (Cothran et al. 2015). However, ecologically equivalent or neutral species need not be closely related: experimental results have shown that the co-occurring barnacles *Jehlius cirratus* and *Notochthamalus scabrosus* also seem to be neutral species, having no demographic advantage when each is made rare (Shinen and Navarrete 2014).

We may have an implicit confirmation bias by focusing our studies on taxa having clear ecological differences, and so we need to more seriously consider the prospects of communities harboring many ecologically equivalent and neutral species. In this chapter, I explore the dynamic signatures of these types of species. However, ecological equivalence and neutrality are not synonymous (McPeek and Siepielski 2019). In fact, some ecologically equivalent sets of species may be coexisting. Here, I consider what is needed for a guild of ecologically equivalent species to be truly neutral species.

ECOLOGICALLY EQUIVALENT AND NEUTRAL SPECIES

In exploring the distributions and abundances of 135 tree and shrub species found in a 13.5 ha plot of dry forest in Guanacaste Province, Costa Rica, Hubbell (1979) suggested that the dominance-diversity patterns seen among these species and for forest tree species around the globe are consistent with a very simple model. In this model, very localized disturbance continually kills off a few individual trees in the forest at random with respect to species identity. The probability of any species replacing one of these dead individuals is proportional to the relative abundance of that species in the forest. In other words, the total number of trees in the forest is regulated, but species replace one another at local sites at random (Hubbell 1979, 2001, Hubbell and Foster 1986).

Hubbell was struck by how well this neutral model of community dynamics produced results that are consistent with broad-scale community patterns, such as rank-abundance curves and distributions of relative species abundances (Hubbell 1979, Hubbell and Foster 1986, Hubbell 2001). In this model, species are assumed to be functionally and so *ecologically equivalent* in their abilities to compete. The resulting community dynamics of species' relative abundances are equivalent to neutral alleles in population genetics: their relative abundances should follow a random walk until only one species is left in the community. Functional equivalence "states that trophically similar species are, at least to a first approximation, demographically identical on a per capita basis in terms of their vital rates, of birth, death, dispersal—and even speciation" (Hubbell 2005, p. 166). Hubbell (1979, 2001) called these dynamics *community drift*. He based his model of neutral community dynamics on Moran's model of genetic drift, in which neutral alleles are equated to species, because it matches the dynamics of local individual death and random replacement as in the forest (Hubbell 2001). However, the Wright-Fischer model of genetic drift can also be used to develop this framework and give identical results (McPeek and Gomulkiewicz 2005).

The publication of Hubbell's (2001) monograph on the "unified neutral theory" sparked a great exploration of whether community patterns were consistent with the assumption that species differences really do not matter, and so for many properties that one may be interested in for a given community, species act as though they are neutral alleles drifting to fixation or extinction (e.g., Volkov et al. 2003, Chave 2004, Etienne and Olff 2005, McGill et al. 2006, Canard et al. 2014). This debate, like many scientific debates, quickly spiraled into an either-or argument: either communities displayed neutral dynamics or they displayed dynamics consistent with niche differentiation; any signature consistent with niche differentiation in these tests was seen as validation to reject neutral dynamics entirely (Volkov et al. 2003, McGill et al. 2006).

However, I think much of the uproar caused by Hubbell's proposals stemmed from his semantics. The term *community drift* implies that all the species in a community are ecologically equivalent, but "probably no ecologist in the world with even a modicum of field experience would seriously question the existence of niche differences among competing species on the same trophic level" (Hubbell 2005, p. 166). Hubbell limited his definition of community to "co-occurring assemblages of trophically similar species" (2005, p. 166). Taken from this more narrow perspective, this implies guilds or functional groups of ecologically equivalent species embedded in the larger food web (Leibold and McPeek 2006, Siepielski et al. 2010, McPeek and Siepielski 2019).

Therefore, in this chapter I explore the community dynamics resulting from having a functional position in a food web occupied by more than one ecologically equivalent species. By "ecologically equivalent," I mean something slightly more stringent than Hubbell: to me, ecologically equivalent means that all the parameters in the model describing population dynamics are identical across the species. It is not difficult to imagine how ecologically equivalent species arise to enter communities, since many modes of speciation affect only characters involved in mate recognition and compatibility and do not alter traits that shape ecological performance of individuals (reviewed in chapter 4 of McPeek 2017b). This is why cryptic species are so prevalent. However, as we will see, having a guild of ecologically equivalent species embedded at some functional position in a food web does not necessarily mean that they will display community drift dynamics. Some constellation of ecological properties can render them as coexisting species, while other constellations of properties render them neutral species. Sorting out these properties is the goal of this chapter.

Neutral Resource Competitors

Begin by returning to the basic model of a single consumer feeding on a single resource. All the same results could be derived using saturating and Beddington-DeAngelis functional responses, but for simplicity and clarity, I focus primarily on the situation with the consumer having a linear functional response. Also, because the species are ecologically equivalent and so have the same parameter values, I drop the parameter subscripts. Model (3.3) then becomes

$$\frac{dN_1}{N_1 dt} = baR_1 - f$$
$$\frac{dR_1}{R_1 dt} = c - dR_1 - aN_1$$

$$(10.1)$$

Again, to coexist in this community with the resource, the consumer must be able to satisfy its invasibility criterion of being able to have a positive per capita population growth rate on the equilibrium abundance of the resource when only the resource is present—that is, satisfy inequalities (3.4) or (3.5). If it can invade, the equilibrium abundances of the two species are

$$R^*_{1(N_1)} = \frac{f}{ab} \quad \text{and} \quad N^*_{1(R_1)} = \frac{cab - df}{a^2 b}.$$

$$(10.2)$$

This equilibrium, if feasible, is also a stable equilibrium. To now, I have simply asserted this based on the literature. However, it is informative here to prove it. Stability of an equilibrium in these models is evaluated using the Jacobian matrix—the matrix of partial differential equations of each total growth rate equation taken with respect to the abundance of each species and evaluated at the equilibrium of interest (May 1973, Strogatz 2015). The Jacobian matrix of model (10.1) is then

$$\begin{bmatrix} \frac{\partial}{\partial R_1}\left(\frac{dR_1}{dt}\right) & \frac{\partial}{\partial N_1}\left(\frac{dR_1}{dt}\right) \\ \frac{\partial}{\partial R_1}\left(\frac{dN_1}{dt}\right) & \frac{\partial}{\partial N_1}\left(\frac{dN_1}{dt}\right) \end{bmatrix}_{R^*_{1(N_1)}, N^*_{1(R_1)}} = \begin{bmatrix} c - 2dR^*_{1(N_1)} - aN^*_{1(R_1)} & -aR^*_{1(N_1)} \\ baN^*_{1(R_1)} & baR^*_{1(N_1)} - f \end{bmatrix} = \begin{bmatrix} -dR^*_{1(N_1)} & -aR^*_{1(N_1)} \\ baN^*_{1(R_1)} & 0 \end{bmatrix}.$$

The stability of the model is determined by calculating the eigenvalues of this matrix. If all the eigenvalues are negative, the equilibrium is stable (May 1973, Hirsch et al. 2012, Strogatz 2015). The eigenvalues are the roots of the characteristic equation of this matrix (which is found by solving the matrix equation $|\mathbf{A} - \lambda\mathbf{I}| = 0$, where $\mathbf{A}$ is the Jacobian matrix and $\mathbf{I}$ is the identity matrix) (Hirsch et al. 2012, Strogatz 2015). After substituting and simplifying, the characteristic equation for this model is

$$\lambda^2 + dR^*_{1(N_1)}\lambda + a^2 bR^*_{1(N_1)}N^*_{1(R_1)} = 0. \tag{10.3}$$

The roots of this polynomial, which are the eigenvalues of the Jacobian matrix, are both negative if the equilibrium abundances of both species are feasible (May 1973).

These results now serve as background against which to evaluate whether a set of ecologically equivalent and neutral consumer species feeding on one resource can coexist in this community. To do this, now consider a model in which two consumers that have identical parameter values feed on the resource:

$$\frac{dN_1}{N_1 dt} = baR_1 - f$$

$$\frac{dN_2}{N_2 dt} = baR_1 - f \tag{10.4}$$

$$\frac{dR_1}{R_1 dt} = c - dR_1 - aN_1 - aN_2$$

Because the dynamical equations for the two consumer species are identical, their isoclines are also identical (fig. 10.1).

The equilibrium abundances for this model are

$$R^*_{1(\sum N_j)} = \frac{f}{ab} \text{ and } (N_1 + N_2)^*_{1(R_1)} = \left(\sum N_j\right)^*_{1(R_1)} = \frac{cab - df}{a^2 b}. \tag{10.5}$$

Here, the resource's equilibrium abundance remains unchanged (cf. equilibria (10.2) and (10.5)). However, an interesting property emerges for the consumers. Neither consumer is regulated to a specific abundance. Instead, the summed total abundance of both consumers is the regulated metric in this community. This is because the intersection of the three isoclines is not a point but rather a line, and that line is the geometric representation of the summed total consumer abundance being regulated (fig. 10.1). This is a stable line of equilibrium. If more than two ecologically identical consumers were present, the same property of their summed total abundance being regulated to this same specific value emerges. Also, the value to which the summed total abundance of ecologically identical consumers is regulated is identical to the value that a single consumer is regulated (again, compare equilibria (10.2) and (10.5)). This is a fundamental property for neutral species (Hubbell 2001). This guild (Root 1967, Simberloff and Dayan 1991) of ecologically identical species operates in the food web as a single functional group.

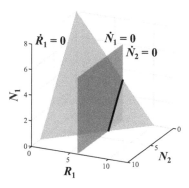

FIGURE 10.1. The isocline configuration for a resource competition module with two ecologically identical consumers that are also neutral species (N_1 and N_2) that feed on one resource (R_1). The consumers have linear functional responses and no direct self-limitation. The isoclines are identified as in figures 3.3 and 3.5. Because the two consumers are ecologically identical, their isoclines are also identical. The intersection of the three isoclines is identified with a thick black line. The parameter values are as follows: $c_1 = 2.0$, $d_1 = 0.2$, $a_{11} = a_{12} = 0.25$, $b_{11} = b_{21} = 0.1$, $h_{11} = h_{21} = 0.0$, $f_1 = f_2 = 0.25$, and $g_1 = g_2 = 0.0$.

The other fundamental property of neutral species is that their abundances will change at random because of demographic stochasticity, but within the constraint that their summed total abundance is regulated to the value specified in equilibrium (10.5). In other words, demographic stochasticity will cause the abundances of the species to move at random along the line of intersection of the three isoclines. This means that the relative abundances of species in this guild act like alleles undergoing genetic drift by changing at random over time, the dynamics that Hubbell called community drift (Hubbell 1979, 2001, Hubbell and Foster 1986). Thus, just like neutral alleles undergoing genetic drift, neutral species in a community undergoing community drift may go extinct at random if they drift to low relative abundance (Chesson and Huntly 1997, Hubbell 2001). Personally, I prefer the term *ecological drift* instead of community drift, because the entire community does not display drift dynamics, just the guild of neutral species. The rate at which their relative abundances change at random is proportional to the inverse of their summed total abundance, and so species will be more prone to local extinction via ecological drift when the summed total abundance of the guild is smaller.

These dynamical properties are also encoded in the stability analysis of this model. The characteristic equation for model (10.4) with two consumers is

$$\lambda \left(\lambda^2 + dR^*_{1(\sum N_j)} \lambda + a^2 b R^*_{1(\sum N_j)} \left(\sum N \right)^*_{j(R_1)} \right) = 0. \qquad (10.6)$$

Note the similarity between this polynomial and the characteristic equation for when a single consumer is present (equation (10.3)): they are identical except for the addition of a zero eigenvalue (i.e., $\lambda = 0$). With q neutral consumer species, the corresponding characteristic equation would have $q - 1$ zero eigenvalues. These zero eigenvalues are the signatures of ecological drift within this guild of species.

The presence of a guild of neutral species occupying a functional position within a food web presents somewhat of a logical conundrum concerning coexistence. With more than one guild member present, none of the guild members are individually coexisting in the community. To see this, consider the dynamics of N_2 invading the community in which N_1 is already present. At the time of invasion, N_2's per capita population growth rate is

$$\frac{dN_2}{N_2 dt} = baR^*_{1(N_1)} - f = ba\left(\frac{f}{ab}\right) - f = 0.$$

If more than one guild member is present, no guild members would be able to satisfy their invasibility criterion, because they all would have a zero per capita population growth rate were they to be removed from the community and then allowed to reinvade. What coexists in the community is the entire guild, because the entire guild of species acts as a single functional group. If the entire guild of neutral species were extirpated, the first member of the guild to reinvade would have a positive per capita population growth rate. Every subsequent guild member to reinvade would have a zero per capita growth rate. Thus, the concept that only one consumer species can coexist at a stable equilibrium if only a single resource limits consumer abundances is preserved.

Figure 10.2 shows examples of the dynamics of a guild of five consumers feeding on a single resource, with the abundance of each consumer being perturbed at random in each iteration of the model to simulate demographic stochasticity. In one example simulation, over the course of 1,000 iterations of the model, four of the five consumers are lost, leaving only one consumer to coexist with the resource (fig. 10.2(a)). At each iteration, the consumers are regulated to $(\Sigma N_j)^*_{1(R_1)} = (cab - df)/(a^2 b)$, so that as one species is lost, the average abundance of each of the remaining consumers increases accordingly. In this simulation, the resource abundance varies because of the variation in consumer abundances. It is very important to remember that, even though consumer species are being lost at random over time, the ecological dynamics of the community are maintaining the entire guild. The consequences of ecological drift within this neutral species guild are the same if the species have saturating functional responses and the community dynamics are a stable limit cycle (fig. 10.2(b)).

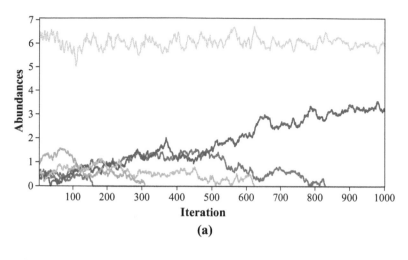

(a)

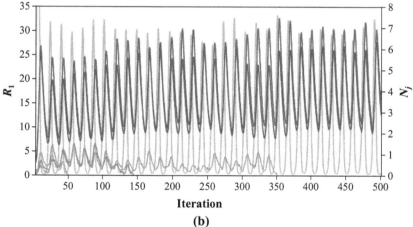

(b)

FIGURE 10.2. The abundance dynamics for five ecologically identical consumers that are also neutral species (N_1, N_2, N_3, N_4, N_5) that feed on one resource (R_1) when (a) the consumers all have linear functional responses and (b) the consumers all have saturating functional responses that result in a stable limit cycle. (a) The abundance of R_1 is the gray trace along the top of the panel, and the abundances of the five consumers are the lower traces in the panel. All but one consumer eventually goes extinct. (b) Abundance traces are shown for the species when the community orbits a limit cycle with stochastic perturbations. The parameter values for both (a) and (b) are as in figure 10.1, with (a) having all $h_{1j} = 0.0$ and (b) having $d_1 = 0.03$ and $h_{1j} = 0.1$.

Neutral Consumers in a Keystone Module

Neutral species can exist at any functional position in a food web. The analysis presented in the previous section is easily translated to multiple top predators that are ecologically identical. This is also true for functional positions at intermediate trophic levels. To see this, first consider a simple three-trophic-level food chain:

$$\frac{dP_1}{P_1 dt} = mnN_1 - x$$

$$\frac{dN_1}{N_1 dt} = baR_1 - mP_1 - f.$$

$$\frac{dR_1}{R_1 dt} = c - dR_1 - aN_1$$

Here again, all the parameters are as defined in chapters 3 and 4. The equilibrium abundances for this model are

$$R^*_{1(N_1,P_1)} = \frac{cmn - ax}{dmn}, \quad N^*_{1(R_1,P_1)} = \frac{x}{mn}, \quad P^*_{1(R_1,N_1)} = \frac{cabmn - a^2bx - fdmn}{dm^2 n}. \quad (10.8)$$

The characteristic equation for this model is

$$\lambda^3 + dR^*_{1(N_1,P_1)}\lambda^2 + \left[m^2 n N^*_{1(R_1,P_1)} P^*_{1(R_1,N_1)} + a^2 b R^*_{1(N_1,P_1)} N^*_{1(R_1,P_1)} \right]\lambda$$
$$+ dm^2 n R^*_{1(N_1,P_1)} N^*_{1(R_1,P_1)} P^*_{1(R_1,N_1)} = 0, \quad (10.9)$$

which has all negative eigenvalues if the three-species equilibrium is feasible (May 1973). Therefore, this is a stable equilibrium.

Now introduce q ecologically equivalent consumers at the intermediate trophic level:

$$\frac{dP_1}{P_1 dt} = mn\sum_{j=1}^{q} N_j - x$$

$$\frac{dN_j}{N_j dt} = baR_1 - mP_1 - f . \quad (10.10)$$

$$\frac{dR_1}{R_1 dt} = c - dR_1 - a\sum_{j=1}^{q} N_j$$

The demographic dynamics of this model again only regulate three abundances, which are identical to those that result with only one consumer present:

$$R^*_{1(\sum N'_{j(R_1,P_1)},P_1)} = \frac{cmn - ax}{dmn}, \quad \sum N^*_{j(R_1,P_1)} = \frac{x}{mn}, \quad P^*_{1(R_1,\sum N'_{j(R_1,P_1)})} = \frac{cabmn - a^2bx - fdmn}{dm^2n}. \quad (10.11)$$

The characteristic equation for this model is also analogous to that of the model when only one consumer is present, except for a series of terms resulting in zero eigenvalues:

$$\lambda^{q-1}\left[\begin{array}{c} \lambda^3 + dR^*_{1(N_1,P_1)}\lambda^2 + \left[m^2nN^*_{1(R_1,P_1)}P^*_{1(R_1,N_1)} + a^2bR^*_{1(N_1,P_1)}N^*_{1(R_1,P_1)}\right]\lambda \\ + dm^2nR^*_{1(N_1,P_1)}N^*_{1(R_1,P_1)}P^*_{1(R_1,N_1)} \end{array}\right] = 0. \quad (10.12)$$

Here again, the ecologically identical consumers will display ecological drift dynamics, within the constraint that their summed total abundance is the quantity being regulated (Siepielski et al. 2010, McPeek and Siepielski 2019). Therefore, consumers should be lost from this guild of neutral species at random over time as their abundances fall to very low values because of demographic stochasticity.

Although I have only considered models that have all species at one trophic level as ecologically equivalent, such a guild of neutral species could be found at any functional position within a trophic level, and their interactions with species outside the guild can involve all the same dynamical complexities that have been considered in previous chapters (e.g., saturating or Beddington-DeAngelis functional responses, point equilibria or stable limit cycles). For example, damselfly genera in eastern North American lakes are different in ways that are completely consistent with coexisting functional groups. *Enallagma* and *Ischnura* compete for small prey in the littoral zones of these lakes, with *Ischnura* being the better resource competitor and *Enallagma* better at avoiding their shared predators (i.e., the better apparent competitor) (McPeek 1998), just as is expected for coexisting consumers in a diamond/keystone predation module from chapter 4. Additionally, *Lestes* is fed upon by the same predators, but feeds on a somewhat different resource base, and so appears to coexist as well (Stoks and McPeek 2003a, 2003b, 2006). Each of these genera has multiple species in each lake, and the congeners within each appear to be neutral species (Siepielski et al. 2010, Siepielski and McPeek 2013).

In fact, it is probably more proper to think of the ecological entities (i.e., species) that have been considered in previous chapters as functional groups and not individual species, because each functional position can harbor a guild of neutral species and the dynamics at the level of the functional positions remain the same (Leibold and McPeek 2006). Functional groups are coexisting. However, when viewed from a taxonomic perspective, species may be continually lost because of ecological drift at each position until only one species remains: this remaining species would then be coexisting.

Because the issues for neutral species are the same, regardless of their functional position in the food web, I defer discussing the empirical testing issues until the end of the chapter.

ECOLOGICALLY EQUIVALENT BUT COEXISTING SPECIES

As I defined at the beginning of this chapter, ecologically equivalent species are those that have identical parameters in the equations describing their abundance dynamics. However, not all ecologically equivalent species are neutral species. Ecologically equivalent species may also be coexisting species.

As community ecologists, we tend to focus on the direct and indirect interactions among species at different functional positions in the food web. Predator eats prey; resource competitor draws down a shared resource that another competitor also needs to survive; a mutualist services its interaction partner. Species that are different either in the structure of the equations describing their population dynamics or in the parameter values in the equations are, by definition, occupying different functional positions within the food web. However, when a guild of ecologically identical species occupying a single functional position is present, the mechanisms of interactions among guild members determine whether they are coexisting or neutral species.

In particular, whether a guild of ecologically identical species is neutral or coexisting depends critically on whether guild members depress their own demographic rates through some mechanism of self-limitation and whether that self-limitation affects conspecific and heterospecific guild members equally. Many different mechanisms can generate such self-limitation among conspecifics: allelopathy (Rice 1984, Thacker et al. 1998, Inderjit and Duke 2003, Loh and Pawlik 2014), competing for mates (Zhang and Hanski 1998, M'Gonigle et al. 2012, Ruokolainen and Hanski 2016), territoriality (Robinson and Terborgh 1995, Grether et al. 2009, Losin et al. 2016), interfering with feeding (Beddington 1975, DeAngelis et al. 1975, Huisman and De Boer 1997, Stillman et al. 1997, Skalski and Gilliam 2001, de Villemereuil and López-Sepulcre 2011, Le Bourlot et al. 2014), cannibalism (Fox 1975, Polis 1981, Rudolf 2007), and stress responses induced by overcrowding (McPeek, Grace, et al. 2001, Glennemeier and Denver 2002, McPeek 2004). The strengths of these intraspecific mechanisms relative to the effects these mechanisms have on heterospecifics within the guild is the determining feature of whether the guild contains neutral species that will display drift dynamics or coexisting species (McPeek and Siepielski 2019). For example, consider a guild of forest birds that feed on the same set of resources and defend breeding territories. The area of suitable habitat limits these populations by defining the number of breeding territories that can be established in a community. Some bird species defend their territories against both conspecifics and closely related heterospecifics, whereas other species defend their breeding territories against conspecifics but ignore heterospecific individuals (Robinson and Terborgh 1995). Remember the quote from MacArthur (1958) about territoriality against conspecifics being all that is needed for coexistence that was cited in chapter 3.

In the examples presented in the two previous sections, guild members expressed no self-limitation, and so they directly affected their own abundances to the same degree as they directly influenced the abundances of other guild members—not at all. To explore how these mechanisms may influence the dynamics within a guild of ecologically identical species, add self-limitation to model (10.4) so that the two consumers may directly limit both their own abundance and the abundance of the other consumer to varying degrees (see box 3.1). The model becomes

$$\frac{dN_1}{N_1 dt} = baR_1 - f - g_{11}N_1 - g_{12}N_2$$

$$\frac{dN_2}{N_2 dt} = baR_1 - f - g_{21}N_1 - g_{22}N_2, \tag{10.13}$$

$$\frac{dR_1}{R_1 dt} = c - dR_1 - aN_1 - aN_2$$

where g_{ij} quantifies the direct demographic limitation effect of species j on species i. These could be two bird species competing for a common resource, and the direct intraspecific and interspecific limitation terms quantify the degree to which territoriality limits their abundances (Robinson and Terborgh 1995). Or they may be damselfly species that compete for the same prey but also fight with one another, sometimes cannibalize one another, and generate stress responses that limit their growth (McPeek and Crowley 1987, McPeek, Grace, et al. 2001, McPeek 2004).

First, consider the situation where each species only limits its own per capita growth rate (i.e., $g_{11} > 0$ and $g_{22} > 0$ but still with $g_{11} = g_{22}$) and has no effect on the other guild member (i.e., $g_{12} = g_{21} = 0$). Here, both consumer species are ecologically identical in their abilities to interact with the resource, and so both their isoclines intersect the R_1 axis at the same point. However, each consumer isocline tilts away from its own axis, so that the two consumer isoclines intersect along a line (fig. 10.3(a)).

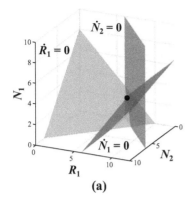

(a)

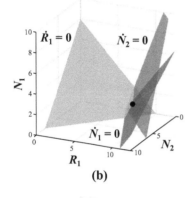

(b)

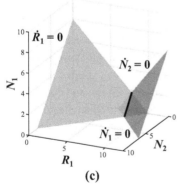

(c)

FIGURE 10.3. Examples of the isoclines for two eco-logically equivalent consumers (N_1 and N_2) that feed on one resource (R_1), when (a) both consumers limit their own abundances but have no effect on the other consumer's abundance, (b) both consumers have a greater direct effect on their own abundance than they have on the other consumer's abundance, and (c) both consumers have identical effects on both conspecifics and heterospecifics. The isoclines are as identified in figure 10.1. In (a) and (b), the stable point equilibrium is identified with a filled black circle. In (c), the line of intersection of the three isoclines is identified with a thick black line. The isoclines are identified as in figures 3.3 and 3.5. All parameters are as specified in figure 10.1, except as follows: (a) $g_{11} = g_{22} = 0.025$ and $g_{12} = g_{21} = 0$; (b) $g_{11} = g_{22} = 0.025$ and $g_{12} = g_{21} = 0.0125$; and (c) $g_{11} = g_{22} = g_{12} = g_{21} = 0.025$.

Consequently, if each consumer can coexist alone with the resource (i.e., they both satisfy the invasibility criterion of inequality (3.4)), this line of intersection itself intersects the R_1 isocline at a single point, and so the two species will each be coexisting in the community with R_1 at a point equilibrium at

$$R^*_{1(N_1,N_2)} = \frac{cg_{11}g_{22} + af(g_{11} + g_{22})}{dg_{11}g_{22} + a^2b(g_{11} + g_{22})},$$

$$N^*_{1(R_1,N_2)} = \frac{g_{22}(cab + fd)}{dg_{11}g_{22} + a^2b(g_{11} + g_{22})}, \quad N^*_{2(R_1,N_1)} = \frac{g_{11}(cab + fd)}{dg_{11}g_{22} + a^2b(g_{11} + g_{22})}$$

(10.14)

and not a line of equilibrium. Given the characteristic equation for this equilibrium,

$$\lambda^3 + \left[dR^*_{1(N_1,N_2)} + g_{11}N^*_{1(R_1,N_2)} + g_{22}N^*_{2(R_1,N_1)} \right]\lambda^2$$

$$+ \left[\begin{array}{c} dR^*_{1(N_1,N_2)}\left(g_{11}N^*_{1(R_1,N_2)} + g_{22}N^*_{2(R_1,N_1)} \right) \\ + a^2bR^*_{1(N_1,N_2)}\left(N^*_{1(R_1,N_2)} + N^*_{2(R_1,N_1)} \right) + g_{11}g_{22}N^*_{1(R_1,N_2)}N^*_{2(R_1,N_1)} \end{array} \right]\lambda \quad , \quad (10.15)$$

$$+ R^*_{1(N_1,N_2)}N^*_{1(R_1,N_2)}N^*_{2(R_1,N_1)}\left(a^2b(g_{11} + g_{22}) + dg_{11}g_{22} \right) = 0$$

it is stable if it is feasible, since all the roots of this polynomial are negative.

This is the most extreme example that results in a stable equilibrium at which the two ecologically equivalent consumers coexist. More generally, any number of ecologically equivalent consumers can coexist here if each limits its own abundance and has no direct effect on limiting the abundances of other guild members. A concrete example of such a scenario would be a guild of species that all have equal capabilities for harvesting resources and avoiding predators, and that are territorial but only defend their territories against conspecifics. Here, each species is its own limiting factor; so one additional consumer could potentially invade and coexist that does not limit its own abundance, is a poorer resource competitor than the guild, and can support a population at the resource abundance to which the guild of identical consumers depresses it. Identical results are also obtained if each consumer experiences feeding interference (i.e., substitute Beddington-DeAngelis functional responses in equations (10.13)) only with conspecifics.

Alternatively, if the two ecologically identical consumers in model (10.13) directly influence both conspecific and heterospecific population dynamics to varying degrees (i.e., all $g_{ij} > 0$ but maintaining $g_{11} = g_{22}$ and $g_{12} = g_{21}$ for ecological equivalence), the potential outcomes mirror exactly the results of the Volterra-Lotka-Gause model of competition (equations (2.6) in chapter 2). The abundances at the three-species equilibrium (i.e., the point where the three isoclines intersect) in this case are

$$N^*_{1(R_1,N_2)} = \frac{(g_{22}-g_{12})(cab+fd)}{d(g_{11}g_{22}-g_{12}g_{21})+a^2b(g_{11}+g_{22}-g_{12}-g_{21})}$$

$$N^*_{2(R_1,N_1)} = \frac{(g_{11}-g_{21})(cab+fd)}{d(g_{11}g_{22}-g_{12}g_{21})+a^2b(g_{11}+g_{22}-g_{12}-g_{21})}. \qquad (10.16)$$

$$R^*_{1(N_1,N_2)} = \frac{c(g_{11}g_{22}-g_{12}g_{21})+af(g_{11}+g_{22}-g_{12}-g_{21})}{d(g_{11}g_{22}-g_{12}g_{21})+a^2b(g_{11}+g_{22}-g_{12}-g_{21})}$$

The first issue is whether these equilibrial abundances are feasible.

Assuming that each consumer can coexist alone with R_1, four possible relationships among the intraspecific and interspecific effects can result, as with the Volterra-Lotka-Gause competition model, and the same outcomes correspond to those relationships (Slobodkin 1961). The characteristic equation for this model is

$$\lambda^3 + \left[dR^*_{1(N_1,N_2)} + g_{11}N^*_{1(R_1,N_2)} + g_{22}N^*_{2(R_1,N_1)} \right]\lambda^2$$

$$+ \left[\begin{array}{c} dR^*_{1(N_1,N_2)}\left(g_{11}N^*_{1(R_1,N_2)} + g_{22}N^*_{2(R_1,N_1)} \right) \\ + a^2bR^*_{1(N_1,N_2)}\left(N^*_{1(R_1,N_2)} + N^*_{2(R_1,N_1)} \right) + N^*_{1(R_1,N_2)}N^*_{2(R_1,N_1)}(g_{11}g_{22}-g_{12}g_{21}) \end{array} \right]\lambda \ . \qquad (10.17)$$

$$+ R^*_{1(N_1,N_2)}N^*_{1(R_1,N_2)}N^*_{2(R_1,N_1)}\left(a^2b(g_{11}+g_{22}-g_{12}-g_{21})+d(g_{11}g_{22}-g_{12}g_{21}) \right) = 0$$

The two consumers will both coexist with R_1 if each has a greater effect on its own abundance than the effect it has on the other consumer (i.e., $g_{11}=g_{22}>g_{12}=g_{21}$), because these relationships make both of their abundances feasible and all the roots of the characteristic equation negative (fig. 10.3(b)). Also, in the context of the issue under consideration now, even though these species are ecologically equivalent in their abilities to engage in interactions with the other species outside the guild and are equivalent in the strengths of their interactions with conspecifics and heterospecifics, they differ in the strengths of their interactions with conspecifics and heterospecifics. This is the key to when ecologically equivalent species can coexist or not. Here and here alone the conclusions drawn from the Volterra-Lotka-Gause model of competition are actually directly relevant: each species limits its own abundance more than it limits the abundances of other guild members.

The models presented here all involve some type of direct density-dependent effect of one species on another to permit ecologically identical species to coexist with one another. However, even this is not a requirement to foster coexistence in such a guild. The mere act of searching and finding mates can accomplish the same mechanistic outcome, if the various guild members segregate spatially when breeding (Zhang and Hanski 1998, M'Gonigle et al. 2012, Ruokolainen and Hanski 2016).

If both consumers have a greater effect on the other species than it has on itself (i.e., $g_{12} = g_{21} > g_{11} = g_{22}$), the priority effect ensues such that the consumer that has an initial abundance advantage will exclude the other consumer; again, just as in the Volterra-Lotka-Gause competition model.

One guild member will deterministically drive another guild member extinct only if the assumption is that all species have identical effects on conspecifics and on heterospecifics ($g_{11} \neq g_{22}$ and $g_{12} \neq g_{21}$). In this case, the guild members are still ecologically equivalent in their interactions with species outside the guild (i.e., the resource species in this case), but not so with respect to other guild members. If N_1 has a greater effect on N_2 than it has on itself, but N_2 has a greater effect on itself than it does on N_1 (i.e., $g_{21} > g_{11}$ and $g_{22} > g_{12}$), N_1 will drive N_2 extinct. Conversely, if N_1 has a greater effect on itself than on N_2, but N_2 has a greater effect on N_1 than on itself (i.e., $g_{21} < g_{11}$ and $g_{22} < g_{12}$), N_2 will drive N_1 extinct. These are the final two possible outcomes that were discussed in chapter 2 for the Volterra-Lotka-Gause model.

However, these species that affect one another directly within the guild may still be neutral species and so display ecological drift dynamics. For this to be true, all guild members must affect conspecifics and all heterospecifics identically. For the two-consumer model in (10.13), this means $g_{11} = g_{21} = g_{12} = g_{22} = g$: the consumers are all ecologically equivalent with all species interactions outside the guild, and have identical demographic effects on conspecifics and all heterospecifics. This recovers the line of equilibrium, which constrains the sum of consumer abundances to be the same as that favored when only one consumer is present (fig. 10.3(c)). The characteristic equation for q consumers in this model is

$$\lambda^{q-1}\left(\lambda^2 + \left(dR^*_{1\left(\sum N_j\right)} + g\left(\sum N_j\right)^*_{(R_1)}\right)\lambda + \left(a^2 b + dg\right)R^*_{1\left(\sum N_j\right)}\left(\sum N\right)^*_{(R_1)}\right) = 0, \quad (10.18)$$

which again shows that the consumers form a guild of neutral species.

Tests for neutral community dynamics often take the form of fitting models based on niche differences and models of neutral species to large data sets describing community properties, such as relative species abundance relationships (e.g., Clark and McLachlan 2003, Volkov et al. 2003, Etienne and Olff 2005, McGill et al. 2006, Canard et al. 2014). One problem with these approaches is that they typically assume that the community contains only coexisting species (i.e., niche structured) or only neutral species (i.e., neutrality) and so do not typically recognize the structure advocated here that communities are made up of coexisting functional groups that each may contain some number of neutral species (Leibold and McPeek 2006).

When taxa are apportioned into putative functional groups, such approaches do suggest that many may harbor guilds of neutral species. For example, Mutshinda et al. (2016) developed a Bayesian approach to explore the drivers of abundance changes in a seven-year time series of abundances for 57 diatom and 17 dinoflagellate species in the English Channel. They found that changes in total functional groups' biomasses strongly correlate with environmental drivers, such as irradiance, temperature, and nutrient concentrations, but the temporal changes within functional groups are most consistent with the random fluctuations of drift dynamics (Mutshinda et al. 2016).

Obviously, such observational studies can be suggestive, but they are not dispositive of mechanism. That requires an experimental approach that can discriminate neutral from coexisting species, regardless of the mechanisms that may foster coexistence or that regulate the abundance of the guild of putative neutral species. Imagine some guild of species like the consumers we have considered in the models in this chapter. If this is a guild of neutral species, manipulating the relative abundances of guild members while holding their total abundance constant should have no effect on their per capita survival, growth, or fecundity rates. However, manipulating the total abundance of all species should cause these rates to change similarly in all putative guild members.

In contrast, if they are all coexisting, each is being regulated to a specific abundance, even if they are ecologically equivalent with how they interact with species outside the guild or not. This means that manipulating their abundance away from this equilibrium should change their demographic rates significantly: they should have higher survival, growth, and fecundity rates in low abundance treatments, and these rates should be low in high abundance treatments. Therefore, these rates should change among relative abundance treatments that all have the same total guild abundance, since a change in relative abundance requires a change in their absolute abundance.

Thus, the appropriate experiment for identifying a guild of neutral species is to manipulate the total and relative abundances of guild members in a cross-factored design (Adler et al. 2007, Siepielski et al. 2010). This design is what plant ecologists would call a de Wit replacement series (de Wit 1960, Rodríguez 1997, Jolliffe 2000). Across treatments in which the summed total abundance of all species is the same, differences in relative abundance for a species across treatments should have no effect on the demographic rates of neutral species. Thus, for neutral species, their per capita demographic rates should only differ across total abundance manipulations (i.e., significant total abundance main effect with no significant relative abundance main effect or interaction between total and relative abundance).

In contrast, if the guild members are coexisting, species should have more favorable demographic rates in treatments at which their absolute abundances are

lower; for example, at lower relative abundances, their absolute abundance is farther from the abundance to which they are being regulated in these conditions.
Therefore, coexisting species should show an interaction between total and relative abundance treatments as well as both significant main effects.

Both total and relative abundance treatments are needed in this experiment to
demonstrate significant effects for either type of species. If only relative abundance treatments are done at a single total abundance, no significant differences
among the relative abundance treatments have two explanations. This result
would occur if the species may truly be neutral, but it may also occur simply
because of insufficient power to detect differences (i.e., a bad experiment). Change
in total abundance should affect the demographic rates of neutral species, and so
a significant total abundance main effect proves that the experiment has sufficient
power to detect demographic differences, which makes a nonsignificant relative
abundance main effect a strong result to identify neutral species (Siepielski et al.
2010). We used this experimental design to show that the 12–15 *Enallagma* species that can be found in lakes with fish across eastern North America appear to be
a guild of neutral species (Siepielski et al. 2010). In contrast, species in different
damselfly genera are ecologically different from one another, and thus are coexisting, because they respond differently to manipulations of both total and relative
abundance (McPeek 1998, Siepielski et al. 2011).

It is important to realize that this experimental design addresses the question of
whether species are coexisting, but it does not address anything about the mechanism fostering that coexistence—*why* are they coexisting? The ecologically identical but coexisting species considered in this chapter would generate exactly the
same results as a guild of resource competitors that are differentiated in their abilities to utilize different resources. This type of experiment must be complemented
with studies described in the previous chapters to show whether species are ecologically different or not and, if so, how they are different in ways that foster the
coexistence of each. Do species that evince the signature of coexistence in this
experiment differ in their abilities to interact with species outside the guild in
question (e.g., resources, prey, predators, mutualists), or are they ecologically
identical to outside species but more strongly limit their own abundances more
than those of other guild members only?

SUMMARY OF MAJOR INSIGHTS

- Ecologically equivalent species are defined as a guild of species that all would
  have the same parameters in models for their per capita growth equations.
  Ecologically equivalent species may be either coexisting or neutral species.

- The guild of ecologically equivalent species will be a guild of neutral species if each species directly affects the demographics of conspecifics and heterospecifics within the guild equally.
- The guild of ecologically equivalent species will be a guild of coexisting species if each species directly restrains the demographics of conspecifics to a greater degree than those of heterospecifics.
- Each member of a guild of neutral species is not coexisting, since each would have a zero per capita population growth rate when it invades. What is coexisting in the community is the entire guild of species.

MacArthur's Recasting Revisited

Invasibility is the most direct test of whether a species is coexisting or not in a community (MacArthur 1972, Turelli 1981, Chesson and Ellner 1989, Law and Morton 1996). However, such tests are logistically nearly impossible to perform in most real communities (Siepielski and McPeek 2010). For example, to test invasibility of a single *Enallagma* damselfly species in a small pond, one would first have to remove all individuals of that species from the pond, wait some number of years while the community in the pond comes to its steady state in that species' absence, and then reintroduce the species at low abundance to see if it will increase, all the while preventing natural immigration of the species. And that is just for one species, to say nothing about the proper experimental replication (i.e., this would have to be done in multiple ponds simultaneously) that would be needed or querying the coexistence status of more than one species.

Thus, we have always needed alternative methodologies. The most scientifically rational, indirect method is to build models of interacting species and develop criteria from those models that would identify which are and are not coexisting. The model developed by Gause (1934) and fully analyzed by Gause and Witt (1935) has served as the primary conceptual framework in which community ecologists have developed these criteria. Over the years, variations on this model have been developed (MacArthur 1969, 1970, May 1973, Abrams 1975, Vandermeer 1975, Chesson 1990, 2000b), but they all result in the same basic criteria (reviewed in chapter 2). However, as discussed previously, the Volterra-Lotka-Gause models are phenomenological and so encapsulate no mechanisms of species interactions.

A model that is mechanism-free to address coexistence is not inherently bad and in fact would have substantial advantages. Primarily, not being tied to any particular mechanism would make the resulting criteria a general test of coexistence. To be useful, those criteria would have to give consistent and correct answers across all possible (or at least most) underlying mechanisms that are actually promoting or retarding the coexistence of the species of interest. This is a very laudable goal, and it is the assumption that underlies the use of the Volterra-Lotka-Gause framework in designing empirical tests.

Two lines of justification have primarily been put forward. One was expounded by May (1973), who recognized the inherent nonlinearity of the functions that drive the demographics of species interactions. The Volterra-Lotka-Gause competition model can be interpreted as a linearized perturbation analysis at the equilibrium point of the community, where a Taylor expansion is made around the equilibrium point and all quadratic and higher terms are discarded. This assumption is also the basis for May's (1973) broader analysis of species interactions based on the *community matrix*. This conceptual justification means that the only empirical scope for the model would be small abundance perturbations when all community members are at or near their equilibrium abundances. Analyses when species are far from their natural abundances (e.g., as when they are invading) would not be consistent with this interpretation of the model.

MacArthur's recasting of consumer-resource models into the Volterra-Lotka-Gause competition form represents the other major line of justification (Schoener 1974a, 1974b, Abrams 1975, Chesson 1990, 2000b, 2012, Chesson and Kuang 2008, HilleRisLambers et al. 2012, Kleinhesselink and Adler 2015, Letten et al. 2017). Schaffer (1981) called this *ecological abstraction* and explored many of its dynamical consequences (see also Levine 1976, Schaffer and Leigh 1976, Lawlor 1979). MacArthur's purpose was to show that the parameters of the Volterra-Lotka-Gause competition equations are simple functions of the more complicated dynamics of resource competition among a guild of consumers. Chesson and Kuang (2008) showed the same to be true for the circuitry of apparent competition mediated through a set of shared predators feeding on these same consumers. Hence, if the assumptions of this recasting are met, evaluating the coexistence of these consumers in the Volterra-Lotka-Gause framework then implies satisfying the coexistence criteria derived for more mechanistic models of resource and apparent competition.

However, just because one model can be converted into another does not mean that the latter will give you the right answer about the former, or that the metrics needed to evaluate the new formulation of the model can be estimated empirically. We know that this recasting is only possible for consumer-resource models if all recast species experience some form of direct self-limitation (see chapter 2). Abrams (1980a) showed that the resulting competition coefficients are not simply a function of the parameters of the model but rather change with species abundances when per capita self-limitation terms in these models are nonlinear. Abrams (1980b) also derived the competition coefficients between two consumers with saturating functional responses feeding on two resources and showed that they change with the resource abundances instead of simply being constants defined by model parameters. Both of these effects are what generate the need to include terms for *higher-order interactions* in Volterra-Lotka-Gause approaches to coexistence (Wilbur 1972, Neill 1974, 1975, Abrams 1980a, 1983b, Case and Bender 1981, Wilbur and Fauth 1990, Billick and Case 1994). The assumption that the subsumed species are always at equilibrium with the abundances of the other species being modeled is also a general property of abstraction and reduced dimensionality (Schaffer 1981) and further limits its applicability to situations where these hidden species have very fast dynamics relative to those being explicitly considered. All of these make comparisons among empirically estimated competition parameters measured in different settings problematic.

However, setting all these matters aside, a primary issue that has not been considered to my knowledge is simply whether the Volterra-Lotka-Gause competition parameterization gives the correct answer for coexistence when the underlying consumer-resource model is specified. In other words, if we take some underlying model of a set of multitrophic-level consumer-resource interactions as "true" and explore the consequences of this approach as an empiricist would do, does evaluating the parameters of the recast model of the Volterra-Lotka-Gause form give the correct answers about their coexistence, even in the best of circumstances where all per capita demographic rates are linear? In addition, what biological interpretations do the parameters in the Volterra-Lotka-Gause form have, and are they empirically quantifiable? The answers to these questions are critical for using this framework as the conceptual basis to design experiments testing coexistence.

RECASTING CONSUMER-RESOURCE MODELS

The conceptual model of Volterra-Lotka-Gause competition is applied to evaluate coexistence of species at all functional positions in the interaction web of a community: plants, herbivores, predators, and so on. However, the parameter interpretations that result when MacArthur's recasting is done for species at different trophic levels may not be the same. In other words, the intrinsic rates of increase and the competition coefficients of equations (2.19) and (2.22) may not have the same interpretations based on the underlying parameters for plants, herbivores, or predators. Moreover, our ability to empirically quantify them may be different for species at different trophic levels. Thus, my goal here is to construct equations of the Volterra-Lotka-Gause form from the basic consumer-resource models we have explored in detail in chapters 3 and 4, and evaluate what conclusions we would draw from the recast competition metrics about their coexistence for species at different trophic levels. The same can be done for food web models including omnivory (chapter 5), but analyses including omnivory only complicate the algebra without adding any fundamental insights, in my experience. Therefore, I ignore omnivory here.

For clarity, begin with the community of species described by equations (3.17) with q consumer species feeding on p resource species. To accomplish the recasting at both trophic levels, all consumers have linear functional responses for feeding on all resources ($h_{ij} = 0$), and all consumers have some degree of self-limitation ($g_j > 0$). With these assumptions, the dynamics of all the species at one trophic level can be subsumed into the equations for species at the other trophic level. Remember, this is accomplished by setting all the equations at one trophic level equal to zero to determine the functions describing their isoclines, and these isocline functions are then substituted into the dynamical equations for the species at the other trophic level. This is what necessitates the assumption that the species that are subsumed are always at equilibrium with the abundances of the species at the other trophic level.

Doing this for equations (3.17) and rearranging into the Volterra-Lotka-Gause form, the resulting set of equations are derived:

$$\frac{dN_j}{N_j dt} = \left[\sum_{i=1}^{p} a_{ij}b_{ij}\frac{c_i}{d_i} - f_j\right] - \left[g_j + \sum_{i=1}^{p} a_{ij}b_{ij}\frac{a_{ij}}{d_i}\right]N_j - \sum_{j'=1(j'\neq j)}^{q}\left[\sum_{i=1}^{p} a_{ij}b_{ij}\frac{a_{ij'}}{d_i}\right]N_{j'}$$

$$\frac{dR_i}{R_i dt} = \left[c_i - \sum_{j=1}^{q} a_{ij}\left(-\frac{f_j}{g_j}\right)\right] - \left[d_i + \sum_{j=1}^{q} a_{ij}\frac{a_{ij}b_{ij}}{g_i}\right]R_i - \sum_{i'=1(i'\neq i)}^{p}\left[\sum_{j=1}^{q} a_{ij}\frac{a_{i'j}b_{i'j}}{g_j}\right]R_{i'} \tag{11.1}$$

Here, a set of equations that explicitly model direct resource limitation at the consumer trophic level and direct predator limitation at the resource trophic level has been transformed into a set of equations where these ecological processes are implicitly embedded in the resulting parameters of the Volterra-Lotka-Gause competition models at each trophic level. Hence, the consumer equations only contain terms with consumer abundances, and the resource equations only contain terms with resource abundances. These equations now allow us to evaluate coexistence among the consumers or coexistence among the resources using the Volterra-Lotka-Gause competition framework without reference to the dynamics of the species at the other trophic level. This is the approach that empirical community ecologists are using when they step into a real community and use the Volterra-Lotka-Gause competition framework to focus exclusively on a set of ecologically similar species at one trophic level.

The same process can be accomplished for a community with any number of trophic levels. For example, Chesson and Kuang (2008) presented recasting of the three-trophic-level food web by only incorporating resource and predator dynamics into the intermediate consumer trophic level. However, the same recasting could be done for each of the three trophic levels. When applied to equations (4.13) with all $h_{ij} = 0$, all $l_{jk} = 0$, all $g_j > 0$, and all $y_k > 0$, the following set of equations describing the dynamics at each trophic level in Volterra-Lotka-Gause form results:

$$
\frac{dP_k}{P_k dt} = \left[\sum_{j=1}^{q} \left(\frac{m_{jk} n_{jk}}{\Theta_j} \left(\sum_{i=1}^{p} a_{ij} b_{ij} \left(\frac{c_i}{d_i} - \sum_{j'=1(j'=j)}^{q} \frac{a_{ij'}}{d_i} N_{j'} \right) - f_j \right) \right) - x_k \right]
$$
$$
- \left[y_k + \sum_{j=1}^{q} m_{jk} n_{jk} \frac{m_{jk}}{\Theta_j} \right] P_k - \sum_{k'=1(k'\neq k)}^{s} \left[\sum_{j=1}^{q} m_{jk} n_{jk} \frac{m_{jk'}}{\Theta_j} \right] P_{k'},
$$

$$
\text{where } \Theta_j = g_j + \sum_{i=1}^{p} a_{ij} b_{ij} \frac{a_{ij}}{d_i};
$$

$$
\frac{dN_j}{N_j dt} = \left[\sum_{i=1}^{p} a_{ij} b_{ij} \frac{c_i}{d_i} - \sum_{k=1}^{s} m_{jk} \left(-\frac{x_k}{y_k} \right) - f_j \right] - \left[g_j + \sum_{i=1}^{p} a_{ij} b_{ij} \frac{a_{ij}}{d_i} + \sum_{k=1}^{s} m_{jk} \frac{m_{jk} n_{jk}}{y_k} \right] N_j
$$
$$
- \sum_{j'=1(j'\neq j)}^{q} \left[\sum_{i=1}^{p} a_{ij} b_{ij} \frac{a_{ij'}}{d_i} + \sum_{k=1}^{s} m_{jk} \frac{m_{j'k} n_{j'k}}{y_k} \right] N_{j'}
$$

$$\quad (11.2)$$

$$
\frac{dR_i}{R_i dt} = \left[c_i + \sum_{j=1}^{q} \frac{a_{ij}}{\Psi_j} \left(f_j + \sum_{j'=1(j'=j)}^{q} \left(\sum_{k=1}^{s} m_{jk} \left(\frac{m_{j'k} n_{j'k}}{y_k} \right) N_{j'} \right) + \sum_{k=1}^{s} \left(m_{jk} \left(-\frac{x_k}{y_k} \right) \right) \right) \right]
$$
$$
- \left[d_i + \sum_{j=1}^{q} a_{ij} \frac{a_{ij} b_{ij}}{\Psi_j} \right] R_i - \sum_{i'=1(i'\neq i)}^{p} \left[\sum_{j=1}^{q} a_{ij} \frac{a_{ij} b_{i'j}}{\Psi_j} \right] R_{i'},
$$

$$
\text{where } \Psi_j = g_j + \sum_{k=1}^{s} m_{jk} n_{jk} \frac{m_{jk}}{y_k}.
$$

The terms in square brackets for each species at each trophic level in both (11.1) and (11.2) underlie the intrinsic rate of increase, the intraspecific competition coefficient, and the interspecific competition coefficients, respectively, from left to right, in the Volterra-Lotka-Gause competition equations:

$$
\frac{dP_k}{P_k dt} = r_k - \alpha_{kk} P_k - \sum_{k'=1(k'\neq k)}^{s} \alpha_{kk'} P_{k'}
$$
$$
\frac{dN_j}{N_j dt} = r_j - \alpha_{jj} N_j - \sum_{j'=1(j'\neq j)}^{q} \alpha_{jj'} N_{j'}. \qquad (11.3)
$$
$$
\frac{dR_i}{R_i dt} = r_i - \alpha_{ii} R_i - \sum_{i'=1(i'\neq i)}^{p} \alpha_{ii'} R_{i'}
$$

With two trophic levels, the intrinsic rates of increase and competition coefficients are all simple functions of the parameters of the underlying model (3.17), but with three trophic levels, only species on the consumer trophic level have intrinsic rates of increase and competition coefficients (i.e., the terms in square brackets) that are purely functions of the parameters of the underlying model (4.13) (cf. corresponding terms in square brackets in (11.1) to (11.2)). It is impossible to remove consumer abundances from the recast resource and predator equations in (11.2). I have also done this recasting for models with more than three trophic levels: with four or more trophic levels, the dynamical equations at every trophic level are comparable to the resource and predator equations in (11.2), where the intrinsic rates of increase depend on species abundances at other trophic levels. This is not really a problem, given that the explicit assumption must be made in any case that the species at the other trophic levels are at equilibrium given the abundances of species at the trophic level of interest. Therefore, these terms can be thought of as part of the intrinsic rates of increase. However, it does mean that these intrinsic rates of increase will change as the abundances of the species of interest change, which will hamper making informed empirical comparisons.

Coexistence in this new framework still requires that the isoclines of the respective species intersect at positive abundances to give a stable equilibrium (see chapter 2). With only two species at one of the trophic levels, coexistence requires

$$\frac{\alpha_{11}}{\alpha_{21}} > \frac{r_1}{r_2} > \frac{\alpha_{12}}{\alpha_{22}}. \tag{11.4}$$

As discussed in chapter 2, this is *the* classic result that has been derived a number of times, both directly (e.g., inequality (6.25) of May 1973, inequality (4) of Vandermeer 1975) and in the present context of MacArthur's recasting (e.g., inequality (2) of Abrams 1975, inequality (13) of Chesson 1990).

This inequality is also the basis for elaborations into what is now called *modern coexistence theory* (e.g., Chesson 2000b, 2012, 2018, HilleRisLambers et al. 2012, Saavedra et al. 2017), but its applicability on a pairwise basis at a trophic level when more than two species are present remains unclear (Strobeck 1973, Goh 1977, Barabás et al. 2016, Saavedra et al. 2017, Singh and Baruah 2020). Chesson and colleagues elaborated this approach, although they used a different form of the phenomenological Volterra-Lotka-Gause model,

$$\frac{dN_j}{N_j dt} = r_j \left(1 - \beta_{jj} N_j - \sum_{j'=1(j' \neq j)}^{q} \beta_{jj'} N_{j'} \right) \tag{11.5}$$

(e.g., see equations (1) in both Chesson 2000b and Chesson 2012), to query coexistence by identifying what Chesson (2000b) called "equalizing" and "stabilizing" effects (see also Chesson and Huntly 1997, Chesson 2000b, 2012, 2018, Adler et al. 2007, Chesson and Kuang 2008, Godoy and Levine 2014). In this form, the competition coefficients are standardized by r_j:

$$\beta_{jj'} = \frac{\alpha_{jj'}}{r_j}.$$

Thus, we have three different formulations that are attributed to Volterra-Lotka-Gause competition: the familiar Gause form of equations (2.6) (Gause 1934, Gause and Witt 1935, Slobodkin 1961); the form of equations (11.3) used by May (1973) and Vandermeer (1975) and resulting from MacArthur's recasting (MacArthur 1969, 1970, Chesson 1990, Chesson and Kuang 2008); and the form of equation (11.5) (Chesson 2000b, 2012). Each, when analyzed in its own right, points to similar aspects of the species' natural histories to foster coexistence. I will try to analogize their notation to that resulting from the MacArthur recasting in equations (11.3), so that comparisons can be made.

The conclusions drawn from analyzing equation (11.5) are that with two species "for stable coexistence, each species must depress its own growth more strongly than it depresses the growth of the other species as it increases in population density" (Chesson 2012, p. 10066). In inequality (11.4), this means that the left ratio is greater than one (i.e., $\beta_{11} > \beta_{21}$) and the right ratio is less than one (i.e., $\beta_{22} > \beta_{12}$) for a pair of coexisting species, and otherwise for a pair of non-coexisting species. This result obtains directly from setting equation (11.5) equal to zero and solving for the conditions that result in a feasible and stable two-species equilibrium. Thus, the metric

$$\rho = \sqrt{\frac{\beta_{12}\beta_{21}}{\beta_{11}\beta_{22}}}$$

is the geometric mean of these ratios between this pair of species and is used as the measure of "niche overlap" to quantify the stabilizing effects contributing to their coexistence (Chesson 2000b, 2012, 2018, Chesson and Kuang 2008, Godoy and Levine 2014).

The equalizing effect on coexistence comes from the ratio of the "long-term low-density growth rates" of the two species κ_j, which is comparable to the middle ratio of intrinsic rates of increase in inequality (11.4); this is a broader concept, because the derivation of these ratios also includes the effects of spatiotemporal environmental effects such as were discussed in chapters 8 and 9 (Chesson 1994, 2000b, 2008, 2012, 2018, Chesson and Huntly 1997, Chesson and Kuang 2008). However, this component of coexistence in this framework does not result from equation (11.5). Coexistence in the framework developed by Chesson and others then requires that

$$\frac{\kappa_1}{\kappa_2}\rho < 1 \tag{11.6}$$

(e.g., Chesson 2012, Godoy and Levine 2014). This condition is essentially the same as inequality (11.4), given that equations (11.3) are not the basic equations used to derive (11.6).

A number of excellent and logistically difficult laboratory and field experimental studies have evaluated coexistence among assemblages of algae and plants in recent years based on the framework used to derive inequality (11.6) (e.g., Levine and HilleRisLambers 2009, Narwani et al. 2013, Godoy et al. 2014, Godoy and Levine 2014, Kraft et al. 2015, Germain et al. 2016, Matías et al. 2018). However, the results of these experiments are curious. For example, in a laboratory study of coexistence among eight algal species that co-occur in lakes across eastern North America, 11 of 25 pairs (44%) satisfied inequality (11.6) (Narwani et al. 2013). Pairwise comparisons were also made among 10 annual plants common to Spanish grasslands, and only 8 of 44 species pairs (18%) satisfied inequality (11.6) when grown in experiments under the environmental conditions for a normal year in their Mediterranean climate, while only 2 of the 44 pairs (5%) satisfied this coexistence criterion when grown under drought conditions (Matías et al. 2018). In an even larger experimental study, pairwise comparisons of 18 plant species in a California grassland found that only 12 of 102 species pairs (12%) satisfied inequality (11.6) (Kraft et al. 2015). These experimental studies used an underlying annual plant demographic model that also characterizes plant interactions phenomenologically to justify the use of inequality (11.6), instead of the consumer-resource framework used by MacArthur (1969, 1970) and Chesson and Kuang (2008) to justify (11.4). However, the two approaches arrive at the same criterion for evaluating coexistence: inequality (11.6) is simply a transformation of inequality (11.4). Taking these results at face value, one would conclude that most co-occurring algae and plant species are not coexisting.

Given these empirical results, consider how well satisfying or not satisfying inequality (11.4) among each pair of species corresponds to coexistence of species in the consumer-resource models. In other words, for predators, consumers, and resources in chapters 3 and 4, what pairwise species relationships emerge for inequalities (11.4) and (11.6) among species that do and do not coexist in these underlying models? I focus here on a few of the examples we have considered thus far.

First, consider the food web depicted in figure 11.1, where four consumers feed on a single resource and are fed upon by one predator, using equations (4.8), where all six species coexist. All pairwise comparisons among the four consumers satisfy their Volterra-Lotka-Gause coexistence criteria. However, species that are poorer resource competitors all have interspecific competition coefficients on

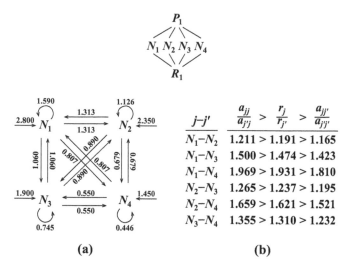

FIGURE 11.1. The Volterra-Lotka-Gause competition parameters for four consumers feeding on one resource and being fed upon by one predator according to equations (4.8). In this case, all six species coexist. The interaction network (a) shows the competition parameters from equations (11.2) for the four consumer species. The intrinsic rates of increase are associated with the straight arrows pointing at each species from no other species, the intraspecific competition coefficients are associated with the circular arrows from each species to itself, and the interspecific competition coefficients are associated with arrows directed from one species to another. The table (b) gives the values for the pairwise coexistence criterion for each species pair (i.e., inequality (11.4)). The parameters are as follows: $c_1 = 2.0$, $d_1 = 0.2$, $a_{11} = 0.5$, $a_{12} = 0.45$, $a_{13} = 0.4$, $a_{14} = 0.35$, all $b_{1j} = n_{jk} = 0.1$, all $h_{1j} = l_{jk} = 0.0$, all $f = 0.1$, all $g_j = 0.025$, $m_{11} = 0.6$, $m_{21} = 0.5$, $m_{31} = 0.4$, $m_{41} = 0.3$, $x_1 = 0.1$, and $y_1 = 0.025$.

superior competitors that are larger than their intraspecific competition coefficients. Therefore, the inference that $a_{jj} > a_{jij}$ is a necessary condition for coexistence is wrong. All of these species comparisons also satisfy inequality (11.6), with all having $\rho < 1$, but ρ is the geometric average of these ratios for the two species and so does not apply to individual species.

In figure 3.13, three different sets of three consumers feeding on three resources are presented, based on model (3.16). In the first set (fig. 3.13(a)–(b)), each consumer is the best resource competitor for one of the resources, and all six species coexist. If this model is recast into the Volterra-Lotka-Gause form of equations (11.1), all pairs of consumers and all pairs of resources satisfy inequalities (11.4) and (11.6). Additionally, if the Volterra-Lotka-Gause isoclines are constructed from these derived parameters, the intersections of the isoclines predict the correct equilibrium abundances (fig. 11.2). However, the system dynamics away

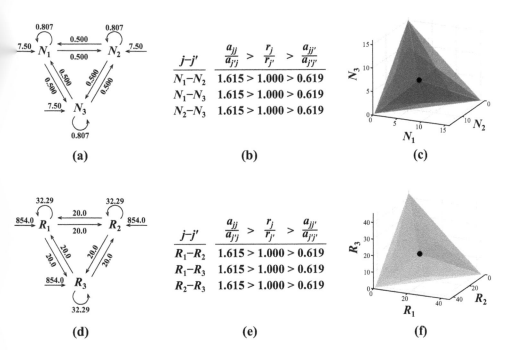

FIGURE 11.2. The Volterra-Lotka-Gause competition parameters for three consumers feeding on three resources, as in figure 3.13(a)–(b). In this case, each consumer is the best resource competitor for one of the resources, and all six species coexist. The interaction network (a) shows the competition parameters from equations (11.1) for the three consumer species. The intrinsic rates of increase are associated with the horizontal arrows pointing at each species, the intraspecific competition coefficients are associated with the circular arrows from each species to itself, and the interspecific competition coefficients are associated with arrows from one species to another. The table (b) gives the values for the pairwise coexistence criterion for each species pair (i.e., inequality (11.4)). The isocline that results from the Volterra-Lotka-Gause parameterization is shown in (c). The same information is given for the three resource species in (d), (e), and (f), respectively. The equilibrium abundances of the six species are identified with a filled black circle in the isoclines ((c) and (f)). In this case, the relationships among the Volterra-Lotka-Gause isoclines identify the correct equilibrium abundances. Compare these to the isoclines for the underlying model in figure 3.13(a)–(b). The parameters are the same as those given for figure 3.13(a)–(b).

from the equilibrium do not behave as they should in relation to the Volterra-Lotka-Gause isoclines (Schaffer 1981). In other words, one cannot deduce changes in abundances at either trophic level by considering the Volterra-Lotka-Gause isoclines other than the position of the equilibrium.

A simulation of the MacArthur recasting module can be accessed at https://mechanismsofcoexistence.com/MacArthurCompetition.html.

In the second set (fig. 3.13(c)–(d)), one consumer is the best resource competitor for two of the resources, another consumer is the best resource competitor for the third resource, and the third consumer is a generalist that is not the best at competing for any resource. In this set, the generalist drives the other two consumers extinct and coexists with the three resources (fig. 11.3). Consistent with this outcome, inequality (11.4) is not satisfied for the pairwise comparisons involving the generalist consumer. However, even though all three resources coexist, one pairwise comparison among the resources is not satisfied, but this is because the extinct consumers' parameters are included in their competition parameters. As a result, the intersection among the resource Volterra-Lotka-Gause

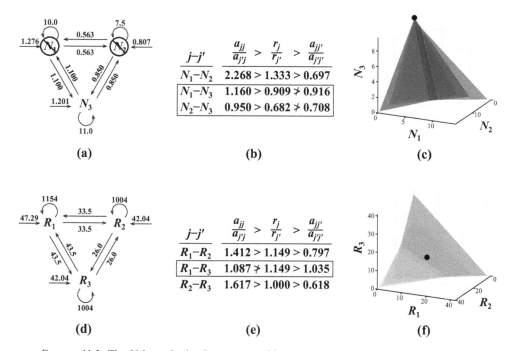

(a)   **(b)**   **(c)**

(d)   **(e)**   **(f)**

FIGURE 11.3. The Volterra-Lotka-Gause competition parameters for three consumer species feeding on three resource species, as in figure 3.13(c)–(d). In this example, N_1 is the best resource competitor for R_1 and R_3, N_2 is the best resource competitor for R_2, and N_3 is a generalist that is the best resource competitor for none of the resources. However, in this case, only N_3 coexists with the three resource species. All competition parameters and isoclines are as described for figure 11.1. Compare the Volterra-Lotka-Gause isoclines presented here to the isoclines for the underlying model in figure 3.13(c)–(d). Note that, in this example, the relationship among the Volterra-Lotka-Gause consumer isoclines does predict the correct equilibrium abundances for the consumers, but the relationship among the Volterra-Lotka-Gause resource isoclines does not. The parameters are the same as those given for figure 3.13(c)–(d).

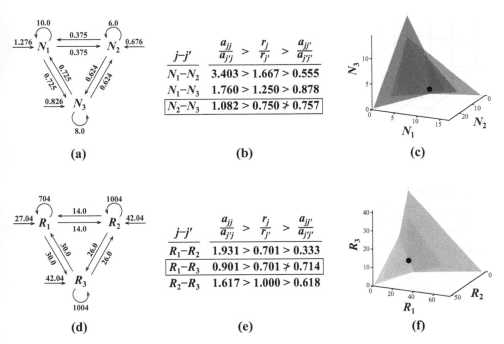

FIGURE 11.4. The Volterra-Lotka-Gause competition parameters for three consumer species feeding on three resource species, as in figure 3.13(e)–(f). In this example, as in figure 11.3, N_1 is the best resource competitor for R_1 and R_3, N_2 is the best resource competitor for R_2, and N_3 is a generalist that is the best resource competitor for none of the resources. However, in this case, all six species coexist. All competition parameters and isoclines are as described for figure 11.1. Here, the relationships among the Volterra-Lotka-Gause isoclines identify the correct equilibrium abundances. Compare the Volterra-Lotka-Gause isoclines presented here to the isoclines for the underlying model in figure 3.13(e)–(f). The parameters are the same as those given for figure 3.13(e)–(f).

isoclines also do not correspond to the true equilibrium abundances. The larger issue this example presents is that the Volterra-Lotka-Gause framework is applicable mainly in retrospect, since the species that will coexist at other trophic levels must be determined before the analysis is conducted.

In the final set (fig. 3.13(e)–(f)), just as in the second set, one consumer is the best resource competitor for two resources, the second consumer is the best competitor for the third resource, and the third consumer is a generalist that is not the best consumer on any resource. However, in this set, all six species coexist. Despite all the species coexisting, one consumer pair and one resource pair do not satisfy either inequality (11.4) or inequality (11.6). However, the intersection among the Volterra-Lotka-Gause isoclines do predict the correct equilibrium

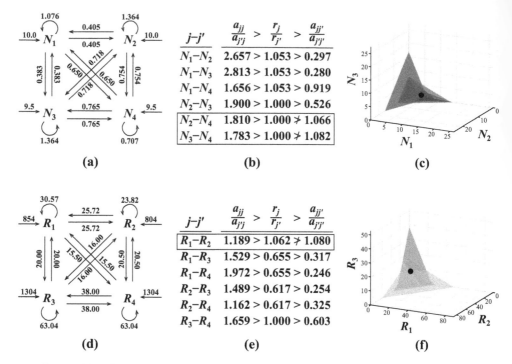

FIGURE 11.5. The Volterra-Lotka-Gause competition parameters for four consumer species feeding on four resource species, as in figure 3.14. In this example, N_1 is the best resource competitor for R_1 and R_2, N_2 is the best resource competitor for R_3, N_3 is the best resource competitor for R_4, and N_4 is a generalist that is the best resource competitor for none of the resources. In this case, all eight species coexist. All competition parameters and isoclines are as described for figure 11.1. In this case, the relationships among the Volterra-Lotka-Gause isoclines identify the correct equilibrium abundances. The parameters are the same as those given for figure 3.14, except all $c_i = 4.0$, all $h_{ij} = l_{jk} = 0.0$, and all $g_i = 0.001$.

abundances for all species, while the community dynamics show no relation to these isoclines away from the equilibrium (fig. 11.4).

Many areas of parameter space in the models of chapters 3 and 4 have coexisting species pairs that do not satisfy their Volterra-Lotka-Gause/Chesson criteria. For example, consider the four consumer species feeding on four resource species in figure 3.14. If all the resources have $c_i = 4.0$, all eight species coexist. However, two consumer species pairs and one resource pair do not satisfy their Volterra-Lotka-Gause/Chesson coexistence criteria (figure 11.5). My experience exploring parameter spaces for these examples suggests that coexisting consumers and predators that are broad feeding generalists are the species that tend to fail some of their pairwise inequalities (11.4) and (11.6) with more specialized feeders (e.g., N_4 in fig. 11.5).

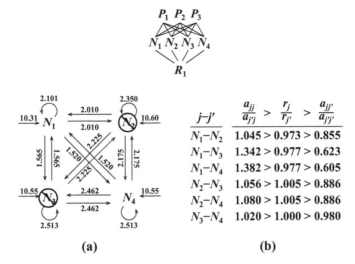

FIGURE 11.6. The Volterra-Lotka-Gause competition parameters for four consumer species that are fed upon by three predators and that feed on one resource species. For the parameters used, only N_1 and N_4 coexist with the three predators and one resource. The parameters are as follows: $c_1 = 4.0$, $d_1 = 0.02$, all $a_{1j} = 0.5$, all $b_{1j} = n_{jk} = 0.1$, all $h_{1j} = l_{jk} = 0.0$, all $f_j = 0.2$, all $g_j = 0.0$, $m_{11} = m_{42} = m_{33} = 0.4$, $m_{21} = m_{32} = m_{23} = m_{43} = 0.3$, $m_{13} = 0.1$, $m_{31} = m_{41} = 0.05$, $m_{12} = 0.01$, all $x_k = 0.02$, and all $y_k = 0.02$.

These are exactly the kinds of species with broad and overlapping diets that are found in real communities, such as the Mo'orea coral reef or the Laikipia savanna.

The opposite relationship is also true. Consider the three-trophic-level community in figure 11.6 with three predators feeding on four consumers, which in turn are feeding on a single resource. From the species pool in this example, only two consumers coexist with the three predators and one resource. However, all pairwise comparisons of the four consumers satisfy both inequalities (11.4) (fig. 11.6(b)) and (11.6). (Remember, more predators than consumers can coexist here because each predator also limits its own abundance in order to do the recasting.)

These examples indicate that many coexisting species fail their Volterra-Lotka-Gause/Chesson coexistence criteria (i.e., inequalities (11.4) and (11.6)), and many species that cannot coexist in a particular community nonetheless satisfy those criteria in that community. These results imply that simply because consumer-resource models can be recast into the form of Volterra-Lotka-Gause competition, these consumer-resource models are no justification for using Volterra-Lotka-Gause competition as the conceptual framework for developing empirical tests of species coexistence that are more mechanistic. Obviously, this may account for the mixed empirical results with algae and plants summarized above.

However, other difficulties are also apparent when the recast equations are considered in detail. For the consumers in the two-trophic-level set of equations (11.1), the derived intrinsic rates of increase and competition coefficients all have solid biological interpretations that are easily quantifiable. Remembering that the equilibrium abundance of each resource in the absence of all consumers is $R^*_{i(R_1,\ldots,R_r)} = c_i/d_i$, each consumer's intrinsic rate of increase (i.e., the term in the first square brackets in the consumer equations of (11.1)) is exactly the per capita growth rate when that consumer is rare and no other consumers are present. These are the empirical conditions under which it would be measured.

The competition coefficients quantify how the effects of resource depletion by one consumer impact the change in per capita growth rate of the other consumer. In the consumer equations of (11.1), the ratios $-a_{ij}/d_i$ describe the change in the abundance of R_i with a change in N_j, which is the slope of the N_j isocline when expressed in terms of R_i's abundance. The total coefficient $a_{ij}b_{ij}(-a_{ij}/d_i)$ is then how this change in resource abundance translates into the per capita growth rate of the consumer in question. Thus, the indirect effects among the consumers mediated through the resources, including the indirect effect of a species on itself, are embedded in these coefficients. The same interpretations apply to the competition coefficients among the resources as mediation of the apparent competitive effects through the consumers. The standard density manipulation experiments can thus be used to estimate these parameters.

The problem arises with the intrinsic rates of increase for the resources. If the consumers experience no direct self-limitation, the consumer dynamics cannot be subsumed into the resource equations (Chesson and Kuang 2008). Remember that the entire technique assumes that the abundances of the subsumed species are at equilibrium with the current abundance of the species of interest. However, with no direct self-limitation, most resource combinations have no consumer equilibrium abundance. To see this, consider the isocline system of a single consumer feeding on a single resource (as in fig. 3.3(a)). Because the consumer isocline is a vertical line, the consumer is at equilibrium only at $R_i = f_j/(a_{ij}(b_{ij} - f_jh_{ij}))$. Only when the consumer experiences some degree of self-limitation (i.e., either $g_j > 0$ or $z_j > 0$) does the consumer have equilibrium abundances at all resource abundances: in other words, the consumer isocline has values across the possible range of each resource's abundance (as in fig. 3.3(d)–(e)). This is why predator self-limitation is required to incorporate the dynamics of species higher in the food web (Chesson and Kuang 2008).

The problem arises because the consumer isocline has negative values at $R_i < f_j/(a_{ij}(b_{ij} - f_jh_{ij}))$. In each resource's intrinsic rate of increase in equations (11.1), the term $-f_j/g_j$ is the intersection point of the N_j isocline with its own axis, which is a negative value. While this component has a perfectly reasonable theoretical interpretation, it is empirically unquantifiable: species cannot have negative abundances in nature. Thus, it is empirically impossible to quantify the appropriate ratio of intrinsic rates of increase for resource pairs in equations (11.1). The same types of terms appear in the intrinsic rates of increase for both consumers and resources in the three-trophic-level system of equations (11.2).

This is a general feature of all these types of recast models for species at all trophic levels that have enemies above them in the food web. Thus, if the recasting of consumer-resource models is the justification, only coexistence of species at the very top of the food web—only species with no enemies above them—can be evaluated empirically using the Volterra-Lotka-Gause coexistence criteria. This may also be why the empirical results for algae and plants imply that most species are not coexisting.

More fundamentally, using Volterra-Lotka-Gause competition as the framework for designing our empirical inquiries substitutes correlation for causation. Empirical studies that measure competition coefficients in this framework are essentially measuring the correlated response of one species to a change in the abundance of another. However, just as quantitative genetic parameters of a population will change as the distributions of alleles at underlying genetic loci, phenotypic traits, and the environment change, the Volterra-Lotka-Gause competition parameters measured for any two species will change as the context of their interaction changes, but with little to no understanding of why.

The language of Volterra-Lotka-Gause competition also blinkers us to focus almost exclusively on resource competition as the defining interaction that must be used to explain how many species can coexist. Remember, in chapter 1, I asked you whether a question about the competitive abilities of species makes you think about the abilities of species to avoid their predators. Nowhere is the importance of the language we use and how we permit it to shape our conceptions more evident than with the paradox of the plankton. Ever since Hutchinson speculated about why so many phytoplankton species appear to coexist in freshwater lakes, the defining focus has been on how a variable environment shifts the resource competitive abilities of the many phytoplankton species to permit their coexistence (e.g., Huisman and Weissing 1999, 2001a, 2002, Litchman and Klausmeier 2001, Litchman et al. 2004, Li and Chesson 2016). This is despite the fact that dozens of consumers feed on those phytoplankton and so are also important limiting factors for their coexistence (e.g., Gliwicz 1975, Leibold 1989, Sarnelle 2005, Leibold et al. 2017).

The language of Volterra-Lotka-Gause competition also creeps into "mechanism." Because criterion (11.4) or (11.6) is the test for coexistence, interpretation of these criteria become the stated mechanism by which the species coexist: namely, a species must limit its own abundance more than it limits the others or, more succinctly, intraspecific competition must be less than interspecific competition. The correspondence between the real mechanism of coexistence and the Volterra-Lotka-Gause interpretation of mechanism is more commensurable for very simple community modules (e.g., two consumers feeding on two resources) (Tilman 1980, 1982). However, the correspondence becomes much more tenuous when more than two trophic levels are present or many species are present at each trophic level.

The Volterra-Lotka-Gause criteria are also what make the issue of coexistence seem to be one of comparison among guild members. Those inequalities are inherently a comparison: can these two species coexist. In contrast, invasibility is a question that only one species can answer: can this species increase when rare and all other species are at their demographic steady states for its absence. By thinking of the problem in Volterra-Lotka-Gause terms, one has removed all the limiting factors that affect species population growth rates from the problem, and so there is nothing about an individual species' performance to evaluate on its own.

The Volterra-Lotka-Gause competition framework for understanding coexistence also is not commensurable with broader issues of community structure or other related levels of inquiry that require an understanding of species interactions. Models of Volterra-Lotka-Gause competition have little to no utility for issues concerning diet breadth (e.g., Mittelbach 1981, Kartzinel et al. 2015, Casey et al. 2019), food web structure (e.g., Paine 1980, Yodzis 1980, 1981, Martinez 1992, Dunne et al. 2002, Allesina et al. 2008, Benke 2018), omnivory (e.g., Diehl 1995, Diehl and Feißel 2000, Vandermeer 2006, Rudolf 2007, Amarasekare 2008a), top-down versus bottom-up effects (e.g., Hairston et al. 1960, Oksanen et al. 1981, Hunter and Price 1992, Menge 1992, Carpenter and Kitchell 1993, Wollrab et al. 2012), and general patterns of species richness and diversity.

Given that tests of invasibility are so hard to conduct in most ecosystems, we should continue to pursue the development of tests of "whether species are coexisting" that are mechanism-free. Perhaps the problem with the Volterra-Lotka-Gause criteria is that they are pairwise comparisons, and stating the criteria simultaneously for three or more species is currently not possible (Saavedra et al. 2017). Or perhaps new methods will be developed that can assess coexistence regardless of mechanism (Maynard, Miller, et al. 2019, Maynard, Wootton, et al. 2019).

SUMMARY OF MAJOR INSIGHTS

- In pairwise comparisons, coexisting species frequently have a greater indirect impact on their own population growth rate than they have on another coexisting species' growth rate (i.e., $\alpha_{ii} > \alpha_{i'i}$ is not true for many coexisting species). Therefore, each species limiting its own abundance more than it limits other species' abundances is not a necessary criterion for Volterra-Lotka-Gause coexistence.

- Even when all per capita demographic rates are linear, the intrinsic rates of increase of species at some trophic levels are functions of the species at other trophic levels. Consequently, empirical estimates of the intrinsic rates of increase will be highly sensitive to the presence and abundances of other species in the community.
- The intrinsic rates of increase in the Volterra-Lotka-Gause framework are empirically inestimable for species that have enemies above them in the food web. Given that almost all species have enemies (i.e., herbivores, predators, pathogens), the intrinsic rates of increase for most species may be inestimable.

Philosophical and Practical Implications

I entered graduate school in the early 1980s as a great debate roiled about the types of data and studies needed to identify the processes and mechanisms that structure biological communities. Competition for limiting resources was the dominant theoretical paradigm for understanding which species could be community members and what properties those species must have to be included (e.g., MacArthur 1969, 1970, May and MacArthur 1972, May 1973, Roughgarden 1974, Schoener 1974a, Vandermeer 1975, Tilman 1982, Abrams 1983a). Consequently, competition for limiting resources was the primary species interaction considered in empirical studies of coexistence and community structure (e.g., MacArthur 1958, Connell 1961, Grant 1966, Brown 1973, Roughgarden 1974, Diamond 1975, Dunham 1980, Hairston 1980). The importance of other types of species interactions and environmental effects, such as predation and disturbance, were seen primarily as modifiers that reduced the strength of competition (e.g., Brooks and Dodson 1965, Paine 1966, 1969, 1974, Harper 1969, Dayton 1971, Murdoch and Oaten 1975, Menge 1976, Menge and Sutherland 1976, Lubchenco 1978, Sousa 1979, Paine and Levin 1981, Hairston 1986).

A major critique to emerge in this debate was that the importance of resource competition was merely assumed and rarely rigorously demonstrated in most systems (Connor and Simberloff 1979, Strong et al. 1979, Connell 1980). Based on this assumption, many of the most prominent studies of the time simply examined phenotypic patterns or diet overlap among co-occurring species to infer the consequences of resource competition for coexistence and community structure (e.g., MacArthur 1958, Schoener 1965, Grant 1966, Brown 1973, Pianka 1974, Roughgarden 1974, Diamond 1975). Surveys of field experiments that manipulated the densities of putative competitors found competition to be a prominent but not ubiquitous interaction (Connell 1983b, Schoener 1983). However, these studies were also open to the criticism that other types of species interactions could generate the same experimental responses (e.g., Holt 1977, Reader 1992).

All this motivated a philosophical debate about the types of studies and data needed to evaluate the causes of coexistence and the resulting structure of communities (Connor and Simberloff 1979, Strong et al. 1979, Quinn and Dunham

1983, Roughgarden 1983, Salt 1983, Simberloff 1983, Strong 1983, Bender et al. 1984). This debate also sparked calls for the explicit study of the mechanisms of competition: identifying and experimentally validating what resources are limiting; determining the morphological, physiological, and behavioral properties of individuals that determine their abilities to acquire and utilize those resources; performing density manipulations of competitors to test for associated changes in the abundances of other competitors and their limiting resources; and exploring patterns of change in competitor abundances along gradients of those resources (Schoener 1986, Tilman 1987).

Listening to this debate at meetings, reading this debate in the literature, and discussing these issues with my colleagues left a lasting impression on me. The motivation for writing this book was that, almost 40 years on, we seem to still be mired in the same general discourse about competition as the primary structuring interaction for all species in communities: the types of data, studies, and experiments have changed, but the central tenets have not. For many, the central theoretical paradigm for understanding coexistence remains the phenomenological Volterra-Lotka-Gause competition model (e.g., Chesson 2000b, Szilagyi and Meszéna 2009, HilleRisLambers et al. 2012, Letten et al. 2017, Saavedra et al. 2017), and the empirical approaches advocated by many to understand and test coexistence are based on this theoretical construct (e.g., Adler et al. 2007, Levine et al. 2008, 2017, Mayfield and Levine 2010, Mayfield and Stouffer 2017). However, as the last chapter demonstrated, much of the underlying justification based on translating mechanistic models of networks of species interactions into Volterra-Lotka-Gause competition cannot be used to support this approach.

Therefore, in this chapter, I would like to broaden the conversation started by Schoener (1986) and Tilman (1987) beyond just resource competition by offering a philosophical perspective on the mechanistic approach of embracing the complexity of communities to understand species coexistence and larger issues of community structure.

PHILOSOPHICAL FRAMEWORK

Species are embedded in complex webs of interactions: the biological community is this complex web. The interactions that each species directly experiences are what determine its abundance and, consequently, whether it can coexist in a community. These direct effects are in turn determined by the effects of other species at more distant positions in the interaction web—the indirect effects that propagate through the interaction web. Despite this reality, our dominant theoretical paradigm for understanding why species can coexist—Volterra-Lotka-Gause competition—ignores this complexity, and examines small groups of species as though they directly interact with one another without any reference to this larger causal network. Moreover, the explanations for why species can or cannot coexist in this abstracted framework (e.g., intraspecific competition is greater than inter-specific competition) are not rooted in the real reasons for each species' success or failure in its interaction web; namely, whether resources and symbionts are abundant enough and enemies are not too abundant for the species to support a population. The many successful empirical dissections of these complex webs of interactions in all types of environments that can be found in the literature—only a few of which are cited in the previous chapters—also show that the complexity of these interaction webs is no excuse for this abstraction: with hard work and ingenuity, scientists can understand this complexity and then apply the knowledge gained to problems of sustainability and conservation.

For me, the pragmatist philosophers of the late nineteenth and early twentieth centuries (Charles Sanders Peirce, William James, F. C. S. Schiller, John Dewey) defined the importance of addressing nature on its own terms and explained why the words, language, and models we use to conceptualize nature define how we see nature and, consequently, what we measure and study about nature. Pragmatism is based on a framework of realism, which assumes that a real causal structure exists to be understood, and that structure can be determined to a substantial degree based on the success of predictions of what practical experiences will be had if one's current conceptions of that structure are true. As William James wrote:

> Pragmatism, on the other hand, asks its usual question. "Grant an idea or belief to be true," it says "what concrete difference will its being true make in any one's actual life? How will the truth be realized? What experiences will be different from those which would obtain if the belief were false? What, in short, is the truth's cash-value in experiential terms?"
>
> The moment pragmatism asks this question, it sees the answer: *True ideas are those that we can assimilate, validate, corroborate, and verify. False ideas are those that we can not.* That is the practical difference it makes to us to have true ideas; that, therefore, is the meaning of truth, for it is all that truth is known-as. (James 1991, pp. 88–89; italics in original)

In the present context, how we conceptualize interactions within communities defines our ability to understand them:

> Consider what effects, which might conceivably have practical bearings, we conceive the object of our conception to have. Then, our conception of these effects is the whole of our conception of the object. (Peirce 1878, p. 293)

or

> Truth, as any dictionary will tell you, is a property of certain of our ideas. It means their "agreement," as falsity means their disagreement, with "reality." Pragmatists and intellectualists both accept this definition as a matter of course. They begin to quarrel only after the question is raised as to what may precisely be meant by the term "agreement," and what by the term "reality," when reality is taken as something for our ideas to agree with. (James 1991, pp. 87–88)

If the language and models we use do not capture the essences of the mechanisms of species interactions and the complexity of the interaction web in which species are embedded, we cannot develop a true understanding of the processes that foster species coexistence and thereby structure communities. Abstracting indirect interactions among species into direct interactions that ignore the actual limiting factors that regulate species' abundances is not studying reality.

Science is the most exacting expression of the pragmatists' experiential nature of understanding. We formulate a conception of how the corner of nature that we study works, and then we formulate expectations for what we should experience in different contexts. Those expectations are the hypotheses we derive, and those experiences are the data we collect in defined observational and experimental contexts to evaluate those hypotheses. The art and creativity of science are expressed in defining those expectations (i.e., hypotheses) and devising experiences (i.e., observations and experiments) to evaluate those expectations.

How do we know something is true about nature from the studies we devise? The philosophy of science tells us that we can never know something with absolute certainty: all we can definitively identify is what is not true (Popper 1959). This is because we cannot be sure we have considered all possible causal structures and features and all the nuances of those causes that may generate the phenomenon in which we are interested. Conversely, the fact that some conception has predictive "cash-value" in some instances does not mean that it is "true." Science is a humbling endeavor, because all scientists know that they are to some degree always wrong. The best we can be is approximately true.

What is the best method for being "approximately true"? In a seminal essay on the subject, T. C. Chamberlin (1890) outlined what he saw as the advantages and traps of various ways of doing science (see also Chamberlin 1897). Chamberlin argued that a major trap for scientists was to have what he called a "ruling theory" that dominates the thinking of scientists because it seduces them into confirmation bias.

> The mind lingers with pleasure upon the facts that fall happily into the embrace of the theory, and feels a natural coldness toward those that seem refractory.

Instinctively there is a special searching-out of phenomena that support it, for the mind is led by its desires. There springs up, also, an unconscious pressing of the theory to make it fit the facts, and a pressing of the facts to make them fit the theory. (Chamberlin 1890, p. 93)

Is the fact that only 12%–44% of co-occurring species are found to satisfy their Volterra-Lotka-Gause competition criteria "approximately true" or a sign that Volterra-Lotka-Gause competition is a "ruling theory"? (See chapter 11.)

To counter the trap of a ruling theory, Chamberlin argued for what he called the method of *multiple working hypotheses*. Platt (1964) characterized Chamberlin's method as *strong inference*.

The effort is to bring up into view every rational explanation of new phenomena, and to develop every tenable hypothesis respecting their cause and history. The investigator thus becomes the parent of a family of hypotheses; and, by his parental relation to all, he is forbidden to fasten his affections unduly upon any one. (Chamberlin 1890, p. 93)

Betini et al. (2017) offered a cogent analysis of the approach and of applying it in ecology and evolution.

However, Chamberlin's argument was not just to have multiple working hypotheses for causes of a single phenomenon: the researcher should construct an entire program of research for how multiple agents of causation may interact to shape a set of interconnected phenomena.

But an adequate explanation often involves the co-ordination of several agencies, which enter into the combined result in varying proportions. The true explanation is therefore necessarily complex. Such complex explanations of phenomena are specially encouraged by the method of multiple hypotheses, and constitute one of its chief merits. We are so prone to attribute a phenomenon to a single cause, that, when we find an agency present, we are liable to rest satisfied therewith, and fail to recognize that it is but one factor, and perchance a minor factor, in the accomplishment of the total result. . . . The problem, therefore, is the determination not only of the participation, but of the measure and the extent, of each of these agencies in the production of the complex result. (Chamberlin 1890, pp. 93–94)

Clearly, this argument applies directly to organizing studies of the complex web of interactions governing species coexistence in a community. This research program requires one to identify and evaluate each possible cause separately and independently, but then to explore how these multiple agents act together in the larger causal network that shapes community structure.

This approach is also augmented dramatically if mutually reinforcing hypotheses that explore each component from *different vantages* are also tested. For example, if nitrate is thought to be a limiting nutrient for some plant species, a number of predictions can be made: (1) individual plants should take up nitrate from the surrounding soil; (2) the abundance of the plant should be higher in areas where soil nitrate availability is higher; (3) increasing nitrate availability to individual plants when grown alone should increase components of their absolute fitness (i.e., increase growth, survival, and fecundity); and (4) increasing the plant's local abundance should decrease nitrate availability in the soil. All of these are predictions derived from the various expected consequences of the actions of this single causal agent, and so represent multiple interlocking examples of the pragmatists' experiences that should prove or disprove the cash value of considering nitrate as a limiting nutrient for this plant. If all these predictions are supported, the conclusion that nitrate is a limiting factor for this plant is much more strongly supported than if only one of these predictions were tested. However, as we have seen, even if nitrate is a limiting nutrient for this plant, not all of these predictions may be upheld. For example, if multiple nutrients limit this and other plants, its abundance may decrease along natural gradients of increasing nitrate availability when in the presence of other plant species if it is not the best competitor for nitrate (see chapter 3). Thus, patterns of support and refutation across multiple predictions about a single agent can add significant insight to not only the efficacy of that single agent but also the broader structure of the causal network of interactions in which that agent is embedded.

Such a research program should utilize every data analytic tool at our disposal. Of course, experimental manipulations are the gold standard for testing predictions that fulfill the pragmatists' methods exactly: if I create experience X, I should see this result, but if I create experience Y, I should see this different result. However, observational studies are more appropriate at the beginning stages when simply identifying the possible players as important. Also, observational studies are more appropriate for testing certain types of hypotheses, for example, covarying changes in species abundances along natural environmental gradients (Betini et al. 2017). New techniques for evaluating the validity of different causal network structures in observational data are being derived in other fields and would be valuable added inferential tools for community ecologists (Pearl 2000, Imbens and Rubin 2015, Morgan and Winship 2015, Pearl and Mackenzie 2018). Additionally, once the causal network is largely organized and verified, parameter estimation of various community metrics (e.g., direct interaction strengths, indirect effects) will become important. As both Schoener (1986) and Tilman (1987) advocated, studies of the phenotypic properties of individual species that influence their ecological performances in the various interactions should also accompany field tests to provide mechanistic context for why species are or are not successful.

Given this rationale and overview, I would like to outline the components of such a research program to explore why species coexist in a complex interaction web. Most of this simply organizes the larger framework of what many have done in the past and are doing now. I do this because it is often hard to see such organization by reading individual papers. For example, Betini et al. (2017) concluded that few ecologists were applying the methodology of multiple working hypotheses in testing ecological theory, but their analysis evaluated this issue based on the contents of single papers. Given that any research program applying the multiple working hypotheses paradigm is exploring a complex network of interactions, it is not surprising that no single paper would be capable of presenting what would appear to be a good example. Such programs are best analyzed as a body of work, because a multiple working hypotheses program of study may represent an entire career's worth of research (e.g., Grant and Grant 2014) or results from a large team of researchers (e.g., Carpenter and Kitchell 1993, Borer et al. 2014). All of the study types discussed below have been more fully described in previous chapters. Here, my goal is to provide a philosophically organized framework for what we have discussed in previous chapters.

Limiting Factors

How can we embrace the complexity of a network of species interactions that constitutes a community and make sense of it, particularly in the context of why various member species are or are not coexisting? The way, in my opinion, is to systematically apply the pragmatists' and Chamberlin's approaches. Most community ecologists are interested in explaining the relationships of particular groups of species—either specific taxa, functional guilds, or trophic levels—and not the entire community. For me, aquatic insects and most specifically odonates have always been a fascination. This necessarily directs my attention to the components of the interaction web that affect those species.

Given a set of species of interest, the factors that directly limit their abundances are the first concern. Thus, the first interesting philosophical question is how to identify limiting factors for species of interest. As we have seen throughout the preceding chapters, a limiting factor is a component of a species' environment that both directly influences one or more of its per capita demographic rates and responds to changes in its abundance (Levin 1970, Leibold 1995). If one were to derive a complete model of the community, each limiting factor would appear as a state variable in the equation of population growth rate for the species of interest and would be represented as a dynamical equation itself in the full community model. For example, to fully characterize the population dynamics of a single phytoplankton species, the equation describing its population dynamics would be a function of the abundances of light, mineral nutrients, and vitamins that it extracts from the water (Hutchinson 1961) and of species that directly inflict mortality on it; and all these factors would also have dynamical equations describing changes in their abundances.

Limiting factors fall into four categories. The first category contains depletable energy sources, materials, and foodstuffs to fuel and support metabolic processes, individual growth, and fecundity: in other words, abiotic and biotic resources. The second general category contains those species that are typically defined as "enemies": herbivores, predators, omnivores, and pathogens and parasites. The third category contains mutualists that influence a species' fitness in a variety of ways. And the final category has the species itself and other similar species that directly impact its demographic rates through mechanisms such as territoriality, feeding interference, allelopathy, and mate competition.

What are not in any of these categories are resource competitors or apparent competitors; those species impart indirect effects, which come later in this research program. To see this, simply examine equations (4.13) for multiple predators feeding on multiple consumers, which are in turn feeding on multiple resources. The consumer equations do not contain terms that include the abundances of any other consumers. They do contain terms that include the resource abundances, because each consumer is *resource limited*: in fact, the definition of resource limitation is that the abundance of resources limits a consumer's per capita population growth rate. They also contain terms that include the predator abundances, because each consumer is also *predator limited*. The interactions among the consumers are indirect effects mediated through the resources and predators. Thus, if one could devise an experimental protocol to hold resource and predator abundances constant while manipulating the abundances of other consumers, the per capita growth rate of the consumer species of interest would not change. Even the consumer's own abundance does not appear in any of the feeding terms, because even intraspecific resource competition and apparent competition are both indirect effects. The consumer's own abundance does appear in equations (4.13), but here it is part of a different mechanism that imparts direct self-limitation, the fourth category of limiting factors.

A somewhat more nuanced philosophical interpretation must be made for the direct effects of consumers feeding on the resources and the predators feeding on the consumers in equations (4.13). The direct effect on each prey in these predator-prey interactions comes from the predator abundances, but their feeding rates are influenced by the abundances of their other prey if the functional responses are saturating. The effect each prey has on the abundance of another prey is a *trait-mediated indirect effect* (Werner and Peacor 2003), where the trait of the intermediary species is the consumer's or predator's feeding rate. Werner and Peacor (2003) did not consider a saturating functional response to be such a trait-mediated indirect effect, but I would respectfully beg to differ. The abundance of one resource only affects the per capita growth rate of another resource species because of its effect on the feeding rate of the consumer, which is a performance metric based on the traits of the consumer. The abundance of the predator and its feeding rate are what determine its impact on any given prey species.

The initial phase of Chamberlin's (1890, 1897) multiple working hypotheses research program is then to design observations and experiments that identify the set of limiting factors for the species of interest. These observational and experimental studies are typically relatively simple, but they are essential. Moreover, they must be grounded in the natural history of the organisms. In Chamberlin's language, "every tenable hypothesis" represents the full possible set of factors that directly limit the species. Does this species eat that species? Does this species pollinate that species? Which parasites infect this species?

After such natural history observations suggest a set of factors, James's cash value of these are the results that are obtained in simple experimental manipulations of each factor to evaluate whether the abundance of each limiting factor directly influences one or more of the species' demographic rates. These are the simple "sledgehammer" experiments that are typically done at the beginning of a research program to identify what the key interactions are, as described in previous chapters. If this predator is removed, does the survival rate of this prey species increase appreciably? If this prey is supplemented, does this predator grow faster, survive better, or produce more offspring? If this pollinator is excluded from the flowers of this plant, are fewer seeds produced? Do infected individuals feed and subsequently grow slower than uninfected individuals? Is this species territorial, and so limits the number of individuals that will reproduce?

The results of such experiments provide the initial characterization of the relative importance of each in the demography of the species of interest. For example, a species may appear in the diets of two predators, but the experimental results might indicate that only one of the predators is a substantial mortality source. These observational and experimental studies are also the initial steps in a *mechanistic natural history* understanding of the community. Obviously, for many systems, much of this information will be known from the work of others. However, thinking expansively about the full range of limiting factors and what previous studies may have missed or not considered will prevent researchers from being ensnared in Chamberlin's ruling theory trap. (Note to graduate students: Know the natural histories of your organisms!)

Enumerating the set of limiting factors for the collection of species of interest also serves two important purposes for future studies. First, these types of studies provide the basic information that will be used for characterizing how the species of interest differ in the relative importance of the various limiting factors. Second, they form the basis of characterizing such species properties as resource and apparent competitive abilities that will be used in later studies to understand the larger network of species interactions. Species that share one or more limiting factors are connected to one another in the circuitry of the interaction web and so cast indirect effects on one another.

As has been repeatedly stressed, the number of limiting factors that regulate the abundances of a guild of species should also be related to the maximum number of species that can coexist in the guild (Levin 1970). Remember, diversity at one trophic level begets diversity at adjacent trophic levels. Thus, to be able to predict and so understand species richness in any given region in the interaction web, it is essential that the number of limiting factors be enumerated for those species. Spatiotemporal variation in the strengths of these direct interactions is also the basis for how relative nonlinearities and the temporal and spatial storage effects that help or hinder coexistence are generated, and so these initial studies provide the mechanistic basis for understanding these processes as well.

Invasibility

Whether a species can satisfy its invasibility criterion is the fundamental test of whether a species is coexisting in a community—can the species increase when rare and the other species are at their demographic steady states in its absence? Obviously, this question can be answered without any knowledge whatsoever of the limiting factors involved. However, experimental tests of invasibility are almost never done, primarily because they are logistically impossible to perform except in the simplest of systems (Siepielski and McPeek 2010).

Consequently, for most species in most ecosystems, the status of a species as coexisting must be inferred from information derived from the studies of its limiting factors and its demographic responses to those limiting factors. Because this is a population dynamics question, such inferences would be best made from estimates of per capita population growth rates derived from studies such as life table response experiments (e.g., Caswell 1989, 1996, 2010, Levin et al. 1996, Fréville and Silvertown 2005, Angert 2006, Davison et al. 2013, Kane et al. 2017) or demographic projections based on other types of observational and experimental results (e.g., Cáceres 1997, McPeek and Peckarsky 1998, Adler et al. 2010). This is also the point where experimental results that quantify changes in various demographic rates along gradients of abundances of a species' limiting factors will be crucial, because these types of analyses will allow the researcher to project per capita population growth rates for various scenarios of its invasion.

Considering the mechanisms promoting invasibility also emphasizes the need for the multiple working hypotheses approach. Even in the simplest models considered in this book, invasibility is the consequence of the balance that a species strikes among demographic forces. For example, a single predator can coexist with a single resource if it satisfies inequality (3.5), which occurs when its foraging return for feeding on the resource is greater than its intrinsic death rate when it invades. However, for species in real communities, this balance is more complicated because dozens of resources, enemies, and mutualists, as well as spatiotemporal variation, are all potentially involved. Moreover, considering coexistence as a comparison among the various species in a guild requires an understanding of how they differentially balance the multiple conflicting demands of the limiting factors they share.

Yet Another Revision of the Niche Concept

The *niche* is a central concept in community ecology, but the word conjures many different definitions to ecologists (Grinnell 1917, Elton 1927, Hutchinson 1958, Leibold 1995). The most widely accepted is Hutchinson's (1958) niche conception derived from set theoretics, in which the environment is described by an N-dimensional axis system. The niche of the species is then defined as the N-dimensional hypervolume containing every point "which corresponds to a state of the environment which would permit the species S_1 to exist indefinitely" (Hutchinson 1958, p. 416); in other words, satisfy its invasibility criterion.

The full set of axes that comprise the environmental description in Hutchinson's niche conception has always been ambiguous. Because of his focus on resource competition, Hutchinson listed environmental factors, like temperature, and biotic factors, like size of food particles, as axes. Interestingly, he did not include resource competitors as axes and instead defined the effects of competitors as the difference in the size of the volume when a competitor is not and is present: the fundamental versus realized niche, respectively (Hutchinson 1958). However, this has always struck many as arbitrary. Why could one not also play the same game with a predator or mutualist of the species in question: its realized niche would be that volume in the state space where the species of interest can exist indefinitely in the presence of the predator or the mutualist (Leibold 1995). These definitions of niches as fundamental or realized also only think of the presence/absence of these other types of species and not their quantitative abundances. As we have repeatedly seen, species may be able to coexist with some species in some communities but not in others, because of changes in the abundances of the other species (e.g., figs. 3.7, 3.9, 4.9, 5.10, 6.14).

All these other species are components of the environment for the species of interest. In other words, all these species could just as easily be included in Hutchinson's environmental description as axes in the N-dimensional environmental space, with the values of the axes being their quantitative abundances. Levin (1970) argued for just such an expansion of Hutchinson's environmental space to include all the abiotic (e.g., light, water, mineral nutrients) and biotic (e.g., food, prey, enemies, mutualists) limiting factors that are important to a species, as well as other types of nondepletable environmental factors (e.g., temperature, humidity, soil texture, salinity). In fact, Elton made the same argument in his description of the niche:

> Animals have all manner of external factors acting upon them—chemical, physical, and biotic—and the "niche" of an animal means its place in the biotic environment, *its relations to food and enemies.* (Elton 1927, pp. 63–64; italics in original)

Thus, considering all these descriptions together, imagine an N-dimensional state space defined by all the nondepletable environmental factors and abiotic and biotic limiting factors that are important to the species of interest. Add to this an axis for per capita population growth rate when the species is rare. In this conception, a species' niche is then the volume in this expanded axis system in which the species has a positive per capita population growth rate when rare.

This niche conception also comports exactly with all the models we have considered. Every abiotic resource and species, including the species itself, that influences the per capita growth rate of the species of interest is an axis in the N-dimensional hyperspace. Axes that describe nondepletable abiotic metrics determine the parameters of the equations. The population dynamics equations of all the species then describe how a community moves through this state space. Axes that quantify the traits of all the species can also be added to this system if desired to understand how the different species may evolve in response to one another (McPeek 2017a, 2017b). Given this expanded view, we have been working with the niches of all these species collectively from the start of chapter 3. Moreover, when conceived in this more expansive framework, the niche becomes the basis for formulating both models of community dynamics and Chamberlin's multiple working hypotheses approach to empirically investigate coexistence and community structure.

Indirect Effects and the Larger Interaction Network

The set of limiting factors are the direct effects that determine whether a species can coexist. The actions and abundances of the other species in the interaction network influence what the abundances of the limiting factors are. Hence, *why* one species can coexist is determined by the actions and abundances of the other species in the interaction web affecting their limiting factors. However, understanding the circuitry and mechanisms of these indirect effects are critical since the outcomes are not always intuitive (Wootton 1994a,1994b, 1994c).

These are the studies in a multiple working hypotheses program that involve species with one or more intermediaries between them. We have already considered multiple examples of the types of nonintuitive results that may obtain when the underlying circuitry of interactions is ignored (e.g., figs. 3.14, 4.9, 5.13, 6.17, 7.4). The most common of these are experiments that manipulate the abundance of one putative (resource or apparent) competitor and follow the responses of other species in their guild (e.g., Connell 1983b, Schoener 1983, Gurevitch et al. 1992, 2000, Holt and Bonsall 2017). However, once multiple species are present at each trophic level, even responses to these simple manipulations can be difficult to predict without understanding the circuitry of species interactions (e.g., Wilbur 1972, 1997, Neill 1974, 1975, Wilbur and Fauth 1990). For example, up to 40% of the net effects between predators and their prey are estimated to be positive, because of the combined direct and indirect effects that propagate through the various food web connections between them (Montoya et al. 2009).

These indirect effects also extend across multiple species connections that may be propagated throughout the food web. The most apparent of these responses are top-down and bottom-up effects that result from changes in composition or abundances at the top and bottom of the food web (e.g., Hairston et al. 1960, Oksanen et al. 1981, McQueen et al. 1986, Carpenter and Kitchell 1993). Rigorous observational and experimental studies have quantified the effects of top predators in generating trophic cascades in multiple ecosystems (Estes and Palmisano 1974, Carpenter and Kitchell 1993, Estes et al. 1998, Shurin et al. 2002, Ripple and Beschta 2012, Ripple et al. 2015) and productivity effects at the bottom (Oksanen et al. 1981, McQueen et al. 1986, Mittelbach et al. 2001). However, the actions of mutualists at the bottom of food webs that support higher trophic levels seem to be much less well appreciated (Bertness and Leonard 1997, Hartnett and Wilson 1999, Smith et al. 1999, Stachowicz 2001, Kula et al. 2004, Stachowicz and Whitlatch 2005).

Employing a multiple working hypotheses framework with hypotheses that consider different vantages is especially important for exploring the indirect effects between two species through multiple circuits of an interaction web. For example, in a diamond/keystone community module, a predator has both a negative direct effect by inflicting mortality on a consumer, a positive indirect effect on the consumer's resource supply by decreasing the abundances of other consumers, and also potentially a negative indirect effect by decreasing that resource supply if the predator is an omnivore (figs. 4.4–4.6, 5.9–5.10) (e.g., Hochberg et al. 1994, Bonsall and Hassell 1998, McPeek 1998, Muller and Godfray 1999, Bety et al. 2002, Cronin 2007, Orrock et al. 2010, Bompard et al. 2013). Another excellent example is the multiple routes of influence that zooplanktivorous fish have on the factors that limit phytoplankton abundances and diversity. Fish feed on the herbivores that feed on the phytoplankton, and because of that feeding, fish excrete phosphorus and nitrogen compounds that nourish the phytoplankton (e.g., Vanni et al. 1997, Vanni 2002, McIntyre et al. 2007). Consequently, the fish have no direct effects on the phytoplankton but multiple routes of indirect effects. Simply measuring the net effect of one species on another provides no understanding of how indirect effects are propagated through the interaction web and so no way to understand why that net effect obtains. Because the net effects will change as all the intervening circuits change in magnitude for a multitude of reasons, we will also have no ability to generalize or predict how or why these effects might change among communities, and no ability to apply such insights to problems.

These are also the studies that require more sophisticated analytical approaches. For example, Wootton (1994b) advocated a multiple working hypotheses approach of performing a series of density manipulations of individual species in the putative causal network and using path analysis to test the strengths of linkages. Structural equation modeling offers a related set of statistical techniques to test the validity of a given causal network and estimate the strengths associated with the various connections (Grace 2006). Other fields are also developing analytical tools for analyzing causal networks that offer new ideas for both analyzing observational data and designing new types of experiments to evaluate causal linkages (Pearl 2000, Woodward 2010, Imbens and Rubin 2015, Pearl and Mackenzie 2018, Hernán et al. 2019). These more sophisticated analytical techniques of causal networks should become important components of the community ecologist's statistical tool kit.

Mechanisms of Interactions

Once the larger network of interactions is mapped and a general basis of the interactions are characterized, more nuanced aspects of the mechanisms involved in species interactions become relevant for study. Some of these features are what generate the need for unpredictable higher-order interactions in the Volterra-Lotka-Gause competition framework and make MacArthur's recasting problematic (see the many citations in previous chapters). However, the details of these mechanisms affect both qualitative and quantitative aspects of invasibility criteria for many different types of species. Therefore, their inclusion in the study of the mechanisms of coexistence is important.

FUNCTIONAL RESPONSES

Trophic interactions are central to the success of every species in a community. All species, from bacteria, to algae and plants, to whales, must forage for food and materials from their environment and must themselves suffer the consequences of foraging enemies. One critical mechanistic component of all these interactions is whether the uptake rates of foragers are approximately linear or saturating. Bacteria taking up molecules from their environment by active transport and plants foraging for mineral nutrients are constrained by the saturating properties of enzyme kinetics (Michaelis and Menten 1913). Herbivores and predators must process and digest consumed food, which also causes saturating rates of foraging return (Holling 1959a, 1959b, 1966, Jeschke et al. 2002). However, little empirical effort has been put into exploring how saturating functional responses may alter the criteria for coexistence as compared to linear functional responses.

A saturating functional response makes resource competitive components of invasibility criteria more difficult to fulfill, but a saturating functional response reduces the indirect effect of one consumer on another that is propagated through a shared resource. Remember that greater handling time causes the maximum feeding rate of a predator to asymptote at a lower value, which causes its isocline to intersect the axes of its prey at higher values (see figs. 3.3, 3.5). This means that a higher prey abundance is required for the predator to coexist, and these effects manifest whether or not the predator itself has enemies (chapters 3–5). However, the beneficial consequence of this for other predators that share prey with this species is that a higher prey abundance will be available for them, because the saturating functional response reduces the ability of each forager to depress prey abundances. Consequently, the indirect effect propagated from one species to another through their shared prey is reduced in magnitude with larger handling times in saturating functional responses.

Saturating functional responses also affect the components of coexistence criteria influenced by interactions with enemies above the species in the food web. If alternative prey are present, the foraging rate of a predator feeding on one prey is diminished because of the saturating effects of feeding on another, and so the invasibility criteria of other prey are easier to satisfy (as in fig. 3.5(b)). Consequently, the indirect effects propagated from one species to another through their shared enemies is diminished because the feeding by enemies becomes saturated as more prey become available (e.g., fig. 6.14).

Saturating functional responses that cause asymptotic benefits to partners are also critical for mutualistic interactions. The primary effect of saturating functional responses causing asymptotic benefits in uni-consumer and bi-consumer mutualisms is to prevent the possibility of the orgy of mutual benefaction, where mutualist abundances increase without bounds (chapter 6). Asymptotic benefits in mutualistic interactions embedded in the interaction network can also increase the overall stability of the entire community (Qian and Akçay 2020).

Saturating functional responses are also what make possible one major form of relative nonlinearities that can foster coexistence with cycling population dynamics (chapter 8) (Hsu et al. 1978b, Armstrong and McGehee 1980).

All of these consequences of saturating functional responses are, to my knowledge, largely unexplored empirically. Much of this is probably due to the fact that coexistence can be evaluated qualitatively (i.e., does this species have adequate resources to increase when rare?) without needing to explore the more subtle effects of feeding satiation (e.g., does a greater handling time for a consumer make it harder for it to successfully invade?). However, given the importance of saturating functional responses to diminishing the strengths of indirect effects being propagated across the interaction web, I believe investigations into these issues would be very profitable.

SPATIOTEMPORAL VARIATION

One of the most difficult empirical problems currently may be assessing and quantifying the role of spatiotemporal variation in fostering or retarding coexistence in the field. Variation per se is not sufficient; only specific forms of spatiotemporal ecological variation will foster coexistence (Chesson and Huntly 1997, Fox 2013). Laboratory experiments have conclusively demonstrated the importance of nonlinearities and covariation between environmental conditions and abundances to fostering coexistence among phytoplankton in the laboratory (Huisman and Weissing 1999, 2002, Litchman and Klausmeier 2001, Litchman et al. 2004, Descamps-Julien and Gonzalez 2005). However, such experiments that effectively manipulate spatiotemporal variation in limiting factors are extremely difficult to accomplish in the field. The standard methods for estimating the storage effect also require a specific model to fit to community interactions (e.g., Angert et al. 2009, Adler et al. 2010, Usinowicz et al. 2012, 2017), and some studies report aggregated estimates of the storage effect across all species of interest—what Chesson (2008) called the community average—instead of estimates for how spatiotemporal variation may help or hinder the invasibility of each species separately (e.g., Angert et al. 2009).

New simulation techniques may make estimating the magnitudes of nonlinearity and storage effect contributions to coexistence easier across a variety of models, but these techniques still require a basic understanding of the network of species interactions to serve as a backbone for the simulations (Ellner et al. 2016). Moreover, the "fluctuation-independent" properties of the interacting species (i.e., what one would consider to promote coexistence under constant environmental conditions based on chapters 3–7) are still required to understand and predict coexistence in a spatiotemporally varying environment. What nonlinearities, environmental and abundance covariation, and the storage effect—fluctuation-dependent processes—do is to boost the per capita growth rates of species that would be on the margins but not over the threshold of coexisting in a constant environment (chapters 8–9).

In fact, this is what has been inferred from a number of studies of spatiotemporal variation and coexistence. In laboratory studies of competition between two freshwater diatoms, *Cyclotella pseudostelligera* and *Fragilaria crotonensis*, Descamps-Julien and Gonzalez (2005) showed these two species coexist in constant temperature environments across a range of temperatures, with *C. pseudostelligera* having a much higher relative abundance than *F. crotonensis*. However, in an environment in which temperature fluctuates sinusoidally, the two species also coexist but with *F. crotonensis* at an abundance nearly identical to *C. pseudostelligera*; temperature variation apparently boosts the pattern of growth rate of *F. crotonensis* (Descamps-Julien and Gonzalez 2005).

Likewise, Adler et al. (2010) fit two different interaction models to abundance data for a shrub (*Artemisia tripartita*) and three grasses (*Pseudoroegneria spicata, Hesperostipa comate, Poa secunda*) in standardized quadrats taken over 31 years at the US Sheep Experimental Station in Idaho, USA. Their simulation analyses suggested that fluctuation-independent ecological differences among these species could fully account for the coexistence of the three grass species, although fluctuation-dependent processes (e.g., relative nonlinearities, storage effects) were estimated to slightly boost their invasion growth rates. In contrast, estimates for *A. tripartita* indicated that fluctuation-dependent processes are necessary in combination with its fluctuation-independent differences from the other species to boost its low density growth rate enough to permit its coexistence (Adler et al. 2010). Thus, in this community, *A. tripartita* is the species that is pushed across its coexistence criterion by spatiotemporal variation.

DIRECT SELF-LIMITATION

MacArthur (1958, p. 609) noted in his dissertation that "as G. E. Hutchinson pointed out in conversation, if each species has its density (even locally) limited by a territorial behavior which ignores the other species, then there need be no further differences between the species to permit them to persist together." For-mulations of territoriality in more mechanistic models show how territoriality can generate population regulation (Both and Visser 2003, López-Sepulcre and Kokko 2005) , and the models considered in chapter 10 show this insight about coexis-tence from Hutchinson and MacArthur to be true. Many other mechanisms can also directly generate negative density dependence within a species that can regu-late populations (Tanner 1966), including despotic habitat filling (Pulliam and Danielson 1991, McPeek, Rodenhouse, et al. 2001), cannibalism (Fox 1975, Polis 1981, Van Buskirk and Smith 1991, Rudolf 2007), opportunity costs and stress responses induced by overcrowding and agonistic interactions (Marra et al. 1995, Lochmiller 1996, McPeek, Grace, et al. 2001, Glennemeier and Denver 2002), interference among foraging individuals (Beddington 1975, DeAngelis et al. 1975, Huisman and De Boer 1997, Stillman et al. 1997, Skalski and Gilliam 2001, de Villemereuil and López-Sepulcre 2011, Le Bourlot et al. 2014), allelopathy (Rice 1984, Thacker et al. 1998, Wardle et al. 1998, Inderjit and Duke 2003), and some forms of mate finding and competition for mates (Zhang and Hanski 1998, Bauer et al. 2005, M'Gonigle et al. 2012, Ruokolainen and Hanski 2016).

As community ecologists, we tend to focus on interspecific interactions. How-ever, species whose population dynamics are influenced by one or more of these various mechanisms generating direct self-limitation add their own limiting factor to the community. Geometrically, all these various mechanisms cause a species' isocline to bend away from its own axis (e.g., figs. 3.3(d)–(e), 3.8(c), 4.2(b)–(c), 4.5(b)). Consequently, a species with some measure of direct self-limitation is reduced in its ability to deplete resources and inflate predator abundances. These mechanisms generally play no role in whether a species can meet its own invasibil-ity criterion for coexistence, but they can strongly influence whether other species can satisfy theirs by making their criteria less severe. Hence, in a multiple working hypotheses framework, analyses of direct self-limitation become important when studying the indirect effects that propagate through the interaction network and exploring what factors shape species richness in a given guild or trophic level.

One category of directly self-limiting processes that do influence a species' invasibility are those that generate Allee effects at very low abundances (Allee and Bowen 1932, Courchamp et al. 2008). An *Allee effect* occurs when the per capita population growth rate of a species increases with its own abundance, par-ticularly at low abundance, and such relationships are apparent in many different types of species (Stephens et al. 1999, Kramer et al. 2009). They are often caused by difficulty in mate finding or the lack of needed social contact and cohesion at low density (Courchamp et al. 2008, Kramer et al. 2009). Allee effects will influ-ence coexistence if per capita population growth rate is near zero or negative at very low abundances—what Kramer et al. (2009) called strong Allee effects.

Mechanisms that generate direct self-limitation also can influence the degree to which ecologically very similar species may be neutral versus coexisting species (chapter 10). If these mechanisms depress the demographic rates of conspecifics more than heterospecifics, they will foster coexistence within a guild of similar species (McPeek and Siepielski 2019). However, if conspecifics and heterospecifics are impacted similarly, these mechanisms of self-limitation will not foster coexistence.

As Schoener (1986) and Tilman (1987) argued, greater understanding will also be developed if all these studies are further complemented by functional studies of how the phenotypic properties of the species shape their abilities to engage in these interactions. Any significant discussion of traits is beyond the scope of this book. However, I know from personal experience that simultaneously exploring both the interaction network in which damselfly species are embedded (e.g., McPeek 1990b, 1998, Stoks and McPeek 2003b, Siepielski et al. 2010, 2011) and how the phenotypic traits of those species determine their performances in those interactions (e.g., McPeek 1990a, 1995, 1999, 2000, 2004, McPeek, Grace, et al. 2001, Stoks and McPeek 2006) creates a synergy that advances both avenues of inquiry far beyond what each could have done separately.

Understanding the traits that shape ecological performance capabilities of individuals and species is another approach that can identify potential mechanisms of connections among species. For example, much of my empirical work has centered on how *Enallagma* and *Ischnura* species coexist in the littoral zones of ponds and lakes. Species in these two genera feed on and are limited by the same resources and are fed upon by the same predators (McPeek 1998, Siepielski et al. 2011). Ecological performance differences in field experiments indicated that they coexist because individuals of the *Ischnura* species grow much faster (i.e., individuals gain mass at a higher rate) but suffer greater mortality from predators than *Enallagma* species (McPeek 1998). Behavioral studies of the two genera indicated that *Ischnura* is much more active than *Enallagma*, which can explain their differences in both of these performance capabilities (Stoks et al. 2003): higher activity makes individuals more conspicuous to visual predators, and higher activity is typically associated with higher foraging returns. However, one experimental result was incommensurate with these interpretations. If *Ischnura* species harvest resources at a higher rate from the environment, they should be better resource competitors than *Enallagma* species. However, my experimental results indicated that species in these two genera are symmetrical competitors (McPeek 1998).

Laboratory behavioral and performance studies provided the answer to reconciling these conflicting field experimental results. Higher activity does result in *Ischnura* species being detected by predators at higher rates, and so explains their higher predator-imposed mortality rates (McPeek 1990a, 1998, Stoks et al. 2003). However, detailed studies of feeding and digestive physiology showed that *Enallagma* and *Ischnura* species eat the same amounts of food over the course of a day, and so they depress their food resources to the same degree. Hence, this explains the field experimental result that they are symmetrical competitors. In contrast, *Enallagma* individuals have substantially lower rates of mass gain in the presence of predators because of stress responses to the presence of predators, whereas *Ischnura* individuals gain mass at similar rates in the presence and absence of predators (McPeek, Grace, et al. 2001, McPeek 2004). These field and laboratory experiments indicate that *Enallagma* and *Ischnura* species differ in ways that cause them to satisfy inequalities (4.9). However, their difference in competitive ability is because they have similar attack coefficients on their prey, but *Ischnura* has a much higher conversion efficiency than *Enallagma* in the presence of predators. Thus, the laboratory results of behavioral and physiological studies provided the mechanistic explanation for the field experimental results as to why these species may coexist.

A parallel focus on traits also opens avenues for exploring the development of community structure over deep time (McPeek and Brown 2000, Webb et al. 2002) and how these ecological interactions generate natural selection to drive the adaptation of species to one another (e.g., McPeek 1997, 2000, 2008a, 2017a, 2019a). I offer these comments also to highlight that applying a multiple working hypotheses framework to the same system over decades will foster a productive and joyful career.

MECHANISTIC NATURAL HISTORY

This framework for exploring coexistence is based on the pragmatists' realist approach for developing ideas about how nature works and expounds Chamberlin's multiple working hypotheses method to test those ideas. The most clear-eyed approach for developing a causal network of species interactions should use the natural histories of the species in the communities as the guide. The natural history of each species identifies the important links to its environment and to other species. It may seem strange to end what is basically a book exploring models of coupled differential equations by invoking natural history, but it is not. In fact, it is critical.

As I have said a few times already, *the words, language, and models we use to conceptualize nature define how we see nature and consequently how we measure and study nature.* How better to conceptualize what fosters or retards species' coexistence and broader processes shaping community structure than to conceptualize the actual components of the natural histories of the species involved. Therefore, the theory we use to see nature must reflect the natural history of the components of nature we are exploring. This approach creates a *mechanistic natural history* of the community.

This approach builds a conceptual and working model of the interaction web being studied that is rooted in the biology of those species. Understanding this web of interactions then reveals the important factors that promote or retard the coexistence of the component species. This mechanistic natural history approach directly addresses the questions about why species are coexisting while simultaneously serving as the foundation for these broader issues of community structure. Who are the players: the limiting factors that permit or prevent a species from satisfying its invasibility criterion? What limiting factors that impinge on one species are directly affected by other species and what are those other species? What then is the resulting network of interactions across multiple species and trophic levels?

This approach also provides the foundation for addressing broader questions about species diversity and community structure, including patterns of diet breadth, diversity and overlap (e.g., Mittelbach 1981, Kartzinel et al. 2015, Casey et al. 2019); broader patterns of food web structure (e.g., Paine 1980, Yodzis 1980, 1981, Martinez 1992, Dunne et al. 2002, Allesina et al. 2008, Benke 2018); the role of omnivory (e.g., Diehl 1995, Diehl and Feißel 2000, Vandermeer 2006, Rudolf 2007, Amarasekare 2008a); top-down versus bottom-up effects (e.g., Hairston et al. 1960, Oksanen et al. 1981, Hunter and Price 1992, Menge 1992, Carpenter and Kitchell 1993, Wollrab et al. 2012); the centrality of mutualisms in food webs (Jordano et al. 2003, Palmer et al. 2003, Bascompte et al. 2006, Lee and Inouye 2010, Jones et al. 2012, Bronstein 2015a, Palmer et al. 2015); the ubiquity of pathogens (Ebert et al. 2000, Decaestecker et al. 2003, Duffy et al. 2005, Lafferty et al. 2006a, Lafferty et al. 2006b, Lafferty et al. 2008, Duffy et al. 2010); and the importance of spatiotemporal heterogeneity (e.g., Huisman and Weissing 1999, 2001a, Litchman and Klausmeier 2001, Huisman and Weissing 2002, Litchman et al. 2004, Li and Chesson 2016). Of course, the models analyzed in this book are to some substantial degree abstractions of more complicated mechanisms, but they form the foundation for building even more mechanistic approaches when needed and when appropriate.

The mechanisms that promote or retard coexistence are knowable. However, these mechanisms differ for various functional positions in the interaction network. Understanding mechanisms provides great insight into a substantially broader range of questions about communities and their application to practical problems of conservation and environmental sustainability. New methods may be able to predict which species are coexisting without reference to the true underlying mechanisms (e.g., Maynard et al. 2019a, Maynard et al. 2019b). However, understanding why all these species are coexisting is the fundamental basis for predicting the responses of species, trophic levels, and entire communities in a changing world.

The next time you watch a harrier gliding over a field in search of its prey or a Cooper's hawk perched in a tree surveying the surroundings for its next meal, I hope this book will motivate you to contemplate the small piece of the complex and wonderful web of species interactions you are witnessing, and spark your imagination for the myriad consequences of that bird obtaining this one meal. Then go out and study it.

Literature Cited

Abdullah, A. S., C. S. Moffat, F. J. Lopez-Ruiz, M. R. Gibberd, J. Hamblin, and A. Zerihun. 2017. Host-multi-pathogen warfare: Pathogen interactions in co-infected plants. Frontiers in Plant Science **8**:1806.

Abrahamczyk, S., and S. S. Renner. 2015. The temporal build-up of hummingbird/plant mutualisms in North America and temperate South America. BMC Evolutionary Biology **15**:104.

Abrams, P. 1975. Limiting similarity and the form of the competition coefficient. Theoretical Population Biology **8**:356–375.

Abrams, P. 1980a. Are competition coefficients constant? Inductive versus deductive approaches. American Naturalist **116**:730–735.

Abrams, P. 1980b. Consumer functional response and competition in consumer-resource systems. Theoretical Population Biology **17**:80–102.

Abrams, P. 1983a. The theory of limiting similarity. Annual Review of Ecology and Systematics **14**:359–376.

Abrams, P. 1983b. Arguments in favor of higher order interactions. American Naturalist **121**:887–891.

Abrams, P. 1984. Variability in resource consumption rates and the coexistence of competing species. Theoretical Population Biology **25**:106–124.

Abrams, P. A. 1976. Niche overlap and environmental variability. Mathematical Biosciences **28**:357–372.

Abrams, P. A. 1992. Adaptive foraging by predators as a cause of predator-prey cycles. Evolutionary Ecology **6**:56–72.

Abrams, P. A. 1993. Effect of increased productivity on the abundances of trophic levels. American Naturalist **141**:351–371.

Abrams, P. A. 1995. Implications of dynamically variable traits for identifying, classifying, and measuring direct and indirect effects in ecological communities. American Naturalist **146**:112–134.

Abrams, P. A. 1999. Is predator-mediated coexistence possible in unstable systems? Ecology **80**:608–621.

Abrams, P. A. 2004. When does periodic variation in resource growth allow robust coexistence of competing consumer species? Ecology **85**:372–382.

Abrams, P. A. 2006. The prerequisites for and likelihood of generalist-specialist coexistence. American Naturalist **167**:329–342.

Abrams, P. A. 2009a. The implications of using multiple resources for consumer density dependence. Evolutionary Ecology Research **11**:517–540.

Abrams, P. A. 2009b. When does greater mortality increase population size? The long history and diverse mechanisms underlying the hydra effect. Ecology Letters **12**:462–474.

Abrams, P. A. 2010. Implications of flexible foraging for interspecific interactions: Lessons from simple models. Functional Ecology 24:7–17.

Abrams, P. A., and X. Chen. 2002. The effect of competition between prey species on the evolution of their vulnerabilities to a shared predator. Evolutionary Ecology Research 4:897–909.

Abrams, P. A., and R. D. Holt. 2002. The impact of consumer–resource cycles on the coexistence of competing consumers. Theoretical Population Biology 62:281–295.

Abrams, P. A., R. D. Holt, and J. D. Roth. 1998. Apparent competition or apparent mutualism? Shared predation when populations cycle. Ecology 79:201–212.

Abrams, P. A., and H. Matsuda. 1996. Positive indirect effects between prey species that share predators. Ecology 77:610–616.

Abrams, P. A., H. Matsuda, and Y. Harada. 1993. Evolutionary unstable fitness maxima and stable fitness minima of continuous traits. Evolutionary Ecology 7:465–487.

Abrams, P. A., and J. D. Roth. 1994. The effects of three-species food chains with nonlinear functional responses. Ecology 75:1118–1130.

Adam, T. C., D. E. Burkepile, B. I. Ruttenberg, and M. J. Paddack. 2015. Herbivory and the resilience of Caribbean coral reefs: Knowledge gaps and implications for management. Marine Ecology Progress Series 520:1–20.

Adler, P. B., and J. M. Drake. 2008. Environmental variation, stochastic extinction, and competitive coexistence. American Naturalist 172:186–195.

Adler, P. B., S. P. Ellner, and J. M. Levine. 2010. Coexistence of perennial plants: An embarrassment of niches. Ecology Letters 13:1019–1029.

Adler, P. B., A. Fajardo, A. R. Kleinhesselink, and N. J. Kraft. 2013. Trait-based tests of coexistence mechanisms. Ecology Letters 16:1294–1306.

Adler, P. B., J. HilleRisLambers, P. C. Kyriakidis, Q. Guan, and J. M. Levine. 2006. Climate variability has a stabilizing effect on the coexistence of prairie grasses. Proceedings of the National Academy of Sciences, USA 103:12793–12798.

Adler, P. B., J. HilleRisLambers, and J. M. Levine. 2007. A niche for neutrality. Ecology Letters 10:95–104.

Adler, P. B., D. Smull, K. H. Beard, R. T. Choi, T. Furniss, A. Kulmatiski, J. M. Meiners, A. T. Tredennick, and K. E. Veblen. 2018. Competition and coexistence in plant communities: Intraspecific competition is stronger than interspecific competition. Ecology Letters 21:1319–1329.

Ågren, G. I. 1988. Ideal nutrient productivities and nutrient proportions in plant growth. Plant, Cell and Environment 11:613–620.

Akçay, E. 2015. Evolutionary models of mutualism. Pages 57–76 in J. L. Bronstein, editor, Mutualism. Oxford University Press, Oxford, UK.

Allee, W. C., and E. S. Bowen. 1932. Studies in animal aggregations: Mass protection against colloidal silver among goldfishes. Journal of Experimental Biology 61:185–207.

Allesina, S., D. Alonso, and M. Pascual. 2008. A general model for food web structure. Science 320:658–661.

Amarasekare, P. 1998. Interactions between local dynamics and dispersal: Insights from single species models. Theoretical Population Biology 53:44–59.

Amarasekare, P. 2002. Interference competition and species coexistence. Proceedings of the Royal Society B-Biological Sciences 269:2541–2550.

Amarasekare, P. 2004a. The role of density-dependent dispersal in source–sink dynamics. Journal of Theoretical Biology 226:159–168.

Amarasekare, P. 2004b. Spatial variation and density-dependent dispersal in competitive coexistence. Proceedings of the Royal Society B-Biological Sciences 271:1497–1506.

Amarasekare, P. 2007. Spatial dynamics of communities with intraguild predation: The role of dispersal strategies. American Naturalist 170:819–831.

Amarasekare, P. 2008a. Coexistence of intraguild predators and prey in resource-rich environments. Ecology 89:2786–2797.

Amarasekare, P. 2008b. Spatial dynamics of foodwebs. Annual Review of Ecology, Evolution, and Systematics 39:479–500.

Amarasekare, P. 2008c. Spatial dynamics of keystone predation. Journal of Animal Ecology 77:1306–1315.

Amarasekare, P., M. F. Hoopes, N. Mouquet, and M. Holyoak. 2004. Mechanisms of coexistence in competitive metacommunities. American Naturalist 164:310–326.

Anderies, J. M., and B. E. Beisner. 2000. Fluctuating environments and phytoplankton community structure: A stochastic model. American Naturalist 155:556–569.

Anderson, R. M., and R. M. May. 1978a. Regulation and stability of host-parasite population interactions: I. Regulatory processes. Journal of Animal Ecology 47:219–247.

Anderson, R. M., and R. M. May. 1978b. Regulation and stability of host-parasite population interactions: II. Destabilizing processes. Journal of Animal Ecology 47:249–267.

Anderson, R. M., and R. M. May. 1980. Infectious diseases and population cycles of forest insects. Science 210:658–661.

Anderson, R. M., and R. M. May. 1981. The population dynamics of microparasites and their invertebrate hosts. Philosophical Transactions of the Royal Society of London B-Biological Sciences 291:451–524.

Anderson, R. M., and R. M. May. 1991. *Infectious Diseases of Humans: Dynamics and Control*. Oxford University Press, New York.

Angert, A. L. 2006. Demography of central and marginal populations of monkeyflowers (*Mimulus cardinalis* and *M. lewisii*). Ecology 87:2014–2025.

Angert, A. L., T. E. Huxman, P. Chesson, and D. L. Venable. 2009. Functional tradeoffs determine species coexistence via the storage effect. Proceedings of the National Academy of Sciences, USA 106:11641–11645.

Armstrong, R. A., and R. McGehee. 1976. Coexistence of species competing for shared resources. Theoretical Population Biology 9:317–328.

Armstrong, R. A., and R. McGehee. 1980. Competitive exclusion. American Naturalist 115:151–170.

Armsworth, P. R., and J. E. Roughgarden. 2008. The structure of clines with fitness-dependent dispersal. American Naturalist 172:648–657.

Arnott, S. E., and M. J. Vanni. 1993. Zooplankton assemblages in fishless bog lakes: Influence of biotic and abiotic factors. Ecology 74:2361–2380.

Auerbach, M., and A. Shmida. 1993. Vegetation change along an altitudinal gradient on Mt. Hermon, Israel—no evidence for discrete communities. Journal of Ecology 81:25–33.

Bachelot, B., and C. T. Lee. 2018. Dynamic preferential allocation to arbuscular mycorrhizal fungi explains fungal succession and coexistence. Ecology 99:372–384.

Bakker, E. S., M. E. Ritchie, H. Olff, D. G. Milchunas, and J. M. Knops. 2006. Herbivore impact on grassland plant diversity depends on habitat productivity and herbivore size. Ecology Letters 9:780–788.

Barabás, G., R. D'Andrea, and S. M. Stump. 2018. Chesson's coexistence theory. Ecological Monographs **88**:277–303.

Barabás, G., M. J. Michalska-Smith, and S. Allesina. 2016. The effect of intra- and interspecific competition on coexistence in multispecies communities. American Naturalist **188**:E1–E12.

Bascompte, J., P. Jordano, C. J. Melián, and J. M. Olesen. 2003. The nested assembly of plant-animal mutualistic networks. Proceedings of the National Academy of Sciences, USA **100**:9383–9387.

Bascompte, J., P. Jordano, and J. M. Olesen. 2006. Asymmetric coevolutionary networks facilitate biodiversity maintenance. Science **312**:431–433.

Bascompte, J., and C. J. Melián. 2005. Simple trophic modules for complex food webs. Ecology **86**:2868–2873.

Bauer, S., J. Samietz, and U. Berger. 2005. Sexual harassment in heterogeneous landscapes can mediate population regulation in a grasshopper. Behavioral Ecology **16**:239–246.

Beals, E. 1960. Forest bird communities in the Apostle Islands of Wisconsin. Wilson Bulletin **72**:156–181.

Beals, E. W. 1969. Vegetational change along altitudinal gradients. Science **165**:981–985.

Becker, J. H., and A. S. Grutter. 2004. Cleaner shrimp do clean. Coral Reefs **23**:515–520.

Beddington, J. R. 1975. Mutual interference between parasites or predators and its effect on searching efficiency. Journal of Animal Ecology **44**:331–340.

Begon, M., R. G. Bowers, N. Kadianakis, and D. E. Hodgkinson. 1992. Disease and community structure: The importance of host self-regulation in a host-host-pathogen model. American Naturalist **139**:1131–1150.

Benard, M. F., and S. J. McCauley. 2008. Integrating across life-history stages: Consequences of natal habitat effects on dispersal. American Naturalist **171**:553–567.

Bender, E. A., T. J. Case, and M. W. Gilpin. 1984. Perturbation experiments in community ecology: Theory and practice. Ecology **65**:1–13.

Beninca, E., B. Ballantine, S. P. Ellner, and J. Huisman. 2015. Species fluctuations sustained by a cyclic succession at the edge of chaos. Proceedings of the National Academy of Sciences, USA **112**:6389–6394.

Beninca, E., J. Huisman, R. Heerkloss, K. D. Johnk, P. Branco, E. H. Van Nes, M. Scheffer, and S. P. Ellner. 2008. Chaos in a long-term experiment with a plankton community. Nature **451**:822–825.

Benke, A. C. 2018. River food webs: An integrative approach to bottom-up flow webs, top-down impact webs, and trophic position. Ecology **99**:1370–1381.

Berestycki, H., and A. Zilio. 2019. Predator-prey models with competition: The emergence of territoriality. American Naturalist **193**:436–446.

Bertness, M. D. 1991a. Interspecific interactions among high marsh perennials in a New England salt marsh. Ecology **72**:125–137.

Bertness, M. D. 1991b. Zonation of *Spartina patens* and *Spartina alterniflora* in New England salt marsh. Ecology **71**:138–148.

Bertness, M. D., and A. M. Ellison. 1987. Determinants of pattern in a New England salt marsh community. Ecological Monographs **57**:129–147.

Bertness, M. D., and G. H. Leonard. 1997. The role of positive interactions in communities: Lessons from intertidal habitats. Ecology **78**.

Beschta, R. L., and W. J. Ripple. 2016. Riparian vegetation recovery in Yellowstone: The first two decades after wolf reintroduction. Biological Conservation 198:93–103.

Betini, G. S., T. Avgar, and J. M. Fryxell. 2017. Why are we not evaluating multiple competing hypotheses in ecology and evolution? Royal Society Open Science 4:160756.

Bety, J., G. Gauthier, E. Korpimäki, and J. F. Giroux. 2002. Shared predators and indirect trophic interactions: Lemming cycles and arctic-nesting geese. Journal of Animal Ecology 71:88–98.

Beverton, R. J. H., and S. J. Holt. 1957. On the dynamics of exploited fish populations. Great Britain Fisheries Investigations, Series II, XIX:553.

Bickford, D., D. J. Lohman, N. S. Sodhi, P. K. Ng, R. Meier, K. Winker, K. K. Ingram, and I. Das. 2007. Cryptic species as a window on diversity and conservation. Trends in Ecology and Evolution 22:148–155.

Billick, I., and T. J. Case. 1994. Higher order interactions in ecological communities: What are they and how can they be detected? Ecology 75:1529–1543.

Binckley, C. A., and W. J. Resetarits. 2008. Oviposition behavior partitions aquatic landscapes along predation and nutrient gradients. Behavioral Ecology 19:552–557.

Blanquart, F., and S. Gandon. 2011. Evolution of migration in a periodically changing environment. American Naturalist 177:188–201.

Boecklen, W., and C. NeSmith. 1985. Hutchinsonian ratios and log-normal distributions. Evolution 39:695–698.

Bolker, B. M. 2003. Combining endogenous and exogenous spatial variability in analytical population models. Theoretical Population Biology 64:255–270.

Bolker, B. M., and S. W. Pacala. 1997. Using moment equations to understand stochastically driven spatial pattern formation in ecological systems. Theoretical Population Biology 52:179–197.

Bolker, B. M., and S. W. Pacala. 1999. Spatial moment equations for plant competition: Understanding spatial strategies and the advantages of short dispersal. American Naturalist 153:575–602.

Bolker, B. M., S. W. Pacala, and C. Neuhauser. 2003. Spatial dynamics in model plant communities: What do we really know? American Naturalist 162:135–148.

Bompard, A., C. C. Jaworski, P. Bearez, and N. Desneux. 2013. Sharing a predator: Can an invasive alien pest affect the predation on a local pest? Population Ecology 55:433–440.

Bond, R. R. 1957. Ecological distribution of breeding birds in the upland forests of southern Wisconsin. Ecological Monographs 27:351–384.

Bonsall, M. B., and M. P. Hassell. 1997. Apparent competition structures ecological assemblages. Nature 388:371–373.

Bonsall, M. B., and M. P. Hassell. 1998. Population dynamics of apparent competition in a host-parasitoid assemblage. Journal of Animal Ecology 67:918–929.

Bonsall, M. B., and R. D. Holt. 2010. Apparent competition and vector-host interactions. Israel Journal of Ecology and Evolution 56:393–416.

Borer, E. T., C. J. Briggs, and R. D. Holt. 2007. Predators, parasitoids, and pathogens: A cross-cutting examination of intraguild predation theory. Ecology 88:2681–2688.

Borer, E. T., E. W. Seabloom, D. S. Gruner, W. S. Harpole, H. Hillebrand, E. M. Lind, P. B. Adler, et al. 2014. Herbivores and nutrients control grassland plant diversity via light limitation. Nature 508:517–520.

Both, C., and M. E. Visser. 2003. Density dependence, territoriality, and divisibility of resources: From optimality models to population processes. American Naturalist **161**:326–336.

Boucher, D. H. 1985. Lotka-Volterra models of mutualism and positive density-dependence. Ecological Modelling **27**:251–270.

Bowers, R. G., and J. Turner. 1997. Community structure and the interplay between inter-specific infection and competition. Journal of Theoretical Biology **187**:95–109.

Braun, E. L. 1935. The vegetation of Pine Mountain, Kentucky: An analysis of the influence of soils and slope exposure as determined by geological structure. American Midland Naturalist **16**:517–565.

Braun, E. L. 1940. An ecological transect of Black Mountain, Kentucky. Ecological Monographs **10**:193–241.

Brett, M. T., and M. M. Benjamin. 2007. A review and reassessment of lake phosphorus retention and the nutrient loading concept. Freshwater Biology **0**:070907013155001.

Briggs, C. J., and E. T. Borer. 2005. Why short-term experiments may not allow long-term predictions about intraguild predation. Ecological Applications **15**:1111–1117.

Bronstein, J. L. 2009. The evolution of facilitation and mutualism. Journal of Ecology **97**:1160–1170.

Bronstein, J. L. 2015a. *Mutualism*. Oxford University Press, Oxford, UK.

Bronstein, J. L. 2015b. The study of mutualism. Pages 3–19 *in* J. L. Bronstein, editor, *Mutualism*. Oxford University Press, Oxford, UK.

Bronstein, J. L., W. G. Wilson, and W. F. Morris. 2003. Ecological dynamics of mutualist/antagonist communities. American Naturalist **162**:S24–S39.

Brooker, R. W., and R. M. Callaway. 2009. Facilitation in the conceptual melting pot. Journal of Ecology **97**:1117–1120.

Brooker, R. W., F. T. Maestre, R. M. Callaway, C. L. Lortie, L. A. Cavieres, G. Kunstler, P. Liancourt, et al. 2007. Facilitation in plant communities: The past, the present, and the future. Journal of Ecology **96**:18–34.

Brooks, J. L., and S. I. Dodson. 1965. Predation, body size, and composition of plankton. Science **150**:28–35.

Brown, B. L., R. P. Creed, and W. E. Dobson. 2002. Branchiobdellid annelids and their crayfish hosts: Are they engaged in a cleaning symbiosis? Oecologia **132**:250–255.

Brown, B. L., R. P. Creed, J. Skelton, M. A. Rollins, and K. J. Farrell. 2012. The fine line between mutualism and parasitism: Complex effects in a cleaning symbiosis demonstrated by multiple field experiments. Oecologia **170**:199–207.

Brown, J. H. 1973. Species diversity of seed-eating desert rodents in sand dune habitats. Ecology **54**:775–787.

Bruno, J. F., J. J. Stachowicz, and M. D. Bertness. 2003. Inclusion of facilitation into ecological theory. Trends in Ecology & Evolution **18**:119–125.

Buhnerkempe, M. G., M. G. Roberts, A. P. Dobson, H. Heesterbeek, P. J. Hudson, and J. O. Lloyd-Smith. 2015. Eight challenges in modelling disease ecology in multi-host, multi-agent systems. Epidemics **10**:26–30.

Burd, M. 1994. Bateman's principle and plant reproduction: The role of pollen limitation in fruit and seed set. Botanical Review **60**:83–139.

Burkepile, D. E., and M. E. Hay. 2010. Impact of herbivore identity on algal succession and coral growth on a Caribbean reef. Plos One **5**:e8963.

Cáceres, C. E. 1997. Temporal variation, dormancy, and coexistence: A field test of the storage effect. Proceedings of the National Academy of Sciences, USA **94**:9171–9175.

Calcagno, V., N. Mouquet, P. Jarne, and P. David. 2006. Coexistence in a metacommunity: The competition-colonization trade-off is not dead. Ecology Letters **9**:897–907.

Calcagno, V., C. Sun, O. J. Schmitz, and M. Loreau. 2011. Keystone predation and plant species coexistence: The role of carnivore hunting mode. American Naturalist **177**:E1–E13.

Caley, M. J., and D. Schluter. 1997. The relationship between local and regional diversity. Ecology **78**:70–80.

Callaway, R. M. 1997. Positive interactions in plant communities and the individualistic-continuum concept. Oecologia **112**:143–149.

Callaway, R. M. 2007. Positive interactions and interdependence in plant communities. Springer, Dordrecht, Netherlands.

Callaway, R. M., R. W. Brooker, P. Choler, Z. Kikvidze, C. L. Lortie, R. Michalet, L. Paolini, et al. 2002. Positive interactions among alpine plants increase with stress. Nature **417**:844–848.

Canard, E. F., N. Mouquet, D. Mouillot, M. Stanko, D. Miklisova, and D. Gravel. 2014. Empirical evaluation of neutral interactions in host-parasite networks. American Naturalist **183**:468–479.

Cantrell, R. S., and C. Cosner. 2001. On the dynamics of predator–prey models with the Beddington–DeAngelis functional response. Journal of Mathematical Analysis and Applications **257**:206–222.

Cantrell, R. S., C. Cosner, and Y. Lou. 2010. Evolution of dispersal and the ideal free distribution. Mathematical Biosciences and Engineering **7**:17–36.

Cantrell, R. S., C. Cosner, Y. Lou, and C. Xie. 2013. Random dispersal versus fitness-dependent dispersal. Journal of Differential Equations **254**:2905–2941.

Caraco, T., and I. N. Wang. 2008. Free-living pathogens: Life-history constraints and strain competition. Journal of Theoretical Biology **250**:569–579.

Carpenter, R. C. 1986. Partitioning herbivory and its effects on coral reef algal communities. Ecological Monographs **564**:345–364.

Carpenter, S. R., and J. F. Kitchell. 1987. The temporal scale of variance in limnetic primary production. American Naturalist **129**:417–433.

Carpenter, S. R., and J. F. Kitchell. 1993. *The Trophic Cascade in Lakes*. Cambridge University Press, Cambridge, UK.

Carpenter, S. R., J. F. Kitchell, and J. R. Hodgson. 1985. Cascading trophic interactions and lake productivity. Bioscience **35**:634–639.

Carson, W. P., and R. B. Root. 1999. Top-down effects of insect herbivores during early succession: Influence on plant biomass and plant dominance. Oecologia **121**:260–272.

Carson, W. P., and R. B. Root. 2000. Herbivory and plant species coexistence: Community regulation by an outbreaking phytophagous insect. Ecological Monographs **70**:73–99.

Casas, J. J., and P. H. Langton. 2008. Chironomid species richness of a permanent and a temporary Mediterranean stream: A long-term comparative study. Journal of the North American Benthological Society **27**:746–759.

Case, T. J. 2000. *An Illustrated Guide to Theoretical Ecology*. Oxford University Press, New York.

Case, T. J., and E. A. Bender. 1981. Testing for higher order interactions. American Naturalist **118**:920–929.

Case, T. J., and R. G. Casten. 1979. Global stability and multiple domains of attraction in ecological systems. American Naturalist **113**:705–714.

Casey, J. M., C. P. Meyer, F. Morat, S. J. Brandl, S. Planes, V. Parravicini, and A. Mahon. 2019. Reconstructing hyperdiverse food webs: Gut content metabarcoding as a tool to disentangle trophic interactions on coral reefs. Methods in Ecology and Evolution **10**:1157–1170.

Caswell, H. 1978. Predator-mediated coexistence: A nonequilbrium model. American Naturalist **112**:127–154.

Caswell, H. 1989. Analysis of life table response experiments. I. Decomposition of effects on population growth rate. Ecological Modeling **46**:221–237.

Caswell, H. 1996. Analysis of life table response experiments. II. Alternative parameterizations for size- and stage-structured models. Ecological Modeling **88**:73–82.

Caswell, H. 2001. *Matrix Population Models: Construction, Analysis, and Interpretation.* 2nd edition. Sinauer Associates, Sunderland, MA.

Caswell, H. 2010. Life table response experiment analysis of the stochastic growth rate. Journal of Ecology **98**:324–333.

Caswell, H., and M. G. Neubert. 1998. Chaos and closure terms in plankton food chain models. Journal of Plankton Research **20**:1837–1845.

Caves, E. M., P. A. Green, and S. Johnsen. 2018. Mutual visual signalling between the cleaner shrimp *Ancylomenes pedersoni* and its client fish. Proceedings of the Royal Society B-Biological Sciences **285**:20180800.

Chamberlin, T. C. 1890. The method of multiple hypotheses. Science **15**:92–96.

Chamberlin, T. C. 1897. Studies for students: The method of multiple working hypotheses. Journal of Geology **5**:837–848.

Charlesworth, B. 1994. *Evolution in Age-Structured Populations.* Cambridge University Press, New York.

Chave, J. 2004. Neutral theory and community ecology. Ecology Letters **7**:241–253.

Chesson, P. 1978. Predator-prey theory and variability. Annual Review of Ecology and Systematics **9**:323–347.

Chesson, P. 1990. MacArthur's consumer-resource model. Theoretical Population Biology **37**:26–38.

Chesson, P. 1991. A need for niches? Trends in Ecology and Evolution **6**:26–28.

Chesson, P. 1994. Multispecies competition in variable environments. Theoretical Population Biology **45**:227–276.

Chesson, P. 2000a. General theory of competitive coexistence in spatially-varying environments. Theoretical Population Biology **58**:211–237.

Chesson, P. 2000b. Mechanisms of maintenance of species diversity. Annual Review of Ecology and Systematics **31**:343–366.

Chesson, P. 2003. Quantifying and testing coexistence mechanisms arising from recruitment fluctuations. Theoretical Population Biology **64**:345–357.

Chesson, P. 2008. Quantifying and testing species coexistence mechanisms. Pages 119–164 *in* F. Valladares, A. Camacho, A. Elosegui, C. Gracia, M. Estrada, J. C. Senar, and J. M. Gili, editors, *Unity in Diversity: Reflections on Ecology after the Legacy of Ramon Margalef.* Fundacion BBVA, Bilbao, Spain.

Chesson, P. 2012. Species competition and predation. Pages 10061–10085 *in* R. A. Meyers, editor, *Encyclopedia of Sustainability Science and Technology*. Springer, New York.

Chesson, P. 2018. Updates on mechanisms of maintenance of species diversity. Journal of Ecology **106**:1773–1794.

Chesson, P., and N. Huntly. 1997. The roles of harsh and fluctuating conditions in the dynamics of ecological communities. American Naturalist **150**:519–553.

Chesson, P., and J. J. Kuang. 2008. The interaction between predation and competition. Nature **456**:235–238.

Chesson, P. L. 1982. The stabilizing effect of a random environment. Journal of Mathematical Biology **15**:1–36.

Chesson, P. L. 1986. Environmental variation and the coexistence of species. Pages 240–256 *in* J. Diamond and T. J. Case, editors, *Community Ecology*. Harper & Row, New York.

Chesson, P. L., and S. Ellner. 1989. Invasibility and stochastic boundedness in monotonic competition models. Journal of Mathematical Biology **27**:117–138.

Chesson, P. L., and R. R. Warner. 1981. Environmental variability promotes coexistence in lottery competitive systems. American Naturalist **117**:923–943.

Clark, C. M., E. E. Cleland, S. L. Collins, J. E. Fargione, L. Gough, K. L. Gross, S. C. Pennings, K. N. Suding, and J. B. Grace. 2007. Environmental and plant community determinants of species loss following nitrogen enrichment. Ecology Letters **10**:596–607.

Clark, C. M., and D. Tilman. 2008. Loss of plant species after chronic low-level nitrogen deposition to prairie grasslands. Nature **451**:712–715.

Clark, J. S., and J. S. McLachlan. 2003. Stability of forest biodiversity. Nature **423**:635–638.

Cleaveland, S., M. K. Laurenson, and L. H. Taylor. 2001. Diseases of humans and their domestic mammals: Pathogen characteristics, host range and the risk of emergence. Philosophical Transactions of the Royal Society of London B-Biological Sciences **356**:991–999.

Clobert, J., E. Danchin, A. A. Dhondt, and J. D. Nichols. 2001. *Dispersal*. Oxford University Press, New York.

Clobert, J., J. F. Le Galliard, J. Cote, S. Meylan, and M. Massot. 2009. Informed dispersal, heterogeneity in animal dispersal syndromes and the dynamics of spatially structured populations. Ecology Letters **12**:197–209.

Coffman, W. P., and C. L. de la Rosa. 1998. Taxonomic composition and temporal organization of tropical and temperate assemblages of lotic Chironomidae. Journal of the Kansas Entomological Society **71**:388–406.

Cohen, J. E. 1970. A Markov contingency-table model for replicated Lotka-Volterra systems near equilibrium. American Naturalist **104**:547–560.

Collinge, S. K., and C. Ray. 2006. *Disease Ecology: Community Structure and Pathogen Dynamics*. Oxford University Press, New York.

Collins, S. L., A. K. Knapp, J. M. Briggs, J. M. Blair, and E. M. Steinauer. 1998. Modulation of diversity by grazing and mowing in native tallgrass prairie. Science **280**:745–747.

Colman, D. R., E. C. Toolson, and C. D. Takacs-Vesbach. 2012. Do diet and taxonomy influence insect gut bacterial communities? Molecular Ecology **21**:5124–5137.

Comita, L. S., H. C. Muller-Landau, S. Aguilar, and S. P. Hubbell. 2010. Asymmetric density dependence shapes species abundances in a tropical tree community. Science **329**:330–332.

Conley, D. J., H. W. Paerl, R. W. Howarth, D. F. Boesch, S. P. Seitzinger, K. E. Havens, C. Lancelot, and G. E. Likens. 2009. Controlling eutrophication: Nitrogen and phosphorus. Science **323**:1014–1015.

Connell, J. H. 1961. The influence of interspecific competition and other factors on the distribution of the barnacle *Chthamalus stellatus*. Ecology **42**:710–723.

Connell, J. H. 1971. On the role of natural enemies in preventing competitive exclusion in some marine animals and in rain forest trees. Pages 298–312 *in* P. J. den Boer and G. R. Gradwell, editors, *Dynamics of Populations*. Centre for Agricultural Publishing and Documentation, Wageningen, Netherlands.

Connell, J. H. 1978. Diversity in tropical rainforests and coral reefs. Science **199**:1302–1310.

Connell, J. H. 1980. Diversity and the coevolution of competitors, or the ghost of competition past. Oikos **35**:131–138.

Connell, J. H. 1983a. Interpreting the results of field experiments: Effects of indirect interactions. Oikos **41**:290–291.

Connell, J. H. 1983b. On the prevalence and relative importance of interspecific competition: Evidence from field experiments. American Naturalist **122**:661–696.

Connell, J. H., and R. O. Slayter. 1977. Mechanisms of succession in natural communities and their role in community stability and organization. American Naturalist **111**:1119–1144.

Connor, E. F., and D. Simberloff. 1979. The assembly of species communities: Chance or competition? Ecology **60**:1132–1140.

Cornell, H. V. 1985. Local and regional richness of cynipine gall wasps on California oaks. Ecology **66**:1247–1260.

Cornell, H. V., and S. P. Harrison. 2013. Regional effects as important determinants of local diversity in both marine and terrestrial systems. Oikos **122**:288–297.

Cornell, H. V., and R. H. Karlson. 1996. Species richness of reef-building corals determined by local and regional processes. Journal of Animal Ecology **65**:233–241.

Cornell, H. V., R. H. Karlson, and T. P. Hughes. 2007. Local–regional species richness relationships are linear at very small to large scales in west-central Pacific corals. Coral Reefs **27**:145–151.

Cornell, H. V., and J. H. Lawton. 1992. Species interactions, local and regional processes, and limits to the richness of ecological communities: A theoretical perspective. Journal of Animal Ecology **61**:1–12.

Cortez, M. H., and P. A. Abrams. 2016. Hydra effects in stable communities and their implications for system dynamics. Ecology **97**:1135–1145.

Côté, I. M. 2000. Evolution and ecology of cleaning symbiosis in the sea. Oceanography and Marine Biology: Annual Review **38**:311–355.

Cothran, R. D., P. Noyes, and R. A. Relyea. 2015. An empirical test of stable species coexistence in an amphipod species complex. Oecologia **178**:819–831.

Cottingham, K. L., and J. M. Butzler. 2006. The community ecology of *Vibrio cholerae*. Pages 105–118 *in* S. K. Collinge and C. Ray, editors, *Disease Ecology: Community Structure and Pathogen Dynamics*. Oxford University Press, New York.

Courchamp, F., J. Berec, and J. Gascoigne. 2008. *Allee Effects in Ecology and Conservation*. Oxford University Press, New York.

Creed, R. P., J. D. Lomonaco, M. J. Thomas, A. Meeks, and B. L. Brown. 2015. Reproductive dependence of a branchiobdellidan annelid on its crayfish host: Confirmation of a mutualism. Crustaceana **88**:385–396.

Cressman, R., and V. Křivan. 2006. Migration dynamics for the ideal free distribution. American Naturalist **168**:384–397.

Cressman, R., and V. Křivan. 2010. The ideal free distribution as an evolutionarily stable state in density-dependent population games. Oikos **119**:1231–1242.

Cressman, R., V. Křivan, and J. Garay. 2004. Ideal free distributions, evolutionary games, and population dynamics in multi-species environments. American Naturalist **164**:473–489.

Crombie, A. C. 1944. On intraspecific and interspecific competition in larvae of graminivorous insects. Journal of Experimental Biology **20**:135–151.

Crombie, A. C. 1945. On competition between different species of graminivorous insects. Proceedings of the Royal Society B-Biological Sciences **132**:362–395.

Cronin, J. T. 2007. Shared parasitoids in a metacommunity: Indirect interactions inhibit herbivore membership in local communities. Ecology **88**:2977–2990.

Crowder, L. B., and W. E. Cooper. 1982. Habitat structural complexity and the interaction of bluegills and their prey. Ecology **63**:1802–1813.

Crowley, P. H. 1979. Predator-mediated coexistence: An equilibrium interpretation. Journal of Theoretical Biology **80**:129–144.

D'Andrea, R., and A. Ostling. 2016. Challenges in linking trait patterns to niche differentiation. Oikos **125**:1369–1385.

Davis, I. C., M. Girard, and P. N. Fultz. 1998. Loss of CD4$^+$ T cells in human immunodeficiency virus type 1-infected chimpanzees is associated with increased lymphocyte apoptosis. Journal of Virology **72**:4623–4632.

Davis, M. P., J. S. Sparks, and W. L. Smith. 2016. Repeated and widespread evolution of bioluminescence in marine fishes. Plos One **11**:e0155154.

Davison, R., F. Nicolè, H. Jacquemyn, and S. Tuljapurkar. 2013. Contributions of covariance: Decomposing the components of stochastic population growth in *Cypripedium calceolus*. American Naturalist **181**:410–420.

Dayton, P. K. 1971. Competition, disturbance, and community organization: The provision and subsequent utilization of space in a rocky intertidal community. Ecological Monographs **41**:351–389.

Dean, A. M. 1983. A simple model of mutualism. American Naturalist **121**:409–417.

DeAngelis, D. L. 1992. Dynamics of nutrient cycling and food webs. Springer, Amsterdam.

DeAngelis, D. L., R. A. Goldstein, and R. V. O'Neill. 1975. A model for trophic interaction. Ecology **56**:881–892.

DeBach, P. 1964. *Biological Control of Insect Pests and Weeds*. Chapman and Hall, London.

DeBach, P., and R. A. Sundby. 1963. Competitive displacement between ecological homologues. Hilgardia **34**:105–166.

Decaestecker, E., A. Vergote, D. Ebert, and L. de Meester. 2003. Evidence for strong host clone-parasite species interactions in the *Daphnia* microparasite system. Evolution **57**:784–792.

de Groot, N. G., and R. E. Bontrop. 2013. The HIV-1 pandemic: Does the selective sweep in chimpanzees mirror humankind's future? Retrovirology **10**:53.

Descamps-Julien, B., and A. Gonzalez. 2005. Stable coexistence in a fluctuating environment: An experimental demonstration. Ecology **86**:2815–2824.

de Roos, A. M., and L. Persson. 2013. *Population and Community Ecology of Ontogenetic Development*. Princeton University Press, Princeton, NJ.

de Villemereuil, P. B., and A. López-Sepulcre. 2011. Consumer functional responses under intra- and inter-specific interference competition. Ecological Modelling **222**:419–426.

de Wit, C. T. 1960. *On Competition*. Centre for Agricultural Publishing and Documentation, Wageningen, Netherlands.

Diamond, J. M. 1975. Assembly of species communities. Pages 342–444 *in* M. L. Cody and J. M. Diamond, editors, *Ecology and Evolution of Communities*. Harvard University Press, Cambridge, MA.

Díaz, S., A. J. Symstad, F. Stuart Chapin, D. A. Wardle, and L. F. Huenneke. 2003. Functional diversity revealed by removal experiments. Trends in Ecology and Evolution **18**:140–146.

Diehl, S. 1995. Direct and indirect effects of omnivory in a littoral lake community. Ecology **76**:1727–1740.

Diehl, S., and M. Feißel. 2000. Effects of enrichment on three-level food chains with omnivory. American Naturalist **155**:200–218.

Dodson, S. I. 1970. Complementary feeding niches maintained by size-selective predation. Limnology and Oceanography **15**:131–137.

Doebeli, M., and G. D. Ruxton. 1997. Evolution of dispersal rates in metapopulation models: Branching and cyclic dynamics in phenotype space. Evolution **51**:1730–1741.

Doncaster, C. P., J. Clobert, B. Doligez, L. Gustafsson, and E. Danchin. 1997. Balanced dispersal between spatially varying local populations: An alternative to the source-sink model. American Naturalist **150**:425–445.

Douglas, A. E. 2015. The special case of symbioses: Mutualisms with persistent contact. Pages 20–34 *in* J. L. Bronstein, editor, *Mutualism*. Oxford University Press, Oxford, UK.

Dowd, E. M., and L. D. Flake. 1985. Foraging habitats and movements of nesting great blue herons in a prairie river ecosystem, South Dakota. Journal of Field Ornithology **56**:379–387.

Droop, M. R. 1974. The nutrient status of algal cells in continuous culture. Journal of the Marine Biological Association of the United Kingdom **54**:825–855.

Drown, D. M., M. F. Dybdahl, and R. Gomulkiewicz. 2013. Consumer-resource interactions and the evolution of migration. Evolution **67**:3290–3304.

Dubinsky, Z., and P. L. Jokiel. 1994. Ratio of energy and nutrient fluxes regulates symbiosis between zooxanthellae and corals. Pacific Science **48**:313–324.

Duffy, M. A., C. E. Cáceres, S. R. Hall, A. J. Tessier, and A. R. Ives. 2010. Temporal, spatial, and between-host comparisons of patterns of parasitism in lake zooplankton. Ecology **91**:3322–3331.

Duffy, M. A., S. R. Hall, A. J. Tessier, and M. Huebner. 2005. Selective predators and their parasitized prey: Are epidemics in zooplankton under top-down control? Limnology and Oceanography **50**:412–420.

Dunham, A. E. 1980. An experimental study of interspecific competition between the iguanid lizards *Sceloporus merriami* and *Urosaurus ornatus*. Ecological Monographs **50**:309–330.

Dunne, J. A., R. J. Williams, and N. D. Martinez. 2002. Food-web structure and network theory: The role of connectance and size. Proceedings of the National Academy of Sciences, USA **99**:12917–12922.

Dupuch, A., L. M. Dill, and P. Magnan. 2009. Testing the effects of resource distribution and inherent habitat riskiness on simultaneous habitat selection by predators and prey. Animal Behaviour **78**:705–713.

Eadie, J. M., L. Broekhoven, and P. Colgan. 1987. Size ratios and artifacts: Hutchinson's rule revisited. American Naturalist **129**:1–17.

Eberhard, W. G. 1985. Sexual selection and animal genitalia. Harvard University Press, Cambridge, MA.

Ebert, D., M. Lipsitch, and K. L. Mangin. 2000. The effect of parasites on host population density and extinction: Experimental epidemiology with *Daphnia* and six microparasites. American Naturalist **156**:459–477.

Edwards, K. F. 2019. Mixotrophy in nanoflagellates across environmental gradients in the ocean. Proceedings of the National Academy of Sciences, USA **116**:6211–6220.

Ehret, A. E., M. O. Westendorp, I. Herr, K. M. Debatin, J. L. Heeney, R. Frank, and P. H. Krammer. 1996. Resistance of chimpanzee T cells to human immunodeficiency virus type 1 Tat-enhanced oxidative stress and apoptosis. Journal of Virology **70**:6502–6507.

Ellner, S. 1989. Convergence of stationary distributions in two-species stochastic competition models. Journal of Mathematical Biology **27**:451–462.

Ellner, S. P., R. E. Snyder, and P. B. Adler. 2016. How to quantify the temporal storage effect using simulations instead of math. Ecology Letters **19**:1333–1342.

Elser, J. J., M. E. Bracken, E. E. Cleland, D. S. Gruner, W. S. Harpole, H. Hillebrand, J. T. Ngai, E. W. Seabloom, J. B. Shurin, and J. E. Smith. 2007. Global analysis of nitrogen and phosphorus limitation of primary producers in freshwater, marine and terrestrial ecosystems. Ecology Letters **10**:1135–1142.

Elser, J. J., and S. R. Carpenter. 1988. Predation-driven dynamics of zooplankton and phytoplankton communities in a whole-lake experiment. Oecologia **76**:148–154.

Elser, J. J., W. F. Fagan, R. F. Denno, D. R. Dobberfuhl, A. Flolarin, A. Huberty, S. Interlandl, et al. 2000. Nutritional constraints in terrestrial and freshwater food webs. Nature **408**:578–580.

Elton, C. S. 1927. *Animal Ecology*. MacMillan, New York.

Elton, C. S. 1942. *Voles, Mice and Lemmings*. Clarendon Press, Oxford, U.K.

Emery, N. C., P. J. Ewanchuk, and M. D. Bertness. 2001. Competition and salt-marsh plant zonation: Stress tolerators may be dominant competitors. Ecology **82**:2471–2485.

Eskola, H. T., and K. Parvinen. 2007. On the mechanistic underpinning of discrete-time population models with Allee effect. Theoretical Population Biology **72**:41–51.

Esposti, M. D., and E. M. Romero. 2017. The functional microbiome of arthropods. Plos One **12**:e0176573.

Estes, J. A., D. O. Duggins, and G. B. Rathbun. 1989. The ecology of extinctions in kelp forest communities. Conservation Biology **3**:252–264.

Estes, J. A., and J. F. Palmisano. 1974. Sea otters: Their role in structuring nearshore communities. Science **185**:1058–1060.

Estes, J. A., M. T. Tinker, T. M. Williams, and D. F. Doak. 1998. Killer whale predation on sea otters linking oceanic and nearshore ecosystems. Science **282**:473–476.

Etienne, R. S., and H. Olff. 2005. Confronting different models of community structure to species-abundance data: A Bayesian model comparison. Ecology Letters 8:493–504.

Fager, E. W., and J. A. McGowan. 1963. Zooplankton species groups in the north Pacific. Science 140:453–460.

Ferrari, F. D., and H. Ueda. 2005. Development of leg 5 of copepods belonging to the calanoid superfamily Centropagoidea (Crustacea). Journal of Crustacean Biology 25:333–352.

Foster, B. L., and K. L. Gross. 1998. Species richness in a successional grassland: Effects of nitrogen enrichment and plant litter. Ecology 79:2593–2602.

Fowler, M. S. 2009. Density dependent dispersal decisions and the Allee effect. Oikos 118:604–614.

Fox, J. W. 2013. The intermediate disturbance hypothesis should be abandoned. Trends in Ecology and Evolution 28:86–92.

Fox, J. W., and D. A. Vasseur. 2008. Character convergence under competition for nutritionally essential resources. American Naturalist 172:667–680.

Fox, L. R. 1975. Cannibalism in natural populations. Annual Reviews of Ecology and Systematics 6: 87–106.

Fretwell, S. D., and H. L. Lucas, Jr. 1969. On territorial behavior and other factors influencing habitat distribution in birds. Acta Biotheoretica 19:45–52.

Fréville, H., and J. Silvertown. 2005. Analysis of interspecific competition in perennial plants using life table response experiments. Plant Ecology 176:69–78.

Fryxell, J. M., J. B. Falls, E. A. Falls, R. J. Brooks, L. Dix, and M. A. Strickland. 1999. Density dependence, prey dependence, and population dynamics of martins in Ontario. Ecology 80:1311–1321.

Funk, W. C., M. Caminer, and S. R. Ron. 2011. High levels of cryptic species diversity uncovered in Amazonian frogs. Proceedings of the Royal Society B-Biological Sciences 279:1806–1814.

Fussmann, G. F., and B. Blasius. 2005. Community response to enrichment is highly sensitive to model structure. Biology Letters 1:9–12.

Gabriel, J. R., F. Saucy, and L. F. Bersier. 2005. Paradoxes in the logistic equation? Ecological Modelling 185:147–151.

Gandon, S. 2004. Evolution of multihost parasites. Evolution 58:455–469.

Gaston, K. J., and T. M. Blackburn. 2000. Pattern and process in macroecology. Blackwell, Malden, MA.

Gatto, M. 1991. Some remarks on models of plankton densities in lakes. American Naturalist 137:264–267.

Gause, G. F. 1932. Experimental studies on the struggle for existence. Journal of Experimental Biology 9:389–402.

Gause, G. F. 1934. *The Struggle for Existence*. Williams and Wilkins, Baltimore, MD.

Gause, G. F., and A. A. Witt. 1935. Behavior of mixed populations and the problem of natural selection. American Naturalist 69:596–609.

Gellner, G., and K. McCann. 2012. Reconciling the omnivory-stability debate. American Naturalist 179:22–37.

Germain, R. M., J. T. Weir, and B. Gilbert. 2016. Species coexistence: Macroevolutionary relationships and the contingency of historical interactions. Proceedings of the Royal Society B-Biological Sciences 283:20160047.

Gilpin, M. E. 1975. *Group Selection in Predator-Prey Communities*. Princeton University Press, Princeton, NJ.

Gilpin, M. E., and F. J. Ayala. 1973. Global models of growth and competition. Proceedings of the National Academy of Sciences, USA 70:3590–3593.

Gilpin, M. E., and M. L. Rosenzweig. 1972. Enriched predator-prey systems: Theoretical stability. Science 177:902–904.

Gleason, H. A. 1926. The individualistic concept of the plant association. Bulletin of the Torrey Botanical Club 53:7–26.

Gleeson, S. K. 1994. Density dependence is better than ratio dependence. Ecology 75:1834–1835.

Glennemeier, K. A., and R. J. Denver. 2002. Role for corticoids in mediating the response of *Rana pipiens* tadpoles to intraspecific competition. Journal of Expermental Zoology 292:32–40.

Gliwicz, Z. M. 1975. Effect of zooplankton grazing on photosynthetic activity and composition of phytoplankton. SIL Proceedings, 1922–2010 19:1490–1497.

Gliwicz, Z. M., and D. Wrzosek. 2008. Predation-mediated coexistence of large- and small-bodied *Daphnia* at different food levels. American Naturalist 172:358–374.

Glynn, P. W. 1993. Coral reef bleaching: Ecological perspectives. Coral Reefs 12:1–17.

Godoy, O., N. J. B. Kraft, and J. M. Levine. 2014. Phylogenetic relatedness and the determinants of competitive outcomes. Ecology Letters 17:836–844.

Godoy, O., and J. M. Levine. 2014. Phenology effects on invasion success: Insights from coupling experiments to coexistence theory. Ecology 95:726–736.

Goh, B. S. 1977. Global stability in many-species systems. American Naturalist 111:135–143.

Goh, B. S. 1979. Stability in models of mutualism. American Naturalist 113:261–275.

Goldberg, D. E., and T. E. Miller. 1990. Effects of different resource additions on species diversity in an annual plant community. Ecology 71:213–225.

Goldberg, D. E., R. Turkington, L. O. Whittaker, and A. R. Dyer. 2001. Density dependence in an annual plant community: Variation among life history stages. Ecological Monographs 71:423–446.

Grace, J. B. 2006. *Structural Equation Modeling and Natural Systems*. Cambridge University Press, New York.

Grace, J. B., and R. G. Wetzel. 1981. Habitat partitioning and competitive displacement in cattails (*Typha*): Experimental field studies. American Naturalist 118:463–474.

Grant, B. R., and P. R. Grant. 1982. Niche shifts and competition in Darwin's finches: *Geospiza conirostris* and congeners. Evolution 36:637–657.

Grant, P. R. 1966. Ecological compatibility of bird species on islands. American Naturalist 100:451–462.

Grant, P. R., and I. Abbott. 1980. Interspecific competition, island biogeography and null hypotheses. Evolution 34:332–341.

Grant, P. R., and B. R. Grant. 2006. Evolution of character displacement in Darwin's finches. Science 313:224–226.

Grant, P. R., and B. R. Grant. 2014. *40 Years of Evolution: Darwin's Finches on Daphne Major Island*. Princeton University Press, Princeton, NJ.

Gray, M. W., G. Burger, and B. F. Lang. 1999. Mitochondrial evolution. Science 283:1476–1481.

Grether, G. F., N. Losin, C. N. Anderson, and K. Okamoto. 2009. The role of interspecific interference competition in character displacement and the evolution of competitor recognition. Biological Reviews 84:617–635.

Grime, J. P. 1977. Evidence for the existence of three primary strategies in plants and its relevance to ecological and evolutionary theory. American Naturalist **111**:1169–1194.

Grinnell, J. 1917. The niche-relationships of the California thrasher. Auk **34**:427–433.

Grover, J. P. 1994. Assembly rules for communities of nutrient-limited plants and specialist herbivores. American Naturalist **143**:258–282.

Grover, J. P. 1997. Resource competition. Chapman & Hall, New York.

Grover, J. P., and T. H. Chrzanowski. 2004. Limiting resources, disturbance, and diversity in phytoplankton communities. Ecological Monographs **74**:533–551.

Grover, J. P., and R. D. Holt. 1998. Disentangling resource and apparent competition: Realistic models for plant-herbivore communities. Journal of Theoretical Biology **191**:353–376.

Gurevitch, J., J. A. Morrison, and L. V. Hedges. 2000. The interaction between competition and predation: A meta-analysis of field experiments. American Naturalist **155**:435–453.

Gurevitch, J., L. L. Morrow, A. Wallace, and J. S. Walsh. 1992. A meta-analysis of competition in field experiments. American Naturalist **140**:539–572.

Hairston, N. G. 1951. Interspecific competition and its probable influence upon the vertical distribution of Appalachian salamanders in the genus *Plethodon*. Ecology **32**:266–274.

Hairston, N. G. 1980. The experimental test of an analysis of field distributions: Competition in terrestrial salamanders. Ecology **61**:817–826.

Hairston, N. G., F. E. Smith, and L. B. Slobodkin. 1960. Community structure, population control, and competition. American Naturalist **94**:421–425.

Hairston, N. G., D. W. Tinkle, and H. M. Wilbur. 1970. Natural selection and the parameters of population growth. Journal of Wildlife Management **34**:681–690.

Hairston, N. G., Sr. 1986. Species packing in *Desmognathus* salamanders: Experimental demonstration of predation and competition. American Naturalist **127**:266–291.

Hall, S. R., M. A. Duffy, and C. E. Cáceres. 2005. Selective predation and productivity jointly drive complex behavior in host-parasite systems. American Naturalist **165**:70–81.

Hanski, I. 1981. Coexistence of competitors in patchy environment with and without predation. Oikos **37**:306–312.

Hanski, I. 1983. Competition of competitors in patchy environment. Ecology **64**:493–500.

Hanski, I. 1998. Metapopulation dynamics. Nature **396**:41–49.

Hanski, I. 1999. *Metapopulation Ecology*. Oxford University Press, Oxford, UK.

Harms, K. E., S. J. Wright, O. Calderón, A. Hernández, and E. A. Herre. 2000. Pervasive density-dependent recruitment enhances seedling diversity in a tropical forest. Nature **404**:493–495.

Harper, J. L. 1969. The role of predation in vegetational diversity. Brookhaven Symposium in Biology **22**:48–62.

Harpole, W. S., J. T. Ngai, E. E. Cleland, E. W. Seabloom, E. T. Borer, M. E. Bracken, J. J. Elser, et al. 2011. Nutrient co-limitation of primary producer communities. Ecology Letters **14**:852–862.

Harpole, W. S., and K. N. Suding. 2007. Frequency-dependence stabilizes competitive interactions among four annual plants. Ecology Letters **10**:1164–1169.

Harpole, W. S., and D. Tilman. 2007. Grassland species loss resulting from reduced niche dimension. Nature **446**:791–793.

Hartnett, D. C., and G. W. T. Wilson. 1999. Mycorrhizae influence plant community structure and diversity in tallgrass prairie. Ecology **80**:1187–1195.

Hassell, M. 1978. *The Dynamics of Arthropod Predator-Prey Systems*. Princeton University Press, Princeton, NJ.

Hastings, A. 1977. Spatial heterogeneity and the stability of predator-prey systems. Theoretical Population Biology **12**:37–48.

Hastings, A. 1978. Spatial heterogeneity and the stability of predator-prey systems: Predator-mediated coexistence. Theoretical Population Biology **14**:380–395.

Hastings, A. 1980. Disturbance, coexistence, history, and competition for space. Theoretical Population Biology **18**:363–373.

Hastings, A. 1982. Dynamics of a single species in a spatially varying habitat: The stabilizing role of high dispersal rates. Journal of Mathematical Biology **16**:49–55.

Hastings, A., K. C. Abbott, K. Cuddington, T. Francis, G. Gellner, Y. C. Lai, A. Morozov, S. Petrovskii, K. Scranton, and M. L. Zeeman. 2018. Transient phenomena in ecology. Science **361**.

Hastings, A., and T. Powell. 1991. Chaos in a three-species food chain. Ecology **72**:896–903.

Hautier, Y., P. A. Niklaus, and A. Hector. 2009. Competition for light causes plant biodiversity loss after eutrophication. Science **324**:636–638.

Hauzy, C., M. Gauduchon, F. D. Hulot, and M. Loreau. 2010. Density-dependent dispersal and relative dispersal affect the stability of predator-prey metacommunities. Journal of Theoretical Biology **266**:458–469.

Hawkins, B. A., and S. G. Compton. 1992. African fig wasp communities: Undersaturation and latitudinal gradients in species richness. Journal of Animal Ecology **61**:361–372.

Hay, M. E. 1986. Associational plant defenses and the maintenance of species diversity: Turning competitors into accomplices. American Naturalist **128**:617–641.

Haygood, R. 2002. Coexistence in MacArthur-style consumer-resource models. Theoretical Population Biology **61**:215–223.

He, Q., and M. D. Bertness. 2014. Extreme stresses, niches, and positive species interactions along stress gradients. Ecology **95**:1437–1443.

Hemp, A. 2005. Continuum or zonation? Altitudinal gradients in the forest vegetation of Mt. Kilimanjaro. Plant Ecology **184**:27–42.

Hempson, T. N., N. A. J. Graham, M. A. MacNeil, D. H. Williamson, G. P. Jones, and G. R. Almany. 2017. Coral reef mesopredators switch prey, shortening food chains, in response to habitat degradation. Ecology and Evolution 7:2626–2635.

Henderson, J. 1927. *The Practical Value of Birds*. MacMillan, New York.

Henry, C. S. 1985. The proliferation of cryptic species in *Chrysoperla* green lacewings through song divergence. Florida Entomologist **68**:18–38.

Henry, C. S., M. L. M. Wells, and C. M. Simon. 1999. Convergent evolution of courtship songs among cryptic species of the Carnea group of green lacewings (Neuroptera: Chrysopidae: Chrysoperla). Evolution **53**:1165–1179.

Hernán, M. A., J. Hsu, and B. Healy. 2019. A second chance to get causal inference right: A classification of data science tasks. Chance **32**:42–49.

Herre, E., N. Knowlton, U. Mueller, and S. Rehner. 1999. The evolution of mutualisms: Exploring the paths between conflict and cooperation. Trends in Ecology and Evolution **14**:49–53.

Hesaaraki, M., and S. M. Moghadas. 2001. Existence of limit cycles for predator-prey systems with a class of functional responses. Ecological Modelling **142**:1–9.

Hespenheide, H. A. 1973. Ecological inferences from morphological data. Annual Review of Ecology and Systematics **4**:213–229.

Hillebrand, H., D. S. Gruner, E. T. Borer, M. E. Bracken, E. E. Cleland, J. J. Elser, W. S. Harpole, J. T. Ngai, E. W. Seabloom, J. B. Shurin, and J. E. Smith. 2007. Consumer versus resource control of producer diversity depends on ecosystem type and producer community structure. Proceedings of the National Academy of Sciences, USA **104**:10904–10909.

HilleRisLambers, J., P. B. Adler, W. S. Harpole, J. M. Levine, and M. M. Mayfield. 2012. Rethinking community assembly through the lens of coexistence theory. Annual Review of Ecology, Evolution, and Systematics **43**:227–248.

HilleRisLambers, J., J. S. Clark, and B. Beckage. 2002. Density-dependent mortality and the latitudinal gradient in species diversity. Nature **417**:732–735.

Hirsch, M. W., S. Smale, and R. L. Devaney. 2012. *Differential Equations, Dynamical Systems, and an Introduction to Chaos*. 3rd edition. Academic Press, New York.

Hochberg, M. E., R. T. Clarke, G. W. Elmes, and J. A. Thomas. 1994. Population dynamic consequences of direct and indirect interactions involving a large blue butterfly and its plant and red ant hosts. Journal of Animal Ecology **63**:375–391.

Hochberg, M. E., and R. D. Holt. 1990. The coexistence of competing parasites. I. The role of cross-species infection. American Naturalist **136**:517–541.

Holland, J. N., and D. L. DeAngelis. 2009. Consumer-resource theory predicts dynamic transitions between outcomes of interspecific interactions. Ecology Letters **12**:1357–1366.

Holland, J. N., and D. L. DeAngelis. 2010. A consumer-resource approach to the density-dependent population dynamics of mutualism. Ecology **91**:1286–1295.

Holland, J. N., D. L. DeAngelis, and J. L. Bronstein. 2002. Population dynamics and mutualism: Functional responses of benefits and costs. American Naturalist **159**:231–244.

Holland, J. N., J. H. Ness, A. Boyle, and J. L. Bronstein. 2005. Mutualisms as consumer-resource interactions. Pages 17–33 *in* P. Barbosa, editor, *Ecology of Predator-Prey Interactions*. Oxford University Press, New York.

Holling, C. S. 1959a. The components of predation as revealed by a study of small mammal predation of the European pine sawfly. Canadian Entomologist **91**:209–223.

Holling, C. S. 1959b. Some characteristics of simple types of predation and parasitism. Canadian Entomologist **91**:385–398.

Holling, C. S. 1966. The functional response of invertebrate predators to prey density. Memoirs of the Entomological Society of Canada **48**:1–86.

Holt, R. D. 1977. Predation, apparent competition, and structure of prey communities. Theoretical Population Biology **12**:197–229.

Holt, R. D. 1984. Spatial heterogeneity, indirect interactions, and the coexistence of prey species. American Naturalist **124**:377–406.

Holt, R. D. 1985. Population dynamics in two-patch environments: Some anomalous consequences of an optimal habitat distribution. Theoretical Population Biology **28**:181–208.

Holt, R. D. 1997. Community modules. Pages 333–349 *in* A. C. Gange and V. K. Brown, editors, *Multitrophic Interactions in Terrestrial Ecosystems*. Blackwell Science, London.

Holt, R. D., and M. B. Bonsall. 2017. Apparent competition. Annual Review of Ecology, Evolution, and Systematics **48**:447–471.

Holt, R. D., and A. P. Dobson. 2006. Extending the principles of community ecology to address the epidemiology of host-pathogen systems. Pages 6–27 *in* S. K. Collinge and C. Ray, editors, *Disease Ecology: Community Structure and Pathogen Dynamics*. Oxford University Press, Oxford, UK.

Holt, R. D., J. Grover, and D. Tilman. 1994. Simple rules for interspecific dominance in systems with exploitative and apparent competition. American Naturalist **144**:741–771.

Holt, R. D., and J. Pickering. 1985. Infectious disease and species coexistence: A model of Lotka-Volterra form. American Naturalist **126**:196–211.

Holt, R. D., and G. A. Polis. 1997. A theoretical framework for intraguild predation. American Naturalist **149**:745–764.

Holt, R. D., and M. Roy. 2007. Predation can increase the prevalence of infectious disease. American Naturalist **169**:690–699.

Holyoak, M., M. A. Leibold, and R. D. Holt, editors. 2005. *Metacommunities: Spatial Dynamics and Ecological Communities*. University of Chicago Press, Chicago.

Horn, H. S., and R. H. MacArthur. 1972. Competition among fugitive species in a harlequin environment. Ecology **53**:749–752.

Horn, H. S., and R. M. May. 1977. Limits to similarity among coexisting competitors. Nature **270**:660–661.

Hsu, S.-B. 1980. A competition model for a seasonally fluctuating nutrient. Journal of Mathematical Biology **9**:115–132.

Hsu, S.-B., K.-S. Cheng, and S. P. Hubbell. 1981. Exploitative competition of microorganisms for two complementary nutrients in continuous cultures. SIAM Journal of Applied Mathematics **41**:422–444.

Hsu, S.-B., S. P. Hubbell, and P. Waltman. 1977. A mathematical theory for single-nutrient competition in continuous cultures of micro-organisms. SIAM Journal of Applied Mathematics **32**:366–383.

Hsu, S.-B., S. P. Hubbell, and P. Waltman. 1978a. Competing predators. SIAM Journal of Applied Mathematics **35**:617–625.

Hsu, S.-B., S. P. Hubbell, and P. Waltman. 1978b. A contribution to the theory of competing predators. Ecological Monographs **48**:337–349.

Hsu, S.-B., S. Ruan, and T.-H. Yang. 2015. Analysis of three species Lotka–Volterra food web models with omnivory. Journal of Mathematical Analysis and Applications **426**:659–687.

Hubbell, S. P. 1979. Tree dispersion, abundance, and diversity in a tropical dry forest. Science **203**:1299–1309.

Hubbell, S. P. 2001. *The Unified Neutral Theory of Biodiversity and Biogeography*. Princeton University Press, Princeton, NJ.

Hubbell, S. P. 2005. Neutral theory in community ecology and the hypothesis of functional equivalence. Functional Ecology **19**:166–172.

Hubbell, S. P., and R. B. Foster. 1986. Biology, chance, and history and the structure of tropical rain forest tree communities. Pages 314–329 *in* J. Diamond and T. J. Case, editors, *Community Ecology*. Harper & Row, New York.

Hughes, T. P. 1994. Catastrophes, phase shifts and large-scale degradation of a Caribbean coral reef. Science **265**:1547–1551.

Huisman, G., and R. J. De Boer. 1997. A formal derivation of the "Beddington" functional response. Journal of Theoretical Biology **185**:389–400.

Huisman, J., A. M. Johansson, E. O. Folmer, and F. J. Weissing. 2001. Towards a solution of the plankton paradox: The importance of physiology and life history. Ecology Letters **4**:408–411.

Huisman, J., and F. J. Weissing. 1999. Biodiversity of plankton by species oscillations and chaos. Nature **402**:407–410.

Huisman, J., and F. J. Weissing. 2001a. Biological conditions for oscillations and chaos generated by multispecies competition. Ecology **82**:2682–2695.

Huisman, J., and F. J. Weissing. 2001b. Fundamental unpredictability in multispecies competition. American Naturalist **157**:488–494.

Huisman, J., and F. J. Weissing. 2002. Oscillations and chaos generated by competition for interactively essential resources. Ecological Research **17**:175–181.

Hunter, M. D. 2016. *The Phytochemical Landscape: Linking Trophic Interactions and Nutrient Dynamics*. Princeton University Press, Princeton, NJ.

Hunter, M. D., and P. W. Price. 1992. Playing chutes and ladders: Heterogeneity and the relative roles of bottom-up and top-down forces in natural communities. Ecology **73**:724–732.

Huntley, J. W., and M. Kowalewski. 2007. Strong coupling of predation intensity and diversity in the Phanerozoic fossil record. Proceedings of the National Academy of Sciences, USA **104**:15006–15010.

Huston, M. 1979. A general hypothesis of species diversity. American Naturalist **113**:81–101.

Hutchinson, G. E. 1951. Copepodology for the ornithologist. Ecology **32**:571–577.

Hutchinson, G. E. 1958. Concluding remarks. Cold Spring Harbor Symposium of Quantitative Biology **22**:415–427.

Hutchinson, G. E. 1959. Homage to Santa Rosalia or why are there so many animals? American Naturalist **93**:145–159.

Hutchinson, G. E. 1961. The paradox of the plankton. American Naturalist **95**:137–145.

Hutchinson, G. E., and R. H. MacArthur. 1959a. Appendix on the theoretical significance of aggressive neglect in interspecific competition. American Naturalist **93**:133–134.

Hutchinson, G. E., and R. H. MacArthur. 1959b. A theoretical ecological model of size distributions among species of animals. American Naturalist **93**:117–125.

Hutson, V., and R. Law. 1985. Permanent coexistence in general models of three interacting species. Journal of Mathematical Biology **21**:285–298.

Hutson, V., and K. Schmitt. 1992. Permanence and the dynamics of biological systems. Mathematical Biosciences **111**:1–71.

Imbens, G. W., and D. B. Rubin. 2015. *Causal Inference for Statistics, Social, and Biomedical Sciences: An Introduction*. Cambridge University Press, New York.

Inderjit, and S. O. Duke. 2003. Ecophysiological aspects of allelopathy. Planta **217**:529–539.

Inouye, R. S. 1980. Stabilization of a predator-prey equilibrium by the addition of a second "keystone" victim. American Naturalist **115**:300–305.

Irigoien, X., J. Huisman, and R. P. Harris. 2004. Global biodiversity patterns of marine phytoplankton and zooplankton. Nature **429**:863–867.

Ivlev, V. S. 1955. *Experimental Ecology of the Feeding of Fishes* (English translation by D. Scott 1961). Yale University Press, New Haven, CT.

Jackson, J. B. C., M. X. Kirby, W. H. Berger, K. A. Bjorndal, L. W. Botsford, B. J. Bourque, R. H. Bradbury, et al. 2001. Historical overfishing and the recent collapse of coastal ecosystems. Science **293**:629–637.

Jaeger, R. G. 1970. Potential extinction through competition between two species of terrestrial salamanders. Evolution **24**:632–642.

Jaeger, R. G. 1971. Competitive exclusion as a factor influencing the distributions of two species of terrestrial salamanders. Ecology **52**:632–637.

James, F. C. 1971. Ordinations of habitat relationships among breeding birds. Wilson Bulletin **83**:215–236.

James, M. R., I. Hawes, M. Weatherhead, C. Stanger, and M. Gibbs. 2000. Carbon flow in the littoral food web of an oligotrophic lake. Hydrobiologica **441**:93–106.

James, W. 1991. *Pragmatism*. Prometheus Books, Amherst, NY. (Original work published by Longmans, Green, New York, 1907.)

Janzen, D. H. 1970. Herbivores and the number of tree species in tropical forests. American Naturalist **104**:501–528.

Janzen, D. H., J. M. Burns, Q. Cong, W. Hallwachs, T. Dapkey, R. Manjunath, M. Hajibabaei, P. D. N. Hebert, and N. V. Grishin. 2017. Nuclear genomes distinguish cryptic species suggested by their DNA barcodes and ecology. Proceedings of the National Academy of Sciences, USA **114**:8313–8318.

Jassby, A. D., and T. Platt. 1976. Mathematical formulation of the relationship between photosynthesis and light for phytoplankton. Limnology and Oceanography **21**:540–547.

Jeppesen, E., J. P. Jensen, M. Søndergaard, T. Lauridsen, and F. Landkildehus. 2000. Trophic structure, species richness and biodiversity in Danish lakes: Changes along a phosphorus gradient. Freshwater Biology **45**:201–218.

Jeschke, J. M., M. Kopp, and R. Tollrian. 2002. Predator functional responses: Discriminating between handling and digesting prey. Ecological Monographs **72**:95–112.

Johnson, C. A., and P. Amarasekare. 2013. Competition for benefits can promote the persistence of mutualistic interactions. Journal of Theoretical Biology **328**:54–64.

Johnson, D. M., and P. H. Crowley. 1980. Habitat and seasonal segregation among coexisting odonate larvae. Odonatologica **9**:297–308.

Jolliffe, P. A. 2000. The replacement series. Journal of Ecology **88**:371–385.

Jones, E. I., J. L. Bronstein, and R. Ferrière. 2012. The fundamental role of competition in the ecology and evolution of mutualisms. Annals of the New York Academy of Sciences **1256**:66–88.

Jordano, P., J. Bascompte, and J. M. Olesen. 2003. Invariant properties in coevolutionary networks of plant-animal interactions. Ecology Letters **6**:69–81.

Kallimanis, A. S., W. E. Kunin, J. M. Halley, and S. P. Sgardelis. 2005. Metapopulation extinction risk under spatially autocorrelated disturbance. Conservation Biology **19**:534–546.

Kane, K., J. S. Sedinger, D. Gibson, E. Blomberg, and M. Atamian. 2017. Fitness landscapes and life-table response experiments predict the importance of local areas to population dynamics. Ecosphere **8**:e01869.

Karban, R., D. Hougeneitzmann, and G. Englishloeb. 1994. Predator-mediated apparent competition between two herbivores that feed on grapevines. Oecologia **97**:508–511.

Karlson, R. H., and H. V. Cornell. 1998. Scale-dependent variation in local vs. regional effects on coral species richness. Ecological Monographs **68**:259–274.

Karlson, R. H., and H. V. Cornell. 2002. Species richness of coral assemblages: Detecting regional influences at local spatial scales. Ecology **83**:452–463.

Karlson, R. H., H. V. Cornell, and T. P. Hughes. 2004. Coral communities are regionally enriched along an oceanic biodiversity gradient. Nature **429**:867–870.

Kartzinel, T. R., P. A. Chen, T. C. Coverdale, D. L. Erickson, W. J. Kress, M. L. Kuzmina, D. I. Rubenstein, W. Wang, and R. M. Pringle. 2015. DNA metabarcoding illuminates dietary niche partitioning by African large herbivores. Proceedings of the National Academy of Sciences, USA **112**:8019–8024.

Kartzinel, T. R., J. C. Hsing, P. M. Musili, B. R. P. Brown, and R. M. Pringle. 2019. Covariation of diet and gut microbiome in African megafauna. Proceedings of the National Academy of Sciences, USA **116**:23588–23593.

Kausrud, K. L., A. Mysterud, H. Steen, J. O. Vik, E. Ostbye, B. Cazelles, E. Framstad, et al. 2008. Linking climate change to lemming cycles. Nature **456**:93–97.

Keeling, M. J., and P. Rohani. 2008. *Modeling Infectious Diseases in Humans and Animals*. Princeton University Press, Princeton, NJ.

Kelly, J. P., D. Straberg, K. Etienne, and M. McCaustland. 2008. Landscape influence on the quality of heron and egret colony sites. Wetlands **28**:257–275.

Kiers, E. T., M. Duhamel, Y. Beesetty, J. A. Menash, O. Franken, E. Verbruggen, C. R. Fellbaum, et al. 2011. Reciprocal rewards stabilize cooperation in the mycorrhizal symbiosis. Science **333**:880–882.

Kingsland, S. 1982. The refractory model: The logistic curve and the history of population ecology. Quarterly Review of Biology **57**:29–52.

Klapwijk, M. J., H. Bylund, M. Schroeder, and C. Björkman. 2016. Forest management and natural biocontrol of insect pests. Forestry **89**:253–262.

Klausmeier, C. A. 2010. Successional state dynamics: A novel approach to modeling non-equilibrium foodweb dynamics. Journal of Theoretical Biology **262**:584–595.

Kleinhesselink, A. R., and P. B. Adler. 2015. Indirect effects of environmental change in resource competition models. American Naturalist **186**:766–776.

Knight, T. M. 2003. Floral density, pollen limitation, and reproductive success in *Trillium grandiflorum*. Oecologia **137**:557–563.

Knowlton, N., and F. Rohwer. 2003. Multispecies microbial mutualisms on coral reefs: The host as a habitat. American Naturalist **162**:S51–S62.

Koch, A. L. 1974a. Coexistence resulting from an alternation of density dependent and density independent growth. Journal of Theoretical Biology **44**:373–386.

Koch, A. L. 1974b. Competitive coexistence of two predators utilizing the same prey under constant environmental conditions. Journal of Theoretical Biology **44**:387–395.

Kohl, M. T., D. R. Stahler, M. C. Metz, J. D. Forester, M. J. Kauffman, N. Varley, P. J. White, D. W. Smith, and D. R. MacNulty. 2018. Diel predator activity drives a dynamic landscape of fear. Ecological Monographs **88**:638–652.

Kohler, S. L., and M. A. McPeek. 1989. Predation risk and the foraging behavior of competing stream insects. Ecology **70**:1811–1825.

Korpimäki, E., P. R. Brown, J. Jacob, and R. P. Pech. 2004. The puzzles of population cycles and outbreaks of small mammals solved? Bioscience **54**:1071–1079.

Kraft, N. J. B., O. Godoy, and J. M. Levine. 2015. Plant functional traits and the multidimensional nature of species coexistence. Proceedings of the National Academy of Sciences, USA **112**:797–802.

Kramer, A. M., B. Dennis, A. M. Liebhold, and J. M. Drake. 2009. The evidence for Allee effects. Population Ecology **51**:341–354.

Krebs, C. J., R. Boonstra, and S. Boutin. 2018. Using experimentation to understand the 10-year snowshoe hare cycle in the boreal forest of North America. Journal of Animal Ecology **87**:87–100.

Krebs, C. J., K. Kielland, J. Bryant, M. O'Donoghue, F. Doyle, C. McIntyre, D. DiFolco, et al. 2013. Synchrony in the snowshoe hare (*Lepus americanus*) cycle in northwestern North America, 1970–2012. Canadian Journal of Zoology **91**:562–572.

Křivan, V. 1998. Effects of optimal antipredator behavior of prey on predator-prey dynamics: The role of refuges. Theoretical Population Biology **53**:131–142.

Křivan, V. 2000. Optimal intraguild foraging and population stability. Theoretical Population Biology **58**:79–94.

Křivan, V. 2003. Competitive co-existence caused by adaptive predators. Evolutionary Ecology Research **5**:1163–1182.

Křivan, V., R. Cressman, and C. Schneider. 2008. The ideal free distribution: A review and synthesis of the game-theoretic perspective. Theoretical Population Biology **73**:403–425.

Křivan, V., and J. Eisner. 2003. Optimal foraging and predator–prey dynamics III. Theoretical Population Biology **63**:269–279.

Křivan, V., and J. Eisner. 2006. The effect of the Holling type II functional response on apparent competition. Theoretical Population Biology **70**:421–430.

Křivan, V., and O. J. Schmitz. 2004. Trait and density mediated indirect interactions in simple food webs. Oikos **107**:239–250.

Křivan, V., and E. Sirot. 2002. Habitat selection by two competing species in a two-habitat environment. American Naturalist **160**:214–234.

Kuang, J. J., and P. Chesson. 2010. Interacting coexistence mechanisms in annual plant communities: Frequency-dependent predation and the storage effect. Theoretical Population Biology **77**:56–70.

Kula, A. A. R., D. C. Hartnett, and G. W. T. Wilson. 2004. Effects of mycorrhizal symbiosis on tallgrass prairie plant-herbivore interactions. Ecology Letters **8**:61–69.

Kuris, A. M., R. F. Hechinger, J. C. Shaw, K. L. Whitney, L. Aguirre-Macedo, C. A. Boch, A. P. Dobson, et al. 2008. Ecosystem energetic implications of parasite and free-living biomass in three estuaries. Nature **454**:515–518.

Lack, D. 1947. *Darwin's Finches*. Cambridge University Press, Cambridge, UK.

Lafferty, K. D., S. Allesina, M. Arim, C. J. Briggs, G. De Leo, A. P. Dobson, J. A. Dunne, et al. 2008. Parasites in food webs: The ultimate missing links. Ecology Letters **11**:533–546.

Lafferty, K. D., A. P. Dobson, and A. M. Kuris. 2006. Parasites dominate food web links. Proceedings of the National Academy of Sciences, USA **103**:11211–11216.

Lafferty, K. D., R. F. Hechinger, J. C. Shaw, K. Whitney, and A. M. Kuris. 2006. Food webs and parasites in a salt marsh ecosystem. Pages 119–134 *in* S. K. Collinge and C. Ray, editors, *Disease Ecology: Community Structure and Pathogen Dynamics*. Oxford University Press, New York.

LaManna, J. A., S. A. Mangan, A. Alonso, N. A. Bourg, W. Y. Brockelman, S. Bunyavejchewin, L.-W. Chang, et al. 2017. Plant diversity increases with the strength of negative density dependence at the global scale. Science **356**:1389–1392.

Lande, R. 1982. A quantitative genetic theory of life history evolution. Ecology **63**:607–615.

Lande, R. 1987. Extinction thresholds in demographic models of territorial populations. American Naturalist **130**:624–635.

Lande, R. 1993. Risks of population extinction from demographic and environmental stochasticity and random catastrophes. American Naturalist **142**:911–927.

Lande, R. 2007. Expected relative fitness and the adaptive topography of fluctuating selection. Evolution **61**:1835–1846.

Lande, R., S. Engen, and B.-E. Sæther. 2003. *Stochastic Population Dynamics in Ecology and Conservation*. Oxford University Press, New York.

Lande, R., S. Engen, and B.-E. Sæther. 2009. An evolutionary maximum principle for density-dependent population dynamics in a fluctuating environment. Philosophical Transactions of the Royal Society of London B-Biological Sciences **364**:1511–1518.

Langton, P. H., and J. J. Casas. 1999. Changes in chironomid assemblage composition in two Mediterranean mountain streams over a period of extreme hydrological conditions. Hydrobiologia **390**:37–49.

Law, R., and R. D. Morton. 1993. Alternative permanent states of ecological communities. Ecology **74**:1347–1361.

Law, R., and R. D. Morton. 1996. Permanence and the assembly of ecological communities. Ecology **77**:762–775.

Lawlor, L. R. 1979. Direct and indirect effects of n-species competition. Oecologia **43**:355–364.

Le Bourlot, V., T. Tully, and D. Claessen. 2014. Interference versus exploitative competition in the regulation of size-structured populations. American Naturalist **184**:609–623.

Lee, C. T., and B. D. Inouye. 2010. Mutualism between consumers and their shared resource can promote competitive coexistence. American Naturalist **175**:277–288.

Leibold, M. A. 1989. Resource edibility and the effects of predators and productivity on the outcome of trophic interactions. American Naturalist **134**:922–949.

Leibold, M. A. 1990. Resources and predators can affect the vertical distributions of zooplankton. Limnology and Oceanography **35**:938–944.

Leibold, M. A. 1991. Trophic interactions and habitat segregation between competing *Daphnia* species. Oecologia **86**:510–520.

Leibold, M. A. 1995. The niche concept revisited: Mechanistic models and community context. Ecology **76**:1371–1382.

Leibold, M. A. 1996. A graphical model of keystone predators in food webs: Trophic regulation of abundance, incidence and diversity patterns in communities. American Naturalist **147**:784–812.

Leibold, M. A. 1999. Biodiversity and nutrient enrichment in pond plankton communities. Evolutionary Ecology Research **1**:73–95.

Leibold, M. A., and J. M. Chase. 2017. *Metacommunities*. Princeton University Press, Princeton, NJ.

Leibold, M. A., S. R. Hall, V. H. Smith, and D. A. Lytle. 2017. Herbivory enhances the diversity of primary producers in pond ecosystems. Ecology **98**:48–56.

Leibold, M. A., and M. A. McPeek. 2006. Coexistence of the niche and neutral perspectives in community ecology. Ecology **87**:1399–1410.

Leirs, H., N. C. Stenseth, J. D. Nichols, J. E. Hines, R. Verhagen, and W. Verheyer. 1997. Stochastic seasonality and nonlinear density-dependent factors regulate population size in an African rodent. Nature **389**:176–180.

Leray, M., A. L. Alldredge, J. Y. Yang, C. P. Meyer, S. J. Holbrook, R. J. Schmitt, N. Knowlton, and A. J. Brooks. 2019. Dietary partitioning promotes the coexistence of planktivorous species on coral reefs. Molecular Ecology **28**:2694–2710.

Letten, A. D., P.-J. Ke, and T. Fukami. 2017. Linking modern coexistence theory and contemporary niche theory. Ecological Monographs **87**:161–177.

Levin, L., H. Caswell, T. Bridges, C. DiBacco, D. Cabrera, and G. Plaia. 1996. Demographic responses of estuarine polychaetes to pollutants: Life table response experiments. Ecological Applications **6**:1295–1313.

Levin, S. A. 1970. Community equilibria and stability, and the extension of the competitive exclusion principle. American Naturalist **104**:413–423.

Levine, J. M., P. B. Adler, and J. HilleRisLambers. 2008. On testing the role of niche differences in stabilizing coexistence. Functional Ecology **22**:934–936.

Levine, J. M., J. Bascompte, P. B. Adler, and S. Allesina. 2017. Beyond pairwise mechanisms of species coexistence in complex communities. Nature **546**:56–64.

Levine, J. M., and J. HilleRisLambers. 2009. The importance of niches for the maintenance of species diversity. Nature **461**:254–257.

Levine, S. H. 1976. Competitive interactions in ecosystems. American Naturalist **110**:903–910.

Levins, R. 1966. The strategy of model building in population biology. American Scientist **54**:421–431.

Levins, R. 1969. Some demographic and genetic consequences of environmental heterogeneity for biological control. Bulletin of the Entomological Society of America **15**:237–240.

Levins, R. 1979. Coexistence in a variable environment. American Naturalist **114**:765–783.

Levins, R., and D. Culver. 1971. Regional coexistence of species and competition between rare species. Proceedings of the National Academy of Science, USA **68**:1246–1248.

Lewontin, R. C., and D. Cohen. 1969. On population growth in a randomly varying environment. Proceedings of the National Academy of Science, USA **62**:1056–1060.

Li, L., and P. Chesson. 2016. The effects of dynamical rates on species coexistence in a variable environment: The paradox of the plankton revisited. American Naturalist **188**:E46–E58.

Lin, S., R. W. Litaker, and W. G. Sunda. 2016. Phosphorus physiological ecology and molecular mechanisms in marine phytoplankton. Journal of Phycology **52**:10–36.

Litchman, E., and C. A. Klausmeier. 2001. Competition of phytoplankton under fluctuating light. American Naturalist **157**:170–187.

Litchman, E., C. A. Klausmeier, and P. Bossard. 2004. Phytoplankton nutrient competition under dynamic light regimes. Limnology and Oceanography **49**:1457–1462.

Lochmiller, R. L. 1996. Immunocompetence and animal population regulation. Oikos **76**:594–602.

Loh, T. L., and J. R. Pawlik. 2014. Chemical defenses and resource trade-offs structure sponge communities on Caribbean coral reefs. Proceedings of the National Academy of Sciences, USA **111**:4151–4156.

López-Sepulcre, A., and H. Kokko. 2005. Territorial defense, territory size, and population regulation. American Naturalist **166**:317–329.

Losey, G. S., A. S. Grutter, G. Rosenquist, J. L. Mahon, and J. P. Zamzow. 1999. Cleaning symbiosis: A review. Pages 379–395 *in* V. C. Almada, R. F. Oliveira, and E. J. Gonçalves, editors, *Behaviour and Conservation of Littoral Fishes*. ISPA, Lisbon.

Losin, N., J. P. Drury, K. S. Peiman, C. Storch, and G. F. Grether. 2016. The ecological and evolutionary stability of interspecific territoriality. Ecology Letters **19**:260–267.

Lotka, A. J. 1912. Quantitative studies in epidemiology. Nature **88**:497–498.

Lotka, A. J. 1920. Analytical note on certain rhythmic relations in organic systems. Proceedings of the National Academy of Sciences, USA **6**:410–415.

Lotka, A. J. 1925. Notes on a meeting on the problem of forecasting city populations with special reference to New York City. Journal of the American Statistical Association **20**:569–573.

Lotka, A. J. 1932a. Contribution to the mathematical theory of capture. I. Conditions for capture. Proceedings of the National Academy of Sciences, USA **18**:172–178.

Lotka, A. J. 1932b. The growth of mixed populations: Two species competing for a common food supply. Journal of the Washington Academy of Sciences **22**: 461–469.

Lowe, E. F., and L. W. Keenan. 1997. Managing phosphorus-based, cultural eutrophication in wetlands: A conceptual approach. Ecological Engineering **9**:109–118.

Lubchenco, J. 1978. Plant species diversity in marine intertidal community: Importance of herbivore food preference and algal competitive abilities. American Naturalist **112**:23–39.

Lucas, F. A. 1893. The food of hummingbirds. Auk **10**:311–315.

Ludsin, S. A., M. W. Kershner, K. A. Blocksom, R. L. Knight, and R. A. Stein. 2001. Life after death in Lake Erie: Nutrient controls drive fish species richness, rehabilitation. Ecological Applications **11**:731–746.

MacArthur, R. 1969. Species packing, and what interspecies competition minimizes. Proceedings of the National Academy of Sciences, USA **64**:1369–1371.

MacArthur, R. 1970. Species packing and competitive equilibrium for many species. Theoretical Population Biology **1**:1–11.

MacArthur, R., and R. Levins. 1964. Competition, habitat selection, and character displacement in a patchy environment. Proceedings of the National Academy of Sciences, USA **51**:1207–1210.

MacArthur, R., and R. Levins. 1967. The limiting similarity, convergence, and divergence of coexisting species. American Naturalist **101**:377–385.

MacArthur, R. H. 1958. Population ecology of some warblers of northeastern coniferous forests. Ecology **39**:599–619.

MacArthur, R. H. 1972. *Geographical Ecology*. Princeton University Press, Princeton, NJ.

Magalhães, A. F. P., P. K. Maruyama, L. A. F. Tavares, and R. L. Martins. 2018. The relative importance of hummingbirds as pollinators in two bromeliads with contrasting floral specializations and breeding systems. Botanical Journal of the Linnean Society **188**:316–326.

Maiorana, V. C. 1978. An explanation of ecological and developmental constants. Nature **273**:375–377.

Manny, B. A., R. G. Wetzel, and W. C. Johnson. 1975. Annual contribution of carbon, nitrogen and phosphorus by migrant Canada geese to a hardwater lake. SIL Proceedings, 1922–2010 **19**:949–951.

Margulis, L. 1970. *Origin of Eukaryotic Cells*. Yale University Press, New Haven, CT.

Marra, P. P., K. Lampe, and B. Tedford. 1995. An analysis of daily corticosterone profiles in two species of *Zonotrichia* under captive and natural conditions. Wilson Bulletin 107:296–304.

Martinez, N. D. 1991. Artifacts or attributes? Effects of resolution on the Little Rock Lake food web. Ecological Monographs 61:367–392.

Martinez, N. D. 1992. Constant connectance in community food webs. American Naturalist 139:1208–1218.

Maskell, L. C., S. M. Smart, J. M. Bullock, K. E. N. Thompson, and C. J. Stevens. 2010. Nitrogen deposition causes widespread loss of species richness in British habitats. Global Change Biology 16:671–679.

Matías, L., O. Godoy, L. Gómez-Aparicio, I. M. Pérez-Ramos, and E. Allan. 2018. An experimental extreme drought reduces the likelihood of species to coexist despite increasing intransitivity in competitive networks. Journal of Ecology 106: 826–837.

Matthysen, E. 2005. Density-dependent dispersal in birds and mammals. Ecography 28:403–416.

May, R. M. 1972. Limit cycles in predator-prey communities. Science 177:900–902.

May, R. M. 1973. *Stability and Complexity in Model Ecosystems*. Princeton University Press, Princeton, NJ.

May, R. M. 1981. Models for two interacting populations. Pages 78–104 *in* R. M. May, editor, *Theoretical Ecology*. Sinauer Associates, Sunderland, MA.

May, R. M. 1992. How many species inhabit the Earth? Scientific American 267:42–49.

May, R. M., and R. H. MacArthur. 1972. Niche overlap as a function of environmental variability. Proceedings of the National Academy of Sciences, USA 69: 1109–1113.

Mayfield, M. M., and J. M. Levine. 2010. Opposing effects of competitive exclusion on the phylogenetic structure of communities. Ecology Letters 13:1085–1093.

Mayfield, M. M., and D. B. Stouffer. 2017. Higher-order interactions capture unexplained complexity in diverse communities. Nature Ecology & Evolution 1:62.

Maynard, D. S., Z. R. Miller, and S. Allesina. 2019. Predicting coexistence in experimental ecological communities. Nature Ecology and Evolution 4:91–100.

Maynard, D. S., J. T. Wootton, C. A. Servan, and S. Allesina. 2019. Reconciling empirical interactions and species coexistence. Ecology Letters 22:1028–1037.

McCallum, H., N. Barlow, and J. Hone. 2001. How should pathogen transmission be modelled? Trends in Ecology and Evolution 16:295–300.

McCann, K., and P. Yodzis. 1994. Biological conditions for chaos in a three-species food chain. Ecology 75:561–564.

McCann, K. S. 2011. *Food Webs*. Princeton University Press, Princeton, NJ.

McCauley, S. J. 2006. The effects of dispersal and recruitment limitation on community structure in artificial ponds. Ecography 29:585–595.

McCook, L., J. Jompa, and G. Diaz-Pulido. 2014. Competition between corals and algae on coral reefs: A review of evidence and mechanisms. Coral Reefs 19:400–417.

McGehee, R., and R. A. Armstrong. 1977. Some mathematical problems concerning the ecological principle of competitive exclusion. Journal of Differential Equations 23:30–52.

McGill, B. J., B. A. Maurer, and M. D. Weiser. 2006. Empirical evaluation of neutral theory. Ecology 87:1411–1423.

McIntyre, P. B., L. E. Jones, A. S. Flecker, and M. J. Vanni. 2007. Fish extinctions alter nutrient recycling in tropical freshwaters. Proceedings of the National Academy of Sciences, USA **104**:4461–4466.

McKew, B. A., G. Metodieva, C. A. Raines, M. V. Metodiev, and R. J. Geider. 2015. Acclimation of *Emiliania huxleyi* (1516) to nutrient limitation involves precise modification of the proteome to scavenge alternative sources of N and P. Environmental Microbiology **17**:4050–4062.

McLellan, B. N., R. Serrouya, H. U. Wittmer, and S. Boutin. 2010. Predator-mediated Allee effects in multi-prey systems. Ecology **91**:286–292.

McManus, J. W., L. A. B. Meñez, K. N. Kesner-Reyes, S. G. Vergara, and M. C. Ablan. 2000. Coral reef fishing and coral-algal phase shifts: Implications for global reef status. ICES Journal of Marine Science **57**:572–578.

McPeek, M. A. 1989. Differential dispersal tendencies among *Enallagma* damselflies (Odonata) inhabiting different habitats. Oikos **56**:187–195.

McPeek, M. A. 1990a. Behavioral differences between *Enallagma* species (Odonata) influencing differential vulnerability to predators. Ecology **71**:1714–1726.

McPeek, M. A. 1990b. Determination of species composition in the *Enallagma* damselfly assemblages of permanent lakes. Ecology **71**:83–98.

McPeek, M. A. 1995. Morphological evolution mediated by behavior in the damselflies of two communities. Evolution **49**:749–769.

McPeek, M. A. 1996. Trade-offs, food web structure, and the coexistence of habitat specialists and generalists. American Naturalist **148**:S124-S138.

McPeek, M. A. 1997. Measuring phenotypic selection on an adaptation: Lamellae of damselflies experiencing dragonfly predation. Evolution **51**:459–466.

McPeek, M. A. 1998. The consequences of changing the top predator in a food web: A comparative experimental approach. Ecological Monographs **68**:1–23.

McPeek, M. A. 1999. Biochemical evolution associated with antipredator adaptation in damselflies. Evolution **53**:1835–1845.

McPeek, M. A. 2000. Predisposed to adapt? Clade-level differences in characters affecting swimming performance in damselflies. Evolution **54**:2072–2080.

McPeek, M. A. 2004. The growth/predation risk trade-off: So what is the mechanism? American Naturalist **163**:E88–E111.

McPeek, M. A. 2007. The macroevolutionary consequences of ecological differences among species. Palaeontology **50**:111–129.

McPeek, M. A. 2008a. The ecological dynamics of clade diversification and community assembly. American Naturalist **172**:E270–E284.

McPeek, M. A. 2008b. Ecological factors limiting the distributions and abundances of Odonata. Pages 51–62 *in* A. Córdoba-Aguilar, editor, *Dragonflies and Damselflies: Model Organisms for Ecological and Evolutionary Research*. Oxford University Press, Oxford, UK.

McPeek, M. A. 2012. Intraspecific density dependence and a guild of consumers coexisting on one resource. Ecology **93**:2728–2735.

McPeek, M. A. 2014. Keystone and intraguild predation, intraspecific density dependence, and a guild of coexisting consumers. American Naturalist **183**:E1–E16.

McPeek, M. A. 2017a. The ecological dynamics of natural selection: Traits and the coevolution of community structure. American Naturalist **189**:E91–E117.

McPeek, M. A. 2017b. *Evolutionary Community Ecology*. Princeton University Press, Princeton, NJ.

McPeek, M. A. 2019a. Limiting similarity? The ecological dynamics of natural selection among resources and consumers caused by both apparent and resource competition. American Naturalist **89**:e01328.

McPeek, M. A. 2019b. Mechanisms influencing the coexistence of multiple consumers and multiple resources: Resource and apparent competition. Ecological Monographs **89**:e01328 01310.01002/ecm.01328.

McPeek, M. A., and J. M. Brown. 2000. Building a regional species pool: Diversification of the *Enallagma* damselflies in eastern North America. Ecology **81**:904–920.

McPeek, M. A., and J. M. Brown. 2007. Clade age and not diversification rate explains species richness among animal taxa. American Naturalist **169**:E97–E106.

McPeek, M. A., and P. H. Crowley. 1987. The effects of density and relative size on the aggressive behaviour, movement and feeding of damselfly larvae (Odonata: Coenagrionidae). Animal Behaviour **35**:1051–1061.

McPeek, M. A., and R. Gomulkiewicz. 2005. Assembling and depleting species richness in metacommunities: Insights from ecology, population genetics and macroevolution. Pages 355–373 *in* M. Holyoak, M. A. Leibold, and R. D. Holt, editors, *Metacommunities: Spatial Dynamics and Ecological Communities*. University of Chicago Press, Chicago.

McPeek, M. A., M. Grace, and J. M. L. Richardson. 2001. Physiological and behavioral responses to predators shape the growth/predation risk trade-off in damselflies. Ecology **82**:1535–1545.

McPeek, M. A., and R. D. Holt. 1992. The evolution of dispersal in spatially and temporally varying environments. American Naturalist **140**:1010–1027.

McPeek, M. A., and B. L. Peckarsky. 1998. Life histories and the strengths of species interactions: Combining mortality, growth, and fecundity effects. Ecology **79**:867–879.

McPeek, M. A., N. L. Rodenhouse, R. T. Holmes, and T. W. Sherry. 2001. A general model of site-dependent population regulation: Population-level regulation without individual-level interactions. Oikos **94**:417–424.

McPeek, M. A., L. Shen, J. Z. Torrey, and H. Farid. 2008. The tempo and mode of three-dimensional morphological evolution in male reproductive structures. American Naturalist **171**:E158–E178.

McPeek, M. A., and A. M. Siepielski. 2019. Disentangling ecologically equivalent from neutral species: The mechanisms of population regulation matter. Journal of Animal Ecology **88**:1755–1765.

McPeek, M. A., L. B. Symes, D. M. Zong, and C. L. McPeek. 2011. Species recognition and patterns of population variation in the reproductive structures of a damselfly genus. Evolution **65**:419–428.

McQueen, D. J., J. R. Post, and E. L. Mills. 1986. Trophic relationships in freshwater pelagic ecosystems. Canadian Journal of Fisheries and Aquatic Sciences **43**:1571–1581.

Melián, C. J., V. Křivan, F. Altermatt, P. Starý, L. Pellissier, and F. De Laender. 2015. Dispersal dynamics in food webs. American Naturalist **185**:157–168.

Menge, B. A. 1976. Organization of the New England rocky intertidal community: Role of predation, competition, and environmental heterogeneity. Ecological Monographs **46**:355–393.

Menge, B. A. 1992. Community regulation: Under what conditions are bottom-up factors important on rocky shores? Ecology **73**:755–765.

Menge, B. A. 1995. Indirect effects in marine rocky intertidal interaction webs: Patterns and importance. Ecological Monographs **65**:21–74.

Menge, B. A., and J. P. Sutherland. 1976. Species diversity gradients: Synthesis of the roles of predation, competition, and temporal heterogeneity. American Naturalist **110**:351–369.

M'Gonigle, L. K., R. Mazzucco, S. P. Otto, and U. Dieckmann. 2012. Sexual selection enables long-term coexistence despite ecological equivalence. Nature **484**:506–509.

Michaelis, L., and M. L. Menten. 1913. Die Kinetik der Invertinwirkung. Biochemische Zeitschrift **49**:333–369.

Miller, A. H. 1951. An analysis of the distribution of the birds of California. University of California Publications in Zoology **50**:531–644.

Mills, N., and W. M. Getz. 1996. Modelling the biological control of insect pests: A review of host-parasitoid models. Ecological Modelling **92**:121–143.

Mittelbach, G. G. 1981. Foraging efficiency and body size: A study of optimal diet and habitat use by bluegills. Ecology **62**:1370–1386.

Mittelbach, G. G., C. W. Osenberg, and M. A. Leibold. 1988. Trophic relations and ontogenetic niche shifts in aquatic ecosystems. Pages 219–235 *in* B. O. Ebenman and L. Persson, editors, *Size-Structured Populations: Ecology and Evolution.* Springer, Berlin.

Mittelbach, G. G., C. F. Steiner, S. M. Scheiner, K. L. Gross, H. L. Reynolds, R. B. Waide, M. R. Willig, S. I. Dodson, and L. Gough. 2001. What is the observed relationship between species richness and productivity? Ecology **82**:2381–2396.

Monod, J. 1949. The growth of bacterial cultures. Annual Review of Microbiology **3**:371–394.

Montoya, J. M., G. Woodward, M. C. Emmerson, and R. V. Solé. 2009. Press perturbations and indirect effects in real food webs. Ecology **90**:2426–2433.

Mora, C., D. P. Tittensor, S. Adl, A. G. Simpson, and B. Worm. 2011. How many species are there on Earth and in the ocean? PLoS Biology **9**:e1001127.

Morgan, S. L., and C. Winship. 2015. *Counterfactuals and Causal Inference: Methods and Principles for Social Research.* 2nd edition. Cambridge University Press, New York.

Moritz, M. A. 1997. Analyzing extreme disturbance events: Fire in Los Padres National Forest. Ecological Applications **7**:1252–1262.

Morris, W. F., J. L. Bronstein, and W. G. Wilson. 2003. Three-way coexistence in obligate mutualist-exploiter interactions: The potential role of competition. American Naturalist **161**:860–875.

Morton, R. D., and R. Law. 1997. Regional species pools and the assembly of local ecological communities. Journal of Theoretical Biology **187**:321–331.

Muko, S., and Y. Iwasa. 2000. Species coexistence by permanent spatial heterogeneity in a lottery model. Theoretical Population Biology **57**:273–284.

Muller, C. B., and H. C. J. Godfray. 1997. Apparent competition between two aphid species. Journal of Animal Ecology **66**:57–64.

Muller, C. B., and H. C. J. Godfray. 1999. Predators and mutualists influence the exclusion of aphid species from natural communities. Oecologia **119**:120–125.

Murdoch, W. W., C. J. Briggs, and R. M. Nisbet. 2003. *Consumer-Resource Dynamics.* Princeton University Press, Princeton, NJ.

Murdoch, W. W., and A. Oaten. 1975. Predation and population stability. Advances in Ecological Research **9**:2–132.

Murrell, D. J., and R. Law. 2003. Heteromyopia and the spatial coexistence of similar competitors. Ecology Letters **6**:48–59.

Muscatine, L., and J. W. Porter. 1977. Reef corals: Mutualistic symbioses adapted to nutrient-poor environments. Bioscience **27**:454–460.

Mutshinda, C. M., Z. V. Finkel, C. E. Widdicombe, A. J. Irwin, and N. Norden. 2016. Ecological equivalence of species within phytoplankton functional groups. Functional Ecology **30**:1714–1722.

Myers, J. H. 2018. Population cycles: Generalities, exceptions and remaining mysteries. Proceedings of the Royal Society B-Biological Sciences **285**:20172841.

Mylius, S. D., K. Klumpers, A. M. de Roos, and L. Persson. 2001. Impact of intraguild predation and stage structure on simple communities along a productivity gradient. American Naturalist **158**:259–276.

Narwani, A., M. A. Alexandrou, T. H. Oakley, I. T. Carroll, and B. J. Cardinale. 2013. Experimental evidence that evolutionary relatedness does not affect the ecological mechanisms of coexistence in freshwater green algae. Ecology Letters **16**:1373–1381.

Nee, S., and R. M. May. 1992. Dynamics of metapopulations: Habitat destructuction and competitive coexistence. Journal of Animal Ecology **61**:37–40.

Neill, W. E. 1974. The community matrix and interdependence of the competition coefficients. American Naturalist **108**:399–408.

Neill, W. E. 1975. Experimental studies of microcrustacean competition, community composition and efficiency of resource utilization. Ecology **56**:809–826.

Neubert, M. G., T. Klanjscek, and H. Caswell. 2004. Reactivity and transient dynamics of predator–prey and food web models. Ecological Modelling **179**:29–38.

Nicholson, A. J., and V. A. Bailey. 1935. The balance of animal populations. Proceedings of the Zoological Society of London **3**:551–598.

Noë, R., and H. Peter. 1994. Biological markets: Supply and demand determine the effect of partner choice in cooperation, mutualism and mating. Behavioral Ecology and Sociobiology **35**:1–11.

North, R. L., S. J. Guildford, R. E. H. Smith, S. M. Havens, and M. R. Twiss. 2007. Evidence for phosphorus, nitrogen, and iron colimitation of phytoplankton communities in Lake Erie. Limnology and Oceanography **52**:315–328.

O'Dea, A., and J. B. C. Jackson. 2009. Environmental change drove macroevolution in cupuladriid bryozoans. Proceedings of the Royal Society of London B-Biological Sciences **276**:3629–3634.

O'Dea, A., J. B. C. Jackson, H. Fortunato, J. T. Smith, L. D'Croz, K. G. Johnson, and J. A. Todd. 2007. Environmental change preceded Caribbean extinction by 2 million years. Proceedings of the National Academy of Sciences, USA **104**:5501–5506.

O'Dwyer, J. P. 2018. Whence Lotka-Volterra? Conservation laws and integrable systems in ecology. Theoretical Ecology **11**:441–452.

Oksanen, L., S. D. Fretwell, J. Arruda, and P. Niemela. 1981. Exploitation ecosystems in gradients of primary productivity. American Naturalist **118**:240–261.

Oksanen, T., M. E. Power, and L. Oksanen. 1995. Ideal free habitat selection and consumer-resource dynamics. American Naturalist **146**:565–585.

Olff, H., and M. E. Ritchie. 1998. Effects of herbivores on grassland plant diversity. Trends in Ecology and Evolution **13**:261–265.

Olson, M. H., M. M. Hage, M. D. Binkley, and J. R. Binder. 2005. Impact of migratory snow geese on nitrogen and phosphorus dynamics in a freshwater reservoir. Freshwater Biology **50**:882–890.

Orrock, J. L., M. L. Baskett, and R. D. Holt. 2010. Spatial interplay of plant competition and consumer foraging mediate plant coexistence and drive the invasion ratchet. Proceedings of the Royal Society B-Biological Sciences **277**:3307–3315.

Orrock, J. L., M. S. Witter, and O. J. Reichman. 2008. Apparent competition with an exotic plant reduces native plant establishment. Ecology **89**:1168–1174.

Packer, C., R. D. Holt, P. J. Hudson, K. D. Lafferty, and A. P. Dobson. 2003. Keeping the herds healthy and alert: Implications of predator control for infectious disease. Ecology Letters **6**:797–802.

Paine, R. T. 1966. Food web complexity and species diversity. American Naturalist **100**:65–75.

Paine, R. T. 1969. A note on trophic complexity and community stability. American Naturalist **103**:91–93.

Paine, R. T. 1974. Intertidal community structure: Experimental studies on the relationship between a dominant competitor and its principal prey. Oecologia **15**:93–120.

Paine, R. T. 1980. Food webs: Linkage, interaction strength and community infrastructure. Journal of Animal Ecology **49**:666–685.

Paine, R. T., and S. A. Levin. 1981. Intertidal landscapes: Disturbance and the dynamics of pattern. Ecological Monographs **51**:145–178.

Palmer, T. M., E. G. Pringle, A. C. Stier, and R. D. Holt. 2015. Mutualism in a community context. Pages 159–180 *in* J. L. Bronstein, editor, *Mutualism*. Oxford University Press, Oxford, UK.

Palmer, T. M., M. L. Stanton, and T. P. Young. 2003. Competition and coexistence: Exploring mechanisms that restrict and maintain diversity within mutualist guilds. American Naturalist **162**:S63–S79.

Palmer, T. M., M. L. Stanton, T. P. Young, J. R. Goheen, R. M. Pringle, and R. Karban. 2008. Breakdown of an ant-plant mutualism follows the loss of large herbivores from an African savanna. Science **319**:192–195.

Park, T. 2000. Taxonomy and distribution of the calanoid copepod family Heterorhobdidae. University of California Press, Berkeley.

Passarge, J., S. Hol, M. Escher, and J. Huisman. 2006. Competition for nutrients and light: Stable coexistence, alternative stable states, or competitive exclusion? Ecological Monographs **76**:57–72.

Pearl, J. 2000. *Causality: Models, Reasoning, and Inference*. Cambridge University Press, Cambridge, UK.

Pearl, J., and D. Mackenzie. 2018. *The Book of Why: The New Science of Cause and Effect*. Basic Books, New York.

Pearl, R., and L. J. Reed. 1920. On the rate of growth of the population of the United States since 1790 and its mathematical representation. Proceedings of the National Academy of Science, USA **6**:275–288.

Peirce, C. S. 1878. How to make our ideas clear. Popular Science Monthly **12**:286–302.

Persson, L. 1999. Trophic cascades: Abiding heterogeneity and the trophic level concept at the end of the road. Oikos **85**:385–397.

Persson, L., and A. M. De Roos. 2003. Adaptive habitat use in size-structured populations: Linking individual behavior to population processes. Ecology **84**:1129–1139.

Pfenninger, M., and K. Schwenk. 2007. Cryptic animal species are homogeneously distributed among taxa and biogeographical regions. BMC Evolutionary Biology 7:121.

Pianka, E. R. 1972. r and K selection or b and d selection? American Naturalist 106:581–588.

Pianka, E. R. 1974. Niche overlap and diffuse competition. Proceedings of the National Academy of Sciences, USA 71:2141–2145.

Pimm, S. L. 1982. *Food Webs*. Chapman and Hall, London.

Pimm, S. L., and J. H. Lawton. 1978. On feeding on more than one trophic level. Nature 275:542–544.

Pintar, M. R., J. R. Bohenek, L. L. Eveland, W. J. Resetarits, and C. Seymour. 2018. Colonization across gradients of risk and reward: Nutrients and predators generate species-specific responses among aquatic insects. Functional Ecology 32:1589–1598.

Platt, J. R. 1964. Strong inference. Science 146:347–353.

Platt, W. J. 1975. The colonization and formation of equilibrium plant species associations on badger disturbances in a tall-grass prairie. Ecological Monographs 45: 285–305.

Platt, W. J., B. Beckage, R. F. Doren, and H. H. Slater. 2002. Interactions of large-scale disturbances: Prior fire regimes and hurricane mortality of savanna pines. Ecology 83:1566–1572.

Platt, W. J., and J. H. Connell. 2003. Natural disturbance and directional replacement of species. Ecological Monographs 73:507–522.

Poethke, H. J., and T. Hovestadt. 2002. Evolution of density- and patch-size-dependent dispersal rates. Proceedings of the Royal Society B-Biological Sciences 269:637–645.

Polis, G. A. 1981. The evolution and dynamics of intraspecific predation. Annual Review of Ecology and Systematics 12:225–251.

Polis, G. A., and R. D. Holt. 1992. Intraguild predation: The dynamics of complex trophic interactions. Trends in Ecology and Evolution 7:151–154.

Polis, G. A., and S. D. Hurd. 1995. Extraordinarily high spider densities on islands: Flow of energy from the marine to terrestrial food webs and the absence of predation. Proceedings of the National Academy of Sciences, USA 92:4382–4386.

Polis, G. A., C. A. Myers, and R. D. Holt. 1989. The ecology and evolution of intraguild predation: Potential competitors that eat each other. Annual Review of Ecology and Systematics 20:297–330.

Polis, G. A., and D. R. Strong. 1996. Food web complexity and community dynamics. American Naturalist 147:813–846.

Poole, P., V. Ramachandran, and J. Terpolilli. 2018. Rhizobia: From saprophytes to endosymbionts. Nature Reviews Microbiology 16:291–303.

Popper, K. R. 1959. *The Logic of Scientific Discovery*. Basic Books, New York.

Proulx, M., and A. Maxumder. 1998. Reversal of grazing impact on plant species richness in nutrient-poor vs. nutrient-rich ecosystems. Ecology 79:2581–2592.

Pulliam, H. R. 1988. Sources, sinks, and population regulation. American Naturalist 132:652–661.

Pulliam, H. R., and B. J. Danielson. 1991. Sources, sinks, and habitat selection—a landscape perspective on population dynamics. American Naturalist 137:S50–S66.

Qian, J. J., and E. Akçay. 2020. The balance of interaction types determines the assembly and stability of ecological communities. Nature Ecology & Evolution 4:356–365.

Quinn, J. F., and A. E. Dunham. 1983. On hypothesis testing in ecology and evolution. American Naturalist 122:602–617.

Ramsey, M. W. 1988. Differences in pollinator effectiveness of birds and insects visiting *Banksia menziesii* (Proteaceae). Oecologia 76:119–124.

Rasher, D. B., S. Engel, V. Bonito, G. J. Fraser, J. P. Montoya, and M. E. Hay. 2012. Effects of herbivory, nutrients, and reef protection on algal proliferation and coral growth on a tropical reef. Oecologia 169:187–198.

Raup, D. M. 1992. *Extinction: Bad Genes or Bad Luck?* W. W. Norton, New York.

Reader, R. J. 1992. Herbivory as a confounding factor in an experiment measuring competition among plants. Ecology 73:373–376.

Resetarits, W. J. 2001. Colonization under threat of predation: Avoidance of fish by an aquatic beetle, *Tropisternus lateralis* (Coleoptera: Hydrophilidae). Oecologia 129:155–160.

Revilla, T. A. 2015. Numerical responses in resource-based mutualisms: A time scale approach. Journal of Theoretical Biology 378:39–46.

Reynolds, S. A., and C. E. Brassil. 2013. When can a single-species, density-dependent model capture the dynamics of a consumer-resource system? Journal of Theoretical Biology 339:70–83.

Rice, E. L. 1984. *Allelopathy.* Academic Press, New York.

Ricker, W. E. 1954. Stock and recruitment. Journal of the Fisheries Research Board of Canada 11:559–623.

Ricklefs, R. E. 1987. Community diversity: Relative roles of local and regional processes. Science 235:167–171.

Ricklefs, R. E. 1989. Speciation and diversity: The integration of local and regional processes. Pages 599–622 *in* D. Otte and J. A. Endler, editors, *Speciation and Its Consequences.* Sinauer Associates, Sunderland, MA.

Ricklefs, R. E. 2008. Disintegration of the ecological community. American Naturalist 172:741–750.

Ricklefs, R. E. 2010. Evolutionary diversification, coevolution between populations and their antagonists, and the filling of niche space. Proceedings of the National Academy of Sciences, USA 107:1265–1272.

Ripple, W. J., and R. L. Beschta. 2012. Trophic cascades in Yellowstone: The first 15 years after wolf reintroduction. Biological Conservation 145:205–213.

Ripple, W. J., R. L. Beschta, and L. E. Painter. 2015. Trophic cascades from wolves to alders in Yellowstone. Forest Ecology and Management 354:254–260.

Ristl, K., S. J. Plitzko, and B. Drossel. 2014. Complex response of a food-web module to symmetric and asymmetric migration between several patches. Journal of Theoretical Biology 354:54–59.

Robinson, S. K., and J. Terborgh. 1995. Interspecific aggression and habitat selection by Amazonian birds. Journal of Animal Ecology 64:1–11.

Roche, B., J. M. Drake, and P. Rohani. 2011. The curse of the Pharaoh revisited: Evolutionary bi-stability in environmentally transmitted pathogens. Ecology Letters 14:569–575.

Rodríguez, D. J. 1997. A method to study competition dynamics using de Wit replacement series experiments. Oikos 78:411–415.

Ronce, O. 2007. How does it feel to be like a rolling stone? Ten questions about dispersal evolution. Annual Review of Ecology, Evolution, and Systematics 38:231–253.

Root, R. B. 1967. The niche exploitation pattern of the blue-gray gnatcatcher. Ecological Monographs 37:317–350.

Rosenzweig, M. L. 1969. Why the prey curve has a hump. American Naturalist 103:81–87.

Rosenzweig, M. L. 1971. Paradox of enrichment: Destabilization of exploitation ecosystems in ecological time. Science 171:385–387.

Rosenzweig, M. L. 1973. Exploitation in three trophic levels. American Naturalist 107:275–294.

Rosenzweig, M. L., and R. H. MacArthur. 1963. Graphical representation and stability conditions of predator-prey interactions. American Naturalist 97:209–223.

Ross, R. 1911. Some quantitative studies in epidemiology. Nature 87:466–467.

Roth, V. L. 1981. Constancy in the size ratios of sympatric species. American Naturalist 119:394–404.

Roughgarden, J. 1974. Species packing and the competition function with illustrations from coral reef fish. Theoretical Population Biology 5:163–186.

Roughgarden, J. 1983. Competition and theory in community ecology. American Naturalist 122:583–601.

Rowan, R. 1998. Diversity and ecology of zooxanthellae on coral reefs. Journal of Phycology 34:407–417.

Roy, M., R. D. Holt, and M. Barfield. 2005. Temporal autocorrelation can enhance the persistence and abundance of metapopulations comprised of coupled sinks. American Naturalist 166:246–261.

Royama, T. 1971. A comparative study of models for predation and parasitism. Researches in Population Ecology 13:1–91.

Rudolf, V. H., and J. Antonovics. 2005. Species coexistence and pathogens with frequency-dependent transmission. American Naturalist 166:112–118.

Rudolf, V. H., and K. D. Lafferty. 2011. Stage structure alters how complexity affects stability of ecological networks. Ecology Letters 14:75–79.

Rudolf, V. H. W. 2007. The interaction of cannibalism and omnivory: Consequences for community dynamics. Ecology 88:2697–2705.

Ruelle, D., and F. Takens. 1971. On the nature of turbulence. Communications in Mathematical Physics 20:167–192.

Runge, J. P., M. C. Runge, and J. D. Nichols. 2006. The role of local populations within a landscape context: Defining and classifying sources and sinks. American Naturalist 167:925–938.

Ruokolainen, L., and I. Hanski. 2016. Stable coexistence of ecologically identical species: Conspecific aggregation via reproductive interference. Journal of Animal Ecology 85:638–647.

Ruse, L. P. 1995. Chironomid community structure deduced from larvae and pupal exuviae of a chalk stream. Hydrobiologia 315:135–142.

Saavedra, S., R. P. Rohr, J. Bascompte, O. Godoy, N. J. B. Kraft, and J. M. Levine. 2017. A structural approach for understanding multispecies coexistence. Ecological Monographs 87:470–486.

Saavedra, S., R. P. Rohr, J. M. Olesen, and J. Bascompte. 2016. Nested species interactions promote feasibility over stability during the assembly of a pollinator community. Ecology and Evolution 6:997–1007.

Sale, P. F. 1975. Patterns of use of space in a guild of territorial reef fishes. Marine Biology 29:89–97.

Sale, P. F. 1977. Maintenance of high diversity in coral reef fish communities. American Naturalist **111**:337–359.

Sale, P. F. 1978. Coexistence of coral reef fishes—a lottery for living space. Environmental Biology of Fishes **3**:85–102.

Salt, G. W. 1983. Roles: Their limits and responsibilities in ecological and evolutionary research. American Naturalist **122**:697–705.

Sarnelle, O. 2005. *Daphnia* as keystone predators: Effects on phytoplankton diversity and grazing resistance. Journal of Plankton Research **27**:1229–1238.

Scavia, D., and S. C. Chapra. 1977. Comparison of an ecological model of Lake Ontario and phosphorus loading models. Journal of the Fisheries Research Board of Canada **34**:286–290.

Schaffer, W. M. 1981. Ecological abstraction: The consequences of reduced dimensionality in ecological models. Ecological Monographs **51**:383–401.

Schaffer, W. M., and E. G. Leigh. 1976. The prospective role of mathematical theory in plant ecology. Systematic Botany **1**:209–232.

Schamp, B. S., and A. M. Jensen. 2019. Evidence of limiting similarity revealed using a conservative assessment of coexistence. Ecosphere **10**:e02840.

Schluter, D., and P. R. Grant. 1984. Determinants of morphological patterns in communities of Darwin's finches. American Naturalist **123**:175–196.

Schmidt, J. H., E. A. Rexstad, C. A. Roland, C. L. McIntyre, M. C. MacCluskie, and M. J. Flamme. 2018. Weather-driven change in primary productivity explains variation in the amplitude of two herbivore population cycles in a boreal system. Oecologia **186**:435–446.

Schmidt, K. A., J. M. Earnhardt, J. S. Brown, and R. D. Holt. 2000. Habitat selection under temporal heterogeneity: Exorcizing the ghost of competition past. Ecology **81**:2622–2630.

Schmitt, R. J. 1987. Indirect interactions between prey: Apparent competition, predator aggregation, and habitat selection. Ecology **68**:1887–1897.

Schmitz, O. J. 2010. *Resolving Ecosystem Complexity*. Princeton University Press, Princeton, NJ.

Schmitz, O. J., and K. B. Suttle. 2001. Effects of top predator species on direct and indirect interactions in a food web. Ecology **82**:2072–2081.

Schoener, T. W. 1965. The evolution of bill size differences among congeneric species of birds. Evolution **19**:189–213.

Schoener, T. W. 1973. Population growth regulated by intraspecific competition for energy or time: Some simple representations. Theoretical Population Biology **4**:56–84.

Schoener, T. W. 1974a. Competition and the form of habitat shift. Theoretical Population Biology **6**:265–307.

Schoener, T. W. 1974b. Some methods for calculating competition coefficient from resource-utilization spectra. American Naturalist **108**:332–340.

Schoener, T. W. 1983. Field experiments on interspecific competition. American Naturalist **122**:240–285.

Schoener, T. W. 1986. Mechanistic approaches to community ecology: A new reductionism. American Zoologist **26**:81–106.

Schoener, T. W., and C. A. Toft. 1983. Spider populations: Extraordinarily high densities on islands without top predators. Science **219**:1353–1355.

Schreiber, S. J., M. Benaïm, and K. A. S. Atchadé. 2011. Persistence in fluctuating environments. Journal of Mathematical Biology 62:655–683.

Seabloom, E. W., O. N. Bjørnstad, B. M. Bolker, and O. J. Reichman. 2005. Spatial signature of environmental heterogeneity, dispersal, and competition in successional grasslands. Ecological Monographs 75:199–214.

Searle, C. L., M. H. Cortez, K. K. Hunsberger, D. C. Grippi, I. A. Oleksy, C. L. Shaw, S. B. de la Serna, C. L. Lash, K. L. Dhir, and M. A. Duffy. 2016. Population density, not host competence, drives patterns of disease in an invaded community. American Naturalist 188:554–566.

Sears, A. L. W., and P. Chesson. 2007. New methods for quantifying the spatial storage effect: An illustration with desert annuals. Ecology 88:2240–2247.

Seo, G., and G. S. K. Wolkowicz. 2018. Sensitivity of the dynamics of the general Rosenzweig-MacArthur model to the mathematical form of the functional response: A bifurcation theory approach. Journal of Mathematical Biology 76:1873–1906.

Shinen, J. L., and S. A. Navarrete. 2014. Lottery coexistence on rocky shores: Weak niche differentiation or equal competitors engaged in neutral dynamics? American Naturalist 183:342–362.

Shmida, A., and S. Ellner. 1984. Coexistence of plant species with similar niches. Vegetatio 58:29–55.

Shurin, J. B., and E. G. Allen. 2001. Effects of competition, predation, and dispersal on species richness at local and regional scales. American Naturalist 158:624–637.

Shurin, J. B., E. T. Borer, E. W. Seabloom, K. Anderson, C. A. Blanchette, B. Broitman, S. D. Cooper, and B. S. Halpern. 2002. A cross-ecosystem comparison of the strength of trophic cascades. Ecology Letters 5:785–791.

Shurin, J. B., J. E. Havel, M. A. Leibold, and B. Pinel-Alloul. 2000. Local and regional zooplankton species richness: A scale-independent test for saturation. Ecology 81:3062–3073.

Sieber, M., and F. M. Hilker. 2012. The hydra effect in predator-prey models. Journal of Mathematical Biology 64:341–360.

Siepielski, A. M., K.-L. Hung, E. E. B. Bein, and M. A. McPeek. 2010. Experimental evidence for neutral community dynamics governing an insect assemblage. Ecology 91:847–857.

Siepielski, A. M., and M. A. McPeek. 2010. On the evidence for species coexistence: A critique of the coexistence program. Ecology 91:3153–3164.

Siepielski, A. M., and M. A. McPeek. 2013. Niche versus neutrality in structuring the beta diversity of damselfly assemblages. Freshwater Biology 58:758–768.

Siepielski, A. M., S. J. McPeek, and M. A. McPeek. 2018. Female mate preferences on high-dimensional shape variation for male species recognition traits. Journal of Evolutionary Biology 31:1239–1250.

Siepielski, A. M., A. N. Mertens, B. L. Wilkinson, and M. A. McPeek. 2011. Signature of ecological partitioning in the maintenance of damselfly diversity. Journal of Animal Ecology 80:1163–1173.

Sih, A., P. H. Crowley, M. A. McPeek, J. Petranka, and K. Strohmeier. 1985. Predation, competition, and prey communities: A review of field experiments. Annual Review of Ecology and Systematics 16:269–311.

Silliman, B. R., and M. D. Bertness. 2003. Shoreline development drives invasion of *Phragmites australis* and the loss of plant diversity on New England salt marshes. Conservation Biology **18**:1424–1434.

Simberloff, D. 1983. Competition theory, hypothesis-testing, and other community ecological buzzwords. American Naturalist **122**:626–635.

Simberloff, D., and W. Boecklen. 1981. Santa Rosalia reconsidered: Size ratios and competition. Evolution **35**:1206–1228.

Simberloff, D., and T. Dayan. 1991. The guild concept and the structure of ecological communities. Annual Review of Ecology and Systematics **22**:115–143.

Singh, P., and G. Baruah. 2020. Higher order interactions and species coexistence. Theoretical Ecology **14**:71–83.

Siu, G., P. Bacchet, G. Bernardi, A. J. Brooks, J. Carlot, R. Causse, J. Claudet, et al. 2017. Shore fishes of French Polynesia. Cybium **41**:245–278.

Skalski, G. T., and J. F. Gilliam. 2001. Functional responses with predator interference: Viable alternatives to the Holling Type II model. Ecology **82**:3083–3092.

Skelton, J., K. J. Farrell, R. P. Creed, B. W. Williams, C. Ames, B. S. Helms, J. Stoekel, and B. L. Brown. 2013. Servants, scoundrels, and hitchhikers: Current understanding of the complex interactions between crayfish and their ectosymbiotic worms (Branchiobdellida). Freshwater Science **32**:1345–1357.

Slatkin, M. 1974. Competition and regional coexistence. Ecology **55**:128–134.

Slatkin, M. 1980. Ecological character displacement. Ecology **61**:163–177.

Slobodkin, L. B. 1953. On social single species populations. Ecology **34**:430–434.

Slobodkin, L. B. 1961. *Growth and Regulation of Animal Populations*. Holt, Rinehart and Winston, New York.

Smith, A. 1776. *Wealth of Nations*. Penguin Books, London.

Smith, M. D., D. C. Hartnett, and G. W. T. Wilson. 1999. Interacting influence of mycorrhizal symbiosis and competition on plant diversity in tallgrass prairie. Oecologia **121**:574–582.

Smith, V. H. 1993. Resource competition between host and pathogen. Bioscience **43**:21–30.

Smith, V. H., B. L. Foster, J. P. Grover, R. D. Holt, M. A. Leibold, and F. deNoyelles, Jr. 2005. Phytoplankton species richness scales consistently from laboratory microcosms to the world's oceans. Proceedings of the National Academy of Sciences, USA **102**:4393–4396.

Smith, V. H., and R. D. Holt. 1996. Resource competition and within-host disease dynamics. Trends in Ecology and Evolution **11**:386–389.

Smith-Gill, S. J., and D. E. Gill. 1978. Curvilinearities in the competition equations: An experiment with ranid tadpoles. American Naturalist **112**:557–570.

Snyder, R. E. 2007. Spatiotemporal population distributions and their implications for species coexistence in a variable environment. Theoretical Population Biology **72**:7–20.

Snyder, R. E. 2008. When does environmental variation most influence species coexistence? Theoretical Ecology **1**:129–139.

Snyder, R. E., and P. Chesson. 2003. Local dispersal can facilitate coexistence in the presence of permanent spatial heterogeneity. Ecology Letters **6**:301–309.

Snyder, R. E., and P. Chesson. 2004. How the spatial scales of dispersal, competition, and environmental heterogeneity interact to affect coexistence. American Naturalist **164**:633–650.

Sousa, W. P. 1979. Experimental investigations of disturbance and ecological succession in a rocky intertidal algal community. Ecological Monographs 49:227–254.

Southward, A. J. 1958. The zonation of plants and animals on rocky sea shores. Biological Reviews 331:137–177.

Sparks, J. S., R. C. Schelly, W. L. Smith, M. P. Davis, D. Tchernov, V. A. Pieribone, and D. F. Gruber. 2014. The covert world of fish biofluorescence: A phylogenetically widespread and phenotypically variable phenomenon. Plos One 9:e83259.

Sprules, W. G. 1972. Effects of size-selective predation and food competition on high altitude zooplankton communities. Ecology 53:375–386.

Srivastava, D. S. 1999. Using local-regional richness plots to test for species saturation: Pitfalls and potentials. Journal of Animal Ecology 68:1–16.

Stachowicz, J. J. 2001. Mutualism, facilitation, and the structure of ecological communities. Bioscience 51:235–246.

Stachowicz, J. J., and R. B. Whitlatch. 2005. Multiple mutualists provide complementary benefits to their seaweed host. Ecology 86:2418–2427.

Stanton-Geddes, J., and C. G. Anderson. 2011. Does a facultative mutualism limit species range expansion? Oecologia 167:149–155.

Staples, T. L., J. M. Dwyer, X. Loy, and M. M. Mayfield. 2016. Potential mechanisms of coexistence in closely related forbs. Oikos 125:1812–1823.

Stenseth, N. C., W. Falck, O. N. Bjørnstad, and C. J. Krebs. 1997. Population regulation in snowshoe hare and Canadian lynx: Asymmetric food web configurations between hare and lynx. Proceedings of the National Academy of Sciences, USA 94:5147–5152.

Stephens, P. A., W. J. Sutherland, and R. P. Freckleton. 1999. What is the Allee effect? Oikos 87:185–190.

Stephenson, T. A., and A. Stephenson. 1949. The universal features of zonation between tide-marks on rocky coasts. Journal of Ecology 37:289–305.

Sterner, R. W., and J. J. Elser. 2002. *Ecological Stoichiometry: The Biology of Elements from Molecules to the Biosphere.* Princeton University Press, Princeton, NJ.

Stewart, F. M., and B. R. Levin. 1973. Partitioning of resources and the outcome of interspecific competition: A model and some general considerations. American Naturalist 107:171–198.

Stillman, R. A., J. D. Goss-Custard, and R. W. G. Caldow. 1997. Modelling interference from basic foraging behaviour. Journal of Animal Ecology 66:692–703.

Stoecker, D. K., P. J. Hansen, D. A. Caron, and A. Mitra. 2017. Mixotrophy in the Marine Plankton. Annual Review of Marine Science 9:311–335.

Stoks, R., and M. A. McPeek. 2003a. Antipredator behavior and physiology determine *Lestes* species turnover along the pond-permanence gradient. Ecology 84:3327–3338.

Stoks, R., and M. A. McPeek. 2003b. Predators and life histories shape *Lestes* damselfly assemblages along a freshwater habitat gradient. Ecology 84:1576–1587.

Stoks, R., and M. A. McPeek. 2006. A tale of two diversifications: Reciprocal habitat shifts to fill ecological space along the pond permanence gradient. American Naturalist 168:S50–S72.

Stoks, R., M. A. McPeek, and J. L. Mitchell. 2003. Evolution of prey behavior in response to changes in predation regime: Damselflies in fish and dragonfly lakes. Evolution 57:574–585.

Strobeck, C. 1973. *N* species competition. Ecology 54:650–654.

Strogatz, S. H. 2015. *Nonlinear Dynamics and Chaos: With Applications to Physics, Biology, Chemistry, and Engineering*. 2nd edition. CRC Press, Boca Raton, FL.

Strong, D. R., Jr. 1983. Natural variability and the manifold mechanisms of ecological communities. American Naturalist **122**:636–660.

Strong, D. R., D. Simberloff, L. B. Abele, and A. B. Thistle. 1984. *Ecological Communities: Conceptual Issues and the Evidence*. Princeton University Press, Princeton, NJ.

Strong, D. R., L. A. Szyska, and D. Simberloff. 1979. Tests of community-wide character displacement against null hypotheses. Evolution **33**:897–913.

Struck, T. H., J. L. Feder, M. Bendiksby, S. Birkeland, J. Cerca, V. I. Gusarov, S. Kistenich, K. H. Larsson, L. H. Liow, M. D. Nowak, B. Stedje, L. Bachmann, and D. Dimitrov. 2018. Finding evolutionary processes hidden in cryptic species. Trends in Ecology and Evolution **33**:153–163.

Stuart, B. L., R. F. Inger, and H. K. Voris. 2006. High level of cryptic species diversity revealed by sympatric lineages of Southeast Asian forest frogs. Biology Letters **2**:470–474.

Stump, S. M. 2017. Multispecies coexistence without diffuse competition; or, why phylogenetic signal and trait clustering weaken coexistence. American Naturalist **190**:213–228.

Szabó, P., and G. Meszéna. 2006. Limiting similarity revisited. Oikos **112**:612–619.

Szilagyi, A., and G. Meszéna. 2009. Limiting similarity and niche theory for structured populations. Journal of Theoretical Biology **258**:27–37.

Tanabe, K., and T. Namba. 2005. Omnivory creates chaos in simple food web models. Ecology **86**:3411–3414.

Tanner, J. T. 1966. Effects of population density on growth rates of animal populations. Ecology **47**:733–745.

Tansley, A. G. 1917. On competition between *Galium saxatile* L. (G. Hercynicum Weig.) and *Galium sylvestre* Poll. (G. Asperum Schreb.) on different types of soil. Journal of Ecology **5**:173–179.

Taper, M. L., and T. J. Case. 1985. Quantitative genetic models for the coevolution of character displacement. Ecology **66**:355–371.

Terborgh, J. 1971. Distribution on environmental gradients: Theory and a preliminary interpretation of distributional patterns in the avifauna of the Cordillera Vilcabamba, Peru. Ecology **52**:23–40.

Terborgh, J. W., and J. Faaborg. 1980. Saturation of bird communities in the West Indies. American Naturalist **116**:178–195.

Tessier, A. J., and M. A. Leibold. 1997. Habitat use and ecological specialization within lake *Daphnia* populations. Oecologia **109**:561–570.

Thacker, R. W., M. A. Becerro, W. A. Lombang, and V. J. Paul. 1998. Allelopathic interactions between sponges on a tropical reef. Ecology **79**:1740–1750.

Thiel, T., and B. Drossel. 2018. Impact of stochastic migration on species diversity in metafood webs consisting of several patches. Journal of Theoretical Biology **443**:147–156.

Thomas, M. J., R. P. Creed, J. Skelton, and B. L. Brown. 2016. Ontogenetic shifts in a freshwater cleaning symbiosis: Consequences for hosts and their symbionts. Ecology **97**:1507–1517.

Thompson, J. N. 1994. *The Coevolutionary Process*. University of Chicago Press, Chicago.

Thompson, J. N. 2005. *The Geographic Mosaic of Coevolution*. University of Chicago Press, Chicago.

Thonicke, K., S. Venevsky, S. Sitch, and W. Cramer. 2001. The role of fire disturbance for global vegetation dynamics: Coupling fire into a dynamic global vegetation model. Global Ecology and Biogeography **10**:661–677.

Thuiller, W., S. Lavorel, G. Midgley, S. Lavergne, and T. Rebelo. 2004. Relating plant traits and species distributions along bioclimatic gradients for 88 *Leucadendron* taxa. Ecology **85**:1688–1699.

Tilman, D. 1977. Resource competition between plankton algae: An experimental and theoretical approach. Ecology **58**:338–348.

Tilman, D. 1980. Resources: A graphical-mechanistic approach to competition and predation. American Naturalist **116**:362–393.

Tilman, D. 1982. *Resource Competition and Community Structure*. Princeton University Press, Princeton, NJ.

Tilman, D. 1987. The importance of the mechanisms of interspecific competition. American Naturalist **129**:769–774.

Tilman, D. 1993. Species richness of experimental productivity gradients: How important is colonization limitation? Ecology **74**:2179–2191.

Tilman, D. 1994. Competition and biodiversity in spatially structured habitats. Ecology **75**:2–16.

Tilman, D., M. Mattson, and S. Langer. 1981. Competition and nutrient kinetics along a temperature gradient: An experimental test of a mechanistic approach to niche theory Limnology and Oceanography **26**:1020–1033.

Tilman, D., R. M. May, C. L. Lehman, and M. A. Nowak. 1994. Habitat destruction and the extinction debt. Nature **371**:65–66.

Titman, D. 1976. Ecological competition between algae: Experimental confirmation of resource-based competition theory. Science **192**:463–465.

Tompkins, D. M., R. A. H. Draycott, and P. J. Hudson. 2000. Field evidence for apparent competition mediated via the shared parasites of two gamebird species. Ecology Letters **3**:10–14.

Towne, E. G., and A. K. Knapp. 1996. Biomass and density responses in tallgrass prairie legumes to annual fire and topographic position. American Journal of Botany **83**:175–179.

Travis, C. C., and W. M. Post. 1979. Dynamics and comparative statics of mutualistic communities. Journal of Theoretical Biology **78**:553–571.

Tuljapurkar, S. 1989. An uncertain life: Demography in random environments. Theoretical Population Biology **35**:227–294.

Tuljapurkar, S. D., and S. H. Orzack. 1980. Population dynamics in variable environments. I. Long-run growth rates and extinction. Theoretical Population Biology **18**:314–342.

Turchin, P. 2003. *Complex Population Dynamics: A Theoretical/Empirical Synthesis*. Princeton University Press, Princeton, NJ.

Turelli, M. 1977. Random environments and stochastic calculus. Theoretical Population Biology **12**:140–178.

Turelli, M. 1978a. Does environmental variability limit niche overlap? Proceedings of the National Academy of Sciences, USA **75**:5085–5089.

Turelli, M. 1978b. A reexamination of stability in randomly varying versus deterministic environments with comments on the stochastic theory of limiting similarity. Theoretical Population Biology **13**:244–267.

Turelli, M. 1981. Niche overlap and invasion of competitors in random environments. I. Models without demographic stochasticity. Theoretical Population Biology **20**:1–56.

Turelli, M., and J. H. Gillespie. 1980. Conditions for the existence of stationary densities for some two-dimensional diffusion processes with applications in population biology. Theoretical Population Biology **17**:167–189.

Urbanowicz, C., R. A. Virginia, and R. E. Irwin. 2018. Pollen limitation and reproduction of three plant species across a temperature gradient in western Greenland. Arctic, Antarctic, and Alpine Research **50**:S100022.

Usinowicz, J., C. H. Chang-Yang, Y. Y. Chen, J. S. Clark, C. Fletcher, N. C. Garwood, Z. Hao, et al. 2017. Temporal coexistence mechanisms contribute to the latitudinal gradient in forest diversity. Nature **550**:105–108.

Usinowicz, J., S. J. Wright, and A. R. Ives. 2012. Coexistence in tropical forests through asynchronous variation in annual seed production. Ecology **93**:2073–2084.

Valdovinos, F. S., P. M. de Espanés, J. D. Flores, and R. Ramos-Jiliberto. 2013. Adaptive foraging allows the maintenance of biodiversity of pollination networks. Oikos **122**:907–917.

Van Buskirk, J., and D. C. Smith. 1991. Density-dependent population regulation in a salamander. Ecology **72**:1747–1756.

Vance, R. R. 1978. Predation and resource partitioning in one predator-two prey model communities. American Naturalist **112**:797–813.

Vance, R. R. 1984. Interference competition and the coexistence of two competitors on a single limiting resource. Ecology **65**:1349–1357.

Vance, R. R. 1985. The stable coexistence of two competitors on one resource. American Naturalist **126**:72–86.

Vandermeer, J. 2006. Omnivory and the stability of food webs. Journal of Theoretical Biology **238**:497–504.

Vandermeer, J., I. G. de la Cerda, D. Boucher, I. Perfecto, and J. Ruiz. 2000. Hurricane disturbance and tropical tree species diversity. Science **290**:788–790.

Vandermeer, J. H. 1970. The community matrix and the number of species in a community. American Naturalist **104**:73–83.

Vandermeer, J. H. 1975. Interspecific competition: A new approach to the classical theory. Science **188**:253–255.

Vandermeer, J. H., and D. H. Boucher. 1978. Varieties of mutualistic interaction in population models. Journal of Theoretical Biology **74**:549–558.

Vanni, M. J. 2002. Nutrient cycling by animals in freshwater ecosystems. Annual Review of Ecology and Systematics **33**:341–370.

Vanni, M. J., C. D. Layne, and S. E. Arnott. 1997. "Top-down" trophic interactions in lakes: Effects of fish on nutrient dynamics. Ecology **78**:1–20.

Van Valkenburgh, B., and C. M. Janis. 1993. Historical diversity patterns in North American large herbivores and carnivores. Pages 330–340 *in* R. E. Ricklefs and D. Schluter, editors, *Species Diversity in Ecological Communities: Historical and Geographical Perspectives*. University of Chicago Press, Chicago.

Verdy, A., and P. Amarasekare. 2010. Alternative stable states in communities with intraguild predation. Journal of Theoretical Biology **262**:116–128.

Vergnon, R., R. Leijs, E. H. van Nes, and M. Scheffer. 2013. Repeated parallel evolution reveals limiting similarity in subterranean diving beetles. American Naturalist **182**:67–75.

Verhulst, P.-F. 1838. Notice sur la loi que la population poursuit dans son accroissement. Correspondance mathématique et physique **10**:113–121.

Volkov, I., J. R. Banavar, S. P. Hubbell, and A. Maritan. 2003. Neutral theory and relative species abundance in ecology. Nature **424**:1035–1037.

Volterra, V. 1926. Variatzioni e fluttuazioni del numero d'individui in specie animali conviventi. Memoria della Reale Accademia Nazionale dei Lincei **6**:31–113.

Wang, Y. S., D. L. DeAngelis, and J. N. Holland. 2011. Uni-directional consumer-resource theory characterizing transitions of interaction outcomes. Ecological Complexity **8**:249–257.

Wang, Y. S., D. L. DeAngelis, and J. N. Holland. 2012. Uni-directional interaction and plant-pollinator-robber coexistence. Bulletin of Mathematical Biology **74**:2142–2164.

Wardle, D. A., M.-C. Nilsson, C. Gallet, and O. Zackrisson. 1998. An ecosystem-level perspective of allelopathy. Biological Reviews **73**:305–319.

Waring, R. H., and J. Major. 1964. Some vegetation of the California coastal redwood region in relation to gradients of moisture, nutrients, light, and temperature. Ecological Monographs **34**:167–215.

Watkinson, A. R., and W. J. Sutherland. 1995. Sinks and pseudo-sinks. Journal of Animal Ecology **64**:126–130.

Webb, C. O., D. D. Ackerly, M. A. McPeek, and M. J. Donoghue. 2002. Phylogenies and community ecology. Annual Review of Ecology and Systematics **33**:475–505.

Weber, M. G., and K. H. Keeler. 2013. The phylogenetic distribution of extrafloral nectaries in plants. Annals of Botany **111**:1251–1261.

Wellborn, G. A., D. K. Skelly, and E. E. Werner. 1996. Mechanisms creating community structure across a freshwater habitat gradient. Annual Review of Ecology and Systematics **27**:337–363.

Welti, E. A. R., and A. Joern. 2018. Fire and grazing modulate the structure and resistance of plant-floral visitor networks in a tallgrass prairie. Oecologia **186**:517–528.

Werner, E. E., and M. A. McPeek. 1994. Direct and indirect effects of predators on two anuran species along an environmental gradient. Ecology **75**:1368–1382.

Werner, E. E., and S. D. Peacor. 2003. A review of trait-mediated indirect interactions in ecological communities. Ecology **84**:1083–1100.

Westfall, M. J., and M. L. May. 2006. *Damselflies of North America*. 2nd edition. Scientific, Gainesville, FL.

Whittaker, R. H. 1956. Vegetation of the Great Smoky Mountains. Ecological Monographs **26**:1–80.

Whittaker, R. H., and C. W. Fairbanks. 1958. A study of plankton copepod communities in the Columbia basin, southeastern Washington. Ecology **39**:46–65.

Whittaker, R. H., and W. A. Niering. 1965. Vegetation of the Santa Catalina Mountains, Arizona: A gradient analysis of the south slope. Ecology **46**:429–452.

Whittaker, R. H., and W. A. Niering. 1975. Vegetation of the Santa Catalina Mountains, Arizona. V. Biomass, production, and diversity along the elevational gradient. Ecology **56**:771–790.

Widder, E. A. 2010. Bioluminescence in the ocean: Origins of biological, chemical, and ecological diversity. Science **328**:704–708.

Wiens, J. A. 1982. On size ratios and sequences in ecological communities: Are there no rules? Annales Zoologici Fennici **19**:297–308.

Wiens, J. A. 1984. On understanding a non-equilibrium world: Myth and reality in community patterns and processes. Pages 439–457 *in* D. R. Strong, D. Simberloff, L. B. Abele, and A. B. Thistle, editors, *Ecological Communities: Conceptual Issues and the Evidence*. Princeton University Press, Princeton, NJ.

Wiens, J. A., and J. T. Rotenberry. 1981. Morphological size ratios and competition in ecological communities. American Naturalist **117**:592–599.

Wilbur, H. M. 1972. Competition, predation, and the structure of the *Ambystoma-Rana sylvatica* community. Ecology **53**:3–21.

Wilbur, H. M. 1997. Experimental ecology of food webs: Complex systems in temporary ponds. Ecology **78**:2279–2302.

Wilbur, H. M., and J. E. Fauth. 1990. Experimental aquatic food webs: Interactions between two predators and two prey. American Naturalist **135**:176–204.

Willig, M. R., and L. R. Walker. 1999. Disturbances in terrestrial ecosystems: Salient themes, synthesis, and future directions. Pages 747–767 *in* L. R. Walker, editor, *Ecosystems of Disturbed Ground*. Elsevier, New York.

Wilson, A. M., T. Y. Hubel, S. D. Wilshin, J. C. Lowe, M. Lorenc, O. P. Dewhirst, H. L. A. Bartlam-Brooks, et al. 2018. Biomechanics of predator-prey arms race in lion, zebra, cheetah and impala. Nature **554**:183–188.

Wilson, S. D., and P. A. Keddy. 1986. Species competitive ability and position along a natural stress/disturbance gradient. Ecology **67**:1236–1242.

Wissinger, S., and J. McGrady. 1993. Intraguild predation and competition between larval dragonflies: Direct and indirect effects of shared prey. Ecology **74**:207–218.

Witman, J. D., R. J. Etter, and F. Smith. 2004. The relationship between regional and local species diversity in marine benthic communities: A global perspective. Proceedings of the National Academy of Sciences, USA **101**:15664–15669.

Witt, J. D., D. L. Threloff, and P. D. Hebert. 2006. DNA barcoding reveals extraordinary cryptic diversity in an amphipod genus: Implications for desert spring conservation. Molecular Ecology **15**:3073–3082.

Witt, J. D. S., and P. D. N. Hebert. 2000. Cryptic species diversity and evolution in the amphipod genus *Hyalella* within central glaciated North America: A molecular phylogenetic approach. Canadian Journal of Fisheries and Aquatic Sciences **57**:687–698.

Wolin, C. L., and L. R. Lawlor. 1984. Models of facultative mutualism: Density effects. American Naturalist **124**:843–862.

Wollkind, D. J. 1976. Exploitation in three trophic levels: An extension allowing intraspecific carnivore interaction. American Naturalist **110**:431–447.

Wollrab, S., A. M. de Roos, and S. Diehl. 2013. Ontogenetic diet shifts promote predator-mediated coexistence. Ecology **94**:2886–2897.

Wollrab, S., S. Diehl, and A. M. De Roos. 2012. Simple rules describe bottom-up and top-down control in food webs with alternative energy pathways. Ecology Letters **15**:935–946.

Woodward, J. 2010. Causation in biology: Stability, specificity, and the choice of levels of explanation. Biology and Philosophy **25**:287–318.

Wootton, J. T. 1994a. The nature and consequences of indirect effects in ecological communities. Annual Review of Ecology and Systematics **25**:443–466.

Wootton, J. T. 1994b. Predicting direct and indirect effects: An integrated approach using experiments and path analysis. Ecology **75**:151–165.

Wootton, J. T. 1994c. Putting the pieces together: Testing the indepedence of interactions among organisms. Ecology **75**:1544–1551.

Worm, B., H. K. Lotze, H. Hillebrand, and U. Sommer. 2002. Consumer versus resource control of species diversity and ecosystem functioning. Nature **417**:848–851.

Wright, D. H. 1989. A simple, stable model of mutualism incorporating handling time. American Naturalist **134**:664–667.

Wyatt, G. A., E. T. Kiers, A. Gardner, and S. A. West. 2014. A biological market analysis of the plant-mycorrhizal symbiosis. Evolution **68**:2603–2618.

Xiao, X., and G. F. Fussmann. 2013. Armstrong-McGehee mechanism revisited: Competitive exclusion and coexistence of nonlinear consumers. Journal of Theoretical Biology **339**:26–35.

Yodzis, P. 1980. The connectance of real ecosystems. Nature **284**:544–545.

Yodzis, P. 1981. The stability of real ecosystems. Nature **289**:674–676.

Young, A. M. 1971. Foraging for insects by a tropical hummingbird. Condor **73**:36–45.

Yu, D. W., and H. B. Wilson. 2001. The competition-colonization trade-off is dead; long live the competition-colonization trade-off. American Naturalist **158**:49–63.

Zaret, T. M. 1980. *Predation and Freshwater Communities*. Yale University Press, New Haven, CT.

Zhan, J., and B. A. McDonald. 2013. Experimental measures of pathogen competition and relative fitness. Annual Review of Phytopathology **51**:131–153.

Zhang, D.-Y., and I. Hanski. 1998. Sexual reproduction and stable coexistence of identical competitors. Journal of Theoretical Biology **193**:465–473.

Zhang, H., Y.-B. He, P.-F. Wu, S.-F. Zhang, Z.-X. Xie, D.-X. Li, L. Lin, F. Chen, and D.-Z. Wanga. 2019. Functional differences in the blooming phytoplankton *Heterosigma akashiwo* and *Prorocentrum donghaiense* revealed by comparative metaproteomics. Applied and Environmental Microbiology **85**:e01425–01419.

Index

MONOGRAPHS IN POPULATION BIOLOGY

SIMON A. LEVIN, ROBERT PRINGLE, AND CORINA TARNITA, SERIES EDITORS

Milton Keynes UK
Ingram Content Group UK Ltd.
UKHW020821230924
448618UK00020B/185